11/94

FUNDAMENTALS OF MACHINE TOOL TECHNOLOGY AND MANUFACTURING PROCESSES

C. Thomas Olivo

 *C. Thomas Olivo Associates*

DEDICATION

This book is dedicated to my wife and constant teacher, Hilda G. Olivo. She shared generously of her unique editorial talents, provided inspiration, and contributed significantly to this work from the initial research, analyses, and organization stages to this final publication.

Library of Congress Catalog Card Number: 89–64001

Olivo, C. Thomas

Fundamentals of machine tool technology and
    manufacturing processes.

Includes handbook tables and index.

1. Machine shop practice.   2. Machining technology.
3. Manufacturing processes.   I. Title.

        1990   621.9′02   89–64001

ISBN 0–938561–00–6
ISBN 0–938561–01–4   (laboratory manual)
ISBN 0–938561–02–2   (instructor's guide)
ISBN 0–938561–03–0   (set: Vol. I, II, III)

Cover credits. . . *Design and composite art*, The Drawing Board
                    Cutting tools, Schuyler Photography
                    Turning (lathe), DoALL Company
                    Laser cutting, Amada Laser Division, ACT, Inc.
(Back cover)   *Indexer milling setup*, SMW Systems, Inc.

PREFACE

Fundamentals of Machine Tool Technology and Manufacturing Processes is a state-of-the-technology, fascinating study of the performance skills and the related technology upon which industry depends for craftspersons and technicians; design, programming, and engineering support personnel. and other high-skill workers.

This newly titled book (formerly, **Fundamentals of Machine Technology**) reflects changes in organization and the extended scope and contents of this third edition. The new dimensions are the end result of three parallel studies, as follows. (1) Analyses of performance skills, technology skills, and related technical subjects skills required by occupationally competent workers, projected to identify evolving industry developments. (2) Studies of courses and programs conducted by educational institutions; industry, labor, and military organizations; and by special training agencies. (3) Evaluations of systems, processes, and products used to measure occupational competency skills.

PRACTICAL APPLICATIONS

This book is crafted for students/trainees and employed workers who relate to programs or occupations such as: machine shop practice; tool and die making; skilled machine operatives; machine and systems design, programming, and industrial engineering personnel; machine tool technology and manufacturing specialists, and others in the machining industry.

This highly practical book is designed for use at the technical institute and community college level; in area vocational-technical schools; in college level industrial engineering courses; in apprenticeship and other on-the-job or cooperative school/industry/military occupational programs; and for adult training and upgrading courses. Such courses/programs may be under public or private administration. The book is also adaptable to directed self study.

SPECIAL FEATURES

The following special features are incorporated within the structure and contents.

√ AUTHENTICITY. The contents are written by a highly qualified, occupationally competent writer. As stated, the contents are based on *valid analyses* of industry requirements, training and *curriculum* development *patterns*, and occupational *competency assessment* systems.

√ COMPREHENSIVENESS. The contents reflect the major areas of training in which an individual must develop performance skills, technology skills, and related technical subjects skills for successful employment and advancement.

√ INTERLOCKED PERFORMANCE SKILLS AND TECHNOLOGY. Equal emphasis is placed on essential manipulative "how to" skills and companion "why-to" technology skills. These skills are further interlocked with problem-solving mathematics skills, applications of physical science skills, and blueprint reading communication skills. . .as required at the workplace.

√ FLEXIBLE UNIT ORGANIZATION. The structure of nine *Parts* and *Sections* and 43 instructional *Units* provides a practical sequence to follow or to rearrange to meet other particular training requirements.

√ STRUCTURE WITHIN UNITS. The instructional contents flow from one Unit to the next higher level of learning. Each Unit starts with an *introductory overview* followed by carefully identified *Unit Objectives*. New *Technical Terms* are defined and applied within the Unit contents and are summarized at the end. The content *Summary* provides another opportunity to bring together all of the learning experiences within each unit.

✓ MODERN, ACCURATE ILLUSTRATIONS. Specially selected, relevant, accurate art work and callouts complement the contents.

✓ HANDBOOK TABLES. Appropriate *Handbook Tables* are content referenced within the units. Additional complete tables appear in the Appendix. Practical test items require the use of handbook data.

✓ CONTINUOUS TESTING. End-of-unit *Review and Self Test Items* (with additional workplace-simulated problems provided in the Laboratory Manual) provide the most comprehensive continuous testing resource materials available.

✓ PERSONAL, MACHINE, AND PRODUCT SAFETY. A positive, planned *Safety Program* threads throughout the contents of each unit. Safety and hygienic practices are developed within the text content with additional emphasis provided in a special "Safe Practices" position in each unit. Test items are contained in each unit.

✓ PRACTICAL AESTHETIC QUALITIES. *Two colors* are used on every page to emphasize important components, features, and concepts within each unit. Page layout, color, and physical properties are planned to enhance learning.

✓ COMPLEMENTARY LABORATORY MANUAL. A student/trainee manual provides *Guidelines for Study* that parallel each textbook unit. Additional shop/laboratory problems are included for individual competency enrichment.

✓ INSTRUCTOR/TRAINER GUIDE. This complementary teacher/teaching resource suggests unit-by-unit *Teaching Plans* and *solutions* to Review and Self Test Items in both the textbook and the Laboratory Manual. The Guide is designed to maximize teaching/learning effectiveness.

*   *   *   *   *

The nine content areas of the book follow. Detailed Section and Unit content information is provided in the Table of Contents.

- 1 Introduction to and Career Opportunities in Machine Tool Technology and Manufacturing Occupations
- 2 Measurement and Layout: Technology and Processes (including Customary and SI Metric Systems)
- 3 Bench and Floor Work: Technology and Processes
- 4 Basic Cut-Off and Grinding Machines; Technology and Processes
- 5 Drilling Machines: Technology and Processes
- 6 Turning Machines: Technology and Processes
- 7 Horizontal and Vertical Milling Machines: Technology and Processes
- 8 Production of Industrial and Cutting Tool Materials and Basic Metallurgy
- 9 New Manufacturing Processes and (Nontraditional) Machine Tools

About the Author

Dr. C. Thomas Olivo ("Dr. Tom") is a distinguished technical writer and foremost authority on institutional and training-within-industry programs related to machine tool technology, machine shop practice, tool and die making, and metal products manufacturing.

After serving an apprenticeship, Dr. Olivo achieved journeyman and master status as a tool and die maker. On the training scene, Dr. Tom served successively responsible positions as a teacher, training supervisor, technical institute director, branch chief, Wharton (England)-American military occupational technical school, and State Director of industrial-technical education.

CONTENTS

Part 1 MACHINE TOOL TECHNOLOGY, MACHINING, AND MANUFACTURING OCCUPATIONS

Section One INTRODUCTION

Part 2 MEASUREMENT: TECHNOLOGY AND PROCESSES

Section One MEASUREMENT AND MEASUREMENT SYSTEMS

Section Two MEASUREMENT AND LAYOUT TOOLS AND INSTRUMENTS

Part 6   TURNING MACHINES: TECHNOLOGY AND PROCESSES

Section One   ENGINE LATHES: DESIGN FEATURES AND OPERATIONS

Section Two   LATHE CUTTING TOOLS AND TOOLHOLDERS

Section Three   LATHE WORK BETWEEN CENTERS

Section Two   TYPICAL MILLING SETUPS AND PROCESSES

Part 8   PRODUCTION OF INDUSTRIAL AND CUTTING TOOL MATERIALS

Section Two HEAT TREATING: TECHNOLOGY AND PROCESSES

Part 9 NEW MANUFACTURING PROCESSES AND MACHINE TOOLS

Section One NONTRADITIONAL TOOLING AND MACHINES: TECHNOLOGY AND PROCESSES

APPENDIX

HANDBOOK TABLES

ACKNOWLEDGEMENTS

All advancements since the machining of single parts on the Wilkinson's lathe of the 1790's to current computer numerically controlled machining centers and totally factory integrated manufacturing processes and systems, are founded on "sustained incremental contributions" of leaders. So, too, the quality, relevance, and authoritativeness of this book depends on resources provided by a broad spectrum of leaders, undergirded by other contributions from the institutions and organizations wherein they serve.

The spectrum of individuals who provided resources include: lead teachers serving in institutional programs, trainers from within industry and union sponsored programs, military occupational specialty training personnel, human resource training supervisors/directors, and others. Marketing directors supplied important up-to-the-minute product information, technical data, and excellent illustrations. The scope represented by this cadre of leaders reached across many different types of training programs and training levels; semiskilled to high skill craftspersons and technicians, to product design, tooling, and programming specialists, to engineering support persons.

The *Courtesy of* line that appears in the text above an illustration sincerely acknowledges each company that provided technical assistance. A personal "thank you" is expressed to each contributing individual from the following companies.

AGF Inc. (formerly American Gas Furnace Company)
Amada Laser Division; ACT, Inc.
AMATROL Inc.
American Iron and Steel Institute
American Precision Museum Association, Inc.
American Saw & Manufacturing Company
American Tool, Inc.
Arcadia Supply, Inc.
(B.C.) Ames Company
Armstrong Brothers Tool Company
Bay State Abrasives, Division of Abrasives Industries, Inc.
Beech Aircraft Corporation
Black & Webster, Inc.
Bridgeport Machines, Division of Textron, Inc.
Brown & Sharpe Manufacturing Company
Buck Supreme, Inc.
Cincinnati Milacron Inc.
Circle Machine Company
Clausing Corporation
Cleveland Twist Drill, an Acme-Cleveland Company
Coherent General
Colt Industries, Elox Division
Command Corporation
Control Laser Corporation
(The) Cooper Group, Nicholson Division

Cushman Industries, Inc.
(The) Desmond-Stephan Manufacturing Company
Darex Corporation
DeVlieg Machine Company, Microbore Division
DoALL Company
Dorsey Gage Company
Enco Manufacturing Company
Federal Products Corporation
General Electric Company: Carboloy Systems Department; Superabrasives Department
Grinding Wheel Institute
Hammond Machinery, Inc.
Hardinge Brothers, Inc.
Illinois/Eclipse: A Division of Illinois Tool Works, Inc.
(The) Jacobs Manufacturing Company
J & A Machinery Company
J & S Tool Company, Inc.
Johnson Gas Appliance Company
Kasto-Racine, Inc.
Kearney & Trecker Corporation
Kenametal, Inc.
Kurt Manufacturing Company
Landis Grinding Machines Company
LeBlond Makino Machine Tool Company
Lewis & Judge Machine Tools Company, Inc.
Lodge & Shipley Company

LSP Industries, Inc
Mahr Gage Company
M&M Precision Systems, an Acme-Cleveland
    Company
Menlo Tool Company, Inc.
Metcut Research Associates, Inc.; Manu-
    facturing Technology Division
Monarch Machine Tool Company: Monarch
    Cortland Division: Monarch Sidney
    Division
Moore Special Tool Company, Inc.
Morse Cutting Tools Division, Gulf +
    Western Manufacturing Company
National Acme, an Acme-Cleveland
    Company
National Twist Drill, Division of Lear
    Siegler, Inc.
W.H. Nichols Company
Northwestern Tool, Inc.
Norton Company, Abrasives Marketing
    Group
Pratt & Whitney Company, Inc.; Precision
    Grinding Division

ReliabOil, Inc.
Science Museum of England
SMW Systems Inc.
South Bend Lathe Company, Inc.
(The) L.S. Starrett Company
Suburban Printers, Inc.
Tapmatic Corporation
TE-CO Tooling Components Company
Thermolyne Corporation
TRW/Geometric Tool Division
TRW/Greenfield Tap & Die Company
TRW/J.H. Williams Division
United States Steel Corporation
Universal Vise & Tool Company
USM Machinery Division, Emhart Corpora-
    tion
Vermont Tap & Die Company
(The) Walton Company
Warner & Swasey Turning Division, A Cross
    & Trecker Company
Wilton Tool & Machinery Division, Wilton
    Corporation
Zagar Incorporated

Special recognition is made of Doyle K. Stewart, Coordinator of Manufacturing Processes, Delaware Technical and Community College, and to Dr. Ralph D. O'Brien, Professor and Chairman, Department of Industrial Technology, Northern Kentucky University, for their capable assessment of organization and scope and meticulous technical review of contents; to Noble Stuart, Coordinator of Machine Trades, Regional State Vocational-Technical School at Bowling Green; to Joe Mazoff, President, J & A Machinery Company, for input on new drill point geometry; and to Joseph Reynolds, Vice President of Marketing and Sales, Delmar Publishers, Inc. for friendly assistance relating to market assessment and product marketing and communications.

Recognition is made to Hilda G. Olivo for developmental assistance and editorial expertise and to Judith Ellen Forward for skillful preparation of the functional Index. Thanks is expressed to Jean A. LeMorta of Thoroughbred Graphics for production expertise and quality composing services and to John R. Orozco of The Drawing Board for the unusually fine layout and preparation of the cover and other art work.

Special commendation is made to Judi S. Orozco recognizing her talents and contributions in converting specifications into excellent product layout designs, mechanicals of the contents, art work, and overall supervision. Finally, appreciation is expressed to Frederick J. Sharer, Director of Manufacturing, Delmar Publishers, for helpful advice and counsel in the areas of production and manufacturing.

The aesthetic, physical properties of this book and the quality, relevance, and value of the contents, are a tribute to each individual and to each organization that gave so generously and shared so graciously.

C. Thomas Olivo

PART 1 Machine Tool Technology, Machining, and Manufacturing Occupations

SECTION ONE

Introduction

UNIT 1

Historical Overview and Career Opportunities

OBJECTIVES

A. IMPACT OF EARLY MACHINE TOOLS AND MANUFACTURING CONCEPTS

After satisfactorily completing this unit, you will be able to:

- Develop a rationale about the contributions of early machine tool developments on mass production.
- Describe the underlying features of Wilkinson's 1794 lathe.
- Interpret the meaning of "sustained, incremental contributions" to the development of machine tools and mass production of interchangeable parts.
- Relate to the emergence of nontraditional machine tools and processes.

B. CAREER OPPORTUNITIES: MACHINE TECHNOLOGY AND MANUFACTURING INDUSTRIES

- Establish bases on which job qualifications and early orientation to industries are established.
- Describe the apprenticeship training system and understand other school and on-the-job training programs.
- Interpret job titles in the context of job levels and areas within an occupation.
- Make decisions about job qualifications and take initial steps in career planning.

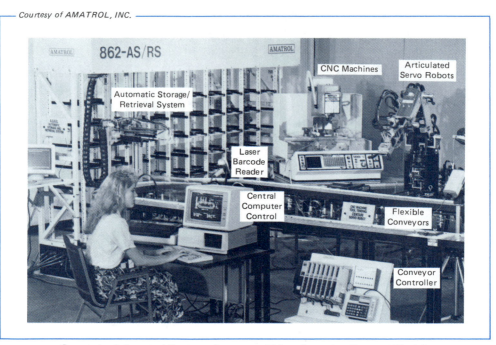

Computer Integrated Factory Automated Manufacturing Systems Model

A. IMPACT OF EARLY MACHINE TOOLS AND CONCEPT OF MASS PRODUCTION

This book relates to the "state-of-the-art" of machine tools, tooling, machining, and mass production processes. The contents provide what Edward A. Battison (Director of the American Precision Museum in Windsor, Vermont) identifies as. . . "the story of sustained, incremental contributions. . .by generations of individuals who stand out for their contributions to the American System of Manufacturing." This endless line of contributors had advanced design concepts beyond Leonardo da Vinci's early drawings of machine elements and mechanical movements to sophisticated, computerized and roboticized equipment.

In "this endless line," it is David Wilkinson of Pawtucket, Rhode Island who provides the reference point to modern machine tool technology and manufacturing processes. Wilkinson applied a carriage he had invented in 1794 to a lathe which he patented in 1798. This lathe advanced the concept of using three bearing points on the underside of a slide rest (carriage). The *main guide* or sliding surface became known as

the *front way* on a lathe bed. This *way* provided a precision guide for two bearing points. The *rear guide* (way) served as a mating surface for the third bearing area of the carriage. Figure 1–1 shows this early design feature.

THE LATHE AS THE DOMINANT MACHINE TOOL

The main guides of the Wilkinson lathe were formed by chisel and file! Precision in turning was achieved by "trial-and-fit" filing to "continue" the rear guide to compensate for variations with the front guide. Other features covered by the Wilkinson patent drawing include a lead screw mounted on centers to permit changeover to a different pitch; a split-nut to move the carriage; and a fixed center in an adjustable tailstock.

From these simple beginnings, the lathe became the dominant machine tool in the manufacturing revolution which followed. In this almost 200 year span, machine elements which were once filed are now machined and held to surface qualities which are measured in micro-inches (μin.), or micrometers (μm) in SI metric.

Courtesy of THE AMERICAN PRECISION MUSEUM

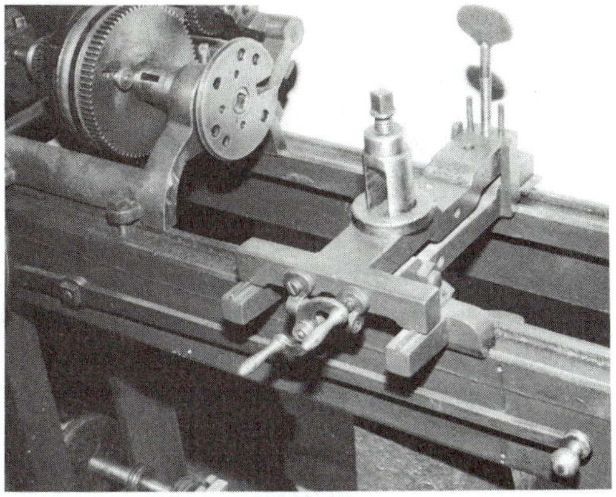

Figure 1–1 Wilkinson application of three bearing surfaces on a carriage at right angles to the bed ways

Tolerances in dimensional, parallelism, concentricity, runout, and other measurements are held to within a thousandth, ten-thousandth, or millionth part of an inch, and equivalent metric measurements.

"INCREMENTAL CONTRIBUTIONS" BY SUCCEEDING GENERATIONS

A parallel lathe development by Henry Maudslay in England incorporated and improved on Wilkinson's major design elements. The same features displayed in Maudslay's early 1800 lathe (Figure 1–2) are still basic components in modern computer-programmed, automated lathes and other turning machines.

In 1845, Stephen Fitch of Middlefield, Connecticut built an "eight station" turret lathe. The turret held and positioned multiple tooling setups. Then, in 1860 Christopher Spencer added a "brain wheel." This feature provided a cam-activated device for controlling the movements of cutting tools.

Following in the path of measuring instruments, Joseph R. Brown applied the vernier principle to a slide caliper. The new instrument enhanced the capability to measure and control dimensions to ±0.001″ and ±0.0001″ limits of accuracy.

Each new instrument and machine tool added to the capability of its predecessor and spurred still other machine tool developments for drilling

Courtesy of BRITISH CROWN COPYRIGHT; SCIENCE MUSEUM; LONDON

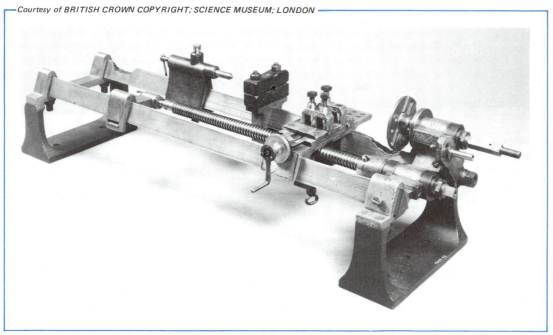

Figure 1–2 Maudslay Lathe of 1800

and hole forming, turning, milling, contouring, boring, grinding, stamping, and other manufacturing processes.

INTERCHANGEABLE PARTS IN MANUFACTURING

Along with machine tool and instrument developments, a concept of using machinery for the manufacture of interchangeable parts emerged. The first successful application of this concept is credited to Simeon North of Middletown, Connecticut. North demonstrated (through on-site examination by government inspectors of his plant, equipment, and manufacturing processes, and the actual fulfillment of a military contract) that it was economically feasible to mass produce interchangeable parts. North used existing machine tools and designed others to replace costly (and comparatively inaccurate) hand processes. North made significant design changes on the milling machine, making it an equally "dominant" production machine to the lathe.

Over the years, the commercial turret lathe of Henry B. Stone; the manufacturing milling machine of Thomas Warner; the profile miller of Frederick Howe; the impact of Thomas Blanchard's cam system for profile turning in woodworking as applied to metalworking machines and processes. . .are examples of. . . "sustained, incremental contributions. . . ." Each generation improved on the work of the preceding generation.

END PRODUCTS OF MANUFACTURING ENTERPRISE

While rooted in the production of guns and military weapons, applications of mass production of interchangeable parts spread rapidly from the mid 1800's to the manufacture of fine parts and mechanisms in watchmaking; business machines like the typewriter; the sewing machine as a household item; and other implements spreading across all economic sectors of society and the world.

NONTRADITIONAL MACHINE TOOLS AND MACHINING

Today's "state-of-the-art" is reflected in the computer-assisted programmed machining center as shown in Figure 1-3 for conventional

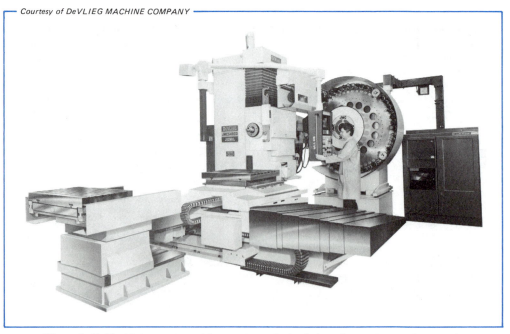

Courtesy of DeVLIEG MACHINE COMPANY

Figure 1–3 Machining center with an automatic tool changing system and computer numerical control requires programming by a skilled technician/craftsperson

machining. Concurrently, since 1940 there has emerged a whole new series of nontraditional machine tools and nontraditional material removal processes. These cope with needs for new alloys and materials which are extremely hard and difficult to fabricate and machine.

Nontraditional machining processes and equipment depend on hydrostatic and sound forces; electrical, electronic, and chemical energy; laser (light) beams converted to thermal energy; and other combined energy systems, using nonconventional tools.

B. CAREER OPPORTUNITIES: MACHINE TOOL, MACHINING, AND MANUFACTURING INDUSTRIES OCCUPATIONS

The industrial sector creates wealth through the production of goods which are needed by each person and are essential to the national well-being of all people, and for national security. This creation of wealth by transforming natural resources into useful, needed products requires a labor force with employable skills.

BASES OF JOB QUALIFICATIONS

The qualifications and duties at each job level are established by analyses of tasks, jobs, occupations, and clusters of jobs which require comparable levels of skill and technology competencies. Analyses reveal the sum-total of learning experiences which interlock mathematics; science; drawing, art, design, blueprint reading, and sketching skills, with required shop and laboratory skill competencies.

It is important for each individual who is planning a career to have the desire to prepare for, meet occupational qualifications, and have the capacity to profit by instruction in the occupation, consistent with labor market needs.

PLANNED EARLY LEARNING DECISIONS

Beginnings of industrial career preparation are rooted in planned early learning experiences in schools in industrial arts and technology courses. In dealing with tools, materials, processes, and products, the student develops an *awareness* and receives an *orientation* to occupations. Occupational *exploration* is provided through

sampling techniques within generalized groups of industries such as transportation, communications, construction, and manufacturing. Assisted by career guidance, counseling, and self-appraisal measures, these early experiences often lead to the selection of an occupation for which preparation within a secondary or post-secondary program leads to an entry level job.

APPRENTICESHIP TRAINING SYSTEM

One common pathway to job preparation is through an apprenticeship. Apprenticeship means that there is a contract (understanding) between the employer and the individual who is to prepare full-time on-the-job to attain the skills and technology essential for a defined trade or technical occupation. Stated are the conditions of training, time and wage increment schedules, conditions of employment, the assessment of occupational competency, and other responsibilities of each party.

Generally, apprenticeships are four years in duration, based on the skill and technology requirements of the occupation. In-school, military occupational specialty, and other previous training experiences may be credited in partial fulfillment of the requirements. Instruction in related subjects is provided in schools and industry training programs as part of the apprenticeship. This instruction deals with related technical communication skills (blueprint reading and sketching), computational skills (applied mathematics), and related science.

After satisfactorily completing and passing all facets of the program each year, the apprentice advances to the first level of journeyman/journeywoman status. With further experiences and continuous updating, upgrading, and retraining, the skilled worker advances to a second level, then a third level and on to master status as the highest level of employment as a trade craftsperson or technician.

OTHER JOB TRAINING SYSTEMS

Limited numbers of trainees are accepted into apprenticeship. Another common practice used by industry is to accept completers of vocational-industrial-technical programs, military occupational specialty training, and other industry/

training agency programs. In many cases, competencies are validated by the satisfactory passing of theory and/or performance skills of a particular occupation. Such persons enter industry as advanced learners for a specific occupation.

With added years of experience, coupled with updating and upgrading experiences, an individual may achieve the equivalent of an apprenticeship and may advance. Importantly, advancement to higher level jobs requires experiences which take place continuously throughout the life span of each worker.

CLASSIFICATION OF JOB TITLES BY LEVEL AND AREA

The classifications: *beginning*, *training*, or *entry*; *intermediate*, and *terminal levels*, are used for identification of *employment levels* (Figure 1–4). Within each of these levels, a number of jobs may be clustered. Each job title in the cluster has common skill and technology requirements.

BASIC JOBS, SUPPLEMENTARY JOBS, AND RELATED JOBS

At each of the three classification levels, there are other job cluster *areas*. For example, in the *basic job area* (in the left column of Figure 1–4), entry into employment and progress vertically is based on preparation for a specific occupation or the jobs within the cluster. An increasingly higher order of skill and technology is required to advance vertically.

Supplementary area jobs, which are identified in the middle column, range throughout the entry, intermediate, and terminal levels. "Supplementary" means that thorough preparation in one occupational field, coupled with additional training, makes it possible to move laterally on a *career ladder* to other jobs that partly depend on the initial skills in the basic area.

Related area jobs within a career ladder are those clusters of jobs that spin off from either basic area or supplementary area jobs and require still further preparation for what may be considered complementary occupations.

In summary, Figure 1–4 shows the structure for classifying jobs at the three levels for each of the three areas. If the nine blocks were filled in with job titles in machine tool technology,

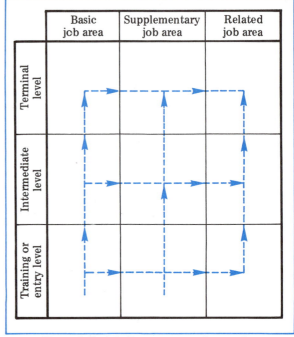

Figure 1–4 Job clusters grouped according to employment levels and areas

machining, and manufacturing, well over 500 different job titles may be included within the career ladder Descriptions, requirements, and other technical information about job titles are provided in standard industrial occupation classification handbooks.

It should be noted that unskilled workers, helpers, single-purpose operators, and low-level skill jobs are not included. These require minimal skills which are usually developed in a short time and on-the-job.

EXAMPLES OF JOB TITLES AT EACH EMPLOYMENT LEVEL

Training, *Learner*, or *Entry Level*. Job titles at this level relate to *semiskilled* craft and technical workers, operatives, and trainees (including apprentices).

Examples of job titles include: machinist apprentice, tool and die maker apprentice; automatic screw machine operator, milling machine operator; materials tester, production heat treater; machine components assembler, parts inspector; technical office personnel.

Courtesy of HARDINGE BROTHERS, INC.

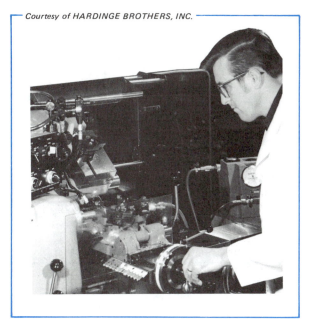

Figure 1–5 Instrument maker operating a Super-Precision® HC-AT chucker and bar machine

Courtesy of MOORE SPECIAL TOOL COMPANY INC.

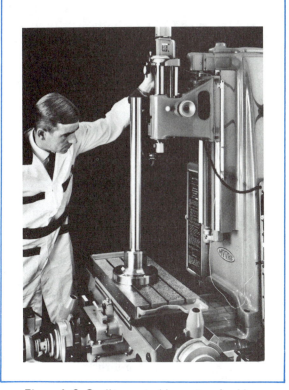

Figure 1–6 Quality control inspector checking the column squareness of a precision machine tool using a cylinder square master

Intermediate Level. This category includes job clusters for *skilled* craftspersons, *semi-professional* technicians and support persons for engineers; sales and marketing representatives; middle management supervisors; industrial trainers.

Examples of job titles include: machinist (grades I, II, III, IV), toolmaker (grades I, II, III), instrument maker (Figure 1–5); metallurgist; CNC, CAD, CAM, CIM system programmer; machine designer, tool designer; parts and gage inspector grade III (Figure 1–6); production set-up mechanic; heavy-duty machine tool operator specialist; all-round heat treater, laboratory technician; production supervisor, in-plant training instructor; sales operations personnel, marketing analyst, technical operations office manager.

Terminal Level. This category includes job titles which depend on the highest levels of decision making and responsibility. Terminal level is the "top level" for manager, director, engineer, researcher, and scientist positions.

Examples of job titles include: mechanical, electronic, metallurgical, tool, industrial, and production engineers; plant manager, plant supervisor; systemwide training director; overall

Courtesy of BEECH AIRCRAFT CORPORATION

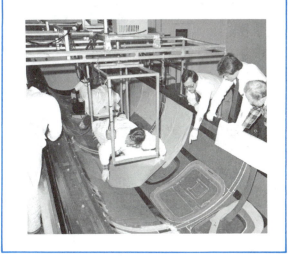

Figure 1–7 Starship component involving a team of workers from three different job levels and responsibilities

computer systems planner or administrator; research and scientific laboratory coordinator.

The relationship of team workers from different levels of job titles and responsibilities is shown in Figure 1–7. Two *CADAM engineers* are checking specifications with a *department chairman* (in sports coat) on a new Starship component. The three *composites technicians* are laying-up epoxy resin-impregnated materials to form the fuselage skin of a new composite airplane.

JOB MOBILITY AND ADVANCEMENT

Job mobility and advancement to intermediate and terminal level positions in the basic, supplementary, and related areas are generally based on the following four groups of qualifications.

- Additional years of successively higher levels of experience and responsibility.
- Continuous updating, upgrading, and extension of skill and technology competencies on-the-job, complemented by special institutes and other training.
- Continuous college level professional experiences (where extablished by occupational analyses). These may be obtained through formalized institutional programs or equivalent advanced degree work through industrial on-the-job professional programs.
- Personal qualities of leadership in group working relationships.

TERMS RELATED TO HISTORICAL MACHINE TOOL DEVELOPMENTS AND CAREER PLANNING

American System of Manufacturing	The result of continuous developments and refinements of machine tools, instruments, cutting tools, control mechanisms, and other devices and materials required to uniformly produce parts and components that are interchangeable.
Trial-and-fit	A method of producing a part by checking it against a mating part or gage each time material is removed.
Incremental contributions (metal products manufacturing)	A succession of new developments of machine tools, instruments, materials, etc., each of which improves and extends the capability to mass produce interchangeable parts to ever-higher levels of quality and cost effectiveness.
Nontraditional machine tools and machining	The application of sound, electrical, electronic, chemical, light, heat, sound, and other sources of energy and nonconventional cutting tools and machine tools and processes in order to remove material.
Career ladder	A succession of job opportunities that permit movement from entry to intermediate to advanced levels of employment in a basic area, a supplementary job area, or a related job area. A series of job changes each of which may require different occupational skill and technology background.
Apprenticeship training	A planned program for developing job skills at the workplace, complemented by competency in applying job-related mathematics, science, and blueprint reading and sketching, and other technology skills. A recognized training program of skill/technology development leading to occupational competency as a skilled craftsperson or technical specialist.
Job mobility	Opportunities and potential for a worker to move from one job to another at a particular job level or to advance to higher levels of employment. Job movement based on successively higher skill, technology, and experience requirements.

UNIT 1 REVIEW AND SELF TEST

A. IMPACT OF EARLY MACHINE TOOLS AND CONCEPT OF MASS PRODUCTION

1. Explain briefly (a) what the main design features were of David Wilkinson's carriage and (b) the impact it had on lathes becoming the dominant machine tool.

2. Identify four basic inspection and measurement tests which are made today in checking the accuracy of machine tools.

3. State two important outcomes on manufacturing which were brought about by refinements in measuring instruments.

4. Interpret what is meant by "producing interchangeable parts in manufacturing."

5. Indicate what effect the parallel developments of machine tools and the manufacture of interchangeable parts had on the growth of the nation.

6. Name two distinct characteristics or uses of nontraditional machine tools.

B. CAREER OPPORTUNITIES: MACHINE TOOL, MACHINING, AND MANUFACTURING INDUSTRIES OCCUPATIONS

1. Interpret briefly what "job clusters" means.

2. Name three major levels and three area groupings which form the matrix of a career ladder.

3. Identify a sequence of planned early in-school experiences which are designed to guide an individual to select an occupational preparation goal.

4. a. Cite four pathways leading to preparation at an entry job level.
    b. Identify a contract plan between an individual and an employer which, upon satisfactory completion, provides mobility to serve as a beginning-level skilled craftsperson.
    c. State three conditions of the contract.

5. Explain the effect on employment of the following two conditions.
    a. Experiences at successively higher skill and technology levels.
    b. Continuous occupational preparation throughout the life span of the individual.

6. a. Layout and label the framework of a career ladder.
    b. Insert at least two appropriate job titles in each of the job level and job area categories.
    c. Contrast (briefly) the difference in job preparation requirements between the entry level jobs in the basic job area and the terminal level jobs in the supplementary job area.

7. State two sets of qualifications which are required for job mobility and advancement to a higher level of employment.

PART 2 Measurement: Technology and Processes

Measurement and Measurement Systems

Accurate interpretation and measurement of dimensions are skills every individual involved in machine technology must have. A thorough understanding of dimensions, units of measure, measurement systems, and measurement terms is necessary for proficiency in machine technology. This section presents the basic principles and practices that are fundamental to accurate dimensional measurement.

UNIT 2

Measurement: Customary and SI Metric Systems

OBJECTIVES

After satisfactorily completing this unit, you will be able to:

- Understand the purposes served by measurement as related to: (a) describing a physical object, (b) constructing a part, and (c) controlling manufacture.
- Take linear measurements with basic measuring tools.
- Relate functions of precision, tolerances, and accuracy to measurements used on industrial drawings.
- Read common graduations in the British inch, decimal-inch, and SI metric measurement systems.
- Use conversion tables to establish equivalent measurements.

DIMENSIONAL MEASUREMENTS

Each measurement is written or expressed as a *dimension*. Whoever designs an object uses a dimension to describe or record a distance or a particular feature. *Dimensional measurement* relates to straight and curved lines, angles, areas, and volumes. Plane surfaces, as well as regular- and irregular-shaped objects, are included. Dimensional measurements, therefore, are the foundation of machine technology and of all industry, business, and commerce.

Dimensional measurements are necessary in designing, building, assembling, operating, and maintaining all parts and mechanisms. Some measurements are computed; others are taken directly with direct-reading measuring rules and other measuring instruments. The degree of accuracy establishes the precision requirements of the measuring tool or instrument to be used.

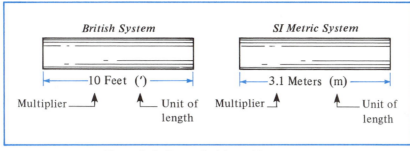

Figure 2-1 Parts of a Linear Measurement

There are three prime reasons for dimensional measurement:

- To describe a physical object;
- To construct a physical object. The object may be a single part, a component in a mechanism, or a complete unit of movable and stationary parts;
- To control the way in which an object is produced by many different people. Controlled measurements are necessary in the manufacture of interchangeable parts.

CHARACTERISTICS OF MEASUREMENT

The most widely used dimensional measurement is *length*, or linear distance. The term used is *linear measurement*. It means that a distance separating two points may be reproduced. Each linear measurement has two parts; a *multiplier* and the *unit of length* (Figure 2–1). The multiplier may be a whole number, a fractional or decimal value, or a combination of the two.

DIMENSIONS, FEATURES, AND MEASUREMENTS

Every line of measurement must have *direction*. The bases of linear measurements are pictured in Figure 2–2. The measurement of an actual dimension begins at a *reference point*. The measurement terminates (ends) at the *measured point*. These two points lie along the *line of measurement*. The two points are also called *references*. Two such points on an object or part define what are called *features*. The features represent the physical characteristics of the part.

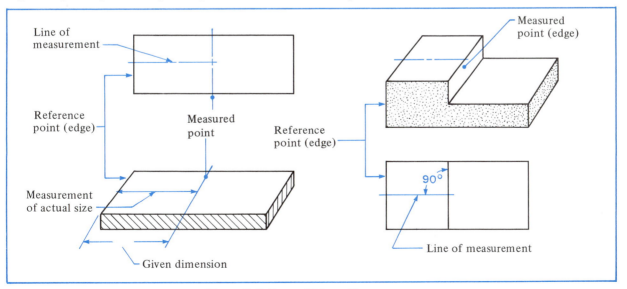

Figure 2-2 Bases of Linear Measurements

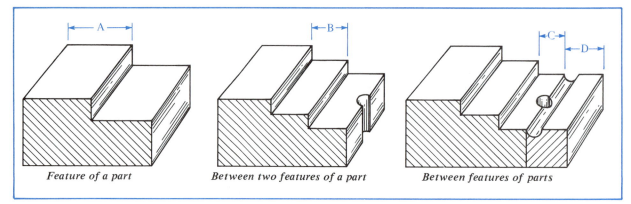

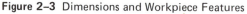

Figure 2–3 Dimensions and Workpiece Features

The notes on drawings, as well as lines that are used for laying out or describing features, are bounded by other lines, edges, ends, or plane surfaces. Three examples of features of different parts are given in Figure 2–3. Features are also bounded by areas. Two shop terms that are used to identify such features are *male* and *female.* These features appear in Figure 2–4.

Measurements are indicated on drawings as dimensions (A, B, C, and D in Figure 2–3). A *dimension* really states the designer's concept of the exact size of an intended feature. By contrast, a *measurement* relates to the actual size. Measurement accuracy depends on the temperature of both the measuring instrument and the workpiece.

PRECISION, TOLERANCES, AND ACCURACY

Three other dimension-related terms must be discussed at this point. *Precision* refers to the degree of exactness. For instance, a part may be dimensioned on a drawing to indicate a precision with *limits* of ±0.001″ or ±0.02 mm. The actual measured dimension, if it *falls within* these limits, known as *tolerances*, may vary by these limits from the size as stated on the drawing. Tolerances thus give the fineness to the range of sizes of a feature. Tolerances are needed to indicate maximum and minimum sizes of parts that will fit and function as designed.

Accuracy deals with the number of measurements that conform to a standard. This number is in comparison to those that vary from the standard. However, in common shop and laboratory practice, the terms *precision* and *accuracy* are used interchangeably.

MEASUREMENT AND INDUSTRIAL DRAWINGS

An object may be described by spoken words or with a photograph or some other form of picture. The sketch of the die block in Figure 2–5 is such a picture. Spoken words, photographs, and simple sketches do not, however, reveal:

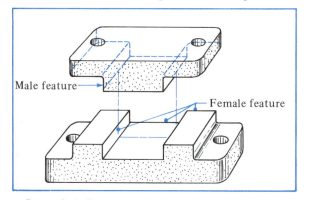

Figure 2–4 Examples of Male and Female Features

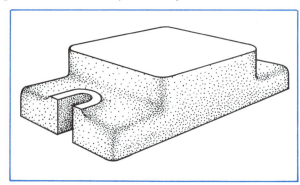

Figure 2–5 Sketch of a Die Block

- Exact dimensions,
- Cutaway sections that show internal details,
- Required machining,
- Additional work processes.

The craftsperson needs to rely on precise technical information. Shape, dimensions, and other important specifications must be furnished on a detailed drawing. Such a drawing conveys identical information to many different people. Some people are involved in the initial design stage; others deal with the same part or mechanism in a final inspection, assembly, or testing stage.

Designers, engineers, technicians, and craftspersons all may work in the same plant or may be separated in many plants throughout the world. They may not speak the same language, except technically. The *industrial drawing* is their universal language. Measurement is an important part of the universal language of an industrial drawing. Measurement is required in the production of every part and device.

INDUSTRIAL DRAWINGS AND BLUEPRINT READING

An *original drawing* (referred to as a *master drawing*) is usually produced with the aid of drafting instruments. While drawings of simple parts sometimes are sketched freehand, other drawings of more complex parts are made on automated drafting machines. Statistical and design data provide input for a computer. The computer then programs and controls automated drafting equipment, and a master drawing may be produced.

Complicated shapes and features may also be drawn directly on a workpiece. The layouts guide the worker in performing hand or machine processes and assembly work.

Original drawings are usually filed as a record. Subsequent design changes are made on these drawings. Reproductions are generally made of original drawings for use in the shop and laboratory and are referred to as *prints of the original*. While few of these prints are *blueprints*, the name still is generally applied among craftspersons.

By developing an ability to *read a blueprint* or sketch, the craftsperson is able to:

- Form a mental picture of a part,
- Visualize and perform different production and assembly processes,
- Make additional drawings that may be needed to machine a part,
- Take necessary measurements.

Reading a print requires an understanding of basic principles of graphic representation and dimensioning. Combined into what is called *drafting*, the principles include:

- The alphabet of lines,
- Views that provide exact information about shape,
- Dimensions and notes (these principles relate to measurements and accuracy),
- Cutting planes and sections (these principles show cross-sectional details of special areas of a workpiece),
- Shop sketching techniques.

THREE BASIC SYSTEMS OF MEASUREMENT

The most important measurement to be considered at this time is the unit of *length* (see Table 2–1). Units of measure for length in the British Imperial System are considered first. Units in the decimal-inch and metric systems follow.

THE BRITISH (U.S.) SYSTEM OF LINEAR MEASUREMENT

The inch is the most widely used unit of linear measure in the *British Imperial System*. The inch is subdivided into a number of equal parts. Each part represents a common fraction or a decimal fraction equivalent. These parts are shown in Figure 2–6.

Inch Common Fractions. The fractions most commonly used on steel rules represent halves,

Table 2–1 Inch-Standard Linear Units of Measure

British Imperial System
Smallest Unit of Measure = International Inch (″)
1 foot (′) = 12″
1 yard (yd) = 3′
1 rod = 5 1/2 yds
1 mile (mi) = 1,760 yds
1 mile (mi) = 5,280 ft
SI Metrics
25.4 millimeters (mm) = 1″

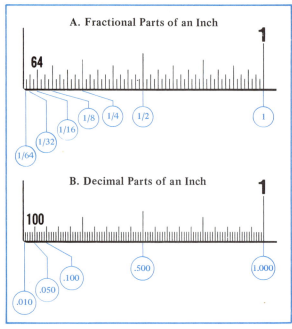

Figure 2–6 Fractional and Decimal Parts of an Inch on Steel Rules

quarters, eighths, sixteenths, thirty-seconds, and sixty-fourths of an inch. The enlarged illustration in Figure 2–6A shows how these fractions are related.

Inch Decimal Fractions. The inch is also divided into tenths, fiftieths, and hundredths of an inch. In the shop and laboratory, the parts are called *decimal fractions*, or *decimal divisions* (Figure 2–6B). Decimals are applied to both British and metric system basic units of measurement. Fractional divisions that are smaller than 1/64″ or 1/100″ are not practical to measure with a rule.

THE DECIMAL-INCH SYSTEM

In the late 1920s the Ford Motor Company promoted a *decimal-inch system* (Figure 2–6B). This system was based on the decimal division of the inch. Dimensions were given on drawings with subdivisions of the inch in multiples of ten. Parts were measured according to the decimal-inch system. The term *mil* was coined to express *thousandths*.

THE METRIC SYSTEM

The units of linear measure in the metric system are similar to units in the decimal-inch system. The difference is in the use of the *meter* instead of the inch. The meter (m) is the standard unit of measure in the metric system. For all practical shop purposes, the *meter is equal to 39.37″*. All other units in the metric system are multiples of the meter. The centimeter equals 0.01m or 0.3937″. The millimeter equals 0.001m or 0.03937″.

LIMITATIONS OF THE INCH AND METRIC MEASUREMENT SYSTEMS

The metric system has the advantage over the British inch system for ease of computation and communication. However, the inch system provides fractional parts of the basic units that are more functional. For example, in the inch system there is a significant range of finely spaced divisions that can be read on steel rules graduated in inches and fractional parts. Readings of 1/64″ and 1/100″ are common. Micrometer measurements in the British system may be taken with a wide range of discrimination: one one-thousandth and one ten-thousandth part of the inch.

By contrast, the millimeter (milli″ = 1/1,000; meter = 39.37″) is 1/1,000 of 39.37″, or almost 0.040″. The next smaller unit in the metric system, 0.1mm, is one-tenth this size, or about 0.004″. It is almost impossible to graduate or read such fractional measurements on a steel rule. Similarly, for micrometer measurements this same 0.1mm (0.004″) is too inaccurate for most shop measurements.

The next submultiple in SI metrics is 0.01mm (the equivalent of almost 0.0004″). A 0.01mm measurement may require a closer tolerance and more expensive machining than is necessary. In practice, metric measurements are rounded off to the nearest whole or half millimeter value. This rounding-off produces larger measurement errors than the inch system does.

COMMON LINEAR MEASURING TOOLS

Linear measurements may be taken with accuracies of 1/64″, 1/100″, and 1/2mm using steel rules. Micrometers, vernier tools, and dial indicators are used on precision measurements.

These measurements are held to one one-thousandth (0.001) and one ten-thousandth (0.0001) part of the inch, or 0.02mm or 0.002mm (almost 0.001″ and 0.0001″), respectively.

Gage blocks provide accuracies in measuring to within 1/100,000th and 1/2,000,000th part of an inch. Similar accuracies in metric measurements may be made with metric gage blocks. Comparator instruments of high amplification and other measuring devices are used to measure to still finer degrees of accuracy.

CONVERSION TABLES

Most trade handbooks contain conversion tables for ready use by skilled craftspersons and technicians. Practical daily applications require the changing of fractional and metric values to decimal equivalents, and vice versa. Table 2–2 shows selected examples of the kind of information that is provided by conversion tables.

Conversion tables with accuracies of measurements rounded to five or six decimal places are widely used in shops and laboratories. Expanded tables now include fractional, decimal, and metric unit equivalents.

Table 2–2 Partial Conversion Table of Standard Measurement Units

Millimeter	Decimal Inch	Fractional Inch	Equivalent Decimal Inch
0.1			0.00394
0.2			0.00787
	0.01		0.01000
0.3			0.01181
0.397		1/64	0.01563
0.4			0.01575
0.5			0.01968
	0.02		0.02000
0.6			0.02362
0.7			0.02756
	0.03		0.03000
0.794		1/32	0.03125
0.8			0.03149
0.9			0.03543
1.0			0.03937
	0.04		0.04000

MEASUREMENT TERMS

Interchangeability	Each part meets dimensional specifications exactly and may be interchanged with every other similar part.
Functions of dimensional measurements	To describe a physical object, to construct an object, and to control the manner in which other individuals produce an object.
Dimensional measurement	The universal language of measuring a dimension.
Linear measurement	Measuring a length between two points along a line of measurement. Expressing a distance in terms of a unit of length and a multiplier.
Reference and measured points	The beginning and termination points of a measurement.
Features	Two definable points from which measurements may be taken. (These points establish the physical characteristics of a part. A feature may be bounded by lines, edges, and ends. The features may be in the same or different planes.)
Dimension	A statement of the exact size of a feature.
Measurement	The actual size of a feature. (A measurement may be affected by temperature change.)
Precision (accuracy)	A stated degree of conformance of a measurement to the standard.
Tolerance	A dimensional range within which a feature or part will perform a specific function and may be interchanged.
SI metric limitations	Measurements with a limited range of discrimination. Maximum degree of accuracy with direct scale and micrometer readings.

Decimal-inch system	Equal divisions of the standard inch in multiples of 10.
Metric system of linear measure	Equal divisions of the meter as the standard unit of linear measure in multiples of 10, 100, 1,000, and so on.
Line-graduated measuring tools	Linear measurement tools on which equal divisions of the inch or meter are indicated by equally spaced lines. (Measurements are read directly from such measuring tools.)

SUMMARY

- Each measurement begins at a reference point and ends at a measured point. Measurements are taken along a line of measurement. The measurement must be related to a measurable characteristic of the feature.
    - A dimension represents the designer's concept of a perfect part. A measurement provides the actual size of a feature.
- Precision refers to the degree a measurement conforms to a stated standard. Tolerances specify a range of acceptable sizes.
    - An industrial drawing graphically provides precise information about size, shape, and other requirements. Measurements are taken based on the dimensions, tolerances, and other specifications found on a drawing.
- The inch is subdivided into fractional parts: 1/64″, 1/32″, 1/16″, 1/8″, 1/4″, and 1/2″.
    - The decimal inch incorporates desirable features of both metric and fractional inch measurements (in multiples of 10, 50, and 100).
- The meter (m) is the basic unit of linear measurement in the SI metric system. The millimeter (mm) is the most commonly used metric unit in the machine and metal products manufacturing industries (1mm = 0.001m or 0.03937″).
    - The steel rule, vernier caliper, and micrometer are practical basic linear measuring tools. The range of precision with these tools is from 1/64″ to 0.0001″ and from 1/2mm to 0.02mm.

UNIT 2   REVIEW AND SELF-TEST

1. a. Cite two examples of direct measurements.
   b. Give two other examples of computed measurements.

2. State three major functions of dimensional measurements.

3. Describe briefly the relationship between producing interchangeable parts and dimensional measurements.

4. Tell how each of the following measurement characteristics relates to dimensional measurement: (a) features, (b) reference point, and (c) measured point.

5. Differentiate between the use of the following measurement terms: (a) precision, (b) tolerance, (c) limits, and (d) accuracy.

6. Give three reasons why a craftsperson must be able to read drawings.

7. Identify the smallest practical unit of length that it is possible to measure with a steel rule in (a) the British (U.S.), (b) decimal-inch, and (c) metric systems.

8. List three instruments or other measuring tools that are used for layout and measuring to accuracies of 0.000001″ (0.000025mm).

SECTION TWO

Measurement and Layout Tools and Instruments

The accuracy with which metal parts are shaped and dimensioned depends, in part, on the craftsperson's ability to correctly manipulate and read a variety of semi-precision and precision measuring instruments. This section presents the principles of and correct practices for using semiprecision line-graduated measuring instruments and related transfer measuring tools.

UNIT 3

Dimensional Measurement: Instruments and Layout Tools

OBJECTIVES

A. LINE GRADUATED RULES

After satisfactorily completing this unit, you will be able to:

- Interpret scales and the use of the steel rule.
- Discriminate in measurement considering the conditions necessary for taking accurate readings.
- Check against worker practices and other factors that result in measurement errors.
- Follow step-by-step procedures for accurately measuring with the steel rule.

B. LAYOUT AND TRANSFER MEASURING TOOLS

- Recognize the functions of the depth gage, combination square, and direct-reading slide caliper and such transfer measuring tools as dividers, calipers, and surface gage for bench and surface plate work.
- Perform each of the following processes:
  - Measure with a Depth Gage and Slide Caliper.
  - Measure and Layout with a Square Head, Center Head, Protractor Head.
  - Set and Measure with Inside and Outside Calipers and Layout with the Hermaphrodite Caliper.
  - Use a Surface Gage.

A. LINE-GRADUATED RULES

The terms *scale* and *rule* are commonly used in industry. Their difference may be explained by comparing an architectural drafting scale with a machinist's rule. Different divisions (graduations) on the drafting scale may represent different linear measurements. For example, a drawing may show a scale of 1″ = 1′ or 1 cm = 1m. By contrast, the graduations on the rule represent actual measurements. The graduations may be in inches and common or decimal fractions. They may also be in millimeters and fractional parts of millimeters in the metric system.

DISCRIMINATION AND MEASUREMENT SCALES

The terms *scale* and *graduation* are used interchangeably in the shop. There are usually four sets of graduations (scales) on a steel rule, one on each edge, front and back. These scales are 1/64″, 1/32″, 1/16″, and 1/8″. The scale to use depends on the *discrimination* that is required—

that is, the degree to which a unit of length is divided and the accuracy of the tool for taking measurements. Figure 3–1 illustrates the discrimination among international inch, decimal-inch, and metric type rules.

For instance, when an inch is divided into 64 equal parts, each division represents 1/64 of an inch. On a metric rule, the centimeter (10mm) is divided into millimeters and half millimeters. The finest reading that can be taken on this rule has a discrimination of one-half millimeter. Thus, common metric rules have a coarser (less accurate) reading than the 1/64″ scale in the inch system.

ALIGNMENT OF GRADUATIONS

The type of rule and the scale to use also depend on the nature of the measurement and the degree of precision required.

The reference points that were described earlier are important in measurement. Measurements with a steel rule must fall along the line of measurement. Either the end of the steel rule or a particular graduation must be aligned with a reference point or edge. *Aligned* means that the

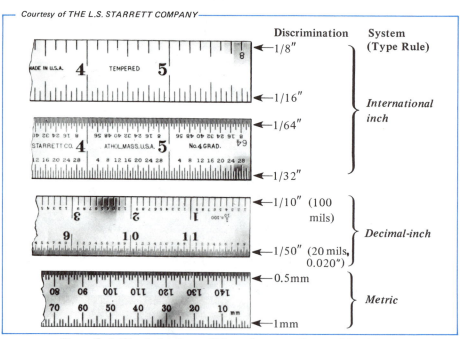

Figure 3–1 Discriminations of Three Common Types of Steel Rules

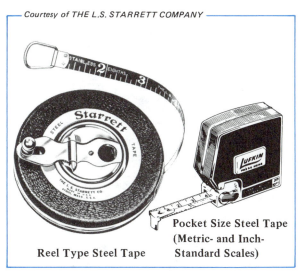

Courtesy of THE L.S. STARRETT COMPANY

Pocket Size Steel Tape
(Metric- and Inch-
Standard Scales)

Reel Type Steel Tape

Figure 3–2 Flexible Tapes Used for Long Measurements

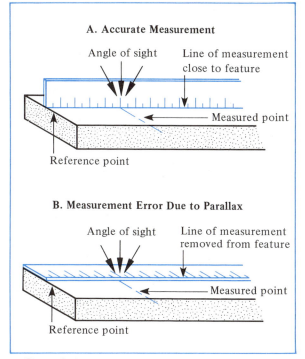

A. Accurate Measurement

Angle of sight

Line of measurement close to feature

Measured point

Reference point

B. Measurement Error Due to Parallax

Angle of sight

Line of measurement removed from feature

Measured point

Reference point

Figure 3–3 Accurate and Inaccurate Measurements

edge or a graduation on a rule falls in the same plane as the measurement reference point or edge. A measurement is then read at the *measured point*.

FLEXIBLE STEEL TAPES

Steel tapes (Figure 3–2) are another form of line-graduated measuring rule. Steel tapes are more practical for making long linear measurements because of their flexibility.

Most *pocket size tapes* are designed with a square head of a specified size. The design permits the tape to be used to measure to a measuring point in a confined area.

When using steel tapes for precision measurements, a flexible tape must be kept on and flush with the line of measurement. Compensation must also be made for expansion and contraction caused by differences in temperature.

ERRORS IN READING MEASUREMENTS

Measurements should be taken as accurately as possible using the *best scale*—that is, the scale selected to give the most accurate reading depending on the requirements of the job. For example, a rule graduated in 20 mils, or 1/64″, or 1/2mm is intended for measurements to the selected degree of accuracy. Errors in reading measurements are caused by (1) inaccurate observation, (2) improper manipulation, and (3) worker bias.

OBSERVATIONAL ERRORS

Whenever possible, a steel rule should be held at right angles to the workpiece and flush along the line of measurement (Figure 3–3A). The position of the observer, then, will not influence the reading and cause *observational errors*.

Parallax Errors. *Parallax* is a form of observational error that relates to the misreading of a measurement. The misreading is caused by the position or angle from which the graduated division is read in relation to the measuring point. The preferred accurate method of measuring a feature is shown in Figure 3–3A. The parallax conditions, as shown in Figure 3–3B, affects the reading and causes an observational error.

Errors in Reading Graduations. Fine graduations, particularly 1/64″ and 1/100″, are another possible source of observational error. In matching such fine graduations with a reference point on a workpiece, the exact dimension may be read incorrectly.

MANIPULATION ERRORS

Errors in measurement also may be caused by *improper manipulation*, or handling of, a steel rule, a workpiece, or both. Sometimes a rule is *cramped*. Cramping is the result of using excessive force. The rule is squeezed, which causes it to bend.

Another common error is to take a measurement when the rule does not lie on the line of measurement.

There is also the possibility of *slippage*, which may occur if a part is heavy or if it takes a comparatively long time to make a measurement. A dimensional error is caused by such slippage, or movement, between the measuring tool and the workpiece.

Good measurement practice requires that a rule be positioned on the line of measurement and held against the reference point using a *light touch*. This light touch is developed with repeated practice.

WORKER BIAS ERRORS

Finally, measurements are affected by *worker bias*, which means that an individual unconsciously reads an error into a measurement. Sometimes this bias is caused by a person's natural tendency to move either the rule or the line of sight to avoid parallax. It is difficult to manipulate a workpiece by hand, keep a rule along the line of measurement and aligned at the reference point, and, at the same time, see the graduation at the measured point.

How to Accurately Measure with the Steel Rule

STEP 1 Select a steel rule with graduations (scale) within the required limits of accuracy.

STEP 2 Remove any rough edges, or *burrs*, from the workpiece. Clean the rule and work. Determine whether the measurement is free of observational, manipulation, and bias errors.

STEP 3 Align the steel rule so that the scale edge is along the line of measurement.

STEP 4 Position the steel rule with the end held lightly against the reference point (Figure 3–4).

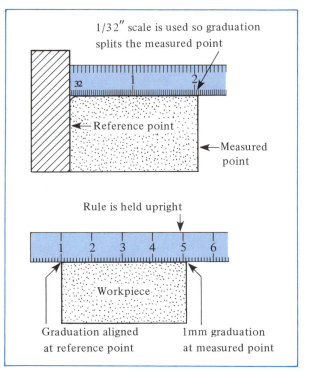

Figure 3–4 Standard Practices for Taking Accurate Linear Measurements

STEP 5 Read the measurement at the measured point (Figure 3–4). For greater precision in measurement, or if a great number of measurements are required, use a magnifier.

Note: If the measurement point does not align with a graduation, turn the rule to another edge having a finer graduation.

STEP 6 Recheck the measurement using standard practices as illustrated in Figure 3–4.

B. LAYOUT AND TRANSFER MEASURING TOOLS

Layout is a common shop term. Layout refers to the preparation of a part so that work processes can be performed according to specific requirements. A layout identifies the actual features. Lines are scribed on a workpiece:

• To show boundaries for machining or performing hand operations,

- To indicate the dimensional limits to which material is to be removed or a part is to be formed,
- To position a part for subsequent operations.

Layout requires the use of three groups of layout, measurement, and inspection instruments:

- Group 1, the steel rule with attachments that make it more functional and precise for layout as well as measurement operations;
- Group 2, related tools for the transfer of measurements;
- Group 3, precision tools, gages, and instruments.

The basic measuring and transfer tools in the first two groups are covered in this unit. Principles and applications of precision micrometers, verniers, dial indicators, and gage blocks in the third group are covered later.

THE DEPTH GAGE

Many machined surfaces require a depth measurement for a groove, hole, recess, or step. Sometimes it is not practical to use a steel rule. The rule may be too wide, or it may be easy to hold the rule on an angle and read it incorrectly.

One of the common adaptations of the steel rule is the attachment of a *head*. The head spans a groove, slot, or other indentation and rests on the reference surface. The steel rule slides through the head and may be made to *bottom* at the measured point. The combination of head and rule is called a *depth gage*.

The rules for this depth gage are narrow and flat. In other depth gages, rods are used for small holes where a measurement with a narrow rule is not possible. The rods may or may not be graduated. Examples of narrow rule depth gages are illustrated in Figure 3–5.

Some depth gages are marked for angles of 30°, 45°, and 60° as shown in Figure 3–5. The blade may be moved to the right or left. A depth gage with an angle adjustment is useful for making a linear measurement to a surface that is at 30°, 45°, or 60° to the reference surface. Depth gages, however, are *not* intended to be used in making angular measurements.

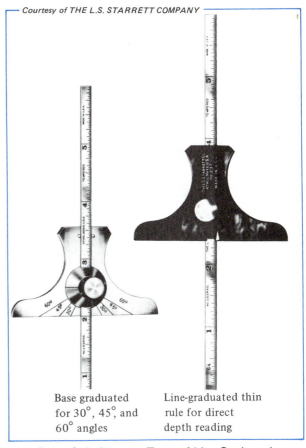

Base graduated for 30°, 45°, and 60° angles

Line-graduated thin rule for direct depth reading

Figure 3–5 Common Types of Line-Graduated Depth Gages

How to Measure with a Depth Gage

STEP 1 Remove burrs and chips from the workpiece and wipe it clean.

STEP 2 Select a depth gage with a blade size that will fit into the groove, hole, or indentation. Then, loosen the clamping screw so that the blade (rule) may be moved by a slight pressure.

STEP 3 Hold the head firmly on the reference surface (Figure 3–6).

STEP 4 Slide the rule gently into the opening until it just bottoms (Figure 3–6). Lock the head and blade.

STEP 5 Release the pressure on the head. Then, try to lightly slide the head on the refer-

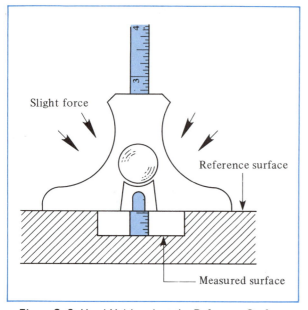

Slight force

Reference surface

Measured surface

Figure 3–6 Head Held against the Reference Surface

STEP ence surface. *Feel*—that is, sense with a light touch—whether the end of the rule is bottomed correctly.

STEP Remove the depth gage and read the
6 depth. If a nongraduated rod is used, place a steel rule next to the rod and read the measurement.

Note: In good shop practice, the measurement steps are repeated. The rule or rod is checked to see that it is properly bottomed and that the measurement is read correctly.

THE COMBINATION SET

The *combination set* is another adaptation of the steel rule. The combination set includes a *square head*, a *center head*, and a *protractor head* (Figure 3–7). A grooved rule is used with each head. The groove permits the rule to be moved into position and clamped.

THE SQUARE HEAD

One feature of the *square head* is that it provides a right-angle reference and thus extends the

use of the steel rule to laying out right-angle or parallel lines. Like the depth gage, the square head and rule may also be used to measure the depth of a hole, indentation, or other linear distance. When a measurement is to be read, the square head covers a portion of one scale beyond the measured point. Although there are four scales on the steel rule, the locking groove on one side permits direct sighting on only two scales. The scale that provides the required discrimination should be selected on the head side. The dimension should be measured at the graduation on the head side.

Another feature of the square head is that one of its surfaces is machined at a 45° angle. This feature permits the layout and measurement of many parts that require a 45° angle. Its application in scribing lines at 45° is illustrated in Figure 3–8.

How to Measure and Lay Out with a Square Head

Preparation

STEP Remove burrs and chips. Wipe the com-
1 bination square, other tools, and the workpiece clean.

STEP Select the scale on the graduated blade
2 that will provide the required accuracy of measurement.

STEP Turn the clamping nut so that the blade
3 may be moved with a slight force.

Measuring a Length

STEP Hold the head firmly against the refer-
1 ence surface.

STEP Move the blade until it exactly splits
2 the measured point. Read the dimension on the blade of the square head.

Laying Out a Line

STEP Move the blade until it extends the re-
1 quired distance from the head. Read the measurement on the scale of the blade on the head side. Lock the blade and the square head.

STEP Position the head firmly against the refer-
2 ence surface. Scribe the required line.

STEP Recheck the dimension for accuracy.
3

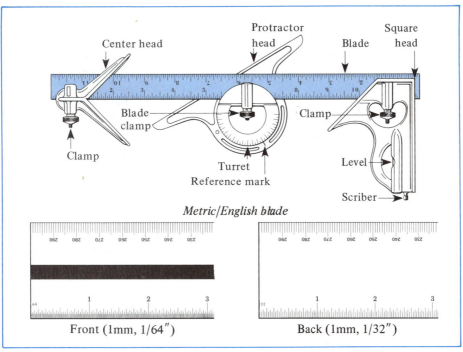

Figure 3-7 Blade and Heads of the Combination Set

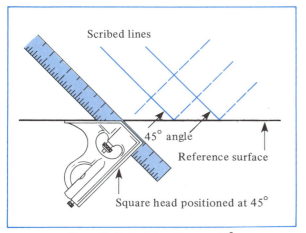

Figure 3-8 Scribing Lines at 45°

THE CENTER HEAD

The *center head* is designed so that the measuring or layout surface of the rule falls along the centerline of a diameter. The rule may be moved in the center head and locked in any desired position.

The center head is used primarily with round stock and cylindrical surfaces. It provides a

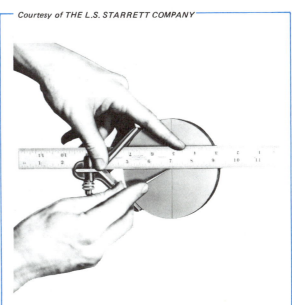

Figure 3-9 Locating a Center Using a Center Head

quick and satisfactory method of measuring a diameter. An application of the center head for laying out centerlines on a round object is shown in Figure 3-9.

THE PROTRACTOR HEAD

The *protractor head* is used to measure angles to an accuracy of 1°. This head contains a turret that is graduated in degrees from 90° to 0° to 90°, a range of 180°.

The base of the protractor head is held against the reference surface (Figure 3–10). The blade (steel rule) is secured to the turret. The turret is turned until the included angle of the blade and the protractor head coincides with the angle to be measured. The number of degrees is read at the reference mark (index line) above the turret.

The protractor head is also useful for measurement and layout work when dimensions are to be indicated on angular surfaces.

Note: A work surface is usually coated with a color-producing, fast-drying fluid. Layout lines may be scribed easily and are clearly visible on such a surface.

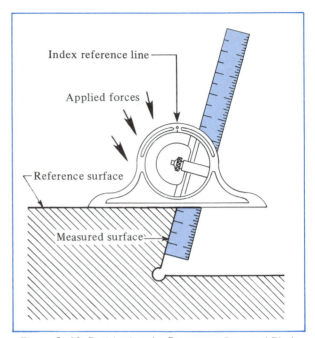

Figure 3–10 Positioning the Protractor Base and Blade

How to Measure and Lay Out Angles with a Protractor Head

Note: The work must be free of oil, grease, and burrs. All tools should be examined in advance for any imperfections that could impair their accuracy.

Measuring an Angle

STEP 1 Remove burrs and chips. Wipe the protractor head, other layout tools, and the workpiece clean.

STEP 2 Insert the blade (rule) so that it extends far enough to measure the reference surface.

STEP 3 Loosen the turret clamp to permit the turret to be rotated when a slight force is applied.

STEP 4 Hold the protractor base firmly against the reference surface of the workpiece (Figure 3–10). Position the protractor head so that the blade and turret may be moved to the desired angle.

STEP 5 Bring the blade down gently to the workpiece and adjust it. The desired angle is reached when there is no light showing between the measured surface of the workpiece and the blade.

STEP 6 Tighten the clamping screw. Read the graduation on the turret that coincides with the reference mark on the protractor head.

Measurements on an Angular Surface

STEP 1 Hold the protractor head securely against the reference surface.

STEP 2 Extend the rule until the end of it splits the measured point.

STEP 3 Read the measurement directly on one of the scales on the rule (Figure 3–10).

TRANSFER INSTRUMENTS

Technically, the *caliper* is a tool that is applied to mechanically *transfer* a measurement that originates at a reference point or surface and terminates at the measured point or surface. The four basic forms of calipers are the (1) *divider*, (2) *inside caliper*, (3) *outside caliper*, and (4) *hermaphrodite (morphy) caliper*. The hermaphrodite caliper combines some of the features of the divider and the other calipers.

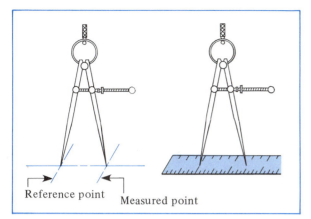

Figure 3–11 Setting the Divider Legs

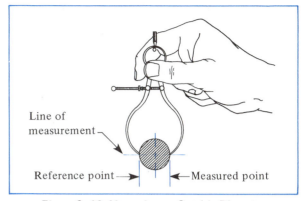

Figure 3–12 Measuring an Outside Diameter

Still other calipers, like the *slide caliper*, are made with scales. Measurements are read *directly*. The slide caliper is described later in this unit. More precise standard and vernier micrometers are considered in Unit 4.

THE DIVIDER

The divider is a simple, accurate tool for transferring or laying out a linear dimension. Parallax in measurement is at a minimum. The sharp points of the divider legs are accurately aligned with measurement lines. The divider legs may be set to a desired dimension with a rule or they may be adjusted to conform to a measurement. The distance between the divider points is then measured on a rule.

It is common practice when setting the divider legs for a measurement to place the point end of one divider leg on a graduation on a rule. Usually, the 1″ or 1cm graduation on the metric scale is used. The distance to the point end of the other leg is read on the rule (Figure 3–11). The difference between the scale readings represents the measurement.

Circles may be scribed by setting the divider legs at the distance representing the radius. One leg is set in the center point; the other leg is swung around. A full circle or an arc is scribed by applying a slight force.

When a number of parts have the same linear measurement, the divider needs to be set only once to duplicate the measurement.

INSIDE AND OUTSIDE CALIPERS

An important point to remember for measurements that are made with inside and outside calipers is that the actual line of measurement is established by *feel* in using the tools. For example, when the legs of an outside caliper are adjusted so that they just slide over a piece of stock, the measurement taken is on the line of measurement (Figure 3–12). If the outside caliper cannot slide over the workpiece with this feel or if it is set too large, an inaccurate measurement results. Accuracy is developed by correct practice in getting the feel of the instrument. Force applied to a caliper causes the legs to *spring* and produces an inaccurate measurement.

Inside and outside calipers are also used to transfer a measurement to or from a particular feature of the workpiece—that is, either the inside or outside caliper is *set* to a particular size. The setting of an outside caliper may be done by measuring the leg distance with a steel rule (Figure 3–13). An inside caliper leg may be set

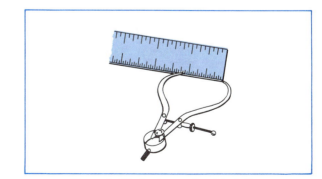

Figure 3–13 Setting an Outside Caliper

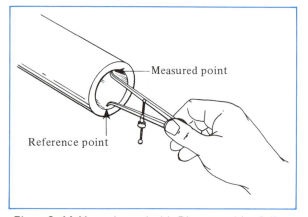

Figure 3–14 Measuring an Inside Diameter with a Caliper

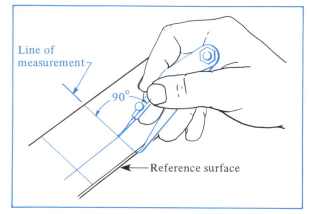

Figure 3–15 Scribing a Parallel Line with a Hermaphrodite Caliper

against another surface feature (reference point, as shown in Figure 3–14), with a micrometer, or with a gage.

If a caliper leg is set against a surface feature, the caliper may then be used to compare a feature against a specific dimension. The size is established by the feel of the caliper legs at the line of measurement. To repeat, if any force is applied, the caliper legs will spring. The lighter the worker feels the caliper measurement, the more accurate the measurement.

Calipers are usually *rocked*—that is, one leg is held lightly or rests on a surface, while the other leg is moved in a series of arcs. The rocking movement helps to establish when the caliper is located at the line of measurement. It is important to be able to manipulate the caliper from a comfortable position.

As a transfer tool, a caliper may be set according to the feature of the workpiece that is

to be measured. The caliper setting then is measured with a measuring tool. The setting also may be compared with a similar feature or part having the desired measurement.

Caution: When measuring a feature of a part which is being machined, stop the machine before calipering the work. There must be no movement of a cutting tool or of the workpiece.

THE HERMAPHRODITE CALIPER

The hermaphrodite caliper (Figure 3–15) is a layout and measurement tool. It has one leg like that of a divider and one like that of a caliper. The caliper leg may be turned 180° for either inside or outside work. The caliper leg usually rests or rides on the reference surface. The divider leg is used to scribe lines parallel to a surface, find the center of a part, or take measurements.

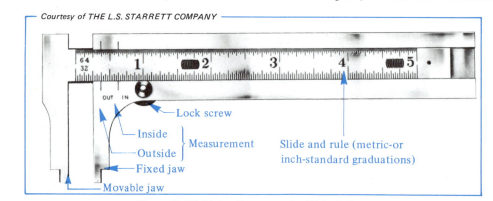

Figure 3–16 Major Features of a Slide Caliper

THE SLIDE CALIPER

The slide caliper combines the features of inside and outside calipers and a rule in one instrument (Figure 3–16). The slide caliper has the advantage of providing a direct reading. A measurement can be locked, or *fixed*, at any position. Locking prevents measurement changes caused by handling. Thus, the direct-reading feature eliminates the need to remember a measurement and is an important feature when a number of cuts are to be taken or a series of operations performed.

However, the slide caliper has disadvantages that must be considered at all times:

- The range of use is less than it is with regular caliper and rule combinations,
- The instrument cannot be adjusted to compensate for wear or the springing of the legs,
- The range of discrimination is limited by the 1/64″ or 1/2mm scale.

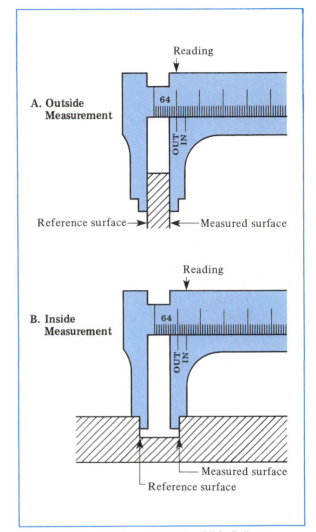

Figure 3–17 Measuring with a Slide Caliper

How to Measure with the Slide Caliper

STEP 1 Check the accuracy of the jaws by bringing them together. Note whether they are parallel and are not sprung. Check the measurement across the jaws with a micrometer or other gage.

STEP 2 Release the lock so that the jaw may be moved easily. Place one leg at the reference point.

STEP 3 Adjust the second leg to the measured point on the line of measurement. Lock the movable jaw and frame. The correct positioning for an outside measurement is shown in Figure 3–17A. The inside measurement position is shown in Figure 3–17B.

STEP 4 Read the measurement at the graduation on the rule. The measurement appears at either the "IN" (inside) or "OUT" (outside) index mark, depending on how the slide caliper is used.

Using the Slide Caliper as a Gage

STEP 1 Set the slide caliper to the required dimension. Lock it in position.

STEP 2 *Gage* the internal or external feature as shown in Figure 3–17A and B. The dimension of the workpiece is correct when the feature *just fits* the caliper setting (the feel between the work and the instrument).

Courtesy of THE L.S. STARRETT COMPANY

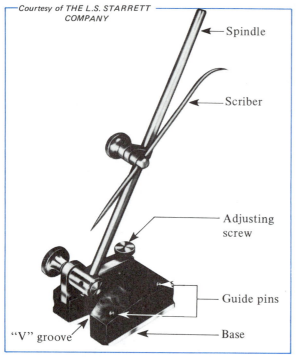

Figure 3–18 Main Parts of a Surface Gage

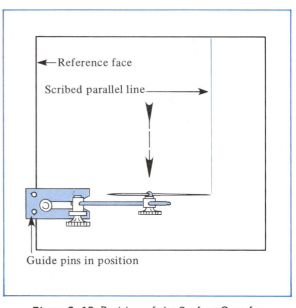

Figure 3–19 Position of the Surface Gage for Scribing Parallel Lines

THE SURFACE GAGE

The surface gage is a layout and transfer measuring tool. Its main parts are illustrated in Figure 3–18. The base is machined with a "V" groove to rest on a flat or round surface. An adjustable spindle is provided with a clamp to hold and position a scriber.

The scriber is set at the approximate required linear measurement. It is then accurately raised and lowered to the required dimension by the adjusting screw. The base also has four *guide pins*. These are pushed down when the surface gage is to be used to scribe lines parallel to a reference surface (Figure 3–19).

When scribing a line, it is easy to spring the scriber if it extends too far from the spindle. Therefore, both the scriber and spindle must be positioned close to the work.

THE SURFACE PLATE

The surface plate is an auxiliary *accessory*. As an accessory, it is combined with other tools, measuring instruments, and test equipment. It is used universally to provide a reference surface. The surface plate is essential to layout and measurement processes. Figure 3–20 provides examples of popular manufactured surface plates.

A surface plate may be a simple, flat plate that has been accurately machined. The term usually refers to a plate that has been machined and scraped to an extreme accuracy or ground to a fine surface finish. The underside is honeycombed with a number of webbed sections that prevent the surface plate from *warpage* and thus provide a permanent flat surface.

The *granite* surface plate is used extensively today and has a number of advantages over the cast-iron type:

• High degree of surface flatness;

• High stability (maintains surface flatness because there is no warpage);

• Capability to remain flat, even when *chipped* (cast-iron surface plates, by contrast, tend to *crater* and require rescraping);

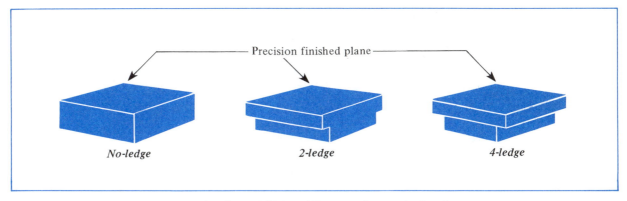

Figure 3–20 General Styles of Precision Granite Surface Plates

• Simplified maintenance (the surfaces are rustproof and may be cleaned easily with soap and water and other readily available cleaners).

The size, construction, and degree of accuracy of surface plates vary. The size naturally is governed by the nature of the layout, measurement, or assembly operations that are to be performed. The degree of precision required determines: (1) the accuracy to which the flat (plane) surface must be machined and scraped and (2) the finish of the surface. The flat surface is used as a reference point or surface.

A. Safe Practices in Dimensional Measurement

• Remove burrs from a workpiece. Burrs cause both inaccuracies in measurement and unsafe handling conditions. Burrs also affect the accurate fitting and movement of mating parts.

• Recheck all precision measurements.

• Consider the possibility of inaccurate measurement readings. These may result from observational, manipulation, and worker bias errors.

• Make it a practice to check and clean measuring tools and the workpiece surface before taking a measurement.

B. Safe Practices in Using Layout and Transfer Measuring Tools

• Examine the workpiece, layout tools, and measuring instruments. Each item should be free of burrs.

• Use a wiping cloth to clean the workpiece and layout and measuring tools. All chips and foreign matter should be removed.

• Handle all tools and parts carefully. Avoid hitting tools or parts together or scraping them against each other.

• Recheck each layout and measurement before cutting or forming. Each dimension should meet the drawing or other technical specification.

• Use a light touch to feel the accuracy of a measurement at the line of measurement. Force applied to a layout or measuring tool may cause the legs to spread and thus produce an inaccurate measurement.

• Replace each layout and measurement tool in a protective container or storage area.

A. LINE GRADUATED MEASURING TOOL TERMS

Scale	The graduations on a rule representing a specific linear dimension.
Steel rule	A common line-graduated measuring tool. A tool used in the shop and laboratory for taking and reading measurements directly from a workpiece.
Discrimination	The degree to which a unit of length is divided and the accuracy of the tool for taking measurements.
Aligned	In relation to a direct measurement, the edge of a steel rule falls on the line of measurement.
Steel tape	A flexible steel rule that may be wound into a container or head.
Best scale	The most appropriate scale to use to obtain a measurement within a desired degree of accuracy.
Parallax error	A condition resulting in a dimensional reading error. An error caused by changing the viewing position of a measurement.
Measurement bias	Conditions affecting measurement accuracy. Measurement errors caused by slippage, reading into or assuming a measurement, or faulty judgment.
Observational error	An error caused by faulty reading (viewing) of a measurement.
Manipulation error	A measurement error caused by improper positioning or handling of a measuring tool or workpiece.

B. TERMS APPLIED TO TRANSFER AND LAYOUT TOOLS

Layout	Scribing and identifying the location, shape, and size of one or more features of a workpiece.
Transfer tools	A group of hand tools and instruments used to indirectly take, transfer, or assist in establishing a measurement. (The measurement is usually read on another measuring tool.)
Depth gage	A tool for measuring the depth of holes, grooves, and slots. A plain or line-graduated depth measurement tool.
Combination set	Additions of square, center, and protractor heads to a grooved, adjustable steel rule.
Divider	An adjustable layout and measuring tool having sharply pointed legs.
Caliper	A transfer measurement tool. The two bent legs are adjustable to the size of a feature. (Inside calipers are used for internal measurements; outside calipers are used for external features.)
Slide caliper	A line graduated instrument that combines the features of inside and outside calipers and a rule.
Feel (as applied to measurement and layout)	The handling of a measuring or layout tool at the line of measurement. (The craftsperson *senses* the physical contact of the tool or instrument at the reference and measured points of the workpiece.)
Surface gage	A layout tool having a base with guide pins, an adjustable spindle, and a scriber used for layout work or the transfer of a measurement.
Surface plate	A true, flat, metal or granite plate. A ribbed cast-iron plate with a perfect plane as a reference surface. A reference plane used for layout, measurement precision, and assembly operations that all require a flat working surface.

SUMMARY

- A steel rule should be selected for linear measurements. The scales must be graduated to the degree of accuracy specified on a drawing. Consideration should be given to the measurement system. Parts with dimensions given in the metric system must be measured with metric scales.

  - Decimal-inch system measurements should be taken with a rule having inch graduations subdivided into decimal values.

- Observational errors in measurement commonly result from parallax errors.

  - Manipulation errors in measurement are largely caused by cramping a steel rule, holding a rule at an angle to the reference point, and not aligning the reading with the measuring point.

- Bias errors are personal errors introduced into a measurement by slippage or the tendency to "read into" a measurement.

  - Workpieces must be clean and free of burrs, which may cause inaccurate measurements and the improper fitting of mating parts.

- All direct-reading measurements should be rechecked for accuracy.

  - The steel rule is one of the basic layout and measuring tools. Attachments to the rule extend its use. The depth gage; combination square, center head, and protractor; and slide caliper are commonly used in shops and laboratories.

- Layout refers to tools, techniques, and processes of marking a work surface. Layout lines show the boundaries of a feature and serve as a guide for hand or machine operations.

  - Layout work requires the use of layout and measuring tools, positioning and clamping devices, and other bench-work accessories.

- The accurate machining or fitting of parts or a feature depends on the ability of the craftsperson to sense, or feel, where the measuring tool accurately touches the workpiece. Contact is at the reference and measured points on the line of measurement.

  - The three general types of calipers include the inside caliper, outside caliper, and hermaphrodite (morphy) caliper. Of the three calipers, the hermaphrodite caliper may be used for both layout work and transferring measurements. Inside and outside calipers are used principally to transfer measurements of internal and external features.

- The surface plate is an accessory used with measurement, layout, and other hand tools. It provides a true, finely machined or scraped-plane layout surface.

  - A slide caliper is used for inside or outside dimensions. The dimension may be preset or the jaws may be adjusted to a dimensional size. The jaws also may be locked in position.

- Accuracy in layout work and measurement depends on feel and judgment. Tools and workpieces must be clean and free of burrs. They must be protected from damage or inaccuracies caused by forcing, hitting, or scraping.

UNIT 3 REVIEW AND SELF-TEST

A. LINE-GRADUATED RULES

1. State what graduated scale on a steel rule should be used to measure each of the following dimensions: (1) 3 1/64", 2 19/32"; (b) 5.01", 14.08"; (c) 75mm, 10.7mm.

2. Indicate three conditions that produce errors in reading measurements on a steel rule.

3. Define the term *aligned* in relation to a direct measurement.

4. State one general practice followed by the craftsperson to check a measurement that has been taken.

5. Give two reasons for removing burrs before measuring and/or fitting two mating parts.

B. LAYOUT AND TRANSFER MEASURING TOOLS

1. List three functions that are served by layout lines.

2. Identify the tool to use to take the following measurements: (a) the diameter of a hole, (b) the distance from the face to the bottom of a hole, (c) the vertical height to the bottom of a dovetail slot, (d) the length of a part that is machined with an angle to within 1°, (e) the diameter of a round bar of stock.

3. Tell what function transfer instruments serve.

4. Describe how a measurement on an inside caliper may be accurately read on a steel rule.

5. Explain the meaning of using a *scale that provides the necessary discrimination*.

6. Cite three examples of layout processes that require the use of a hermaphrodite caliper.

7. Describe briefly how to take an inside measurement with a slide caliper.

8. Identify three layout work processes requiring the use of a surface gage.

9. Cite three advantages of granite surface plates over cast-iron surface plates.

10. State three safety precautions to take to ensure accuracy in laying-out processes.

Precision Measurement: Standard and Vernier Micrometers

Dimensional measurements that must be more accurate than 1/100″ or 1/2mm require the use of precision measuring instruments. Such instruments include standard and vernier micrometers; vernier calipers, depth gages, height gages, and bevel protractors. The principles of and correct practices for manipulating and reading these instruments are examined in this unit.

OBJECTIVES

STANDARD MICROMETERS AND VERNIER MICROMETERS

After satisfactorily completing this unit, you will be able to:

- Understand design features of inch-standard and metric standard outside, inside, and depth micrometers.
- Perform the following processes:
  - Measure with Inch-Standard and Metric Micrometers,
  - Measure with the Vernier Micrometer,
  - Measure with the Depth Micrometer,
  - Measure with an Inside Micrometer.
- Apply standard terminology associated with these instruments.

STANDARD AND VERNIER MICROMETER PRINCIPLES AND MEASUREMENTS

Many parts or features of a part require a degree of accuracy that is ±0.001″ or ±0.0001″ (±0.02mm or ±0.002mm for metric dimensioning). A *standard micrometer* is widely used for precision measurements to within 0.001″ (or 0.02mm). More precise measurements to within 0.0001″ (or 0.002mm) are taken with a *vernier micrometer*. In the shop and laboratory, the term *mike* is used for micrometer and for micrometer measuring practices.

Micrometers are made for direct hand use. They may also be designed as part of another measuring instrument. Their functions parallel the functions of calipers. Of the many different types of micrometers, the three most commonly used types—outside, depth, and inside—are covered in this unit.

Among the advantages of the micrometer are the following:

- Easier and clearer readability than rules or calipers;

- Consistently accurate measurements with close tolerances;

- No observational error due to parallax;

- Portability, comparative ease of handling, fast operation, and reasonable cost;

- Design features that include a built-in adjustment to correct for wear.

MAJOR DESIGN FEATURES OF THE MICROMETER

The major parts of the micrometer are identified in Figure 4–1. The micrometer *anvil* and *spindle* are made of a hardened alloy steel. The ends, or measuring surfaces, are finely lapped so that they are perfectly parallel to each other. As illustrated, a thread is cut directly on the spindle.

A ratchet stop is attached to the end of the spindle. Spring action in the ratchet provides a constant amount of pressure (force). This action makes it possible to obtain the correct feel for all measurements.

Many micrometers are provided with a spindle lock. Rotating the *lock nut* in one direction *locks* the spindle from turning. Slight rotation in the opposite direction releases the pressure. The spindle is then free to turn. Outside micrometers are available in a wide range of sizes from 1/2″ to 60″ (12mm to 1,500mm) capacity. The maximum range of measurements that may be taken with a micrometer is usually 1″ or 25mm.

The micrometer design provides for wear on the anvil and spindle faces and for adjustments due to frame conditions. The *thimble* and spindle are made in two parts. They may be loosened and the thimble scale turned and reset to zero at the index line. The two parts are then secured as a single part. An *adjusting nut* may be turned to take up the clearance caused by wearing of the threads.

MICROMETER PRINCIPLE AND MEASUREMENT

All micrometers operate on the principle that a *circular movement* of a threaded spindle produces an *axial movement*. Thus, as the spindle assembly is turned, the end (face) of the spindle moves toward or away from the fixed anvil. The distance (axial movement) per revolution (circular movement) depends on the pitch of the threaded spindle. The *pitch* refers to the number of threads either per inch or in 25mm.

For correct micrometer measurement a workpiece is calipered along its line of measurement. The actual measurement is the distance between the reference and measured points. Figure 4–2 shows the positions of the micrometer and workpiece in taking a correct measurement. The accuracy of the measurement depends on the feel—that is, the contact between the instrument, the work, and the operator in a controlled environment.

Inch and metric outside micrometers are used for similar measurements. The instruments are handled in the same manner. Direct readings are made on the *barrel* and *thimble scales.*

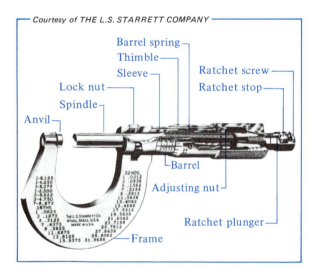

Figure 4–1 Cutaway View of an Inch-Standard Outside Micrometer

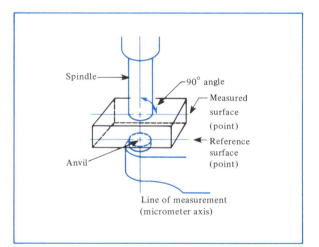

Figure 4–2 Positions of the Micrometer and Workpiece for Correct Measurement

THE INCH-STANDARD MICROMETER

The screw thread on the spindle of the inch-standard outside micrometer has 40 threads per inch. Its pitch, then, is 1/40″, or 0.025″. As the spindle is turned one complete revolution, it moves 0.025″ (Figure 4–3A).

Attached to the spindle is the thimble. The thimble is graduated with 25 graduations around its circumference (periphery). By dividing the movement of the spindle (0.025″) for each revolution by 25, the distance represented by each graduation on the thimble is 0.001″ (Figure 4–3B).

For greater readability the lengths of some of the lines that represent the graduations on the thimble and barrel are varied (Figure 4–4). For example, since every fourth graduation on the barrel represents 0.100″ (0.025″ × 4), each 0.100″ is represented by a long graduation line. The long lines on the barrel are numbered 0, 1, 2, 3, and so on to 10. Each graduation on the thimble represents 0.001″. The long lines on the thimble mark every fifth graduation and are numbered 5, 10, 15, 20, and 0 (for 25). There usually is a series of smaller-sized numbers for the intermediate graduations.

A micrometer measurement is obtained from the readings on the barrel and the thimble. The thimble reading is taken at the point where the

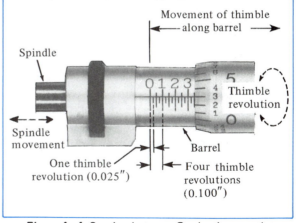

Courtesy of THE L.S. STARRETT COMPANY

Figure 4–4 Spacing between Graduations on the Micrometer Barrel (each represents 0.025″) and on the Thimble (each represents 0.001″)

index line on the barrel crosses a graduated (or fraction) line on the thimble (Figure 4–5). The barrel and thimble readings are added and then represent a micrometer measurement.

Through correct practice, micrometer measurements are read directly and almost automatically. Starting at the index line, the readings on the barrel and thimble are read as a continuous numerical value.

Control of Measurement Conditions

In every measurement situation, four conditions must be controlled:
  — *Control of the part* (this control requires manufacturing controls to produce pre-

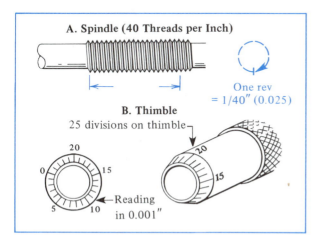

Figure 4–3 Relationships between (A) Micrometer Spindle Pitch (40 threads per inch); Spindle/Thimble Movement (1/40″, or 0.025″, per revolution); and (B) Graduations on the Thimble of an Inch-Standard Micrometer

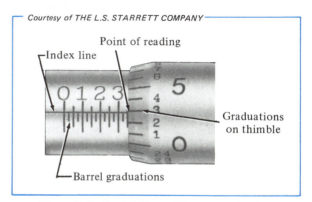

Courtesy of THE L.S. STARRETT COMPANY

Figure 4–5 Establishing a Micrometer Reading

cisely machined surfaces from which accurate measurements may be taken);
— *Built-in design features* (such features permit accurate readings to be taken consistently with the measuring instrument);
— *Craftsperson's expertise* in taking and reading a measurement;
— *Environment* (precision measurements require adequate temperature controls of the workpiece and measuring instruments and the elimination of vibration, dust, and other environmental conditions that impair accuracy).

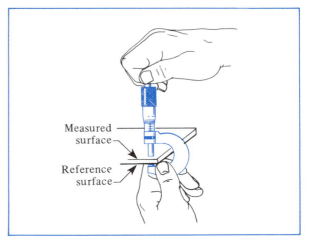

Figure 4–6 Correct Finger and Micrometer Positions for Taking an Accurate Measurement

Principles for Using a Micrometer

— Stop the machine if a mounted piece of work is to be measured. Be sure there is no movement of the work.
— Check the workpiece. Remove burrs, dirt, abrasives, and other foreign particles with a clean cloth. Recheck the surface by slowly and lightly moving the hand across the faces to be measured.
— Clean the measuring surfaces of the micrometer. Bring the anvil and spindle faces lightly together on a piece of paper. Gently pull the paper through. Blow away any remaining paper fibers from the faces.
— Check the accuracy of the micrometer. The thimble reading should be *zero*. Any variation requires adjustment. The cap holding the thimble and spindle are unlocked with a micrometer spanner wrench. The thimble is moved gently until a zero reading is obtained. The cap is then locked. The thimble is secured at zero.
— Use the ratchet to equalize the minute pressure that is applied to obtain uniform measurements.
— Feel the measurement at the line of measurement. The feel is established as the anvil and spindle faces are brought to the reference and measured points and are moved on the work.
— Carefully place the micrometer, as well as all other measuring instruments, in a location where they will not be hit, dropped, or otherwise damaged.

How to Measure with the Inch-Standard Micrometer

Taking a Measurement

STEP 1 Adjust the micrometer opening so that it is slightly larger than the feature to be measured.
STEP 2 Hold the frame of the micrometer with the fingers of one hand. Place the micrometer on the work so that the anvil is held lightly against the reference point.
STEP 3 Turn the ratchet (or the thimble directly if there is no ratchet) with the fingers of the other hand. Continue turning until the spindle face just touches the work (measured point). The correct position of the hands and micrometer are illustrated in Figure 4–6.

Reading a Measurement to within 0.001″

STEP 1 Lock the spindle with the lock nut. Note the last numeral that appears on the graduated barrel (A in Figure 4–7).
STEP 2 Add 0.025″ for each additional line that shows on the barrel (B in Figure 4–7).
STEP 3 Add the number of the graduation on the thimble (C in Figure 4–7).
STEP 4 Read the micrometer measurement in *thousandths*. The measurement is the *sum* of the barrel and thimble readings (A, B, and C in Figure 4–7).

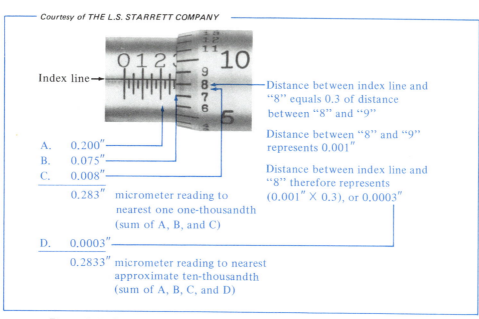

Courtesy of THE L.S. STARRETT COMPANY

Index line →

Distance between index line and "8" equals 0.3 of distance between "8" and "9"

Distance between "8" and "9" represents 0.001″

Distance between index line and "8" therefore represents (0.001″ × 0.3), or 0.0003″

A.   0.200″
B.   0.075″
C.   0.008″
_____
     0.283″   micrometer reading to nearest one one-thousandth (sum of A, B, and C)

D.   0.0003″
_____
     0.2833″ micrometer reading to nearest approximate ten-thousandth (sum of A, B, C, and D)

Figure 4–7 Reading a Standard, or Nonvernier, Micrometer to the Nearest Approximate Ten-Thousandth of an Inch (0.0001″)

Note: Where a micrometer larger than 1″ is used, the total measurement equals the whole number and the decimal fraction read on the micrometer head. For example, with an 8″ micrometer the reading would be 7″ plus whatever decimal is shown on the micrometer.

Estimating to Finer than 0.001″

STEP 1 Estimate the distance between the last graduation on the thimble and the index line. For instance, in Figure 4–7 the distance between "8" and "9" is about one-third (0.0003″).

STEP 2 Add the decimal fractional value of this distance (D in Figure 4–7) to the thousandths reading (the sum of A, B, and C in Figure 4–7, or 0.283″). The answer gives an approximate four-place decimal value (0.2833″ in Figure 4–7).

THE VERNIER MICROMETER AND 0.0001″ (0.002mm) MEASUREMENT

The vernier micrometer is used to make precise measurements that are within the *one ten-thousandth of an inch*, 0.0001″, range (or the metric system equivalent, 0.002mm). The addition of the vernier scale to the barrel of a micrometer extends its use in making these precise measurements. The vernier principle dates back to 1631 and is named for its inventor, Pierre Vernier.

The vernier principle is comparatively simple. There are ten graduations on the top of the micrometer barrel that occupy the same space as nine graduations on the thimble. The difference between the width of one of the nine spaces on the thimble and one of the ten spaces on the barrel is one-tenth of one space.

Each graduation on the thimble represents one-thousandth of an inch. Thus, the difference between graduations on the barrel and the thimble is one-tenth of one-thousandth, or one ten-thousandth part of an inch (Figure 4–8A).

The vernier reading is established by sighting across the graduated lines on the barrel and the divisions on the thimble. The line that coincides represents the *ten-thousandths vernier reading*. This reading is added to the three-place (one-thousandths) reading. The result is a four-place (ten-thousandths) vernier micrometer reading (Figure 4–8B) of 0.2970″.

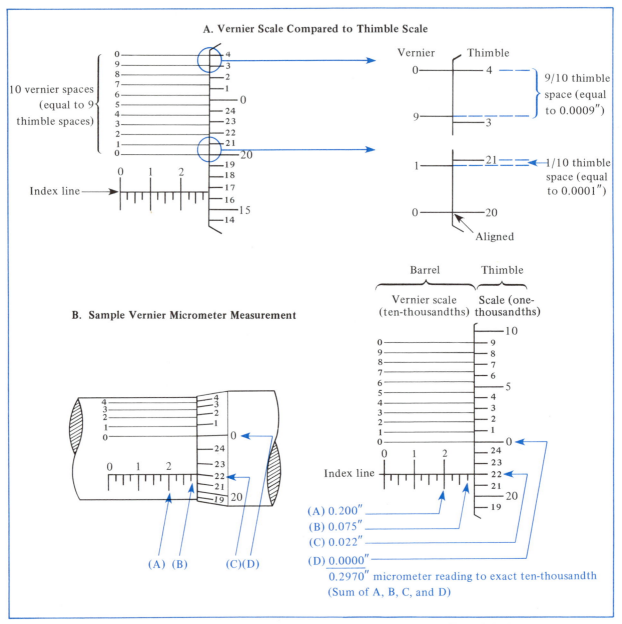

A. Vernier Scale Compared to Thimble Scale

B. Sample Vernier Micrometer Measurement

(A) 0.200″
(B) 0.075″
(C) 0.022″
(D) 0.0000″
0.2970″ micrometer reading to exact ten-thousandth
(Sum of A, B, C, and D)

Figure 4–8 Vernier Micrometer Scales and a Sample Measurement to Exact 0.0001″

THE METRIC MICROMETER

The principles and the applications of inch and of metric micrometers are similar. The difference lies in the basic unit of measure and the discriminating ability of the instruments. For example, standard inch micrometers can be used to measure to .001″ (0.02mm) and vernier micrometers, to .0001″ (0.002mm). By comparison, metric micrometers have discriminations of 0.01mm and 0.002mm. The 0.01 mm measurement is equal to 0.0004″, which is a higher degree of accuracy than is normally required for some work. The 0.002mm measurement (equal

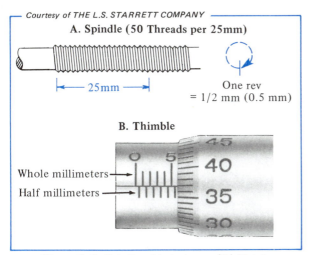

A. Spindle (50 Threads per 25mm)

25mm

One rev
= 1/2 mm (0.5 mm)

B. Thimble

Whole millimeters

Half millimeters

Figure 4–9 Relationship between (A) Metric Micrometer Spindle Pitch (50 threads in 25mm) Spindle/Thimble Movement (1/2mm, or 0.50 mm, per revolution); and (B) Spacing between Graduations on the Barrel (whole and half millimeters)

to approximately 0.00008″) is beyond the range of accuracy of most workpieces and the instrument itself.

METRIC MICROMETER GRADUATIONS AND READINGS

The screw of the metric micrometer has 50 threads in 25mm (Figure 4–9A). Each turn of the spindle produces an axial movement of 1/2mm (about 0.020″). Two turns moves the spindle 1mm (almost 0.040″). In contrast to the inch micrometer, there are two rows of graduations on the metric micrometer barrel (Figure 4–9B). The top row next to the index line represents whole millimeters. They start at 0, and every fifth millimeter is numbered. The graduations on the lower row represent half millimeters (0.50mm).

There are 50 graduations on the thimble (Figure 4–10). One complete revolution of the thimble moves it 1/2mm (or 0.50mm). The difference between two graduations on the thimble is 1/50 of 0.50mm, or 0.01mm.

The metric micrometer reading is established by adding three values (Figure 4–10):

- The last whole number of millimeters that is visible in the top row near the thimble (A in Figure 4–10),

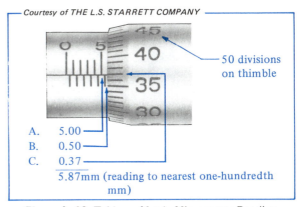

50 divisions on thimble

A. 5.00
B. 0.50
C. 0.37

5.87mm (reading to nearest one-hundredth mm)

Figure 4–10 Taking a Metric Micrometer Reading

- The half-millimeter graduation in the lower row that shows (B in Figure 4–10),
- The graduation on the thimble at the index line (C in Figure 4–10).

Like the inch micrometer, the metric micrometer may be used to estimate any fractional part that falls between two graduations. Estimating adds to the accuracy of the measurement.

THE DEPTH MICROMETER

The depth micrometer (Figure 4–11) combines the features of a micrometer head and base. The same operations are performed as with a depth gage having a steel rule blade. The precision of the depth micrometer is within the 0.001″ range on the standard inch micrometer and 1/2mm (0.50mm) on the metric micrometer head. While vernier graduations on the head would increase the possible range of accuracy of measurements to 0.0001″, such measurements would not be consistently reliable.

The usual range of measurements for depth micrometers is 1″ or 25mm. This range is increased by the addition of interchangeable *measuring rods*. As with all measuring tools, care must be taken to properly seat the measuring rods. The accuracy of the zero setting of the depth micrometer may be easily checked by holding the base against a flat surface. The face (end) of the measuring rod is then brought into contact with the surface by using the ratchet. Any variation from zero should be adjusted following the steps used with a regular micrometer.

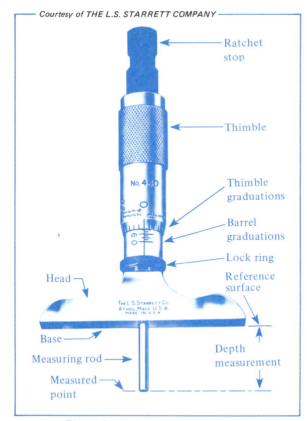

Ratchet stop

Thimble

Thimble graduations

Barrel graduations

Lock ring

Reference surface

Head

Base

Measuring rod

Measured point

Depth measurement

Figure 4–11 Functional Features of the Depth Micrometer

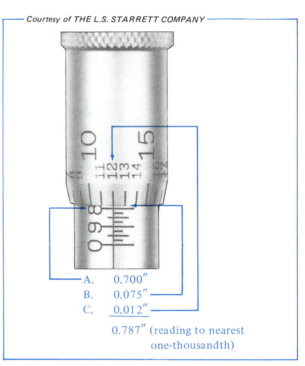

A.   0.700″
B.   0.075″
C.   0.012″

0.787″ (reading to nearest one-thousandth)

Figure 4–12 Reading a Depth Micrometer to the Nearest One-Thousandth of an Inch (0.001″)

Gage blocks of specified sizes may be used to check the accuracy along the depth range of the instrument.

An important point to remember with the depth micrometer is that it measures *in reverse* from other micrometers. (1) The zero reading of the depth micrometer appears when the thimble is at the *topmost* position. (2) The depth micrometer graduations are in *reverse* order. (3) The thimble reads *clockwise*.

How to Measure with the Depth Micrometer

STEP 1 Prepare the workpiece. Select the appropriate length measuring rod. Prepare and test the depth micrometer for accuracy. Then, adjust the measuring rod so that it is almost to the depth to be measured.

STEP 2 Position and apply a slight pressure to hold the micrometer base against the reference surface.

STEP 3 Lower the measuring rod. Use the ratchet to bottom the rod.

STEP 4 Note the numbered graduation that appears in full at the index line on the barrel. The first number of the reading is less than the full numeral that is exposed (A in Figure 4–12, 0.700″).

STEP 5 Read the last (0.025″) graduation that is covered (Figure 4–12). The reading becomes the second and third decimal place reading (B in Figure 4–12, 0.075″).

STEP 6 Read the graduation on the thimble at the index line (Figure 4–12). It represents thousandths of an inch (C in Figure 4–12, 0.012″).

STEP 7 Combine the readings at the index line of the graduations on the barrel and the thimble. The sum represents the depth in thousandths (0.787″ in Figure 4–12).

> **Note:** Where the extension rods are used, the length of the rod is added to the measurement on the micrometer head.

DEPTH MEASUREMENTS IN METRIC

Follow the same procedures with the metric depth micrometer as described for the standard inch micrometer. The measurements will be in millimeters and fractional parts of a millimeter.

THE INSIDE MICROMETER

The inside micrometer (Figure 4–13) consists of a micrometer head, a chuck, and two contact points. Precision extension rods and a spacing collar extend the range of measurements possible with a single micrometer head. A handle is provided for use when small holes or distances are

to be measured. A micrometer head with an extension rod is illustrated in Figure 4–13.

The graduations on the micrometer head may be in thousandths of an inch, in hundredths of a millimeter, or finer. The instrument, like that of all other precision measuring instruments, must be checked against an accurate standard and adjusted for accuracy.

APPLICATION OF THE INSIDE MICROMETER

The features and correct alignment of an inside micrometer for taking an accurate inside measurement are shown in Figure 4–13. A measurement is taken by closing the inside micrometer to almost the size of the feature to be measured. The micrometer thimble is turned until contact is established at the measured point along the line of measurement. The micrometer is removed from the work. The reading is made directly by combining the barrel and thimble readings.

TELESCOPING AND SMALL-HOLE GAGES

The micrometer head size limits its use to minimum internal measurements of close to 2″ or 50mm. For smaller measurements it is necessary to use transfer measuring tools. Among these tools are the *telescoping* and *small-hole gages*. Measurements with these gages extend down to 1/8″ (3mm).

The telescoping gage consists of two tubes (legs) that are spring loaded. The legs are locked at the reference points of a measurement by a locking nut or screw in the handle (Figure 4–14). The measurement is then transferred to and read on a micrometer or other measuring instrument.

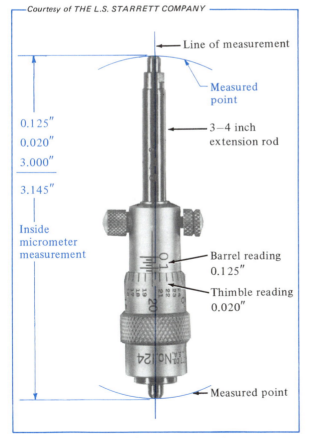

Courtesy of THE L.S. STARRETT COMPANY

Figure 4–13 Taking an Inside Diameter Measurement

Courtesy of THE L.S. STARRETT COMPANY

Figure 4–14 Application of Telescoping gage on an Inside Diameter Measurement

MEASUREMENT TERMS FOR STANDARD AND VERNIER MICROMETERS

Basic micrometers (mikes) — Three basic measuring tools referred to as the *outisde mike*, *inside mike*, and *depth mike*. (The name identifies the measurements that may be taken with each instrument.)

Micrometer reading — A precise measurement representing the distance between two reference points at the line of measurement. A direct reading that is the total of the barrel, thimble, and vernier scale readings. (Where applicable, the length of the extension rod is added.)

Estimating a fractional part (0.0001″ or 0.002mm) — Observing the fractional part of a 0.001″ or 0.02mm division on the thimble at the index line. The addition of the fractional (distance) value to the whole number value on the thimble.

Index line — A horizontal line on the barrel. A fixed point from which micrometer measurements are read.

Telescoping gage — A nongraduated transfer measuring tool. A set of spring-loaded legs that are adjustable for internal measurements. (The legs are locked at a correct measurement.)

Scale (barrel and thimble) — The graduations on the barrel and thimble of the micrometer head. (The barrel scale represents the pitch of the spindle thread. The thimble scale represents 1/25 or 1/50 of the spindle thread pitch—0.001″ or 0.02mm, respectively.)

Control conditions — Control of workpiece production, reliability of the instrument, worker expertise, and environment for precision measuring.

Micrometer head — A precisely machined and graduated adjustable measuring unit. A measuring tool that consistently produces precise measurements. (The addition of a frame, a base, or extension rods extends the use. Inside, outside, depth, and other precise linear measuring instruments are examples.)

SUMMARY

- Micrometers are basic precision measuring tools. Standard micrometers are graduated to measure within 0.001″ or 0.02mm.
  - A vernier scale may be added to the micrometer barrel. The vernier scale makes it possible to read measurements to within 0.0001″ or its metric equivalent, 0.002mm.
- Provision is made in the design of the micrometer to compensate for wear of the contact measuring surfaces and the threaded spindle.
  - Micrometer measurements are based on the principle that the circular motion of a screw thread produces an axial linear movement.

■ The pitch of the metric micrometer screw thread is 1/2mm (approximately 0.020″) as contrasted to the 1/40″ pitch (0.025″) for the inch micrometer.

   ■ The barrel of the metric micrometer has two scales: an upper scale (1mm) and a lower scale (0.50mm). The inch micrometer has a single row of graduations: each graduation represents 0.025″.

■ The accuracy of micrometer measurements (in fact, of all measurements) depends on four controls:
   — Production of the workpiece,
   — Accuracy built into the design features of the instrument,
   — Expertise of the craftsperson,
   — Environment.

   ■ The thimble of the standard metric micrometer is graduated in two one-hundredths of a millimeter, 0.02 mm; the inch micrometer thimble, in one one-thousandth of an inch, 0.001″.

■ A standard micrometer measurement combines the readings on the barrel and thimble. Where extension rods are used, the additional length is added.

   ■ Small inside measurements are usually transferred to a micrometer with small-hole gages or a telescoping gage.

UNIT 4 REVIEW AND SELF-TEST

STANDARD AND VERNIER MICROMETERS

1. Cite three advantages of using micrometers over steel rules and calipers.

2. Tell why measurements taken for the inch micrometer with a ratchet stop are consistently more accurate than measurements made with a standard micrometer.

3. State three safe practices that must be observed when using a micrometer.

4. Indicate how a 0.001″ micrometer may be used to estimate a measurement finer than 0.001″.

5. Describe briefly the vernier principle as applied to an outside micrometer.

6. State how the spindle graduations of a depth micrometer differ from the spindle graduations on an outside micrometer.

7. Explain how to take an accurate measurement with an inside micrometer.

8. Cite an example of where telescoping and small-hole gages must be used instead of inside micrometers or transfer measuring tools.

PART 3 Bench and Floor Work: Technology and Processes

SECTION ONE

Bench Work and Noncutting Hand Tools

Work processes performed with hand-manipulated tools and instruments and with portable power equipment are generally categorized as bench work or floor work. Among these processes are layout, fitting, finishing, assembling, and various other noncutting, cutting, and forming operations. This section introduces bench and floor work.

UNIT 5

Preparation for Bench Work Processes

OBJECTIVES

After satisfactorily completing this unit, you will be able to:

- Understand the functions of a work order, production plan, and blueprint.
- Trace the step-by-step procedures for preparing a workpiece for a required sequence of operations.
- Recognize and apply good work habits and attitudes as they affect worker performance, efficiency, and a wholesome working environment.
- Apply basic *Terms* used in bench and floor work.
- Follow *Safe Practices* in bench and floor work.

SCOPE AND REQUIREMENTS OF BENCH AND FLOOR WORK

Bench and floor work involve the use of hand tools and small portable equipment, layout tools, cutting and forming tools, and measuring instruments. The tools and instruments are all manipulated by hand. Accurately performing a bench work process requires the development of hand skills and a working knowledge of technology related to the process. The term *bench work* refers to setting up, laying out, and checking dimensions and measurements. Bench work also includes filing, sawing, threading and reaming operations, and assembling and fitting processes. Two typical bench work processes are shown in Figure 5–1.

A work area larger than a bench is sometimes required because of the size of a workpiece or unit. Bench work operations are then performed on the floor as *floor work*. Identical tools and instruments are used.

Bench and floor work require the craftsperson and technician to:

- Read and communicate in the technical language of the trade;
- Make necessary computations using technical manuals and handbooks;
- Relate machine and other forming processes to bench finishing operations;
- Plan a work procedure in which each operation is performed in proper sequence;
- Check the accuracy at each step of an operation and make whatever adjustments are needed;
- Use tools properly;
- Observe safe practices to prevent personal injury or machine, tool, or workpiece damage.

BASIC BENCH WORK PROCESSES

Bench and floor work involve a wide variety of hand tools, measuring instruments, layout tools, and accessories, as well as a great range of operations. The applications of these tools, instruments, and accessories vary with each job. It should be noted that many of them are also used in machining operations.

Bench and floor work require the use of hand tools for basic operations such as:

- Hammering,
- Holding,
- Turning and fastening,
- Punching and shearing,
- Sawing,
- Reaming,
- Threading,
- Filing and burring,
- Polishing.

Common hand tools and their proper use for performing these operations are covered in succeeding units. The remainder of this unit deals with:

- Specific information provided by a work order,
- Preparation of surfaces for accurately laying out a workpiece and performing bench operations,
- Work habits and attitudes,
- Safety rules.

Courtesy of THE L.S. STARRETT COMPANY

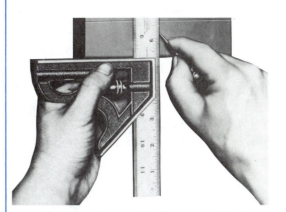

Scribing a "square line" on a coated surface

Hand tapping

Figure 5-1 Typical Bench Work Processes

WORK ORDER, PRODUCTION PLAN, AND BLUEPRINT OR SKETCH

The *work order* and/or *production plan* provide the following types of information about a particular part or unit: name, function, quantity, work processes, schedule, and so on. The production plan provides technical information about tool requirements, the machining sequence, tolerances, sizes, machine setups, cutting speeds and feeds, and other production conditions. The work order and production plan are usually used with a mechanical drawing, freehand sketch, or blueprint.

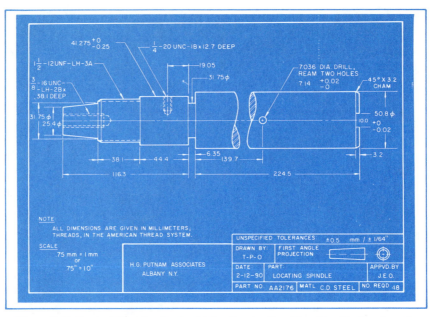

Figure 5–2 Typical Drawing from which a Production Plan May Be Developed

The blueprint graphically describes parts and mechanisms. The blueprint description is complete and accurate so that a part may be made of the same materials, machined to the same degree of accuracy and surface finish, and precision assembled anywhere in the world. Figure 5–2 is an example of a typical working drawing from which a production plan and work order may be prepared.

The blueprint and the work order provide all essential information for the performance of a bench work process. This information includes the technical details of form, sizes, quantity, material, finish, and a part's relation to other fitted or mated parts. The same technical information is conveyed to the designer, draftsperson, engineer, mechanic, technician, and consumer. Each individual is then able to:

- Form a mental picture of the shape, features, material used, and size of each part;
- Visualize the fabricating and manufacturing processes required to make a part;
- Plan the sequence of steps for performing each process;
- Translate the lines, symbols, views, sections, dimensions, and notes found on drawings and sketches in laying out and performing

other bench and floor work processes and machining operations.

The blueprint and the work order or production plan are the starting point. The worker must be able to secure materials, produce the workpiece, and perform efficiently whatever bench and assembly operations are needed.

CONDITIONING THE WORKPIECE

It is important for grease, dirt, and other foreign particles to be removed from a workpiece before layout processes are begun. Therefore, castings are usually machine tumbled to remove sand and rough projections. Forgings are ground to remove rough edges and fins—that is, excess material formed in the area where die sections meet. Steel rods and bars that have protective coatings are immersed in a cleansing bath to remove film, grease, and dirt.

Equally important is the need to continuously check the work to be sure that all burrs are removed. Burrs are a safety hazard because they can cause cuts and bruises. Burrs also can cause damage to parts in assembly and inaccuracies in layout, machining, or finishing operations. Burrs can be removed from hardened surfaces by grinding or stoning the edge or surface with a small abrasive sharpening stone. Burrs on internal surfaces such as holes or fillets are sometimes cut away with a hand scraper. Other edges are usually broken by filing them with a smooth-cut file.

Cleaning, burring, and removing foreign matter from a workpiece, then, are important steps in conditioning the workpiece. The work surface must be clean if a layout dye or other coating is to be applied.

PREPARATION OF THE WORKPIECE

One of the unique functions of all machine and metal trades is the production of accurately assembled parts of precise shape, size, and surface finish. The sizes, shapes, and other characteristics of materials drawn from supply sources should always be checked against the corresponding size, shape, and other physical specifications in the work order. The material should be sufficiently oversized to permit the worker to perform the hand and machine processes necessary to produce a particular part.

The material from which a part is to be made may be a forging, casting, or rolled or otherwise preformed or shaped bar. It is often necessary when working with such material to prepare the work surface. Preparation of the work surface permits layout lines that are clearly visible and easy to follow to be drawn. A precision layout using a vernier height gage is shown in Figure 5-3.

The most practical and universally used of any surface coloring agents are layout dyes. These dyes may be used with ferrous metals and with some nonferrous metals such as aluminum, brass, bronze, and copper. After the layout surface is burred and cleaned to remove any oil film, the dye solution is brushed or sprayed on.

Layout dye dries rapidly and provides a deep blue, lavender, or other-colored surface. Finely scribed lines stand out sharply and clearly. Figure 5-4 shows a cleaned surface being sprayed with a layout dye.

The surfaces of rough castings and forgings may be painted with a thin white pigment that is mixed with alcohol. This white pigment spreads easily and dries rapidly. It produces an excellent coating for layout operations. Ordinary white chalk is used sometimes. However, the chalk

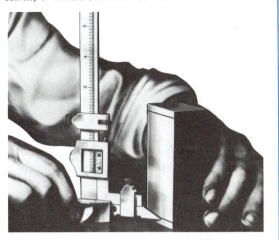

Figure 5–3 Precision Layout Lines Clearly Visible on a Coated Work Surface

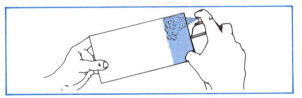

Figure 5–4 Spraying Layout Dye on a Cleaned Surface

coating may wear or be brushed away before operations are completed. The layout must then be repeated. Therefore, chalked coatings are not as desirable or practical as dyes are.

WORK HABITS AND ATTITUDES

The skill of the worker is affected by work habits and attitudes. Concern for the safe handling and storage of all tools is essential. Safe practices as applied to individual bench tools and processes are examined at the time when each tool and its applications are considered.

The skilled craftsperson works in an orderly manner. Attention is paid to the condition, placement, and handling of all layout, cutting, or other hand tools and instruments. The work place and the part or parts are all considered in terms of safety and accuracy.

A number of hand tools have cutting edges or are heavy. These tools must be placed where they do not rest on or scrape against finished surfaces or precision measuring tools. Rough, unmachined surfaces must also be positioned so that they are not in direct contact with finely finished surfaces or precision instruments. Accuracy and personal safety require that separate places on the work bench be planned for placing and working with different tools.

Scraping produces scratched or scored surfaces. The skilled worker, therefore, avoids scraping one surface against another. Instead, parts are gently picked *up* and carefully placed in position. Surfaces of different degrees of finish that are to be placed together should have a thin metal sheet or even paper placed between the surfaces to protect them. Protecting surfaces is particularly important for parts that must be layed out before machining operations are performed.

Safe Practices in Preparing Work for Bench or Assembly Processes

- Clean the work place and remove grease and dirt from all tools.

- Remove abrasive and foreign particles and burrs from all workpieces.

- Place cutting tools and rough hand tools side by side in one area of the work bench.

- Check the accuracy of all measuring tools and the condition of all cutting tool and workpiece edges.

- Place a clean cloth or a soft-surfaced material under all precision measurement and layout tools.

- Pick *up* each tool to avoid sliding one tool on another. Sliding can cause scratching.

- Return each tool to its appropriate place on the work bench.

- Wipe off each tool and return to its proper container or holder.

TECHNICAL TERMS USED IN BENCH AND FLOOR WORK

Bench and floor work	A series of hand or portable machine processes performed at a bench or on the floor. (Examples include laying out, cutting, forming, finishing, fitting, and assembling one or more parts.)
Technical trade language	Terms, symbols, techniques, and devices used by craftspersons and technicians to communicate complete specifications about a part or mechanism.
Hand tools	Tools that are manipulated by hand. (Common hand tools include tools used for layout processes, holding, hammering, fastening, punching, shearing, sawing, filing, threading, burring, and polishing.)
Work order	An information sheet used by a worker to produce a part or mechanism; shape, size, quantity, material, and other specifications.
Production plan	The sequence of steps and the routing of a workpiece through a shop or laboratory.
Workpiece	A mass (material) having a particular shape (features), size, composition, and surface finish. (The workpiece may be cast, forged, preformed, welded, rolled, or machined.)
Preparing the workpiece	The preparation of one or more surfaces of a workpiece by cleaning, burring, and removing foreign particles.
Surface treatment	The application of a coating on a work surface to make layout lines clearly visible and to establish the limits of a dimension.
Coloring agents	Whiting, multi-colored dye, copper sulphate, and other solutions used in shops and laboratories to color a work surface.

Shop layout	Processes of marking the locations and shape of different features of a workpiece. Locations for positioning a workpiece as a guide for hand and machine operations.
Work habits and attitudes	Qualities and attitudes developed by skilled craftspersons. Desirable performance standards on the job.

SUMMARY

- Machining processes are usually preceded by hand and other layout processes. These hand and layout processes are followed by checking, fitting, and assembly operations at a bench or on the floor.
    - Bench and floor processes may be adapted to machine operations.
- Layout and measuring tools and some cutting hand tools are also used on lathes, milling machines, or other machine tools.
    - The craftsperson must be able to:
        - Interpret specifications on drawings,
        - Make calculations,
        - Apply science and trade skill and technical knowledge to produce accurately and efficiently a required part or mechanism.
- In drafting and on shop layouts, lines and surfaces are projected from one view to another. Projection lines are used to show relationships from one view to another.
    - A work order provides written directions about the requirements of a job. The information on a work order may be accompanied by technical information on a production plan. The plan gives the steps to produce a part, the routing of the part through several departments, degrees of accuracy, and other pertinent information.
- Work surfaces must be prepared to receive a coloring agent. Quick-drying paints and layout dyes are used to coat a work surface. Coated surfaces simplify layout operations and make all lines and dimensions clearly visible.
    - Orderly work habits are the hallmark of the skilled craftsperson. The protection of tools and workpieces and their safe handling to prevent personal injury are essential practices.
- The condition of the workpiece affects the accuracy of layout, fitting, assembling, and other bench, floor, and machine operations. Burrs, grease, and other foreign particles must be removed.

UNIT 5 REVIEW AND SELF-TEST

1. List six common bench work processes.
2. Give four reasons why the worker depends on a blueprint and/or work order (production plan).
3. Explain briefly what is meant by (a) the condition and (b) the preparation of a workpiece.
4. a. Describe a projection line.
   b. State the purpose of projection lines on a drawing.
5. Set up a series of four conditions that constitute good work habits and attitudes.

Basic Noncutting Hand Tools

Bench hand tools fall into two groups: (1) *noncutting tools* and (2) *cutting and shaping tools*. Each group of tools relates to a number of bench work processes as follows:

- Noncutting hand processes such as
  - Hammering,   – Clamping,
  - Fastening,   – Positioning.

- Cutting and shaping hand processes such as
  - Sawing,
  - Chiseling,
  - Punching and driving,
  - Filing,
  - Threading,
  - Reaming.

OBJECTIVES

NONCUTTING HAND TOOLS

After satisfactorily completing this unit, you will be able to:

- Identify features and functions of noncutting hand tools grouped for hammering, fastening and gripping, turning, and clamping processes.
- Select proper size and type of straight or offset screwdrivers for slotted and recessed-head screw applications.
- Know different applications of the bench vise, clamps and solid and adjustable wrenches.
- Perform each of the following processes.
  - Use Noncutting Tools (hammers, screwdrivers and wrenches).
  - Use Clamping and Gripping Devices (vises, wrenches, clamps).
  - Select a comfortable and appropriate set of industrial goggles or safety shield.
- Apply new *Terms* related to cutting and noncutting hand tools and common abrasive cutting processes.
- Follow *Safe Practices* in using noncutting hand tools.

NONCUTTING HAND TOOLS

HAMMERING TOOLS AND PROCESSES

BALL-PEEN HAMMERS

One of the everyday tools used in the machine and metal trades is the *ball-peen* hammer. Two other tools that are not as common are the *straight-peen* and the *cross-peen* hammers. One end of each hammer head is shaped as a cylindrical solid. The face is ground slightly crowned and the edge is beveled. This face is used to deliver a flat blow (force) or to shape and form metal. The other specially shaped end of each head permits the operator to deliver a blow in a more confined area or to shape and form metal in a somewhat different manner than with the face. For example, a surface may be *peened* or shaped through a series of cupped indentations with a ball-peen hammer.

Hammers come in many sizes. The size used depends on the work size and required force. Hammer weights range up to three pounds. Ball-peen hammers weighing just an ounce are used in fine instrument work and for small, precise layout operations. The 12-ounce ball-peen hammer is one of the most practical sizes for general machine work. A large force may be delivered by a hammer at a desired point by firmly grasping the handle near the end and delivering a solid, sharp blow.

50

How to Use Hammers

General Hammering

STEP 1 Select a hammer weight suitable for the size of the job. Inspect the handle to see that it is securely fastened to the head. Check the faces to be sure that they are not chipped.

STEP 2 Grasp the hammer near the end of the handle. Hold the hammer firmly but not rigidly.

STEP 3 Start with a light, sharp blow to get the feel and the correct distance for hammering.

STEP 4 Continue and strike heavier, solid blows.

Caution: Watch the point of action and not the hammer head.

Riveting

STEP 1 Strike a blow at a slight angle on the edge of the rivet head.

Note: Sometimes a head is riveted by peening. Peening consists of a series of indentations around the periphery (edge) of a pin or other part. The indentations are produced by striking the rivet head with the ball end of the hammer.

STEP 2 Continue around the entire edge of the rivet head. Peen or flatten the edge as required.

Driving with Soft-Face Hammers

STEP 1 Select a soft-face hammer that has softer head material than the part.

STEP 2 Secure the part or mechanism in a vise or other holding fixture.

Note: The vise must provide adequate support to permit applying the necessary force to drive the part. Check to see that the driven part will clear the vise jaws.

STEP 3 Apply a thin film of lubricant between the mating parts that are to be driven into position.

STEP 4 Deliver sharp, firm blows to separate or to drive the parts together.

SOFT-FACE HAMMERS

There are many assembly and disassembly operations in which machined parts fit tightly together. These parts must be driven into place without damage to any part or surface. In these operations, therefore, the face of a hammer head must be softer than the workpiece. The pounding faces of a soft-face hammer may be made of rawhide, plastic, wood, hard rubber, lead, brass, or other soft metal or material. Some makes of soft-face hammers are designed so that the faces may be replaced when worn.

INDUSTRIAL GOGGLES AND FACE SHIELDS

Although industrial goggles and face shields are not classified as tools, they are introduced here because their use is essential in preventing eye injuries. They are especially important in bench and floor work and in machining operations because of the ever-present danger of injuries from flying particles and other foreign matter.

The worker should obtain a personal pair of goggles or an appropriate shield. Goggles should be venitlated to prevent fogging but should still protect the eyes against grit and other small particles. Goggles that fit over vision-correcting eyeglasses are available.

Face shields are preferred as a lightweight and cooler eye protection device. Shields are practical in operations that do not involve dust, abrasives, and other fine particles.

FASTENERS AND FASTENING TOOLS

Screws, nuts, and bolts are called *fasteners*. Fasteners are turned with two major groups of hand tools: screwdrivers and wrenches. The construction and design of these tools vary depending on the application. For instance, small jewelers' screwdrivers are used for fine precision work. By contrast, some screwdrivers for heavy-duty service have a square shank to which a wrench

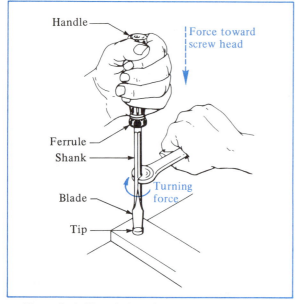

Figure 6–1 Wrench Attached to the Square Shank of a Heavy-Duty Flat-Head Screwdriver for Added Turning Force

can be attached to gain added leverage (force) as shown in Figure 6–1.

SCREWDRIVERS

There are many types of screwdrivers. The types most often used in the machine shop are the flat-head, Phillips-head, and offset screwdrivers.

FLAT- AND PHILLIPS-HEAD SCREWDRIVERS

A screwdriver usually combines four basic parts: handle, ferrule, shank, and blade (head).

A good screwdriver shank and blade are made of a high-quality steel that has been forged to shape, hardened and tempered. Screwdriver sizes are designated by the length of the shank; typical sizes are 6″, 8″, and 10″. The size of the tip is proportional to the blade length.

There are two basic shapes of tips: *flat head* and *Phillips head*. The flat-head and Phillips-head screwdrivers are similar, except for the shape of their blades and tips. The flat-head shape, ground to a slight wedge and flat across the tip, fits screw heads that are slotted completely across the head diameter. The worker should select a tip size that is almost the width

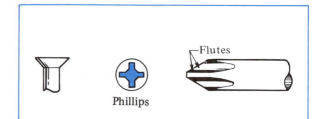

Figure 6–2 Features of a Phillips-Head Screwdriver Blade

and length of the screw slot. The tip of the blade should be flat so that it can be held firmly against the bottom of the slot.

The blade of the Phillips-head screwdriver has four *flutes* that cross at the center of the cone-shaped tip (figure 6–2). It fits screws and bolts that have two slots that cross in the center of the head. The crossed slots extend across the screw head but stop short of the outer edge.

OFFSET SCREWDRIVERS

The third type of screwdriver is the *offset* screwdriver. This screwdriver is made from one solid piece of metal and usually has a flat, wedge-shaped tip on each end. One tip is at a right angle to the "handle" (body of the screwdriver). The other tip is parallel to the handle. Two common offset screwdrivers are illustrated in Figure 6–3. Offset screwdrivers also come with Phillips and Reed and Prince type tips.

The offset screwdriver is used to hold and turn fasteners in spaces that are inaccessible to straight-shank screwdrivers.

HAND WRENCHES

Like screwdrivers, hand wrenches are also considered to be fastening tools. Therefore, hand wrenches are made of a high-quality alloy steel that combines strength, toughness, and durability.

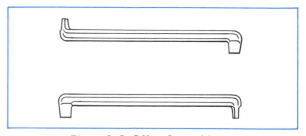

Figure 6–3 Offset Screwdrivers

Wrenches also are heat treated to prevent wear and to gain other desirable physical properties.

Wrenches derive their names from (1) their shape—for example, *open-end* wrench; (2) their construction—for example, *adjustable* wrench; or (3) the requirements of the job—for example, *pipe* wrench. Four common types of wrenches used in bench work are described in this unit. They are:

- Solid nonadjustable wrenches,
- Adjustable wrenches,
- Allen wrenches,
- Special spanner wrenches.

SOLID NONADJUSTABLE WRENCHES

The solid wrench derives its name from its construction. It is made of one piece. The size of the jaw opening is, therefore, permanent (nonadjustable). A solid wrench is designed so that the overall length is proportional to (1) the size of a bolt or nut and (2) the force normally required for tightening or loosening operations.

The three general types (shapes) of solid nonadjustable wrenches are: (1) *solid open-end*, (2) *box*, and (3) *socket*. Each of these wrench types is available in metric or inch sizes and singly or in sets.

Solid Open-End Wrenches. Some solid wrenches are open ended—that is, they are designed with either one open end or two open ends with different sizes. The open end (head) permits the wrench jaws to be slid onto the sides of a bolt or nut from either the top or side and for force to then be applied from whatever position is most advantageous.

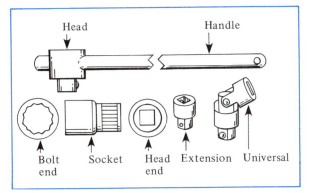

Figure 6-4 Parts and Features of a Socket Wrench Set

Box Wrenches. Other solid nonadjustable wrenches, box wrenches, have closed heads (ends). Each head accommodates a particular size of hexagon bolt or nut. Box wrenches are used whenever space permits the heads to be fitted *over* bolts or nuts.

Socket Wrenches. Still other solid wrenches, socket wrenches, have hollow, fluted sockets. Hexagon sockets fit over hexagon-head nuts and bolts. Square sockets are designed to fit over square-head nuts and bolts.

Socket wrenches provide flexibility. They may be used with a ratchet handle, which permits turning nuts and bolts in confined spaces. They also may be used with flexible adapters and extension bars, which make it possible to turn bolts and nuts that are inaccessible with other wrenches. The parts of a typical socket wrench set are shown in Figure 6-4.

Socket and box wrenches should be used whenever possible in preference to open-end and adjustable wrenches. The proper size and shape of a socket or box wrench *nests*, or *houses*, the nut or bolt to prevent slippage. Thus, a greater force may be applied, and manipulating in confined places is easier.

Torque Wrenches. The proper functioning of many parts requires that the contact surfaces be secured together with a uniform (specified) force. The *torque* (tightening, or turning, force) must be controlled to prevent overstressing and stripping of the threads or shearing off of the bolt or nut.

Tension or torque wrenches are used to measure and control turning force (Figure 6-5). These wrenches have either a graduated dial or other type of scale that indicates the amount

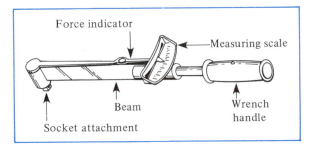

Figure 6-5 Torque Wrench

of applied turning force in foot- and/or inch-pounds.

ADJUSTABLE WRENCHES

Adjustable wrenches are adaptable to a wide range of nut, bolt, or workpiece sizes and shapes. Three common adjustable types are the *adjustable open-end* wrench, the *monkey* wrench, and the *pipe* wrench. One adjustable wrench can often replace many solid (fixed-size) wrenches. The sizes of monkey and pipe wrenches are designated by their overall length— for example: a 10″ monkey wrench and a 12″ pipe wrench. The size of an adjustable open-end wrench is indicated by the maximum dimension of the jaw opening.

Monkey, pipe, and adjustable open-end wrenches require that the direction of pull (turning force) be *toward* the adjustable jaw. For personal safety and to avoid damage to the workpiece or part, these wrenches must be checked and adjusted repeatedly during the operation to maintain a close fit between the jaws and the work. The pipe wrench is self-adjusting to a limited extent. The jaws move together when a force is applied.

The pipe wrench is used for turning objects that are round or irregularly shaped. It's two jaws have teeth. As pressure (force) is applied in the turning direction, the teeth grip the work. The turning force is transmitted to the part and causes it to turn. The teeth, however, leave indentations and burrs, which might cause the workpiece to be rejected.

ALLEN WRENCHES

Hollow or socket type setscrews and other bolts are turned by a hexagon-shaped offset wrench known as an *Allen* wrench. Figure 6–6

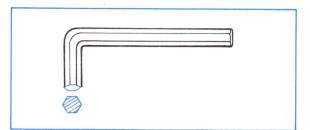

Figure 6–6 Allen Wrench

shows one of these wrenches. Allen wrenches are made of alloy steel and are heat treated to withstand great forces. These wrenches are available individually or in sets and in either millimeter or fractional-inch sizes.

SPANNER WRENCHES

Figure 6–7 shows two types of spanner wrenches. The *face spanner* has pins that fit into notches or holes. The *single-end* (hook) *spanner* is used to turn a special nut or bolt that is notched.

Spanner wrenches are widely used on machine tools. The wrenches turn large nuts that draw and hold a chuck or other device securely to a tapered machine spindle. A reverse turning process unlocks the chuck so that it may be removed.

CLAMPING AND GRIPPING DEVICES
BENCH VISES

Bench vises are devices that can be adjusted quickly to clamp work between their jaws. The vise is usually positioned on a bench at a height that permits the worker to comfortably perform bench work operations.

The bench vise top section can be turned horizontally on a simple swivel arrangement. A swivel lock secures or releases the vise top section from the base, which is attached to the bench.

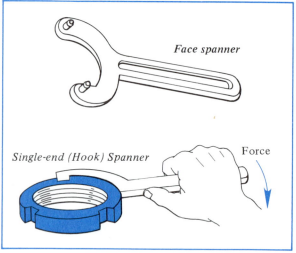

Figure 6–7 Face and Single-End (Hook) Spanner Wrenches

Regular vise jaws are *serrated* (grooved) to provide a good gripping surface. Some workpiece surfaces, however, must be protected against the scratches, burrs, and indentations caused by direct contact with the jaws. Therefore, *soft jaws*, which fit over the serrated vise jaws, are used to prevent scoring of such workpiece surfaces.

PLIERS

Pliers are used for gripping, turning, and drawing wire and small parts and for cutting metal wires and pins. The two most commonly used pliers are *slip-joint* pliers and *needle-nose* pliers.

CLAMPS

Clamps are used for holding work in place. The *C-clamp* is usually used for comparatively heavy-duty clamping. The C-clamp also permits the clamping of parts that are wider than can be accommodated with other (parallel) clamps.

Parallel clamps have two jaws that provide parallel clamping surfaces. The distance between the two jaws is controlled by two screws. When the jaws are parallel with the work faces, sufficient force may be exerted to securely hold the pieces. It is important that as much of the full surface of each parallel jaw be in contact with the work as is possible.

Figure 6–8 shows a simple toolmaker's parallel clamp. As holding devices, a set is generally used for clamping workpieces with finished surfaces during layout, inspection, and basic machining operations. Because the clamping faces of each jaw are ground smooth, they will not mar a finished surface.

How to Use Screwdrivers

STEP 1 Select the appropriate screwdriver type, blade, and tip size for the job.

STEP 2 Inspect the tip to be sure the blade (flat-head or Phillips-head) will enter the screw head. The tip flutes must seat properly in the slot(s) in the screw head.

STEP 3 Grasp the handle in one hand. Guide the tip into the screw slot(s) with the other hand (Figure 6–9).

STEP 4 Apply force toward the screw head. At the same time, turn the screwdriver handle clockwise to tighten a regular right-hand screw. Use a counterclockwise turning force to loosen such a screw.

STEP 5 Tighten all screws by applying uniform turning force. If uniform appearance is required, all screw slots should be aligned in the same relative position.

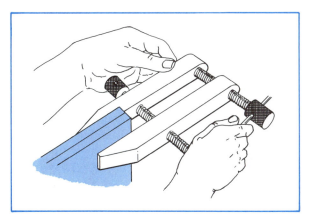

Figure 6–8 Adjusting Clamp Jaws to Be Parallel for Applying Force

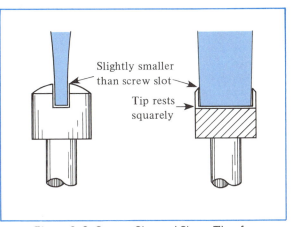

Figure 6–9 Correct Size and Shape Tip of a Flat-Head Screwdriver Blade

How to Use Wrenches

Solid Open-End Wrenches

STEP 1 Select a solid open-end wrench whose jaws fit the nut or bolt.

STEP 2 Place the wrench in the position in which the greatest leverage can be applied.

STEP 3 Pull the wrench handle clockwise to tighten a right-hand bolt or nut. Reverse the direction to loosen this type of bolt or nut. Gradually apply turning force. Use either one or both hands, depending on the job requirements.

Note: When great turning force is needed, be sure that the wrench jaws do not spring out of shape. The fit between the wrench jaws and work should be checked repeatedly.

STEP 4 Move (force) the wrench as far as space permits. Turn over the wrench so that the part can be turned an additional amount. Then, turn over and reposition the wrench on the part. Repeat the turning procedure until the part is properly tightened or has been removed.

Box and Socket Wrenches

STEP 1 If possible, select a correct size of box or socket wrench in preference to an open-end wrench.

STEP 2 Determine what socket wrench accessories, if any, are needed to provide the required leverage with the greatest efficiency.

Note: A solid wrench or a ratchet handle may be used with the socket head (Figure 6-10). Other applications, however, might require the use of an adapter, universal joint, or extension bar.

STEP 3 Place the wrench in the position that will provide maximum leverage. Apply a small amount of lubricant to the threaded parts. Gradually apply force to the end of the wrench handle (Figure 6-10). A clockwise force tightens and a counterclockwise force loosens a right-hand thread.

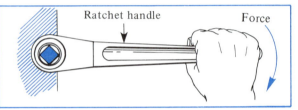

Figure 6-10 Applying Turning Force with a Ratchet-Handle Socket Wrench

Note: The amount of force to use depends on the size and requirements of the job. Accurate judgment by the worker is important. Excessive force can *round over* (smooth) the serrations in the wrench end or socket, damage the bolt head or nut, or break the part.

Torque Wrenches

STEP 1 Use a torque wrench to uniformly tighten bolts and nuts. Pull the wrench handle until the amount of force read on the scale meets the job specifications.

Note: Parts should be assembled by first turning the bolts and nuts by hand. Then use a conventional ratchet or socket until all bolts and nuts are uniformly seated. When two or more nuts are to be tightened, they should be turned alternately and uniformly.

Adjustable Open-End and Monkey Wrenches

STEP 1 Select an adjustable open-end or monkey wrench. The wrench size must be proportional to the size of the nut, bolt, or part to be tightened or loosened.

STEP 2 Adjust the wrench jaws to the approximate size of the part to be turned and place the wrench on the part. The adjustable jaw should be on the side *toward* which the turning force is to be applied. Adjust the jaws to fit the part.

STEP 3 Apply force slowly in the direction of the movable jaw. Continue to increase the force until the part begins to move.

Note: Repeatedly check the jaws and work to ensure that a close fit is maintained as increased force is applied.

Allen Wrenches

STEP 1 Remove dirt and other foreign particles from the screw head.

STEP 2 Select the appropriate Allen wrench that fits the screw head.

STEP 3 Grip the Allen wrench in one hand and apply appropriate force in the direction required for tightening or loosening. Judgment must be used in exerting force appropriate for the size and construction of the part.

Spanner Wrenches

STEP 1 Select a hook spanner or a face (pin) spanner wrench that fits the particular part.

STEP 2 Apply the necessary force near the end of the handle. Reverse the hook spanner to turn a nut in the opposite direction.

How to Use Clamping and Gripping Devices

Bench Vises

STEP 1 Clean the vise. Regularly oil the moving parts.

STEP 2 Determine whether the workpiece may be clamped directly. Use soft jaws or other appropriately soft material to protect finished surfaces.

STEP 3 Position the work between the vise jaws.

> Note: Assembly and other hand operations should be performed as close as possible to the vise jaws. Make sure the surfaces to be clamped are strong enough to withstand the compressing force of the vise jaws.
>
> Place a block below the work to prevent it from moving during hand operations.

STEP 4 Bring the vise jaws against the work. Turn the vise handle to apply sufficient force to securely hold the workpiece in position.

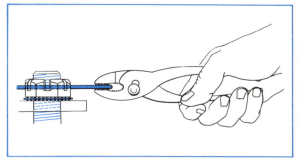

Figure 6–11 Gripping and Applying Force with Pliers

Pliers (Noncutting Operations)

STEP 1 Select the proper type and size of pliers for the job.

STEP 2 Position the workpiece between the nose of the pliers. Then apply a gripping force to the handles (Figure 6–11).

STEP 3 Hold, pull, insert, bend, or twist the part as required.

C-Clamps and Parallel Clamps

STEP 1 Remove burrs and clean the workpieces and clamps. Apply a drop of oil to the moving parts of the clamps.

STEP 2 Open the clamp so that the jaws are slightly wider than the thickness of the parts to be clamped. Use soft metal inserts between the clamp jaws and work surfaces that must be protected.

STEP 3 Position the clamp over the work so that bench or machine operations can be performed without interference. Clamps are often used in sets to hold parts more securely and accurately.

STEP 4 Move the clamp jaws together. Turn the front adjusting screw of a parallel clamp with the fingers to set the jaws to the width of the work. Bring the jaws parallel by turning the outside screw.

STEP 5 Insert a pin or round metal object in the hole of the tightening screw.

STEP 6 Turn this screw to bring the jaws parallel. Once the jaws are in position, apply an additional clamping force to the two jaws.

> Note: When properly positioned and secured, the jaws should be parallel and tightly forced against the workpieces.

Safe Practices for Using Bench Tools

Hammering Tools

- Examine the hammer each time *before* it is used. The handle must be securely wedged in or bonded to the head. The flat (slightly crowned) face must not be chipped.
- Use a soft-face hammer if the work surface must not be dented by the hammer blow.

Fastening Tools

- Firmly press the screwdriver tip into the screw slot or cross-slots while applying turning force. Applying force helps to prevent the tip from slipping out of the slot(s), which can cause burring of the screw head slot(s), and personal injury.
- Turning force should be applied to an adjustable wrench only in the direction of the *movable* jaw.
- Apply turning force gradually, while carefully monitoring the position of the wrench and the work to detect signs of slippage. Excessive force can cause the jaws of any open-end wrench to spring.
- Protect the surfaces of a workpiece from indentations produced by pipe wrenches.

Clamping Devices

- Use soft jaws or other soft material when holding finished work surfaces in a bench vise to prevent marring of the finished surface.
- Position the workpiece in a vise so that it does not overhang into an aisle or other area where a person might accidentally brush against it.

NONCUTTING HAND TOOL TERMS

Ball-peen hammer	A common form of hammer used in machine and metal trades. The hammer head has one end that is a cylindrical solid with a slightly crowned face. The opposite end is shaped like a ball.
Soft-face hammer	A hammer whose end faces are softer than the workpiece. A hammer having crowned faces made of rawhide, plastic, or other relatively soft material.
Fasteners	Screws, nuts, bolts, keys, pins, and other devices that fasten or secure parts together.
Screwdriver tip (flat or Phillips)	The end of the screwdriver blade that conforms to the shape and size of the screw slot(s) into which it is inserted.
Offset screwdriver	A single-piece metal screwdriver with two formed ends (tips). (The ends may be on the same side or at right angles to the body.)
Solid open-end wrench (nonadjustable)	A one-piece steel wrench consisting of a body and one or two open ends. (Each end is machined to fit a particular nut or bolt head size. The ends are open on the front side to permit sliding them into position on the nut or bolt.)
Box wrench	A solid wrench with an opening machined into one or both ends. (Each opening fits around a particular size and shape of bolt, nut, or part.)
Socket wrench	A form of solid wrench. (One end of each socket fits over a bolt or nut of a standard size and shape. The other end of each socket has a square hole into which fits the square shank of a ratchet or straight handle or an accessory such as an extension, adapter, or universal joint.)
Torque wrench	A calibrated inch- and foot-pounds torque measuring device.
Adjustable wrenches	A group of wrenches whose jaw openings adjust to fit different sizes and shapes of nuts and bolts and round or irregularly shaped parts.
Allen wrench	A metal piece whose size and cross-section fit the hollow or socket-type screw heads. (One end is offset to form a handle.)

Spanner wrench	A special form of wrench that fits around large, round, threaded collars. (The collars have either slotted edges or holes in their faces. Single- and double-pin spanner wrenches fit these collars.)
Pliers (general purpose)	A gripping tool for turning, twisting, bending, shearing, or manipulating wires, pins, and similar small parts.
Clamps	Work holding clamping devices of C-shaped or parallel forms.
Eye protective devices	Face shields and industrial goggles of nonshatterable materials and enclosures for protecting the eyes against flying particles or splattering materials.

SUMMARY

- Bench tools are broadly grouped to serve such functions as: hammering, fastening, clamping, forming, sawing, filing, chiseling, threading, and reaming.
  - The ball-peen hammer is used to peen an edge or to direct a blow to a limited area.
- The common fastening tools used in bench work and for other machining operations include flat-head, Phillips-head, and offset screwdrivers.
  - Wrenches are of the nonadjustable type—solid, open-end, box, socket—and the adjustable type—adjustable open-end, monkey wrench, pipe wrench. Other types include the Allen wrench, spanner wrench, and setscrew wrench.
- Two common types of clamping devices are the bench vise and C- and parallel clamps.
  - Safe work practices must be followed in bench work to prevent damage to finished work surfaces, avoid personal injury from sharp or burred edges or parts extending into aisle areas, and protect tools.

UNIT 6 REVIEW AND SELF-TEST

NONCUTTING HAND TOOLS

1. List four noncutting and four cutting and shaping bench processes.
2. Identify (a) one type of hammer that is used to drive a pin and (b) another type that is used for assembling mating parts.
3. a. Describe a flat- and a Phillips-head screwdriver.
   b. State two advantages of the Phillips-head screwdriver over the flat-head screwdriver.
4. Recommend a hand wrench to use in each of the following applications:
   a. Tightening a hexagon nut in a confined recessed area;
   b. Tightening large-size, square-head lag screws;
   c. Assembling hexagon bolts in areas that are inaccessible to open-end wrenches and where each bolt must be uniformly tightened;
   d. Assembling and disassembling large diameter hexagon coupling nuts;
   e. Turning hollow or socket-type setscrews;
   f. Tightening (1) a shouldering collar that is fitted with a single slot on the periphery and (2) another collar that is fitted with two slots on the face.
5. Describe how holding workpieces with a C-clamp differs from gripping with a set of parallel clamps.
6. State two factors that should guide the worker in selecting industrial goggles or a protective face shield.
7. List four general safety precautions to observe when selecting and using hammers, screwdrivers, and wrenches.

SECTION TWO

Cutting and Shaping Hand Tools

Bench and floor work processes that involve cutting and shaping with hand tools are: hand sawing; chiseling, punching, and driving; filing; scraping, burring, and polishing; internal and external threading; and reaming. The principles of and correct practices for performing these hand processes are described in this section.

UNIT 7

Basic Cutting Tools and Processes

OBJECTIVES

After satisfactorily completing this unit, you will be able to:

A. HAND SAWING PROCESSES AND TECHNOLOGY

- Understand the properties and design features of hand hacksaw blades and hacksaw frames and manufacturer's specifications.
- Apply information about cutting speed, force, and the holding of irregular and special sizes of materials, including safety precautions.
- Use the hand hacksaw for cutting.

B. CHISELING, PUNCHING, AND DRIVING TOOLS AND PROCESSES

- Understand basic types of hand chisels, hollow and solid hole punches, and centering punches.
- Perform the following processes.
  - Use Hand Chisels.
  - Use Punches and Drifts.
  - Use Center and Prick Punches for Layout Work.

C. HAND FILE CHARACTERISTICS AND FILING PROCESSES

- Name and identify the characteristics of standard and precision files and filing processes.
- Perform the following processes.
  - Select Appropriate Shape and Size File.
  - Cross File and Check for Flatness and Squareness.
  - Drawfile and Polish a Finished Surface.
  - File a Rounded Corner.
  - Remove Burrs by Filing.
- Follow *Safe Practices* for caring and using files.

D. BURRING AND HAND SCRAPING
- Recognize the hazards of burrs and their removal for personal safety, work appearance, and accuracy.
- Tell the important functions served by scraping and flaking processes.
- Identify flat, half-round, and three-corner scrapers.
- Describe how hand flaking and scraping processes are carried on.

E. COATED ABRASIVES AND HAND FINISHING PROCESSES
- Deal with burring for layout and measuring, for personal and tool safety, and for appearance.
- Interpret technical information about natural and artificial abrasives.
- Use *Terms* related to properties and applications of coated abrasives, forms, backing and bonding materials, grain density, and sizes.
- Perform Hand Polishing with Coated Abrasives.

Hand sawing of metals and other hard materials is an everyday shop process. The hand tool used for this sawing is the *hacksaw*. Correct use of the hacksaw enables the mechanic to efficiently and safely cut material and avoid breakage and unnecessary wear of a saw blade.

A. HAND SAWING PROCESSES AND TECHNOLOGY

Two sets of related factors should be considered when selecting the correct hacksaw frame and blade:

- Shape, size, and kind (properties) of material to be cut;
- Features of a particular saw blade (pitch, set of the teeth, and length; the metal of which the blade is made; and its heat treatment).

HACKSAW FRAMES

Selection of the correct hacksaw frame is an important consideration in the hand sawing process. Hacksaw frames are named according to the shape of their handles—for example, *pistol-grip* and *straight-handle*. Each of these frames is available as either (1) a solid-steel nonadjustable frame that fits one length of blade or (2) an adjustable frame that accommodates various lengths

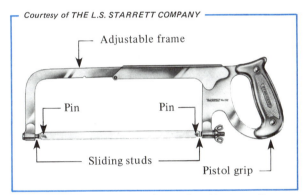

Courtesy of THE L.S. STARRETT COMPANY

Figure 7-1 An adjustable Pistol-Grip Hacksaw Frame

of blades. An adjustable pistol-grip hacksaw frame is shown in Figure 7–1. This particular adjustable frame accommodates hacksaw blade lengths ranging from 8″ to 10″ (200mm to 250mm).

As shown, the front and handle ends of the frame are each equipped with a *sliding stud*. The studs may be turned to and locked in any one of four different positions. They permit turning the blade to 90°, 180°, 270°, and the normal vertical position.

There is a pin on each sliding stud. The frame is set to the blade length. With the teeth pointed forward, the blade is seated against the flat part of each stud. Tension is applied to hold the blade tightly in the frame.

HACKSAW BLADES

Another important consideration in the hand sawing process is the selection of the correct hacksaw blade. Hacksaw blades are made from high-grade tool steel, tungsten alloy steel, tungsten high-speed steel, molybdenum steel, and molybdenum high-speed steel. Tungsten adds *ductile* strength to steel and thus relieves some of its brittleness. Molybdenum adds to its wearing qualities.

All blades are hardened and tempered. There are three hardness classifications: all hard, semi-flexible, and flexible back. Each classification describes the heat treatment of a particular blade. *All hard* indicates that the entire blade is hardened. Although the all-hard blade retains a sharp cutting edge longer, it is relatively brittle and may be broken easily. *Flexible back* means that only the teeth on the blade are hardened. Therefore, in contrast to the all-hard blade, the flexible-back blade can *flex* (bend slightly) during the cutting process without breaking.

Length is also a factor in selecting a hacksaw blade. *Length* refers to the center-to-center distance between the holes in the ends of the blade. Various lengths of blades are available. The most common lengths are 8″, 10″, and 12″ (200mm, 250mm, and 300mm). Blade width is standardized at approximately 1/2″ or 12mm. Hand hacksaw blades are 0.025″ (0.6mm) thick.

Two other important factors also enter into the selection of a blade. They are the *number* of teeth and the *set* of the teeth.

Hacksaw blades are made with a specified number of teeth per inch. This number is called the *pitch* of the blade. For example, a blade with 14 teeth per inch is a 14 pitch blade. Blades are made with pitches that vary from coarse (14 pitch) to medium (18 or 20) to fine (24 to 32).

The pitch to select depends on the hardness, the shape, and the thickness of the part to be cut. For large sections or mild materials, a 14 pitch provides good chip clearance. Harder materials like tool steel, high carbon and high speed steels require an 18 pitch blade. A 24 pitch is used on structural iron shapes, brass, copper, iron pipe, and electrical conduit. A 32 pitch is used for thin tubing and sheet metals.

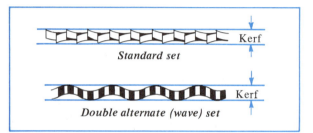

Figure 7-2 Standard and Double Alternate Set Patterns

In addition to being specified in number of teeth per inch (pitch), hacksaw teeth are *set* (positioned) to permit a *kerf* (groove) to be cut freely. The kerf is slightly wider than the thickness of the blade to prevent the blade from binding. A *standard* set means that one tooth is offset to the right and the next to the left, and so on. Some fine-pitch blades have a *double alternate* set, or a *wave* set (Figure 7–2).

CUTTING SPEEDS AND FORCE (PRESSURE)

Blade manufacturers recommend a cutting speed ranging from 40 strokes a minute for tough materials like stainless steels and tool steels to 55 strokes a minute for mild materials like aluminum and brass. At these speeds, the worker is able to control (relieve) the force on the teeth during the return stroke and saw without tiring.

The amount of force to apply when cutting depends, as with other hand tools, on the skill and judgment of the operator. Enough force must be applied on the *forward* cutting stroke to permit the teeth to cut the material. The forces are released and the blade is raised slightly from the work surface to prevent the teeth from scraping during the return stroke.

HOLDING IRREGULAR SHAPES AND THIN-WALLED MATERIALS

A general principle in hand hacksawing is to always engage at least two teeth with the work to prevent digging in or stripping of the teeth or bending of the part. Thin, flat materials should be held between blocks of wood. Tubing and other circular shapes should be nested in a simple form made of soft material. A wider area of clamping surface is thus provided.

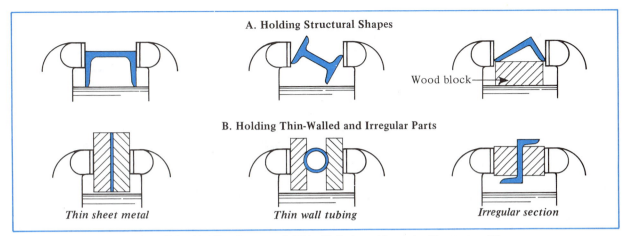

A. Holding Structural Shapes

Wood block⟶

B. Holding Thin-Walled and Irregular Parts

Thin sheet metal   *Thin wall tubing*   *Irregular section*

Figure 7-3 Holding Structural, Thin-Walled, and Irregular Shapes in a Vise

On thin-gage tubing a wooden plug may be inserted in the center. Sometimes the hacksaw cut is made through both the wood support and the workpiece to permit a greater number of teeth to be engaged. Breaking through or damaging the part is thus avoided.

Common practices for holding structural shapes in a vise are illustrated in Figure 7–3A. Similarly, Figure 7–3B shows how thin-walled and irregularly shaped parts should be nested and held.

How to Use the Hand Hacksaw

Selecting the Correct Frame and Blade

STEP 1 Examine the work order or blueprint. Determine the nature and shape of the material to be cut.

STEP 2 Select a hacksaw frame appropriate for the job.

STEP 3 Select the length, pitch, set, and hardness of the hacksaw blade according to the job requirements.

STEP 4 Place the sliding studs in the correct position for cutting. Then insert the blade in the frame. The cutting teeth must point *forward*, away from the operator and toward the front of the frame (Figure 7–4).

STEP 5 Adjust the tension of the blade in the frame. It should be tight enough to prevent the blade from buckling and drifting during use.

Note: Excessive tightening may cause the frame to bend, the blade to break, or the stud pins to shear.

Work Position and Posture While Sawing

STEP 1 Position the workpiece in a bench vise. Use soft jaws, if necessary, to protect the work surface.

Note: Thin-walled parts and irregularly shaped parts should be nested between soft materials. They then should be clamped in position for cutting.

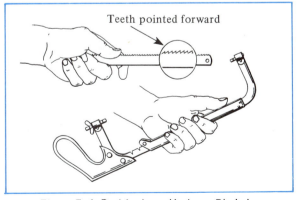

Teeth pointed forward

Figure 7–4 Positioning a Hacksaw Blade in a Hand Hacksaw Frame

Note: The workpiece area to be cut should be positioned as close to the vise jaws as is possible to prevent springing of the work or flexing of the blade. If a long cut is required, it may be necessary to frequently reposition the part nearer the vise jaws.

STEP 2   Grasp the front of the hacksaw frame with one hand. Hold the handle in the other hand.

STEP 3   Take a position near the vise. One foot should be pointed toward the bench. The other foot should be to the side and back. This position gives balance when sawing.

STEP 4   Move the body on both the forward cutting and return strokes. Lean forward from the hips on the forward stroke. Return to the original position at the end of the return stroke.

Starting the Cut

STEP 1   Break any sharp corner that may strip the saw teeth by *notching* (filing) a small groove, or "V" indentation, close to the layout line or a measured point.

STEP 2   Place the front of the blade in the indentation and *guide* it to the cutting line.

STEP 3   Apply a slight, steady force. Push the saw forward at a slight angle across the surface of the work. (The arrows in Figure 7–5 show the force that should be applied during the cutting stroke.)

Note: The force applied depends on the thickness of the material and its composition. For example, a thick part made of machine steel requires a greater force than does a softer metal like brass or aluminum.

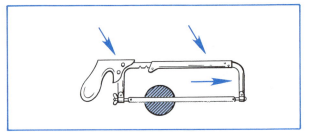

Figure 7–5 Forces Applied During Cutting Stroke

STEP 4   Release the force during the return stroke. Raise the frame and blade. Draw the hacksaw straight back to the starting point. The dotted arrows in Figure 7–6 indicate the releasing of the forces.

STEP 5   Make each stroke as long as the blade, frame, and work area permit. Maintain long, steady strokes. Avoid the tendency to speed up.

Finishing the Cut

STEP 1   Slow down the cutting action near the end of the cut. Reducing speed and force is a safety precaution. Abruptly breaking through the cut, losing balance, and having the free end of the work tear away from the portion held in the vise result in personal injury, damage to the work, or breaking of the blade.

STEP 2   Clean the chips from the blade. Loosen the tension on the blade and frame. Remove the burrs from the workpiece.

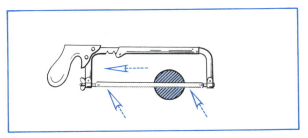

Figure 7–6 Forces Released (Blade Raised)
on Return Stroke

TABLE 7-1 Manufacturers' Recommended Pitches for Hand Hacksawing

Material Types and Recommended Pitches	Correct Pitch	Incorrect Pitch
For large sections and mild materials; 14 pitch	Good chip clearance	Pitch too fine No chip clearance Result: clogged teeth
For harder materials like tool steel, high-carbon, and high-speed steels; 18 pitch		
For angle iron, brass, copper, iron pipe, and electrical conduit; 24 pitch	At least two teeth on a section	Pitch too coarse One tooth on a section Result: stripped teeth
For thin tubing and sheet metals; 32 pitch	 At least four to six teeth on a section	

CHISELS

Many workpieces require the cutting of metal or other hard material by a hand process called *chiseling*. Chiseling usually refers to two operations: (1) *chipping*, which is the process of removing or cutting away material, and (2) *shearing*, which refers to the cutting apart of metals or other materials.

Cold chisels are used in both bench and floor work for chipping metals and other materials.

Cold chisels are also used in shearing operations. The solid jaw of a vise may be used as one blade for shearing; the cold chisel is considered to be the other shearing blade. A scissor-like action of these blades shears the material.

The three main parts of a cold chisel are: (1) the cutting end (edge), (2) the body, and (3) the head. The body shape of the chisel may be hexagonal, octagonal, rectangular, square, or round. Cold chisels are usually forged to shape, hardened, and tempered. Cold chisels will cut materials that are softer than the hardness of the chisel itself. Cold chisels may be driven

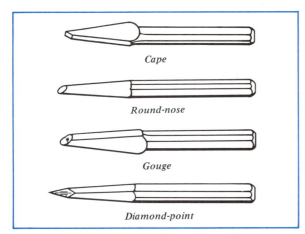

Figure 7-7 Common Types of Cold Chisels

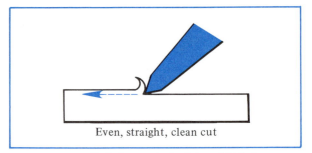

Even, straight, clean cut

Figure 7-8 Correctly Positioned Chisel

(forced) with a hand hammer or a pneumatic (air) hammer.

COMMON TYPES OF COLD CHISELS

Cold chisels are named according to the shape of or the shape made by their cutting edge. The chisel that is most common is the *flat* chisel. This chisel is used to chip (remove) material from a surface. The flat chisel also shears (cuts through) rivets and other parts. The cutting faces of this chisel are ground, in general practice, to an included angle of 60° to 70°. The cutting edge is slightly crowned.

Other common types of cold chisels are shown in Figure 7-7. The *cape* chisel is the narrowest of the chisels in width. The cape chisel is, therefore, especially useful for cutting narrow grooves, slots, and square corners. The cape chisel is also used to cut shallow grooves in work surfaces that are wide. The flat chisel is then used to remove the *stock* (material) between the grooves.

The *round-nose* chisel has a round cutting edge (Figure 7-7). It is designed for chiseling round and semicircular grooves and inside, round *filleted* corners. The *gouge* chisel is used for chipping circular surfaces. The *square* or *diamond-point* chisel is used for cutting square corners and grooves.

HOLDING, POSITIONING, AND STRIKING CHISELS

The holding, positioning, and striking of all types of chisels is important. Skillful cutting

requires that the chisel be held steady in one hand, with the finger muscles relaxed. For a chipping operation the cutting edge is positioned for the depth of cut. The chisel is positioned for each successive rough cut and again for the final finish cut (Figure 7-8).

For a shearing operation the cutting edge of the chisel is set at the cutting line. On mild ferrous metals a 1/16″ or 2mm cut is considered a roughing cut; 1/32″ or 1 mm, a finish cut. The depth of the roughing cut may be increased to 3/32″ on softer nonferrous materials.

Quick, sharp hammer blows on the head of the chisel are most effective for cutting. For personal safety it is important to *follow the cutting edge* and not to watch the head of the chisel. The depth of cut depends on the angle at which the chisel is held. The sharper the angle between the chisel and the workpiece, the deeper the cut. After each cut the chisel is positioned for the next cut. The cutting process is then repeated. The chisel point may be lubricated with a light machine oil for easier driving and faster cutting.

Note: When chiseling cast parts like cast iron, brass, or aluminum, the corners may break away

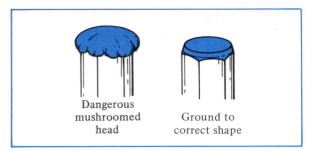

Dangerous mushroomed head

Ground to correct shape

Figure 7-9 Corrected Shape of a Mushroomed Chisel Head

below the desired layout line. The chiseling of cast parts is started from the outside edges. The cut is then finished near the center of the workpiece.

Caution: Put on a pair of safety goggles or a protective shield.

Caution: A *mushroomed* head must be reground for safety. The corrected shape of a mushroomed head is illustrated in Figure 7–9.

HAND PUNCHES

Hand punches (Figure 7-10) are used in bench work for four general purposes:

- To make indentations;
- To drive pins, bolts, or parts;
- To align parts;
- To punch holes.

Punches are known by the shape and diameter of their point, or shank end. Each type of punch is designed for a specific use. Solid, pin, and taper punches are made in a range of sizes. In practice, where parts are to be driven in or removed, the diameter of the punch should be as large as possible to permit the worker to deliver a solid, driving force with a hammer.

The *hollow* punch is used to cut holes in *shim* (very thin) stock or other thin materials. Hollow punches are made in sets from 1/4″ (6mm) to 4″ (100mm) in diameter.

PRICK AND CENTER PUNCHES

Two similar punches are used extensively in center layout work for making cone-shaped indentations: the *prick* punch and the *center* punch.

The center punch has a 90° included angle. This punch produces a centering location as a guide for positioning a drill.

The prick punch is smaller than the center punch and therefore is not subjected to as heavy a striking force. The included angle of the point is 30°. Primarily, the prick punch is used to:

- Form a small center hole from which circles may be laid out with dividers;

- Make small, easily identifiable, cone-shaped depressions along scribed lines. The prick-punched holes assist the worker during machining operations;
- Mark the location of parts for proper identification and assembly.

DRIFTS

Drifts are used to drive parts and are usually round, soft metal pieces. Drifts must be slightly smaller than the diameter of a shaft, pin, stud, or other part that is to be driven. Brass rods are often used as drifts. Brass is softer than steel and other alloys and will not score or mark the part that is being driven.

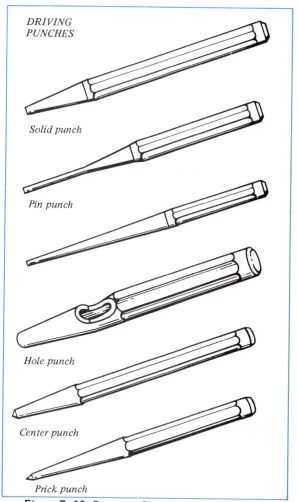

DRIVING PUNCHES

Solid punch

Pin punch

Hole punch

Center punch

Prick punch

Figure 7–10 Common Shapes of Hand Punches

How to Use Punches and Drifts

Driving with Solid and Pin Punches

STEP Grip the workpiece. Position it so that
1 the pin may be driven out.

Note: Sometimes the part is positioned on a solid anvil. The pin is placed over one of the holes in the anvil. At other times, the workpiece is placed on two parallel pieces of metal resting on another solid metal plate.

STEP Select a solid punch. Its size must be a
2 little smaller than the diameter of the pin.

Note: Check the head of the punch. If it is mushroomed, dress (grind) it before using.

STEP Place the punch on the pin and center it.
3 Hold the punch squarely with the fingers of one hand.

STEP Select a hammer suitable for the size of
4 the job. Judge the amount of force needed to start and to drive the pin. Strike a solid blow.

Note: Check to see if the head of the pin or part is mushroomed. A rounded pin head may need to be ground away before attempting to drive the pin.

STEP Select a pin punch for finish driving, if
5 necessary. The use of a long pin punch to finish driving a pin or stud is illustrated in Figure 7–11.

STEP Place the pin punch squarely on the driven
6 pin. Continue to apply force to drive the pin out. The hammer blows should be lightened when the pin begins to move a slight distance each time it is hit.

Driving with a Drift

STEP Select a drift large enough for the size of
1 the job and the amount of force to be used. The drift material should be softer than the part.

Note: The face of the drift should be square. Check the head end. Remove any excess material caused by mushrooming.

STEP Hold the workpiece in a vise, other suit-
2 able holding fixture, or against a surface or part.

STEP Position the drift squarely over the gear,
3 shaft, hardened pin, or other part that is to be removed.

STEP Use solid blows to start driving the drift.
4 Reduce the force as the parts separate.

Aligning Parts

STEP Select the kind of taper punch that is
1 appropriate for the job.

Note: The punch may be tapered to a point, slightly tapered, or of the same size as the pin or stud.

STEP Align the parts by sight. Insert the punch
2 through the holes. Move the parts so that they are aligned more closely.

STEP Tap the taper punch lightly. A tapping
3 action will move the parts into final alignment.

Note: The parts may need to be clamped to hold them in alignment. The taper punch is removed after aligning.

Punching Holes with a Hollow Punch

STEP Mark the center of the hole with a prick
1 punch.

STEP Scribe a circle of the required diameter.
2

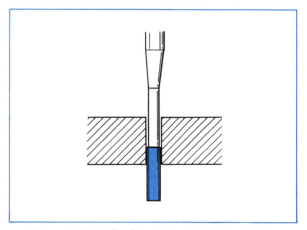

Figure 7–11 A Pin Punch Used to Finish Driving a Pin through a Part

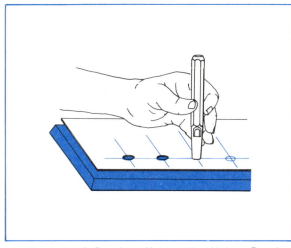

Figure 7-12 Punching Holes with a Hollow Punch

STEP 3 Place the material so that it rests flatly on a block. The hard wood or other flat material should be softer than the workpiece.

STEP 4 Hold the hollow punch square with the workpiece. The diameter of the punch is centered with the scribed circle (Figure 7-12).

STEP 5 Strike a sharp blow to produce an impression.

STEP 6 Raise the punch. Examine the impression. Be sure the impression is centered. If it is off-center, position the punch so that it is concentric with the scribed circle. Then strike a blow to get a correctly centered impression.

STEP 7 Return the hollow punch to the correct impression. Continue to strike hard, sharp blows until the hole is punched.

STEP 8 Place the workpiece upside down on a flat metal plate. Use a soft-face hammer or mallet to flatten the metal part.

Caution: Remove all sharp edges and burrs. The edge (circumference) of the hole should be ground, filed, or scraped.

Punching Holes with a Solid Punch

Note: Small diameter holes are usually punched directly.

STEP 1 Lay out the center of each hole. Scribe a circle of the required diameter.

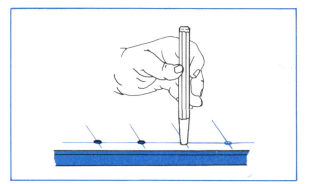
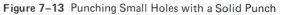

Figure 7-13 Punching Small Holes with a Solid Punch

STEP 2 Select a solid punch.

STEP 3 Place the workpiece on a flat, solid block.

STEP 4 Center the solid punch. Punch the hole (Figure 7-13). Follow previous steps 4 through 8.

C. HAND FILE CHARACTERISTICS AND FILING PROCESSES

The *file* is a cutting tool. Files are used extensively in bench and other hand operations to:

- Alter the shape of a part,
- Reduce the size of a part,
- Remove burrs and edges,
- Finish surfaces that have tool marks,
- Touch up surfaces that require fitting in assembly.

CHARACTERISTICS OF FILES
PARTS OF A FILE

Terms that are commonly used to designate the parts of a file are shown in Figure 7-14.

The cutting edges of a file consist of a number of teeth. These teeth are cut in parallel rows diagonally across the face. Since files are made of high-grade tool steel and are hardened and tempered, the teeth are hard. A file, therefore,

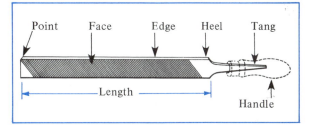

Figure 7-14 Parts of a File

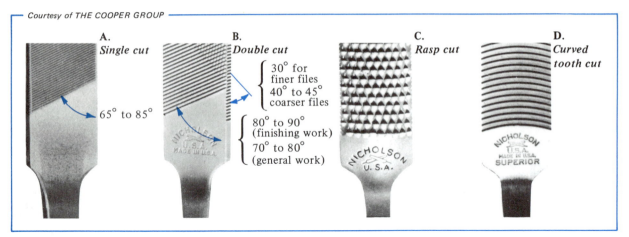

A.
Single cut

65° to 85°

B.
Double cut

30° for finer files
40° to 45° coarser files

80° to 90° (finishing work)
70° to 80° (general work)

C.
Rasp cut

D.
Curved tooth cut

Figure 7–15 Four Different Cuts of Files

can cut softer materials like unhardened steels, cast iron, brass, and nonmetallic products.

FILE CUTS

Files are produced with different *cuts* (parallel rows of teeth). A file with a single series of cuts is called a *single-cut* file. A *double-cut* file has two series of cuts. The two series of parallel rows of teeth cross each other at an angle with the edge. The angle of the cut on single-cut files varies from 65° to 85° with the edge (Figure 7–15A). Double-cut files for general metal work have one set of cuts at a 40° to 45° angle from the edge and a second crossing set at an angle from 70° to 80° (Figure 7–15B). The angles of the two cuts of finer double-cut files for finishing operations are 30° and 80° to 90°, respectively. Two other cuts of files are the *rasp cut* (Figure 7–15C) and the *curved tooth* (Figure 7–15D).

COARSENESS OF CUTS

The distance between each parallel row of cuts denotes the *coarseness* of a file. The coarsest cut on files larger than 10″ is known as *rough cut*. The finest cut is *dead smooth*. In between there are degrees of coarseness from a *coarse cut*, a *bastard cut*, a *second cut*, to a *smooth cut*.

Numbers are used to indicate the cut and relative coarseness of files that are smaller than 10″. There are ten different cuts: #00, #0, #1, #2, #3, #4, #5, #6, #7, and #8. The #00 cut is the coarsest, and #8 is the finest. Again, because

of the relation between coarseness of cut and size of file, a #1 cut 8″ file is coarser than a #1 cut 4″ file.

Each cut of file is used for a particular hand filing process. The most commonly used files in machine shops are: the bastard cut, for rough filing; the second cut, for reducing a part to size; and the smooth cut, for finish filing.

FILE NAMES

The name of a file is associated with either its cross section, particular use, or general shape. For example, a circular shape file is called a *round file*. A file with a rectangular cross section is generally designated as a *flat file*. However, if it is used for mill filing, it is identified as a *mill file* and if the edges are ground as safe filing edges the file is known as a *pillar file*. A cross section of a file having a curved and a flat face is called a *half-round* file.

PRECISION FILES

The files just described are generally classified as *machinists'* files. Three other groups of files, known as *precision* files, are next described from among the many designs of files. Precision files include the *Swiss pattern*, *needle-handle*, and *special-purpose* files.

SWISS PATTERN FILES

Swiss pattern files are used to accurately file and finish small precision parts. They are used commonly by tool and die makers, instrument makers, jewelers, and other highly skilled crafts-persons. The files have cross sections similar to machinists' files.

The major differences between Swiss pattern and machinists' (American pattern) files are that:

- The points of Swiss pattern files are smaller,
- The tapers of Swiss pattern files toward their points and at their tangs are longer and sharper,
- The cuts of the Swiss pattern teeth are finer,
- The specifications of Swiss pattern files are more exacting.

Six of the most common Swiss pattern files —usually called just *Swiss* files—include half round, narrow pillar, knife, round, square, and three square (triangular) shapes. There are over 100 different shapes. The teeth are double cut on the faces. The teeth on the edges of the knife and three-square files are single cut. Some of these files are manufactured with safe edges (no teeth).

NEEDLE-HANDLE FILES

Needle-handle files are another group of small precision files. They are normally 4″ to 6″ long. They are used for fine, precision filing where a small surface area is to be formed or fitted with another part.

Needle-handle files also are designated by their shape, cut, and length. The characteristics and common shapes of some of these files are illustrated in Figure 7–16.

SPECIAL-PURPOSE FILES

Special-purpose files produce different kinds of cutting actions and surface finishes on various materials. They are designed for greater efficiency than general-purpose files. For example, although a mill file is widely used in lathe work, the special-purpose, long-angle, lathe file has a number of advantages:

- The teeth, cut at a longer angle, provide an efficient shearing action, in contrast to the tearing and scoring action of general-purpose files;
- Chatter marks are prevented;
- Clogging of the teeth is reduced;
- Materials are cut away faster;
- A finer surface finish is produced.

Soft metals, plastics, stainless steels, and other materials each require special-purpose files. These files permit maximum cutting efficiency, dimensional accuracy, and high-quality surface finishing. For very soft metals a special single-cut coarse file with a small angle provides an efficient shearing action (Figure 7–17).

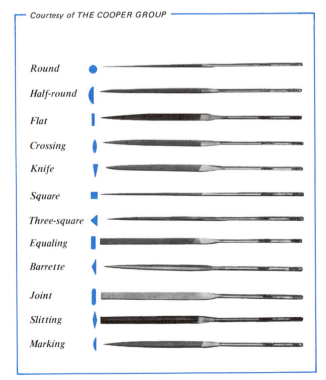

Round
Half-round
Flat
Crossing
Knife
Square
Three-square
Equaling
Barrette
Joint
Slitting
Marking

Figure 7–16 A Set of Assorted Needle-Handle Files

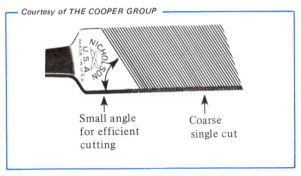

Small angle for efficient cutting

Coarse single cut

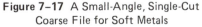

Figure 7–17 A Small-Angle, Single-Cut Coarse File for Soft Metals

FEATURES OF FILES

There are three features of files that are especially important in filing. They are a *safe edge*, *convexity*, and *taper*.

SAFE EDGE

A safe edge is a file edge that has no teeth. The safe edge protects the surface against which the file is moved from any cutting action. The safe edge also permits filing into a sharper corner.

CONVEXITY

Convexity of a file means that the faces are slightly convex and bellied along their length. The faces curve so that they are smaller at the point end of the file. Convexity is important for five reasons:

- A perfectly flat face would require a tremendous downward force to make the teeth bite and a great forward force to produce a cutting action;
- Control of a perfectly flat file would be difficult;
- A flat surface could only be produced if every part of the file were perfectly straight;
- File warpage, resulting from the hardening process, would produce an undesirable concave face;
- Convexity makes it possible for the skilled worker to control the position and amount of cutting.

TAPER

Most files taper slightly in width. The taper on round, square, and three-square (triangular) files is more severe than the taper of other types. The cross-sectional area is reduced considerably from about the middle of the body to the point. A file is termed *blunt* if it has a uniform cross section along its entire length.

THE CARE OF FILES

One common problem in hand filing is *pinning*. Pinning occurs when a *pin* (small particle of the material being filed) gets wedged (pinned) on the lead edge of the file teeth. The pins, as they move on the file across the face of the work, tear into and scratch the filed surface.

Pinning is caused by:

- Bearing too hard (applying too much force) on a file, particularly on a new file;
- Failing to regularly clean the file teeth.

Pinning may be avoided by:

- Brushing the workpiece with a file brush to remove the filings,
- Brushing the file teeth with a wire file card,
- Pushing a metal scorer or a soft metal strip through the grooves of the teeth,
- Rubbing ordinary chalk across the file teeth to chalk them.

FILE HANDLES

Part of the skill in using a file depends on the size, shape, and feel of the handle. A file may be controlled better and hand fatigue may be avoided if the handle is easy to grasp and hold. The handle should be selected for shape and size. The handle (1) must feel comfortable and (2) must give balance to the file.

FILING PROCESSES

CROSS FILING

Cross filing, often referred to as just plain *filing*, is the process of diagonally crossing file strokes. First, the file face is forced against the work and pushed forward on the cutting stroke. The force is relieved on the return stroke. Then, after a few strokes, when a complete surface area has been filed, the position of the file and the filing operation are changed. The strokes are crossed. The procedure continues until the part has been *brought down* (filed) to a layout line or dimension. As much of the file face as is possible should be used during each cutting stroke. Checks for flatness and squareness of filed surfaces are shown in Figure 7–18A and B.

The surface of most materials may be filed directly. However, cast-iron parts require a special procedure. The *scale* (outer surface) of a cast-iron casting is often *file hard*. If a file is used directly on such a surface, the teeth will dull rapidly. The condition may be overcome either by rough grinding the surface or by chipping below the depth of the scale. Once the scale is removed, the cast iron is soft enough to be rough or finish filed.

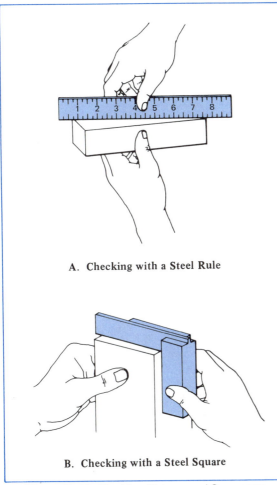

A. Checking with a Steel Rule

B. Checking with a Steel Square

Figure 17-18 Checking Flatness and Squareness

DRAWFILING

Drawfiling is a process of *graining* a workpiece—that is, producing a fine-grained surface finish. For drawfiling, a single-cut file is held with both hands, as illustrated in Figure 7–19. The file is then forced down onto the workpiece as it is moved over the surface. A cutting action is produced by *drawing* (pushing and pulling) the file alternately in each direction. Particular care must be taken to use a sharp file and to keep it clean.

Drawfiled surfaces are sometimes polished by holding a strip of abrasive cloth tightly against the bottom face of the file. The file is then drawn over the work to produce a polished surface.

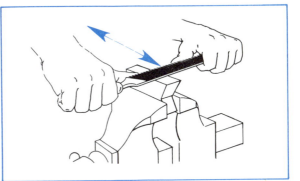

Figure 17-19 Positions of a File, Hands, and Work in Drawfiling

SOFT METAL FILING

The files and processes described thus far have dealt primarily with the filing of steels and other hard metals. These same files may not be suitable for soft materials because filings can clog and wedge between the teeth.

The curved-tooth file, mentioned earlier, is specially designed for efficiently filing soft metals like brass, copper, aluminum, and lead. This file has chisel-edged teeth that produce an excellent finished surface. Clogging of the teeth is reduced by the shape of the grooves between the rows of teeth. The curved-tooth file is also used for finish filing of soft metal parts turned on a lathe.

How to Cross File

Holding the File

STEP 1 Position the work in a vise or other holding device so that there will be no interference while filing. The height of the surface to be filed should be level with the worker's elbow.

STEP 2 Select a file that will produce the fastest cutting action. Selection depends on the material, the size of the part, and the nature of the filing process.

STEP 3 Check the tightness of the file handle. Clean the file teeth with a file brush, card, or metal scorer, as needed (Figure 7–20).

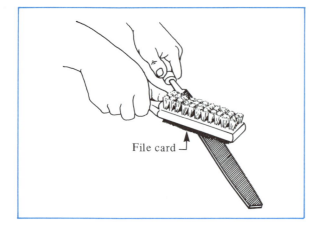

Figure 7-20 Cleaning File Teeth with a File Card

STEP Hold the point end of the file in one hand,
4 with the thumb and palm pushing on the top face. Curve the remaining fingers on the underface (Figure 7-21). Downward force for heavy filing can now be applied by pressing the thumb, palm, and fingers into the file point.

STEP Grasp the file handle in the other hand.
5 The palm should rest against the handle, with the thumb on top. The fingers are curled around the handle (Figure 7-21).

Positioning the Body

STEP Stand close to the bench, with one foot
1 pointed forward. Place the other foot from 8″ to 12″ to the side and about 16″ back.

STEP Lean the upper part of the body forward
2 for most of the forward stroke. Bend the knees slightly. Apply force by pushing with the arms.

STEP Follow through with the arms to the end
3 of the stroke.

STEP Bring the body back to the original posi-
4 tion. Return the file to the start of the next stroke by moving it lightly across the work.

Note: When surfaces are wider than the width of the file, the file is moved to the next area to be filed during the return stroke. Each cut should overlap the preceding cut.

STEP Cross the stroke during rough filing. Cross
5 filing helps to rest the arms and check the evenness of filing. Limit the file strokes to about 50 per minute in order to control the file and to avoid dragging it on return strokes.

Checking the Filed Surface

STEP Test the work surface for flatness and
1 straightness. Use a steel rule or the blade of a square. Check the work surface crosswise, lengthwise, and diagonally.

STEP File the *high spots* (raised areas) down to
2 the specified dimensions. The convexity of the file makes it possible to pinpoint the filing area.

STEP Check the depth and shape of the filed
3 surface either against the layout lines or by direct measurement.

Checking Squareness

STEP Remove burrs. Clean the workpiece and
1 the square.

STEP Hold the solid base of the square against
2 one face of the work. Gently slide the base until the blade contacts the filed surface.

STEP Hold up the workpiece and square to the
3 light. The presence of light between the blade and filed area indicates low places where the surface was filed too much.

STEP Continue to file the high places. Recheck
4 for squareness and correct size.

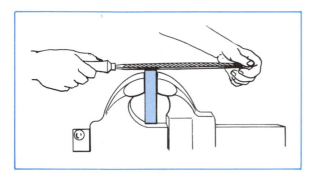

Figure 7-21 Correct Hand Positions for Heavy Filing

How to File a Rounded Corner

STEP 1 Clamp the workpiece so that the layout line is above the vise jaws.

STEP 2 Begin by rough filing the top surface to within 1/64″ of the layout line.

STEP 3 Continue to rough file at about a 45° angle. File to within 1/64″ of the layout line.

STEP 4 File away the excess material on the corner to be rounded. File at a series of angles. (The sequence of file cuts is shown in Figure 7–22A.)

STEP 5 File away the crests formed by each previous angular cut with a finer cut file (Figure 7–22B).

STEP 6 Finish filing by using a rounding, rocking motion. Continue filing until the surface conforms to the required curvature (Figure 7–22C).

Note: After a few strokes, check the filed surface for high or low spots. Checking

may be done against a layout line or with a template of the correct size and shape.

D. DEBURRING AND HAND SCRAPING

A burr is the turned-up or razor-sharp projecting edge of a material. Burrs generally are produced by sawing, drilling, punching, shearing, and most cutting processes. Burrs are also raised through careless handling, hammering, dropping, and nicking. Damaged or spoiled work parts often result from failure to remove burrs before clamping or performing hand or machine operations.

Burred edges, nicks, dents, and ragged or unwanted sharp corners are usually removed by hand filing.

How to File Burrs

STEP 1 Select either a smooth-cut hand file or a smooth-cut mill file.

STEP 2 Mount the workpiece in a vise.

Note: When the workpiece is small, the skilled worker holds the part in one hand, as an easy way of removing burrs from all edges.

Caution: Burrs can be razor sharp. Scraping against them or rubbing with the fingers or hand can produce cuts or bruises.

STEP 3 Hold the file at a slight angle to the burred edge. Take light cuts to *break the corner* and remove the burrs.

STEP 4 Move to the adjoining surface and repeat the filing process.

STEP 5 Continue to remove the burrs from all sides and surfaces of the workpiece.

The sharp edges and burrs on hardened steel pieces may be removed easily with a small oilstone. The burrs on curved surfaces may be scraped with the three-cornered scraper. In summary, parts are burred to:

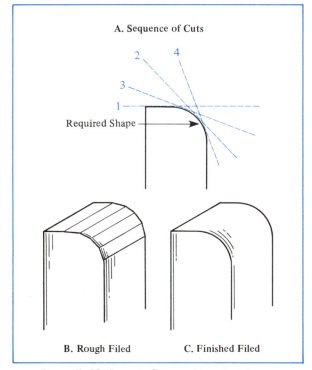

Figure 7–22 Steps in Filing a Rounded Corner

- Ensure dimensional accuracy,
- Provide for the exact fitting of mating parts,
- Permit safe handling,
- Avoid damage to tools and machines,
- Improve the finish and appearance of the part.

BURRING FOR APPEARANCE

One of the hallmarks of a skilled craftsperson is a properly burred workpiece. The rounding of corners adds to the finished appearance of a job. Industrial practices often require finishing in this manner. Many industrial blueprints include notations such as **BREAK EDGES** or **REMOVE ALL UNNECESSARY SHARP CORNERS.**

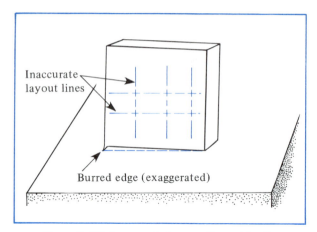

Figure 7–23 Inaccurate Layout Caused by Burr under the Workpiece Edge

BURRING FOR SAFETY

To reemphasize: Parts with burred edges, ridges, and sharp corners are not safe to handle. These razor-sharp edges cut, and even tear, metal against which they may be brushed or rubbed. Consequently, where possible, remove all unnecessary sharp corners and burrs to prevent personal injury.

There are applications in which sharp corners are necessary, such as in mold making and die work. Sharp corners must be retained in such applications.

BURRING FOR ACCURACY

Improper burring is a frequent source of inaccurate layout work. It is impossible to accurately lay out a job if it rests on a burred edge or on a surface that is not clean. The raised surface area (mass) produced by a burr will not seat solidly against a finished surface. This added mass has the same effect as a shim of thin material placed between the surfaces.

Work that is being positioned for layout and machining should be free of burrs. The work must rest evenly on parallels, in a vise, on a table, or in any other holding device. A burr under one edge is sufficient to throw the job out of alignment, as illustrated in Figure 7–23. The result is inaccurate, damaged, or spoiled work.

Measurements taken over burrs are another source of trouble. Burrs prevent holding a steel rule or other measuring tool sufficiently close to a measured point. The presence of burrs makes it difficult to obtain an exact reading. Therefore, all measurements should be taken on surfaces that are free of burrs.

Burrs frequently cause problems when fitting parts together. Burrs may tear a surface against which they are moved. The projections also produce an oversize (larger than specified) condition. In all bench work processes, then, sharp corners, turned-over edges, and machine-produced burrs must be removed for parts to fit accurately.

SCRAPING TOOLS AND PROCESSES

FUNCTIONS OF SCRAPING

Scraping is a bench work process that involves a cutting action. Small amounts of material are removed during each process.

Scraping produces extremely accurate flat or circular bearing surfaces and special surface finish effects. Scraping is not as common a practice today as it was years ago. Modern grinding techniques and improved machining processes produce surface finishes of high quality and precise limits of accuracy. However, the scraping process is still performed when:

- The surface produced by a shaper, planer, milling or other machine has tool marks (these marks may not be suitable as a close-grained, smooth bearing surface);

- A surface slightly deviates from being flat and true (the deviation can be corrected by hand scraping);
- A curved shaft on which another shaft, spindle, or circular part must ride is only a few thousandths of an inch away from a perfect fit and must be scraped into position;
- A surface is to be frosted or flaked (*frosting* is a surface finish design produced by hand scraping).

FUNCTIONS OF FROSTING

Frosting (flaking) refers to a repetitive design. This design is produced by scraping in a particular pattern of strokes. Sometimes the patterns are circular and scroll shaped. At other times the scraping follows straight-line patterns in which strokes are alternated at right angles. There are many combinations.

A surface is frosted to:

- Improve the sliding quality between two mating surfaces and help in eliminating sticking,
- Improve appearance,
- Provide pockets for lubricating fluid.

FLAT SURFACE SCRAPERS

The tools used for scraping and producing surface finishes are simple. Two common forms of scraper, the *flat* and the *hook*, are shown in Figure 7–24. The cutting edge on the end of the flat scraper is square with the faces and edges. However, instead of being perfectly straight, the edge is slightly convex. This convex end permits the operator to control cutting action and position the scraper on high spots and prevents scoring or scratching during the scraping process. The corners of the cutting end often are rounded with a hand stone as a precaution against scoring.

HALF-ROUND (BEARING) SCRAPERS

Another form of scraper is called a *bearing*, or *half-round*, scraper (Figure 7–25). This specially shaped scraper has an offset cutting end that is spoon shaped. The cutting edges on the two sides extend along the lips of the spoon shape. They come to a sharp point. A half-round bearing scraper is suited for scraping and finishing round and curved surfaces that require precision fitting and/or a special surface finish.

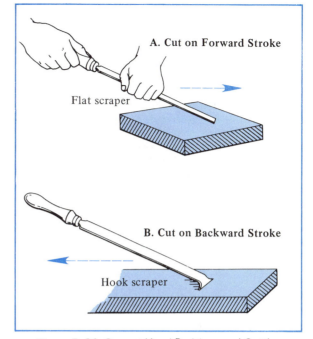

Figure 7–24 Correct Hand Positions and Cutting Angle for Using Scrapers

THREE-CORNERED (TRIANGULAR) SCRAPER

The form of scraper that is most common is called a *three-cornered* (triangular) scraper. Frequently, it is made from an old three-square file. The length is shortened, the teeth are ground off, and the sides are formed into a sharp point. The three-cornered scraper is a handy, practical cutting tool. It is used to remove sharp edges from holes, fillets, or other curved surfaces. These sharp edges may be the result of other hand or machining operations like drilling, reaming, or boring.

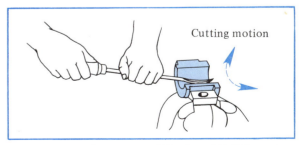

Figure 7–25 Positioning of Hands and Half-Round Scraper for Scraping a Curved Surface

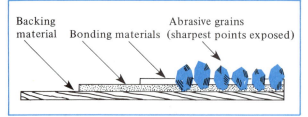

Figure 7–26 Prime Materials Used in the Manufacture of Coated Abrasives

Scrapers are also produced with carbide tips. Many scrapers are moved by hand. Others are held-in-hand power tools that produce the necessary movement for scraping.

E. COATED ABRASIVES AND HAND FINISHING (POLISHING)

Abrasives and their applications to machine cutting and surface finishing processes are covered in detail in the units on grinders and grinding operations. Only basic information on coated abrasives that relates to bench and floor work is considered in this unit.

Coated abrasives are used in hand operations in machine shops. These abrasives produce a smooth or polished surface and remove very slight amounts of material. The edges of their abrasive grains, with their ability to fracture and form new sharp edges as they wear, make coated abrasives excellent for polishing.

STRUCTURE OF COATED ABRASIVES

Coated abrasives are produced by bonding a particular type and quantity of abrasive grains on a suitable flexible backing material. The relationship of the *coating*, the *bonding*, and the *backing* is shown in Figure 7–26.

The abrasive grains are electrostatically embedded in a bond in the backing material. The sharpest points of the grain are thus positioned outward and exposed.

The grains are spaced evenly as either an *open* or *closed* coating. The two basic open and closed coatings are shown in Figure 7–27. An open-coated abrasive cloth has only part of the backing surface covered with abrasive grains (Figure 7–27A). The open spaces between grains prevent the grains of the abrasive material from filling with chips or clogging.

By contrast, in closed coating, all of the backing surface is covered (Figure 7–27B). Closed-coated abrasive cloths, belts, and discs are used in heavy cutting operations that require a high rate of stock removal.

FORMS OF COATED ABRASIVES

Coated abrasives are produced in a number of forms. The most common forms are shown in Figure 7–28. These forms include *abrasive belts*, *grinding discs*, and *abrasive cloth* in sheet or roll form. Abrasive belts provide an efficient cutting material and technique for polishing sheet or strip steel. Grinding discs are used for smoothing welds and for removing surface

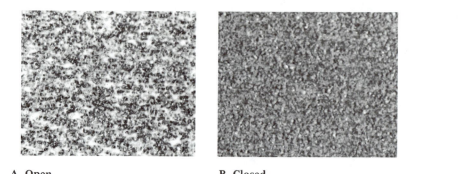

A. Open   B. Closed

Figure 7–27 Two Basic Abrasive Coatings

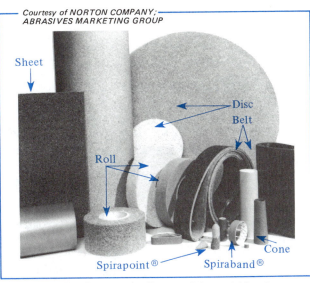

Sheet

Disc

Belt

Roll

Cone

Spirapoint®   Spiraband®

Figure 7–28 Common Forms of Coated Abrasives

irregularities, projections, and burrs. Abrasive cloth is used primarily for hand polishing and surface finishing.

When selecting a coated abrasive, consideration must be given to two sets of factors:

- The type of material in the part and the required surface finish,
- The qualities of the coated abrasive that are needed to cut efficiently and to produce the desired finish.

CHARACTERISTICS OF COATED ABRASIVES

The cutting efficiency of coated abrasives and surface finish depend on the:

- Abrasive material,
- Bonding and backing materials,
- Type of coating,
- Grain size.

ABRASIVE MATERIAL

An *abrasive* is a hard, sharp-cutting, crystalline grain. Abrasives are applied in machine, metal, and other industries to grind, polish, buff, and lap metal surfaces to specific sizes and surface finishes. Abrasives may be natural or artificial.

All abrasives possess at least the three following properties:

- *Penetration hardness*, which is the cutting ability due to the hardness of the grain;
- *Fracture resistance*, which is the ability of the grain to fracture when the cutting edge has started to dull and thereby produce a new sharp cutting edge;
- *Wear resistance*, which is the ability of the grain to resist rapid wear and to maintain its sharpness (wear resistance is affected by the chemical composition of the grains in relation to that of the work being ground).

Natural Abrasives. The natural abrasives that are still used in industry include *garnet*, *flint*, *crocus*, *diamond*, and *emery*. Coated garnet and flint abrasive papers and cloths are used in the woodworking industry. Crocus (red iron oxide) cloth and rouge (flour-sized powder) are used for polishing and producing a very fine finish. Extremely small amounts of metal are removed with rouge.

Artificial Abrasives. *Aluminum oxide*, *silicon carbide*, and *boron carbide* are synthetic (artificial) abrasives. The hardness of each of these abrasives is higher than the hardness of the natural abrasives (except diamond).

Silicon carbide is made by fusing silica sand and coke in an electric furnace at high temperatures. The sand provides the silica; the coke, the carbon. Silicon carbide is used for coated abrasives, grinding wheels, and abrasive sticks and stones. Silicon carbide has a high grain fracture. It is, therefore, desirable for grinding some of the softer metals like cast iron, bronze, and copper. Nonmetallic materials that are ground with silicon carbide include marble, pottery, and plastic.

Aluminum oxide is manufactured from bauxite ore. The ore is heated to extremely high temperatures, at which it is purified to crystalline form. The addition of titanium produces an abrasive that has added toughness qualities. With toughness, resistance to shock, and grain fracturing properties, aluminum oxide abrasives are suitable for grinding both soft and hard steels, wrought and malleable iron, and hard bronze.

Artificial abrasives with ever-higher hardness qualities are being developed. The important fact is that artificial abrasives are harder and have greater impact toughness than the natural abrasives. Artificial abrasives have the following advantages over natural abrasives:

- Faster cutting characteristics,
- Ability to retain their cutting properties longer,
- Uniformity of quality,
- Economical manufacture in a wider range of grain sizes.

BACKING MATERIALS

Paper, *cloth*, *fiber*, and combinations thereof provide the backing material for coated abrasives. Cloth and fiber backings are the most durable. They are used for hand and machine cutting, buffing, and polishing operations.

The most common forms of cloth-backed coated abrasives for hand use are sheets and rolls. The sheets are 9″ × 11″ (225mm × 275mm). The rolls range from 1/2″ to 3″ (12mm to 75mm) wide. Other forms of coated abrasives for high-powered grinding machines include belts, discs, spiral points, and cones.

BONDING MATERIALS AND COATINGS

Bonding material serves as an adhesive. This material holds the grain particles to the backing and indirectly contributes to their cutting properties. Since coated abrasives may be used on dry or wet processes, a combination of bonds is employed. *Glues* provide a bond for dry, light, hand grinding and polishing. *Synthetic resins* are practical for wet grinding with waterproof cloth. There are also modifications of glues, resins, and varnishes. The resin-bonded combina-

Table 7–2 Coarseness Designations and Grain Sizes

Relative Coarseness	Grain Sizes
Extra coarse	12, 16, 20, 24, 30, 36
Coarse	40, 50
Medium	60, 80, 100
Fine	120, 150, 180
Extra Fine	220, 240, 280, 320, 360, 400
Flour	440, 500, 600

tions, in particular, provide moisture- and heat-resistant properties. These properties are required for rough grinding and polishing applications for toughness.

GRAIN SIZES

The abrasive grain sizes for coated abrasives range from extra coarse to flour. The extreme and the intermediate degrees of coarseness are indicated by numbers that range from 12 to 600. Coarseness classifications and their number designations are given in Table 7–2.

In trade language the terms *grain sizes* and *grits* are used interchangeably. Usually coarser grain sizes (grits) are used for heavy stock removal; finer grain sizes, for highly polished surface finishes. For hand polishing it is good practice to start with as coarse a grain as is possible. The grain size must be consistent with the required surface finish. Reducing the grain size from coarser to finer during a series of final finishing operations speeds the cutting action.

How to Hand Polish with Coated Abrasives

Selecting the Coated Abrasive

STEP 1 Determine the abrasive grain size that is appropriate for the material to be polished and the required quality of surface finish. Select the coarsest grade possible. Move to finer grades as the process progresses.

STEP 2 Tear a strip slightly wider than the width of the file or backing object. The length should permit the abrasive cloth to be held tautly.

Note: The grain runs lengthwise on cloth backing sheets and rolls. It is easy to tear an even strip running with the grain. (The direction of tear is illustrated in Figure 7–29.)

STEP 3 Grip the work securely in a vise or other holding device. Use soft jaws or pieces of soft stock to protect the finished surfaces.

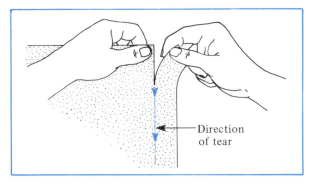

Figure 7-29 Tearing an Abrasive along the Grain of the Fabric

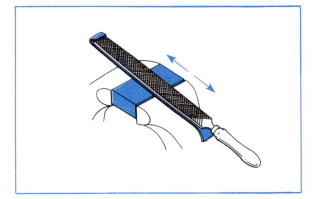

Figure 7-31 Polishing Flat Work Surfaces

STEP 4 Check the surface of the backing block to see that it is free of burrs and is flat and smooth. A file, a hardwood block, or other suitable material may serve as a backing.

Hand Polishing

STEP 1 Fold one end of the abrasive strip over the file point or backing block (Figure 7-30). Hold the strip in position with one hand. Bring the other end along the under face of the file or block. Pull it taut with the other hand.

STEP 2 Place the abrasive strip on the work surface. Apply a slight downward pressure. Move the abrasive strip in a straight line over the work surface. Take as long a stroke as possible. Cut on both the forward and return strokes (Figure 7-31).

STEP 3 Replace the abrasive strip. Change from a coarser grain to a finer grain for fine finishing.

Note: A few drops of oil applied to a metallic surface helps the cutting action.

A. Safe Practices in Hand Hacksawing

- Position the workpiece area where the cut is to be made as close to the vise as is possible to prevent springing, saw breakage, and personal injury.
- Nest thin-walled tubing and other easily damaged parts in suitable forms. Otherwise, they may be collapsed or distorted by the clamping force.
- Select the blade pitch and set that is most suitable for the material and the nature of the cutting operation. Stripped teeth often result when the pitch is too coarse and fewer than two teeth are in contact with the workpiece at one time.
- Apply force only on the forward (cutting) stroke. Relieve the force on the return stroke.
- Reduce the speed and the force at the end of a cut. Slowing down prevents the teeth from digging into the small remaining section of the material. It also prevents the cut portion from breaking away.
- Start a new blade in another place when a blade breaks during a cut to prevent binding and blade breakage. Remember that the groove produced by a blade becomes narrower as the balde wears.
- Saw straight. If the cut *runs* (strays from the correct line), reposition the workpiece a quarter turn and start a new cut. Starting a new cut prevents cramping of the blade

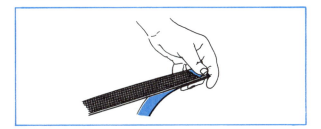

Figure 7-30 Correct Method of Holding an Abrasive Strip on a File Face

and also helps guide the second cut in the correct line.

- Cut a small groove with a file in sharp corners where a saw cut is to be started. The groove permits accurate positioning of the saw and also prevents stripping of the teeth.

B. Safe Practices for Chiseling, Punching Holes, and Driving Parts

- Wear safety goggles or a protective shield when chipping and driving parts. Flying chips and other particles may cause eye injury.

- Place a chipping guard or frame between the work area where a part is being chipped and other workers.

- Grind off any overhanging metal (mushroom) that may form on the head of a chisel, punch, or drift. Fragments of a mushroomed head may break away and can fly off at high speed during hammering.

- Limit the depth of a roughing cut to 1/16" (2mm). Allow 1/32" (1mm) for a finish cut. Heavier roughing cuts require excess force, which may cause a chisel to break or the cutting edge to dull quicker.

- Avoid rubbing the edges of holes that have been punched in metal.

C. Safe Practices in Filing and Polishing

- Avoid hitting a file directly, scraping files together, or placing one file on another.

- Remove pins that cling to and filings that clog the file teeth.

- Use a file with a properly fitted, tight handle.

- Avoid handling parts with burrs or fine, sharp projections. Remove burrs as soon as is practical.

- Remove the outer scale on a cast-iron casting by grinding or chiseling before filing.

- Use an eye protective device for operations involving the use of abrasive materials.

D. Safe Practices in Burring and Scraping

- Hold the cutting face of a flat scraper flat with the workpiece. Incline the edge at an appropriate cutting angle to prevent gouging into or scratching the finished surface.

- Manipulate the three-cornered scraper using a properly secured handle. The cutting face should be moved along the contour of the edge that is to be removed.

- Steady the hand against the part when using a three-cornered scraper. This provides support in applying a slight force, guiding the scraper, and preventing the sharp point and edges from slipping away from the work to cause personal injury.

- Remove only small amounts of material either to precision finish parts or to produce a desired surface finish design.

- Remove sharp projecting edges and burrs that produce inaccuracies in layout, measurement errors, and improper fits. Burrs also cause personal injury and tool damage. Burrs should be removed carefully and as soon as is practical.

E. Safe Practices for Coated Abrasive Polishing

- Select the abrasive grain size required to produce the desired quality of surface finish. Coarse grains cut deeply. They may produce surface scratches that are not removable within the dimensional limits of the finished part.

- Move a coated abrasive in one direction to polish a work surface. Crossing of strokes produces scratch marks. These marks may be objectionable.

BASIC TERMS USED WITH HAND CUTTING BENCH TOOLS AND PROCESSES

A. HAND HACKSAWING

Pitch	The number of teeth per inch or metric equivalent. (The general range of pitch for hand hacksaw blades is 14 to 32.)
Set	The alternate positioning of hacksaw blade teeth to cut a kerf, or groove, wider than the blade body thickness.
Flexible blade	A hacksaw blade that is heat treated. (The teeth are harder than the body, and the blade may be flexed.)
All-hard blade	A heat-treated hacksaw blade whose body and teeth are uniformly hard.
Cutting speed	The number of strokes per minute recommended by hacksaw blade manufacturers. The speed at which maximum performance and control of the cutting process may be obtained.

B. CHISELING, PUNCHING HOLES AND DRIVING TOOLS AND PROCESSES

Flat, round-nose, gouge, cape, and diamond-point chisels	Five common types of cold chisels. (Cold chisels are named according to the shape of their cutting edge or the chip, surface, or groove that is produced.)
Angle of cutting edge	The included angle between the two adjacent sides of the cutting point of a cold chisel.
Chiseling	A general term denoting two processes: chipping and shearing.
Mushroomed head	The flowing over (unsafe condition) of metal on a chisel, punch, or drift head.
Hand punches	Straight- or tapered-shank or hollow-center, round punches. Punches used for four basic hand operations: to make indentations, align parts, drive parts, and blank holes.
Solid, pin, taper, center, and hollow punches	Basic types of hand punches. (The names partially describe the shape and function of each punch. Punches are obtainable in sets of various diameters.)
Drift	A metal bar that is softer than the part to be driven.

C. FILES AND FILING

Single cut	A single series of parallel rows of file teeth. (The teeth are cut diagonally across the file faces.)
Double cut	Two series of parallel rows of teeth. (The rows cross each other diagonally. The double cut produces sharp-pointed cutting edges as file teeth.)
Coarseness of cut (files 10″ and longer)	A relative measurement of distance between two parallel rows of file teeth. (The range of coarseness from coarsest to finest is: rough cut, coarse cut, bastard cut, second cut, smooth cut, and dead smooth.)
Coarseness of cut (files shorter than 10″)	A range of cuts from coarse (#00) to fine (#8). (Intermediate cuts are designated from #0 to #1 through #7.)
File names	A name associated with a particular cross section (shape) or use. (The most common names are: flat, mill, hand, pillar, half-round, round, square, and three-square.)
Needle-handle files	A group of files 6'' and smaller in length with a round handle as the body. The toothed portion tapers sharply to almost a point.

Swiss pattern files	A group of precision files generally applied in instrumentation and toolmaking. (These files have longer tapers, finer cut teeth, and smaller point ends.)
Safe edge	A file edge without teeth.
Pinning	The wedging of filings (pins) on the file face or in the grooves of file teeth.
Cross filing (filing)	Filing materials by diagonally crossing the file strokes.
Drawfiling	A process for graining a workpiece. Drawing a single-cut file or a second-cut mill file across a work surface and cutting on both the forward and return stroke.
Soft metal file and filing	A deeper cut and differently shaped groove that prevents filings from clogging the file teeth. Filing soft materials with a curved-tooth file.
High spots	Areas of a surface that are higher than the specified dimension. File areas requiring the additional removal of material.

D. SCRAPING AND POLISHING TOOLS AND MATERIALS

Frosting (flaking)	Producing a repetitive design of scrolls, circles, or straight lines on a finished work surface.
Scraper/Scraping	A hand cutting tool for precision finishing flat, round, and other curved forms; fitting mating parts; and producing surface finish designs.
High spots	Surface markings produced by moving a workpiece to be fitted on a perfectly flat surface or mating part. Markings that reveal places on a surface that are high in relation to surrounding areas.
Breaking an edge	The removal of an unnecessary sharp edge by filing, scraping, or stoning.
Abrasive	A natural or synthetically produced hard material. (The grains have properties of penetration hardness to cut softer materials, fracture resistance to expose new sharp cutting edges without losing an entire grain, and wear resistance to maintain sharpness.)
Artificial abrasive	A synthetic material having the properties of hard-cutting abrasive grains. Three broad categories of manufactured abrasives: aluminum oxide, silicon carbide, and boron carbide.
Coated abrasive	A cutting product consisting of abrasive grains with special cutting properties, fixed grain quantities, distributed over an area, a bonding agent, and a backing material.

─────────────── SUMMARY ───────────────

A. HAND HACKSAWING

- Hacksaw frames are adjustable to accommodate different lengths of hacksaw blades. The blade may be held securely in any one of four frame positions.
    - Hacksaw blades may be all hard, semiflexible, or flexible back. All blades are designated according to the (1) material of which they are made, (2) pitch, (3) set, (4) length, and (5) heat treatment.
- Blades are positioned correctly in the frame with the teeth pointed toward the front of the frame.
    - The range of pitch for hand hacksaw blades is from 14 to 32. This range permits cutting solid, tubular, and thin sheet metals.
- Cutting speeds of from 40 to 50 strokes per minute are recommended.

- Cutting action is produced by applying a steady forward and slightly angular downward force during the cutting stroke.
- The force is released and the blade is raised slightly on the return stroke.
  - Tool safety precautions are necessary to prevent personal injury and damage to tools and materials. Attention must be paid to:
    - Starting a cut on narrow-sectioned material,
    - Positioning and holding the workpiece,
    - Replacing a worn blade,
    - Cutting action at the end of the cut.

B. CHISELING, PUNCHING HOLES AND DRIVING TOOLS AND PROCESSES

- The cold chisel is a cutting hand tool. The most commonly used cold chisel for chipping and shearing is the flat chisel. The cape, gouge, round-nose, and diamond-point chisels are other common cold chisels. These chisels are used to cut flat, square, "V", and round grooves and surfaces.
  - The recommended depth of cut for rough chipping of ferrous metals is 1/16". A 1/32" cut usually is taken when finish chipping.
- Chisels, hand punches, and drifts should be grasped by the fingers of one hand. To avoid fatigue, the hand should rest on a surface, if possible.
  - Sharp solid blows are needed in chiseling, punching, and driving. The operator must watch and guide the cutting edge, not the tool head.
- Six common types of hand punches are the solid punch, the pin or drift punch, the taper pin punch, the hollow punch, the prick punch, and the center punch.
  - The prick punch and the center punch form indentations. Prick-punched marks provide a guide for layout operations; center-punched marks, for drilling operations.
- Small holes may be produced with the pin or taper punch.
  - Drifts are usually handmade pieces of soft material. A drift conforms in size and shape to the part to be driven.
- Solid and taper punches are used to drive pins, studs, bolts, and other parts and to align holes in mating parts.
  - Chipping operations are dangerous. Safety goggles or a protective shield and a screen are required to prevent injury.

C. FILES AND FILING PROCESSES

- Filing is a cutting process. Filing alters the shape and size and the fit of a part. A particular surface finish may be produced by filing.
  - Swiss pattern and needle-handle files are used for fine precision filing of small surfaces.
- Safe edges on files protect the surface against which the file edge rubs from being filed.
  - The angles of single- and double-cut files vary for different materials.
- The coarseness of a cut relates to the distance between the parallel rows of cuts. Rough, coarse, bastard, second, smooth, and dead smooth cuts are common degrees of coarseness.
  - The coarseness of files that are less than 10" long ranges from #00 (coarsest) to #8 (finest).
- Eight common shapes of files are flat, mill, hand, pillar, half-round, round, square, and three-square (triangular).
  - File faces are produced with convexity, which makes it possible to posi-

tion and control the file to produce flat surfaces. Convexity makes it easier to file.

- The wedging of pins (filings) on the lead edge of file teeth causes a scratching action that can mark or score a surface.
    - A file should be fitted with a handle whose size and shape feels comfortable and balanced.
- Cross filing is the process of diagonally crossing the position of the file cuts.
    - Drawfiling is the fine graining of a work surface. Drawfiling requires the drawing of a single-cut smooth file (using a slight force) across the work surface.
- Filed surfaces are checked for squareness and straightness by using a square and/or other straight-edged tool. High spots are sighted and then filed until the surface is accurate.

D. SCRAPING, COATED ABRASIVES, AND POLISHING

- Scraping produces surfaces that are close-grained, smooth, and perfectly flat and true fitting.
    - Frosted and flaked surface finishes are also scraped. Scroll or other straight-line surface designs are formed by scraping minute amounts from the surface. Frosting improves the sliding quality and appearance of a surface and provides lubrication pockets.
- The three common hand scrapers are the flat, bearing or half-round, and triangular scrapers. The first two produce precision-fit flat and curved surfaces. The triangular scraper is a practical tool for burring and for rounding over edges.
    - Coated abrasives are produced in forms such as belts, discs, abrasive cloths, and powders (flour).
- Cutting action and surface finish depend on the nature of the abrasive grains (size and material), bonding material (glues, resins, and varnishes), coatings (open and closed), and backing material (paper, cloth, and fiber).
    - Abrasives may be natural (emery, garnet, flint, crocus, diamond) or artificial (aluminum oxide, silicon carbide, boron carbide).
- The polishing process should be started with the coarsest grain possible, consistent with the final surface finish required. Successively finer grain sizes (grits) should be used until the final finish is reached.
    - A burr is an unnecessary jagged edge produced by hand or machine processes.
- Burrs may cause inaccurate layouts, improper fitting of parts, damage to and poor appearance of a workpiece. Burrs can scratch and mark tools and machines and produce safety hazards.

UNIT 7 REVIEW AND SELF-TEST

A. HAND SAWING PROCESSES

1. Explain the design features of a hand hacksaw frame that permit the hacksaw blade to be positioned straight or at 90°, 180°, and 270°.
2. Identify four materials from which hacksaw blades are manufactured.
3. List the kinds of information that must be supplied in ordering hand hacksaw blades.
4. State one principle to follow for hand hacksawing thin-sectioned materials and structural shapes.

5. Define cutting speed as applied to hand hacksaw blades.
6. Tell how to start a cut at a sharp corner that may strip the saw teeth.
7. State a safe practice to follow in hand hacksawing when a blade breaks before the cut is completed.

B. CHISELING, PUNCHING, AND DRIVING TOOLS AND PROCESSES
1. a. Name three common hand chisels.
   b. Describe the main design features of the cutting end of each chisel named.
   c. List an application of each chisel named.
2. Tell what effect changing the angle of a chisel has on chiseling.
3. State three guidelines to follow when using a hand chisel for a shearing operation.
4. Distinguish between the design and use of a solid punch in comparison to a hollow punch.
5. Indicate two characteristics of drifts that differ from the characteristics of solid punches.
6. List the steps for aligning holes in two mating parts.
7. State three safety precautions for personal protection against injury during chiseling.

C. HAND FILE CHARACTERISTICS AND FILING PROCESSES
1. Identify (a) the types of cuts and (b) the angles of a general metal work file.
2. State how coarseness is designated for files under 10″ long.
3. a. List three files that are used to produce a flat surface.
   b. Give one application of each file listed.
   c. Describe at least one distinguishing design feature of each file listed.
4. Give reasons why Swiss pattern files are preferred by many precision mechanics in place of American pattern files.
5. Enumerate three advantages of using a long-angle, lathe file to mill file lathe work.
6. Cite four reasons why files are produced slightly convex along their length.
7. State what corrective steps may be taken to avoid pinning.
8. Describe drawfiling.
9. List four safety precautions to follow in filing.

D. BURRING AND HAND SCRAPING
1. State a major function that is served by hand scraping a machined surface.
2. Describe briefly what "frosting (flaking)" a surface means.
3. Explain what is meant by the term "high spots" as related to hand scraping.
4. Cite two reasons why a hand scraper is ground with a slightly convex face.

E. COATED-ABRASIVES AND HAND FINISHING (POLISHING)
1. Contrast open-coated with closed-coated abrasive cloths.
2. Name three properties of all abrasives.
3. State three advantages of using artificial abrasives in place of natural abrasives.
4. Give two guidelines used to finish a machined surface by hand polishing.

UNIT 8

Hand Reamers and Reaming Processes

OBJECTIVES

After satisfactorily completing this unit, you will be able to:

- Compare drilling and hand reaming processes.
- Identify the features of hand reamers and allowances for reaming.
- Understand the functions and applications of straight and spiral-fluted solid and expansion reamers.
- Apply technical information about conditions which produce chatter when reaming.
- Determine cutting fluids for common ferrous and nonferrous metals.
- Perform or simulate step-by-step procedures for hand reaming using solid hand reamers, taper reamers, adjustable blade reamers, and expansion reamers.
- Analyze and follow *Safe Practices* for the use and care of hand reamers.
- Use *Terms* that describe hand reamers and hand reaming.

HAND REAMING

Reaming is the process of shearing away small amounts of material from a previously prepared hole. The process produces a hole that is:

- Perfectly round,
- Held to a specified diameter within +0.001"/ −0.000" (or to a closer tolerance),
- Smooth and has a high-quality surface finish,
- Straight.

Reaming is performed by a multiple-fluted cutting tool known as a *reamer*. Cutting action takes place by rotating and feeding the cutting edges on each flute of the reamer against the periphery (surface) of a hole. The cutting on many hand reamers is done along a tapered starting end. The amount of material removed in hand reaming ranges from 0.001" to 0.003" and occasionally up to 0.005".

COMPARISON OF DRILLING AND REAMING PROCESSES

Many machined holes must be held to close tolerances for roundness, straightness, quality

of surface finish, and dimensional accuracy. Drilling and other machining processes produce holes that have comparatively rough surfaces. These holes are larger in diameter than the drill size. They may not be perfectly straight and round. A drilled hole may vary up to +0.005" in diameter for drill sizes up to 1/2". The variation may be as high as 0.020" for 1" and larger sizes of drills. When a drill is guided by a drill bushing in a fixture, greater accuracy is possible.

Whereas a drill is used to produce a hole in a solid mass, the reamer is a *finishing* tool. The reamer reams a hole precisely to size. Parts may be hand reamed or machine reamed. Only a limited amount of material is removed by a finishing cut with a hand reamer. By contrast, a greater amount may be removed by machine reaming.

This unit is limited to hand reamers and hand reaming processes. *Solid* (fixed), *adjustable*, and *expansion* types of reamers are covered. Consideration is given to *straight*, *spiral*, and *taper* forms. The features of hand reamers are illustrated to help explain the shearing action. A few commonly used cutting fluids are considered. These fluids are related to cutting action and surface finish.

FEATURES OF HAND REAMERS

The main design features of the cutting end of a spiral-fluted hand reamer are identified in Figure 8–1. The hand reamer, whether straight or spiral (helical), has a number of *flutes*. These flutes are cut into and along the body. The flutes are machined so that the flute face is *radial* —that is, if the edge of a steel rule were placed on the front face of the reamer, it would pass through the center.

Each flute has a *land*. Each land has a small *margin*, or circular width behind the cutting edge of the reamer. The margin extends from the cutting face to a *relieved* area. Beyond the margin, the land is relieved to prevent the reamer from binding in the hole.

Shearing (cutting) action is produced by the *starting taper*. Beginning at the point of the reamer, the flutes are ground at a slight taper. This taper merges into the outside surface (periphery) of the reamer. The taper usually extends back for a distance slightly longer than the diameter of the reamer. Additional features of a hand reamer are illustrated in Figure 8–2, which shows a straight-fluted hand reamer.

The *shank* of each reamer is smaller than its outside (basic) diameter. In reaming it is desirable to turn the reamer clear through the workpiece whenever possible to avoid scoring the inside finished surface. The end of the shank is square to permit turning a reamer with a tap wrench. Both ends of the reamer have center holes for two purposes:

- To provide a bearing surface for grinding and sharpening (the bearing surface is concentric with the periphery of the reamer);

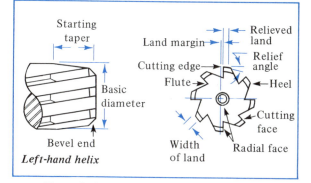

Figure 8–1 Main Features of the Cutting End of a Spiral-Fluted Hand Reamer

- To position the reamer on the machine (where the hole may have been drilled or rough reamed), with a machine center to help to align the reamer with the line of measurement of the workpiece.

Hand reamers are usually made of carbon steel and high-speed steel. Reamers are manufactured in standard dimensional sizes. Special multiple diameters are designed for applications that require the simultaneous reaming of more than one diameter of a hole.

BASIC TYPES OF HAND REAMERS

SOLID HAND REAMERS (STRAIGHT HOLES)

Solid hand reamers may have flutes that are *straight* (parallel) or *helical* (spiral and at an angle). These reamers produce straight holes of a specified diameter, with good surface finish, and to close tolerances. Straight and helical reamers may be purchased in inch-standard and

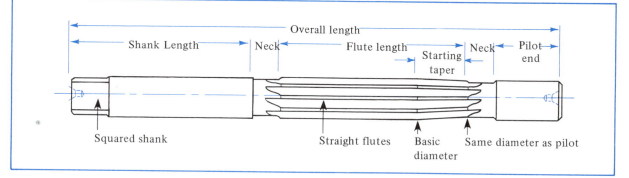

Figure 8–2 Unique Features of a Straight-Fluted Hand Reamer

metric-standard sets. Reamer sizes are in increments of 1/64″ and 1/32″ and metric equivalents. Reamers are also furnished in special sizes. The outside diameter is usually checked with a micrometer.

The outside diameter of an old reamer may be ground to within 0.003″ to 0.005″ of a standard size. Such a reamer is a *roughing* reamer. A roughing reamer is used to correct inaccuracies caused by the cutting action of a drill. A *finishing* reamer is of the exact size as the finished diameter and is used as a *second* reamer to produce an accurately reamed hole.

Spiral-fluted reamers are machined with a *left-hand helix* design. This design permits the cutting edge to cut as the reamer is turned and fed into the work. If the helix were right-hand, the reamer would thread into the work, score the surface, and produce an inaccurate hole. Spiral-fluted reamers are also used for reaming holes with keyways.

SOLID HAND REAMERS (TAPERED HOLES)

The flutes of reamers may be tapered for reaming tapered holes. There are three basic forms of taper reamers: *pin*, *socket*, and *burring*.

Taper Socket Reamer. Taper socket reamers produce tapered holes. These reamers conform to Morse, Brown & Sharpe, or metric-standard tapered socket, tool shank, and other spindle standards. Since these reamers must remove a considerable amount of material, they are produced in a set of two reamers. One reamer is for roughing. A series of shallow grooves is cut across each flute of a roughing reamer and in a staggered pattern (Figure 8–3A). The grooves relieve some of the force required to turn the reamer because they reduce the area over which the cutting takes place. A straight-flute finishing taper reamer is illustrated in Figure 8–3B. Figure 8–3C also illustrates a finishing reamer that has spiral flutes.

Burring Reamer. Burring reamers are not considered as finishing tools in terms of producing a precision surface finish or accurate diameter. A burring reamer has a steep taper. This reamer is used to burr tubing and pipes and to enlarge the diameter of a hole in thin metal. A burring reamer has straight or helical flutes.

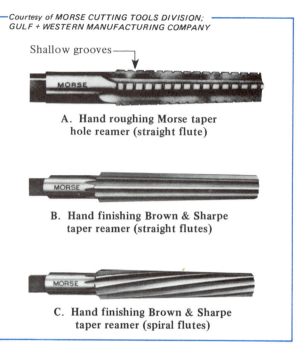

Shallow grooves —

A. Hand roughing Morse taper hole reamer (straight flute)

B. Hand finishing Brown & Sharpe taper reamer (straight flutes)

C. Hand finishing Brown & Sharpe taper reamer (spiral flutes)

Figure 8–3 Taper Socket Hand Reamers

EXPANSION HAND REAMERS

The body of an expansion reamer is bored with a tapered hole. This hole passes through a portion of its length. A section of each flute is slotted lengthwise into the tapered hole. The hole is threaded at the pilot end to receive a tapered plug. As this plug is turned into the tapered portion of the body, the flutes are expanded. The limit of expansion varies from 0.006″ for 1/4″ diameter reamers, to 0.012″ for 1″ to 1 1/2″ reamers, to 0.015″ for 1 9/16″ to 2″ reamers.

The pilot on the expansion reamer is undersized. It serves to align the cutting edges concentric with the hole.

The expansion reamer may be increased in size only to produce oversized holes. These holes require the removal of a small amount of material over a nominal size.

Expansion hand reamers may be straight fluted or left-hand helical fluted. The helical-fluted expansion reamer operates on the same principle of adjustment as the straight-fluted reamer. The helical flute is especially adapted to reaming holes that are grooved or in which the surfaces are interrupted (partly cut away).

ADJUSTABLE HAND REAMERS

Adjustable hand reamers are practical for reaming operations that require a diameter that is larger or smaller than a basic size. A complete set of adjustable hand reamers makes it possible to produce holes over a range of 3″, starting at 1/4″ diameter. Each reamer may be expanded to produce intermediate diameters, which overlap the next size of adjustable reamer.

CHATTER MARKS RESULTING FROM REAMING

Chatter marks are uneven, wave-like imperfections on reamed surfaces. These surfaces have undesirable high and low spots. Chatter marks that result from reaming may be caused by any one or combination of the following conditions:

- The speed at which a reamer is being turned is too fast;
- The distance a reamer is advanced each revolution (feed) is too little;
- A workpiece is not held securely, and there is too much spring to the tool or the work;
- A reamer is unevenly forced into a workpiece;
- Chips are not being removed and are clogging the flutes;
- The drilled or rough-reamed hole is incorrect in terms of size, surface condition, roundness, or straightness.

The remedy is to correct the particular condition and make whatever changes are necessary in speed, feed, proper preparation of the hole to be reamed, and so on. Chatter marks may also be avoided by using a left-hand spiral reamer that may be fed evenly. Some reamers are increment cut; their cutting edges are unevenly spaced. This spacing permits overlapping during the cutting process and thus prevents a ragged surface from being formed.

CUTTING FLUIDS

The use of a cutting fluid on certain materials makes the hand reaming process easier. Less force is required and a higher quality of surface finish is possible. Most metals except cast iron are reamed using a cutting fluid. Generally, for nonferrous metals like aluminum, copper, and brass, a soluble oil or lard oil compound is recommended. Cast iron is reamed dry. Soluble oils and sulphurized oils may be used on ordinary, hard, and stainless steels.

Recommended cutting fluids for various ferrous and nonferrous metals are listed in Appendix Table A-18. While recommended primarily for machine reaming, the cutting fluids may also be used for hand reaming.

How to Use Hand Reamers

Solid Hand Reamers—Straight Hole Reaming

STEP 1 Measure the outside diameter of the machined part. Check the drawing. Establish what tolerance is allowed between the part size and the reamed hole.

STEP 2 Select the correct size of solid hand reamer. Check the outside diameter. (In Figure 8–4 the outside diameter of a reamer is being measured with a micrometer.)

STEP 3 Remove burrs from the holes and other surfaces. Check the condition and size of the hole. The allowance for hand reaming should range from 0.002″ to 0.005″ (0.05mm to 0.12mm) maximum.

Note: If the hole is smaller than the hand reaming allowance, it may be necessary to rough ream with a roughing reamer.

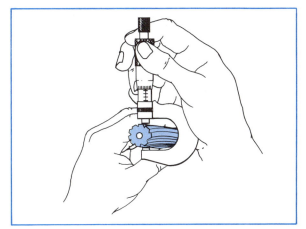

Figure 8–4 Checking the Outside Diameter of a Solid Hand Reamer

STEP 4 Examine the areas of the workpiece that are to be clamped. Be sure they can withstand the force of the vise jaws without bending or cracking.

Note: Sometimes it is more practical to hold the workpiece in a vise so that the hole is reamed in a horizontal position. At other times the reamer may be positioned vertically in the vise. The workpiece is then turned on the reamer.

STEP 5 Select a suitable tap wrench. Start the tapered cutting edge in the hole to center the reamer.

STEP 6 Turn the reamer clockwise slowly while feeding the reamer into the workpiece.

Note: If the reamer is not centered and is turned too fast, the cutting edges may produce a corrugated, chatter-marked surface.

STEP 7 Apply a cutting fluid where required between the cutting edges and the inside surface of the hole.

STEP 8 Continue to turn the reamer clockwise. Feed it into and through the workpiece.

Note: Where very close tolerances of +0.0005″ (+0.01mm) must be held, the reamer is sometimes removed when it cuts to its full diameter. The size of the reamed portion is checked. If necessary, the reamer is changed.

STEP 9 Remove the chips from the hole and reamer flutes. Examine the cutting edges and lands. Remove any burrs. Replace the reamer in its proper storage container.

STEP 10 Inspect the workpiece for quality of surface finish. Insert the mating part into the reamed hole. Check the parts against the job specifications.

Solid Hand Reamers—Taper Reaming (Taper Socket)

STEP 1 Check the drawing or measure the tapered mating part to establish the required size of taper reamer.

STEP 2 Check the diameter of the hole to be reamed.

Note: There should be an adequate amount of material left to permit reaming at the small end of the taper.

STEP 3 Determine how far the reamer must enter the hole to produce the required diameter at the large end of the taper.

STEP 4 Insert the roughing taper reamer in the hole. Turn it clockwise while applying a slight downward force. Use a steady motion.

Note: Taper reamers tend to feed into the work. No undue force should be applied.

STEP 5 Apply a cutting fluid if the material being reamed requires it.

STEP 6 Remove the taper reamer frequently. Clear away the chips.

Note: Continue to turn the reamer in a clockwise direction when removing it.

STEP 7 Check the tapered hole against the sleeve, socket, or other mating part. Leave enough material for the finishing reamer.

STEP 8 Change from a roughing to a finishing reamer. Continue to ream to the required depth.

STEP 9 Remove the burrs at the large and small diameters of the tapered hole.

STEP 10 Wipe the reamer clean. Check and remove any burrs. Oil the cutting edges of the reamer. Store the reamer in a suitable container.

Solid Hand Reamers—Taper Reaming (Taper Pin)

STEP 1 Check the mating parts to see that the holes are aligned. The parts should be held securely in the position in which they are to be joined with a taper pin.

STEP 2 Check the reamer condition and size. Start the reamer and clean the chips periodically. (The taper pin reaming of two mating parts is illustrated in Figure 8–5).

Caution: Taper reamers are hard and fragile. If a reamer draws into the work, release all force. Continue to turn the

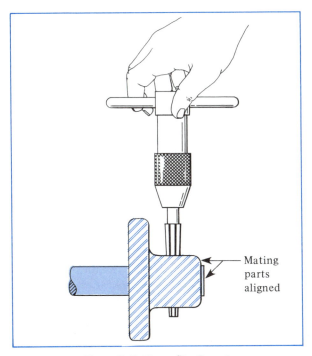

Mating
parts
aligned

Figure 8-5 Taper Pin Reaming

reamer slowly in the cutting direction while withdrawing it. Then carefully start the reaming process again.

STEP Clean the reamer and the part. Insert
3   the taper pin. Check its depth against the job requirements.

STEP Repeat the reaming process until the
4   taper pin fits accurately.

Adjustable Blade Hand Reamers—Adjusting for Size

STEP Measure the diameter of the reamer with
1   a gage or micrometer.

STEP Mount and secure the reamer in a vise.
2

STEP Adjust the reamer diameter. Turn the
3   lower adjusting nut a quarter turn. Tighten the upper adjusting nut the same amount. (The position of a reamer and the turning of the top adjusting nut are shown in Figure 8-6.)

Note: The blades are moved back from the reamer point to enlarge the diameter and vice versa.

STEP Continue the adjusting process until the
4   correct reamer size is reached.

Note: A quarter turn of the adjusting nut is recommended to prevent the blades from riding out of the slots and to prevent chips from entering under the blades.

Adjustable Blade Hand Reamers—Checking the Workpiece and Reaming

STEP Check the hole size. Mount the work (or
1   the reamer if the workpiece is to be turned on the reamer).

STEP Start reaming. Continue part way until the
2   reamer starts to cut to its full diameter.

STEP Continue to turn the reamer while with-
3   drawing it. Clean the workpiece. Be careful of the sharp burred edge.

STEP Check the partially reamed hole for size.
4   Adjust the reamer if necessary.

STEP Continue to ream through. Remove burrs.
5   Clean the workpiece. Test for size and surface finish.

Note: It may be necessary to readjust the blades and to take another very light finishing cut.

STEP Loosen the top adjusting nut a quarter
6   turn. Wipe the reamer clean. Examine and oilstone any burrs. Oil the reamer. Replace it in a container.

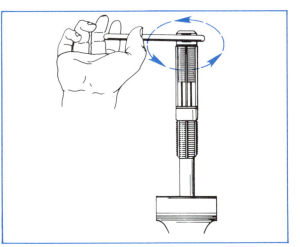

Figure 8-6 Setting the Blades of an Adjustable Hand Reamer to a Required Diameter

Expansion Hand Reamers—Setting the Hand Reamers

STEP 1   Check the cutting edges, lands, and pilot. See that all are free of burrs.

STEP 2   Check the diameter of the predrilled or rough-reamed hole.   There must be enough material left for reaming.

STEP 3   Measure the outside diameter of the reamer at several places.   Turn in the tapered screw to enlarge the reamer diameter. Back out the screw to reduce the diameter.

Caution: The reamer should be expanded only within its design limits. Otherwise, the flutes may be damaged.

STEP 4   Mount the workpiece, position the reamer, and start and finish the reaming process.

Note: Check the reamed diameter as soon as the reamer has cut to the full depth. Adjust the reamer if necessary.

Note: Back off the tapered screw after reaming to protect the flutes and make it possible for the reamer to adjust to its normal undersized position.

Safe Practices in the Care and Use of Hand Reamers

• Examine the cutting edges and lands of reamers for nicks, burrs, or other surface irregularities. These irregularities produce surface scratches and imperfections.   Burrs are removed by hand with an oilstone.

• Turn reamers in a right-hand (clockwise) direction only. Where possible, move the reamer completely through the workpiece while turning. Otherwise, keep turning the reamer clockwise while pulling outward on the tap wrench.

• Leave not more than 0.005″ or 0.1mm of material for hand reaming. Holes that must be held to tolerances of +0.0005″/–0.0000″ should be rough reamed to within 0.001″ to 0.003″ and then finish reamed.

• Examine each drilled hole before reaming it. The hole must not be tapered or bell-mouthed (out of round) or have a rough finished surface. Any of these conditions may cause tool breakage and an inaccurately reamed hole.

• Use a left-hand spiral-fluted reamer for reaming holes in which there is an intermittent cut.

• Rotate the reamer clockwise and slowly. Allow it to align itself in the hole. The feed should be continuous.

HAND REAMER AND HAND REAMING TERMS

Reaming	A shearing/cutting process for producing precision holes. Cutting holes that are dimensionally accurate, round, straight, and have a high-quality surface finish.
Starting taper	The cutting edges of straight hole reamers. A tapered end that begins with a diameter that is smaller than a basic size. (The taper extends along the lands until it merges with the outside [basic] diameter of the reamer.)
Taper pin reamer	A straight- or helical-fluted reamer whose cutting edges are tapered. (The taper conforms to the standard dimensions of tapered pins.)
Taper socket reamer	A straight- or helical-fluted reamer having tapered flutes. (The taper of the flutes meets a specific standard. The standard relates to sockets, tapered spindles, or other tapered mating parts.)
Expansion hand reamer	A hand reamer having flutes. (These flutes may be expanded by a tapered screw to enlarge the reamer size slightly.)
Adjustable blade hand reamer	A hand reamer that has a series of blades. (These blades are adjustable above or below a basic size.)
Burring reamer	A straight- or spiral-fluted and sharply tapered reamer. A reamer for burring tubes and pipes. A reamer used to roughly enlarge holes in thin-gage metals.

Chatter (reaming)	An imperfect reamer cutting action that produces inaccurate high and low surface areas.
Left-hand helix (reamer)	A reamer with flutes cut at an angle to the body and counterclockwise. The left-hand direction of the flutes. (This direction is the reverse of the cutting direction.)

SUMMARY

- Parts are reamed to produce holes that are dimensionally accurate, straight, and have good surface finish.
    - Three common types of hand reamers are: solid, adjustable, and expansion. These types of reamers are available in straight, spiral (helical), and taper forms.
- The cutting action of hand reamers is produced by the cutting edges of the flutes and blades. The starting taper merges with the outside diameter of the reamer.
    - Taper pin reamers accommodate the full range of standard taper pins. Reamer sets provide for size overlapping.
- Taper socket reamers are designed for standard Morse, Brown & Sharpe, and other tapers. Taper reamers are used for sockets, tapered shafts, shanks, and other holders.
    - The expansion hand reamer may be expanded only above a basic dimension. An undersized pilot end positions the blades concentrically with a hole. The pilot guides the reamer so that it cuts evenly.
- Helical-fluted reamers are recommended for reaming holes that are interrupted.
    - Chatter is an undesirable wave-like unevenness of a surface. Chatter marks are caused by excessive speed, too slow a feed, poor mounting, an improperly prepared hole, and various other causes.
- Cutting fluids, when required for a particular material, improve the smoothness of the cutting action, reduce the force required for reaming, and increase the accuracy of the reamed hole.
    - Safe practices must be followed when checking for and removing burrs and turning, feeding, cleaning, and storing reamers.

UNIT 8 REVIEW AND SELF-TEST

1. Compare the variation from the nominal hole size for 12.5mm and 25mm diameter holes that are produced by (a) drilling and (b) reaming.
2. Describe each of the following design features of a hand reamer: (a) radial flute face, (b) margin, and (c) starting taper.
3. State why it is common practice to turn a straight hand reamer through a workpiece.
4. Tell what function is served by the series of shallow grooves cut into each flute in a roughing taper socket reamer.
5. Indicate the limit of expansion on the following diameter ranges of expansion hand reamers: (a) up to 6mm, (b) 1″ to 1 1/2″, and (c) 39mm to 50mm.
6. Cite three advantages of adjustable blade reamers over expansion hand reamers.
7. Give three common corrective steps that are taken to eliminate chatter marks during hand reaming.
8. List four cautions that must be observed to avoid personal injury and/or damage to hand reamers.

Characteristics, Tools, and Hand Tapping Processes of Internal Threading

This unit introduces important milestones (since 1918) in screw thread design and standardization. Thread forms and threading processes are then related to the American National, Unified, and International Metric (ISO) Thread Systems.

These basic understandings are followed by:

- Functions and characteristics of screw threads,
- General cutting tools (taps), holders, and tap extractors for internal threading (tapping) by hand,
- Tap drill sizes to meet specific threading requirements,
- Actual hand tapping processes.

Later units cover more advanced technology and precise machine methods, tools, thread fits, and thread forms. Threading formulas and other computations are treated as they relate to machine set-ups and processes.

OBJECTIVES

After satisfactorily completing this unit, you will be able to:

- Tell the features and characteristics of American National, Unified, ISO Metric, and American National Straight and Taper Pipe Threads
- Describe the nature and functions of screw threads and basic production methods.
- Use technical *Terms* and apply information about tap sets, tap wrenches, and tap (thread) extractors.
- Perform the following processes.
  - Tap Internal Threads by Hand (selecting the tap and wrench, checking for squareness, and tapping holes)
  - Extract a Broken Tap or Threaded Part
- Use formulas to solve tap drill size problems and check with standard tables.

DEVELOPMENT OF THE AMERICAN NATIONAL AND UNIFIED SCREW THREAD SYSTEMS

In the early 1900's, the United States Standard (USS) and the Society of Automotive Engineers (SAE) provided standards for form, pitch, and outside diameters. Nevertheless, further standarization and improvements were needed. The National Screw Thread Commission of 1918, together with the SAE, the American Standards Association (ASA), the American Society of Mechanical Engineers, and other professional groups developed the original *American National Screw Thread Standards.*

The new standards included different tolerances, clearances, and other improvements.

They provided flexibility in thread forming and manufacturing. American National Screw Thread Standards included the coarse-thread series of the USS and the fine-thread series corresponding to the SAE. By 1935 the American Standards Association approved these standards, which later became the National Coarse (NC) and National Fine (NF) series.

Although efforts to standardize were carried on notably by Great Britain, Canada, and the United States, threaded parts among these countries were not interchangeable. The British Whitworth thread had a 55° standard thread angle; the United States, a 60° standard. Finally, in December 1948 the systems of these countries were standardized into the *Unified Screw Thread System.*

The Unified thread system is a compromise between the British and the American systems. The design of the new thread form (shape) is such that parts threaded to conform in shape to either the National Coarse or National Fine series are interchangeable with British parts. The British preference for the feature of a rounded crest and root is incorporated in the Unified system. This feature is important in mass production and the life of a screw thread.

INTRODUCTION TO SCREW THREADS

BASIC THREAD PRODUCING METHODS

Eight basic methods are employed to produce threads. The threads meet high standards of form, fit, accuracy, and interchangeability within a thread system. The eight methods are:

- Hand cutting internal threads with taps and external threads with dies;
- Machine tapping using formed cutters;
- External and internal machine thread cutting with single-point cutting tools;
- Machine threading using multiple-point cutting tools;
- Machine milling with rotary cutters;
- Machine grinding with formed abrasive wheels;
- Machine rolling and forming;
- Casting, using sand, die, permanent mold, shell casting, and other techniques (one unique advantage of the casting method is that parts of some machines—for example, sewing and vending machines, typewriters, and toys—have internal threads cast in place).

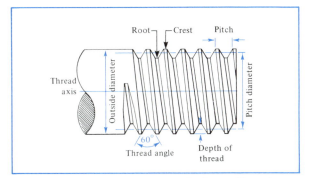

Figure 9-1 Basic Features of a Screw Thread

THE NATURE OF SCREW THREADS

A *screw thread* is defined as a helical (spiral) ridge of uniform section (form and size). This ridge is formed on a cylindrical or cone-shaped surface. The basic features of a screw thread are illustrated in Figure 9–1. Screw threads that are produced on an outer surface, like the threads on bolts and threaded studs, are called *straight external* threads. Screw threads that are cut on cone-shaped surfaces, like pipe threads, are *tapered external* threads. Threads that are cut parallel on inner surfaces, as are the threads of an adjusting nut or a hexagon nut fastener, are known as *straight* or *parallel internal* threads. External and internal parallel threads are referred to in shop language as just *outside* threads or *inside* threads.

Some threads may be advanced (moved) or tightened into a part by turning them clockwise. These threads are classified as *right-hand* threads. Threads that advance when turned counterclockwise are *left-hand* threads. Right- and left-hand threads each have special applications. Unless otherwise indicated on a drawing, the thread is assumed to be right-hand.

AMERICAN NATIONAL AND UNIFIED SCREW THREADS

Each thread has a shape, or profile. This shape is called the *thread form*. Although the size of a thread form varies according to the dimensions of a threaded part, the shape remains constant. The symbol ∪ preceding a thread form on a drawing indicates that the shape conforms to Unified system standards and is accepted by Great Britain, Canada, and the United States.

There are four basic thread series and one special thread series in both the Unified thread system and the American National thread system. The four basic thread series are:

- *Coarse-thread series*, with threads designated as UNC in the Unified system and NC (National Coarse) in the American National system;
- *Fine-thread series*, with threads designated UNF and NF;
- *Extra-fine-thread series*, UNEF and NEF;
- *Constant-pitch series*, 8 UN, 12 UN, and 16 UN for the Unified system and 8 N, 12 N, and 16 N for the American National system.

There are a number of pitch series in the constant-pitch series. There is the 8-pitch, 12-pitch, and 16-pitch series, and others. *Constant-pitch* means there are the same number of threads per inch for all diameters that are included in the series. For example, the 12 N series always has 12 threads per inch regardless of the outside thread diameter. The constant-pitch series is used when the other three thread series fail to meet specific requirements.

There are also designations for special conditions that require a nonstandard or special thread series. Some drawings show the symbol UNS, which indicates that the threads are Unified Special, or NS, which indicates National Special threads.

THE INTERNATIONAL (ISO) METRIC THREAD SYSTEM

Paralleling developments in the Unified and American National systems were international movements. These movements were designed to modify the many European metric thread systems into an internationally accepted, single metric system. The result was the international metric thread form established by the *International Organization for Standardization*, known as *ISO*. This metric standard thread form is used in Europe and in many countries throughout the world.

The ISO thread form is similar to the Unified and the American National forms, except for three basic differences:

— The depth of the thread is greater,
— The sharp-V thread form is flattened at the crest to one-eighth of the thread depth.
— The root is flattened (with a rounded root preferred) to one-eighth of the thread depth.

FUNCTIONS SERVED BY SCREW THREADS

Screw threads serve four prime functions:
— As fasteners to hold two or more parts securely in a fixed position and to permit assembly and disassembly;
— As simple machines (in a physical science context) to transmit and increase force or power (a machine feed screw and a common screw jack are examples of this function);

— As design features on measuring instruments (such as a threaded spindle of a micrometer) to produce motion. (Precise measurement standards may be established by this motion);
— As mechanical devices to change a rotary motion to a straight-line motion (feed screw being turned to raise or lower a machine table).

ISO METRIC THREADS

The ISO metric threads for all general threading and assembling purposes are identified with the *ISO Metric Coarse series*. The *ISO Metric Fine series* is employed in fine precision work. Table 9–1 shows twelve common ISO metric thread sizes ranging from 2–24mm. These sizes cover about the same range as the fractional inch sizes in the National Coarse series up to 1″ outside diameter. There are additional standard sizes in the ISO Metric Coarse series beyond the 2–24mm range. The nine common thread sizes in the ISO Metric Fine series threads cover an equivalent range of fractional inch sizes in the National Fine series up to 1″.

In Table 9–1, for example, the coarse series shows that an ISO metric threaded part with an 8mm outside diameter has a pitch of 1.25mm. The information would appear on an industrial print or drawing as M8–1.25 for the tapped hole it represents. The tap drill diameter is 8mm - 1.25mm, or 6.75mm, which is rounded off to 6.8mm for the tap drill size.

Table 9–1 Selected ISO Metric Thread Coarse Series (2–24mm)

Outside Diameter (mm)	Pitch	Tap Drill Diameter
2	0.4	1.6
2.5	0.45	2.1
3	0.5	2.5
4	0.7	3.3
5	0.8	4.2
6	1.0	5.0
8	1.25	6.8
10	1.5	8.5
12	1.75	10.3
16	2.0	14.0
20	2.5	17.5
24	3.0	21.0

More complete tables cover the whole range of sizes and series. These tables are contained in trade handbooks, Standards Association papers, manufacturers' technical manuals, and other printed manuals.

CUTTING INTERNAL THREADS WITH HAND TAPS

The process of cutting internal threads is referred to as *tapping*. Internal threads may be formed with a cutting tool called a *tap*. Taps are used to meet three general tapping requirements. Some holes are tapped *through* a workpiece. Other holes are tapped part of the way through and are called *blind* holes. Some blind holes are *bottomed* with a thread (Figure 9–2).

FEATURES OF HAND TAPS

A tap is a specially shaped, accurately threaded piece of tool steel or high-speed steel. After a tap has been machined almost to size and form, its threads are *relieved* (reduced in size) to produce cutting edges. The tap is then hardened and tempered. A fine tap is ground to size and shape.

The general features of a hand tap are represented in Figure 9–3. A number of *flutes* (grooves) are milled into the body for the length of the threaded portion. These flutes help to form the

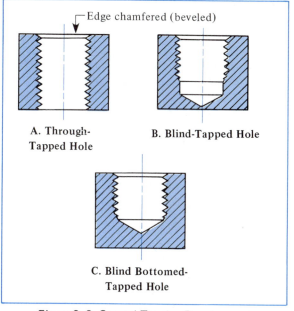

A. Through-Tapped Hole   B. Blind-Tapped Hole

C. Blind Bottomed-Tapped Hole

Figure 9–2 General Tapping Requirements

cutting edges. The flutes also provide channels so that chips can move out of the workpiece along the flutes during the cutting process. Regular taps have straight flutes. Some taps are *spiral fluted* to produce a different cutting action in which the shearing that takes place and the spiral shape push the chips ahead of the tap.

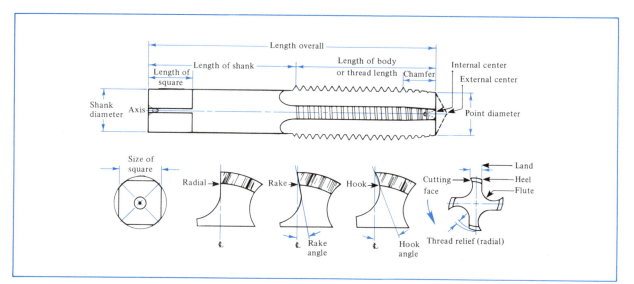

Figure 9–3 Principal Features of a Hand Tap

The *shank* on some taps is smaller in diameter than the tap drill size. This difference in diameter permits the last teeth on the tap to be employed in cutting a thread. Also, the tap may be turned until it completely moves through the workpiece. Besides being smaller in diameter, the shank end is square. The square shape provides a good bearing surface for a tap wrench and for turning the tap. Sometimes there is a *center hole* on the shank end. The center hole is used to align the tap at the line of measurement. Alignment is essential in starting the hand tap and in tapping on a machine.

TAP SETS

Standard Sets. Taps that are larger than 1/4″ usually come in standard size sets for each thread size and pitch. A set consists of three taps: a *taper* tap, *plug* tap, and *bottoming* tap.

The taper tap is sometimes called a *starting* tap. It may be used to start the hand tapping process, particularly when threading a blind hole, or to tap a hole that is drilled completely through a part. The taper tap has the first eight to ten threads ground at an angle (backed off). The taper begins with the tap drill diameter at the point end. It extends until the full outside diameter is reached. These angled threads start the tap and do the rough cutting. Because the taper tap has a greater number of angled teeth engaged in cutting the thread than plug and bottoming taps, less force is required. Also, the taper tap is easier to align with the line of measurement, and there is less likelihood of breakage.

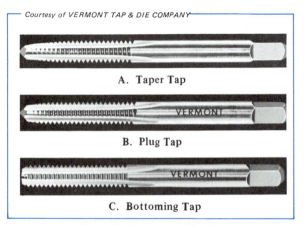

Courtesy of VERMONT TAP & DIE COMPANY

A. Taper Tap

B. Plug Tap

C. Bottoming Tap

Figure 9–4 Standard Tap Set

The plug tap has the first three to five threads tapered. The plug tap is usually the second tap that is used for tapping threads (the taper tap being the first). It may be used for tapping through a part or for tapping a given distance. Certain blind-hole tapping is done with a plug tap if the last few threads need not be cut to the total depth.

The bottoming tap is *backed off* (tapered) at the point end from one to one and one-half threads. This tap is the third tap that should be used when a hole is to be blind tapped and is to *bottom* (cut a full thread) at the end of the tap drill hole. The bottoming tap should be used after the plug tap. Extra care must be taken near the end of the thread to see that the tap is not forced (jammed) against the bottom of the hole.

Machine-Screw Taps. Tap sizes that are smaller than 1/4″ are designated by whole numbers rather than by fractions. For example, a designation of 10–32 means that the outside diameter (OD) of a machine-screw tap is equal to the number of the tap (#10) times 0.013 plus 0.060″. For the 10–32 tap the outside diameter, calculated by using the formula, equals (10 × 0.013) + 0.060″, or 0.190″. The "32" refers to the number of threads per inch. Thus, a 10–32 tap is one that has an outside diameter of 0.190″ and cuts 32 threads per inch.

On smaller size taps the amount of material to be removed in cutting the thread form is limited. The use of two or three taps in a set is therefore not required.

Serial Set Taps. There are times when deep threads must be cut by hand in tough metals. Taps that cut deep threads have a different design than standard taps. These taps are known as *serial* taps. They differ in the amount of material each tap cuts away.

The #1 tap cuts a shallow thread. The #2 tap cuts the thread deeper. The #3 tap is the final sizing tap. It cuts to the required depth to correctly form the teeth.

Serial taps must be used in sets. A combination of these taps reduces the amount of force exerted on any one tap and helps avoid tap breakage.

OTHER TYPES OF HAND AND MACHINE TAPS

Gun Taps. Gun taps derive their name from the action caused by the shape at the point end of the tap. The cutting point is cut at an angle. The design of the point causes chips to shoot out ahead of the tap. The cutting action is shown in Figure 9–5. Gun taps are applied primarily in tapping through holes in stringy metals whose chips tend to bunch up and lodge in straight-fluted taps. With these metals, tapping requires that a greater force be exerted on a tap. This force increases the probability of tap breakage.

Some tap manufacturers produce a *spiral-point design* tap (Figure 9–5 at (B) to meet the same threading requirements as gun taps.

Gun taps are production taps. To withstand the greater forces required in machine tapping, there are fewer flutes on gun taps and the flutes are cut to a shallower depth than standard taps.

Gun taps are furnished in three general forms. Because each form performs a function that is different from the function of a standard tap,

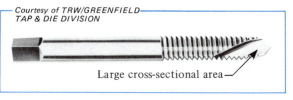

Large cross-sectional area

Figure 9–6 A Gun Flute Only Tap

the names vary from the names of standard taps. The three forms of gun taps are the *plug gun*, *bottoming gun*, and *gun flute only* tap.

The plug gun tap is used for through tapping where holes are open and chips may shoot out.

The bottoming gun tap is used for bottoming tapping operations.

The gun flute only tap is used on soft and stringy metals for shallow through tapping. The design features of this tap appear in Figure 9–6. A large cross-sectional area gives added strength and permits gun flute only taps to be powered with higher-force turning devices than regular machine taps.

Spiral-Fluted Taps. *Low-angle spiral-fluted* taps are especially adapted for tapping soft and stringy metals. Such metals include copper, aluminum, die-cast metals, magnesium, brass, and others. Instead of straight flutes, these taps have spirals. The spiral flutes provide a pathway along which chips may travel out of a workpiece (Figure 9–7).

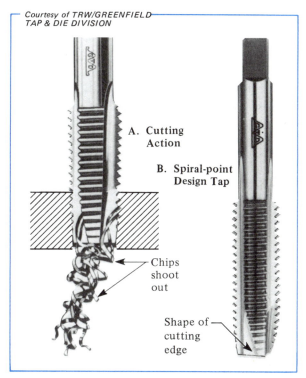

A. Cutting Action

B. Spiral-point Design Tap

Chips shoot out

Shape of cutting edge

Figure 9–5 Cutting Action of a Plug Gun (Spiral-Point Design) Tap

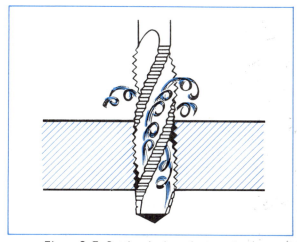

Figure 9–7 Cutting Action of a Low-Angle Spiral-Fluted Tap

Figure 9–8 High-Angle Spiral-Fluted Tap

High-angle spiral-fluted taps are excellent for tapping (1) tough alloy steels, (2) threads that are interrupted and then continue in a part, and (3) blind holes (Figure 9–8).

Pipe Taps. Three common forms of pipe taps are used to cut American National Form Pipe Threads. These threads include (1) standard taper pipe threads (Figure 9–9A), (2) straight pipe threads, and (3) dryseal pipe threads. The processes of producing pipe threads are similar to processes used for tapping any other American National Form threads. The *taper pipe* tap forms NPT (National Pipe Taper) threads. The *straight pipe* tap cuts NPS (National Pipe Straight) threads. The *dryseal pipe* tap (Figure 9–9B) produces American National Standard Dryseal Pipe Threads (NPSF).

Standard taper pipe threads and straight pipe threads are the same shape as the American National Form. Their thread angle is 60°. Their crests and roots are flattened. Their differences lie in diameter designation and pitch. Pipe threads taper 3/4″ per foot.

Parts cut with tapered pipe threads may be drawn together to produce a rigid joint. When a pipe compound seal is used on the threads, the line is made gas or liquid leakproof.

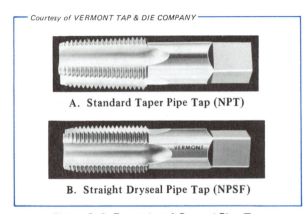

A. **Standard Taper Pipe Tap (NPT)**

B. **Straight Dryseal Pipe Tap (NPSF)**

Figure 9–9 Examples of General Pipe Taps

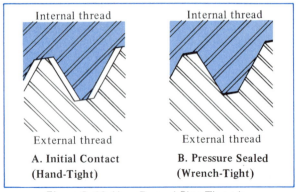

A. Initial Contact (Hand-Tight)	B. Pressure Sealed (Wrench-Tight)

Figure 9–10 How Dryseal Pipe Threads Produce a Pressure-Tight Seal

Straight pipe threads are used in couplings, low-pressure systems, and other piping applications where there is very limited vibration.

Dryseal pipe threads are applied in systems requiring pressure-tight seals. These seals are supplied by making the flats of the crests the same size or slightly smaller than the flats on the roots of the adjoining (mating) thread. Figure 9–10A shows the initial contact between crests and roots as a result of hand turning. A pressure-tight seal produced when parts are further tightened by wrench is illustrated in Figure 9–10B.

FLUTE DEPTH AND TAP STRENGTH

While referred to as hand taps, a number of taps also are used for machine tapping. The difference in some cases is the material from which a tap is made. Hand taps are usually manufactured from high-carbon tool steel; machine taps, from high-speed steel.

Tap sizes that are 1/2″ or larger normally have four flutes. Sizes smaller than 1/2″ may have two, three, or four flutes. For example, machine-screw taps usually have three flutes. The three flutes are cut deeper than flutes of large tap sizes and provide extra space for chips in blind-hole tapping or in tapping deep holes where chips are stringy.

TAP DRILL SIZES AND TABLES

A *tap drill* refers to a drill of a specific size. A tap drill produces a hole in which threads may be cut to a particular thread depth.

Table 9–2 Tap Drill Sizes (75% thread) UNC/NC Series (Partial Table)

Thread Size and Threads per Inch	Major Outside Diameter (Inches)	Tap Drill Size	Decimal Equivalent of Tap Drill
1/4–20	0.2500	#7	0.2010
5/16–18	0.3125	F	0.2570
3/8–16	0.3750	5/16	0.3125
7/16–14	0.4375	U	0.3680
1/2–13	0.5000	27/64	0.4219
9/16–12	0.5625	31/64	0.4844

TAP DRILL SIZES

The *tap drill size* may vary from the root diameter required to cut a theoretical 100% full thread to a larger diameter. The larger diameter leaves only enough material to cut a fraction of a full-depth thread. According to laboratory tests, a 50% depth thread has greater holding power than the strength of the bolt. In other words, the bolt shears before the threads of the tapped hole strip.

The material in a part, the length of thread engagement, and the application of the thread itself are considerations for establishing the percent of the full-depth thread that is required. Finer threads for precision instruments may be cut close to their full depth. By contrast, bolts, nuts, and deep-tapped holes in the NC, UNC, or ISO Metric coarse series may require a smaller percentage of the full-depth thread. The possibility of tap breakage is reduced when holes are threaded to a fraction of the full depth.

When the tap drill size is to be computed, values may be substituted in the following formula:

$$\begin{array}{c}\text{Tap Drill Size}\\(\text{75\% of}\\\text{Total Depth})\end{array} = \begin{array}{c}\text{Outside}\\\text{Diameter}\end{array}\left(-\,0.75 \times \frac{1.299}{\begin{array}{c}\text{Number of Threads}\\\text{per Inch}\end{array}}\right)$$

If a thread depth other than 75% is specified, substitute the required percent in place of 0.75. The nearest standard size drill is selected as the tap drill.

TAP DRILL TABLES

The Appendix contains a series of thread tables. One table is titled *Hole Sizes for Various*

Percents of Thread Height and Length of Engagement. Two other tables relate to various thread dimension in the UNC/NC and UNF/NF series.

Table 9–2 provides just a few thread sizes to illustrate tap drill size tables. The sizes all relate to a thread depth of approximately 75%. Note, as an example, that a 1/2–13 UNC/NC thread requires a 27/64″ tap drill to produce a 75% thread.

TAP WRENCHES

Taps are turned by specially designed *tap wrenches.* Two general forms of wrenches are the *T-handle* (Figure 9-11) and the *straight-handle.* Both of these wrenches have adjustable jaws that permit the use of a single tap wrench handle to accommodate a number of taps with different sizes of square heads. However, the range of tap sizes is so great that tap wrenches are made in a number of sizes.

The size of the square head of a tap is proportional to the tap size. Care must be taken to select the appropriate size of tap wrench. Many taps are broken because excessive force is exerted through the leverage of a wrench that is too large. While designed principally for use as tap wrenches, these wrenches are also used with other square-shanked cutting tools for hand turning operations.

How to Tap Internal Threads by Hand

Selecting the Tap and Tap Wrench

STEP 1 Examine the thread specifications on the work order, print, or shop sketch.

STEP 2 Determine the most practical tap to use (that is available). The correct tap depends on the material, length of thread, and percent of thread depth required. Select an appropriate size and type of tap wrench.

STEP 3 Refer to a table of tap drill sizes for the thread size and form series. Determine the tap drill size that meets the job specifications.

Note: Shop prints normally carry a notation on the size of drill and depth, particularly for blind holes.

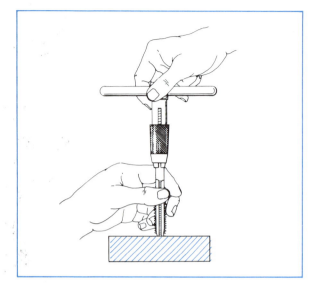

Figure 9–11 Practices in Hand Tapping with a T-Handle Tap Wrench

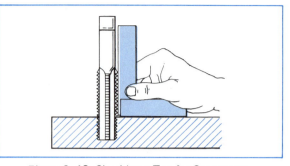

Figure 9–12 Checking a Tap for Squareness

STEP Check the drilled hole for correct tap
4   drill size and depth. Remove all chips. If possible, the drilled hole should be countersunk slightly. Countersinking helps in starting a thread and prevents throwing up a burr.

Starting the Tap

STEP Mount the workpiece securely in a vise.
1   Check the hole to see that it is in a vertical position.

STEP Start with a taper tap. Tighten the shank
2   end in the wrench jaws. Grasp the tap and tap wrench in one hand. Place the tap in the tap hole. Guide the tap with the other hand so that it is vertical. (Figure 9–11 shows the correct position of a tap at the start.)

STEP Apply a slightly downward force while
3   turning the tap clockwise at the same time. Continue to turn with one hand. Apply a constant force for two or three turns. This force should cause the tap teeth to start to form partial threads.

Check for Squareness

STEP Remove the tap wrench carefully. Clean
1   the work surface and check for burrs at the beginning of the thread. Check the tap for squareness (Figure 9–12). Bring the blade of a square against the tap shank in two positions at 90° to each other.

Note: If the tap is not square, back it off by reversing the direction. Restart the tap by turning it clockwise. Apply a limited force in the direction from which the tap leans.

Note: Care and judgment are needed. A tap is brittle. Any sharp jarring or excess force at an angle may cause a tap to snap and break.

STEP Turn the tap two or three turns. Repeat
2   the checking and adjusting steps until the tap is square with the workpiece.

STEP Apply a small quantity of an appropriate
3   cutting fluid to the cutting edges of the tap.

Tapping a Through Hole

STEP Turn the tap with both hands on the tap
1   wrench handles. As the tap takes hold, the downward force may be released.

Note: The two handle ends of small tap wrenches should be held and turned with a steady, gentle, even force. Larger straight-handle tap wrenches and taps are turned by grasping one end of the handle with one hand and the other end with the other hand.

STEP 2 Back off the tap after each series of two or three turns. Backing off is done by reversing the direction of the tap. The reversing action causes the chips to break. They then can move through the flutes and out of the work. Backing off also prevents tap breakage.

STEP 3 Continue to turn the tap, repeating the steps for cutting, lubricating, and backing off. The starting taper tap may either be moved through the workpiece or removed by reversing the direction.

Note: It is good practice to follow the taper tap with a plug tap. The plug tap then serves as a finishing or sizing tap.

TAP EXTRACTORS

The name *tap extractor* indicates that the tool is used to extract parts of taps that have broken

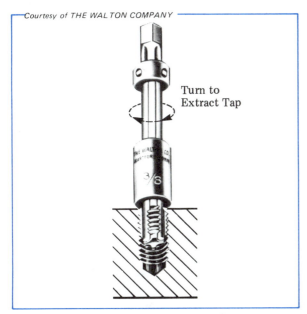

Courtesy of THE WALTON COMPANY

Turn to Extract Tap

3/8

Figure 9-13 Application of a Tap Extractor

off in a hole (Figure 9-13). The extractor has prongs. These prongs fit into the flute spaces of a broken tap. The prongs are pushed down into the flutes. The steel bushing on the extractor is moved down as far as possible to the work. The bushing thus provides support for the prongs. The extractor is turned counterclockwise with a tap wrench to remove a broken right-hand tap. After the broken tap is removed, the tapped hole should be checked to see that no broken pieces remain in the hole.

Safe Practices with Threading Tools and Processes

- Reverse a standard hand tap during the cutting process. A reversing action breaks the chips, makes it easier to move them out of the tapped hole, and reduces the force exerted on the tap.

- Select a size and type of tap wrench that is in proportion to the tap size and related to the nature of the tapping process. Excessive force applied by using oversized tap wrenches is a common cause of tap breakage.

- Exercise care with bottoming taps, particularly in forming the last few threads before bottoming. Avoid excessive force or jamming the point of the tap against the bottom edge of the drilled hole.

- Back out a tap that is started incorrectly. Restart it in a vertical position, square with the work surface. Any uneven force applied to one side of a tap handle may be great enough to break the tap.

- Use caution when using an air blast. Chips should only be blown from a workpiece when (1) each operator in a work area uses an eye protection device and (2) the workpiece is shielded to prevent particles from flying within a work area.

- Handle broken taps carefully. Taps fracture into sharp, jagged pieces. Workpieces with parts of broken taps should be placed so that no worker may brush up against them. Use only a hand tool, such as a pair of pliers, to grasp or try to move a broken tap.

SCREW THREAD AND HAND TAPPING TERMS

Unified thread form	Thread form standards adopted by Great Britain, Canada, and the United States. Standards that ensure interchangeability of threaded parts conforming to the national standards of these countries. A 60° included angle thread form with crest and rounded root. A thread form that incorporates features of the British Standard Whitworth and the American National Form.
National Coarse and National Fine	Two basic thread series built upon the American National Form. (Fine series (NF) threads are used on parts and instruments requiring finer, more precise threads.)
International (ISO) metric thread system	A metric thread system accepted as an international standard. A thread form having a 60° thread angle. (The crests and roots are flattened at 1/8th of the depth. A rounded root is recommended. Drawings are dimensioned with metric thread specifications.)
Pitch-diameter series	A thread system proposed by the Industrial Fasteners Institute. (The system is intended to reduce and replace current Unified and ISO Metric pitch-diameter combinations.)
Constant-pitch series	A threaded series in which the pitch is the same for all diameters that are included in the series—for example, 8 pitch, 24 pitch, 32 pitch.
ISO metric thread designation	Specifications of a metric thread giving the outside diameter and pitch in millimeters—for example, M20-2.5 means an outside diameter of 20mm and a pitch of 2.5mm.
Flutes	Channels running parallel to a tap axis and cut below the thread depth. Grooves that form the teeth of a tap. Channels into which chips flow and are removed during the cutting process.
Relieved	The reduction in size of the mass behind a shape. The removal of material behind a cutting edge. (The removal produces the cutting edge and permits a cutting tool to be turned with minimum resistance, drag or friction.)
Bottoming thread	A common threading term used to indicate that as close to a full thread as possible must be produced at the bottom of a hole.
Tap drill	A standard drill corresponding in size to the diameter to which a hole must be drilled. (The tap drill permits cutting a thread to a specified depth.)
Tap extractor	A pronged tool inserted in the flutes of a broken tap and turned to extract the tap.

SUMMARY

- There are eight basic methods of producing threads: hand cutting, machine cutting with formed cutters, machine cutting with single-point and multiple-point cutting tools, milling, grinding, rolling and forming, and casting.
  - Screw threads serve four functions: to fasten parts, to change force, to produce motion, and to change motion.
- Screw threads are classified as right- or left-hand, straight or tapered, and internal or external.

- Five common thread series in the Unified and American National thread systems are:
  - — Coarse: UNC/NC
  - — Fine: UNF/NF
  - — Extra fine: UNEF/NEF
  - — Constant pitch: UN/N
  - — Special: UNS/NS
- ISO metric coarse and fine series threads are identified on drawings by the letter **M** followed by the outside diameter and pitch size—for example, **M12-1.75**.
  - Standard hand tap sets include taper (for starting), plug (for through and deep-hole tapping), and bottoming taps.
- A cutting fluid, appropriate for the material being tapped, helps the cutting action, improves the quality of the finished threads, and reduces the force required for tapping.
  - Machine-screw taps under 1/4″ are designated by number. Usually a single tap is used to cut the thread to a required percent depth.
- Serial set taps are designated for cutting tough materials. Taps #1, #2, and #3 each cut a portion of the full thread depth. The #3 tap is the final finishing (sizing) tap.
  - The plug gun, bottoming gun, and gun flute only (spiral-point design) are used for stringy metals whose chips adhere to straight-fluted taps.
- The low-angle spiral-fluted tap is designed for tapping soft metals. The high-angle spiral-fluted tap is for tough alloy steels and blind holes.
  - Straight (NPS), standard taper (NPT), and American National dryseal are three common forms of pipe threads.
- There are usually fewer flutes (two and three) on small-sized taps. These flutes are cut deeper for greater chip clearance. These taps may be sheared and fractured easily.

UNIT 9 REVIEW AND SELF-TEST

1. List four basic hand and machine (lathe cutting) methods of producing threads.
2. Differentiate between the pitch to diameter relationship of threads in the UNC series and the constant-pitch series (UN) for the Unified system.
3. Show how information about ISO metric threads with a 25mm outside diameter and a pitch of 4mm is represented on an industrial drawing.
4. Identify when each of the following taps is used: (a) taper, (b) plug, and (c) bottoming.
5. Tell when (a) low-angle spiral-fluted and (b) high-angle spiral-fluted taps are used.
6. Give one design feature of an American National standard dryseal pipe thread.
7. State why general-purpose threading to 75% of full depth is an accepted practice.
8. Explain how part of a tap that breaks in a hole may be removed safely.
9. List four safety practices to follow to avoid personal injury or tap breakage when threading.

Drawings, Dies, and External Threading

Historical and basic technical information about standards for thread forms and thread series, hand cutting tools, and tapping processes were covered for internal threads in Unit 9. This unit provides additional details that relate principally to external threads. Techniques are described for representing and specifying screw threads on shop drawings, prints, and sketches.

Today, it is still necessary and practical to cut threads by hand. While hand-cut threads may not be as precise as threads produced by machining processes, hand-cut threads do meet general requirements. The tools and processes used in bench work for cutting external threads and for measuring pitch are also described in this unit. More advanced technology and processes are covered in later units that deal with different forms and precision methods of cutting internal and external threads.

OBJECTIVES

After satisfactorily completing this unit, you will be able to:

- Apply screw pitch gages and the steel rule in measuring the pitch of American National, Unified, and ISO Metric Threads.
- Identify design features of solid and split-adjustable regular and pipe thread dies.
- Understand the functions of screw extractors.
- Perform the following external threading processes.
    - Cutting Threads with a Hand Threading Die.
    - Extracting a Broken Threaded Part.
- Translate drafting room techniques of representing and dimensioning threads.
- Utilize technical data from thread tables.
- Interpret each new threading *Term.*
- Follow each recommended *Safe Practice.*

SCREW THREAD SPECIFICATIONS AND DRAWING REPRESENTATION

The designer's concept of an exact part is communicated to the craftsperson through the medium of a drawing. The craftsperson must be able to accurately interpret design features, dimensions, and the relationship of one part with another part. Each line, view, dimension, and note on a drawing has a particular meaning for the worker.

To review, threads may be external or internal. A thread may extend entirely through or over a given length. The thread may also be bottomed at the depth of a drilled hole, or it may go only a specified distance. Similarly, an external thread may be cut to a shoulder or threaded part way.

All screw thread specifications are usually shown graphically (represented) on drawings and sketches in one of three ways:

— Pictorial representation,
— Schematic representation,
— Simplified representation.

To provide all essential screw thread details, a drawing must include full specifications for cutting and measuring threads. Technical information about thread sizes, depth or length, class of threads (fits), and surface finish appear near the representation of the threaded portion. Sometimes the thread length is specified by a dimension on the part. Other instructions may be stated as notes.

Dimensions for internal and external threads are added to the part drawing. Simplified drawings of threads with dimensioning codes and dimensions are provided in Figure 10–1. Each drawing includes a series of encircled numerals. The type of dimension represented by a particular numeral is explained by a code (Figure 10–1C). Specifications of an external thread are given on a drawing as shown in Figure 10–1A. Details of a blind threaded hole (internal thread) appear on a drawing as illustrated in Figure 10–1B.

MEASURING SCREW THREAD PITCH

Pitch is the distance at the same point on a thread form between two successive teeth. Two simple screw thread measurement techniques may be used to establish the pitch of a particular threaded part. The first technique requires the use of a screw pitch gage; the second technique, a steel rule.

SCREW PITCH GAGE TECHNIQUE

A *screw pitch gage* has a series of thin blades. Each blade has a number of teeth. The teeth match the form and size of a particular pitch. Each blade is marked for easy reading.

Pitch is checked by selecting one gage blade. The blade teeth are placed in the threaded grooves of an actual part. The blade is sighted to see whether the teeth match the teeth profile of the workpiece. Different blades may need to be tried until a blade is found that conforms exactly to the workpiece teeth. The required pitch is thus established.

STEEL RULE TECHNIQUE

Screw pitch may also be measured by placing a steel rule lengthwise on a threaded part. An inch graduation is usually placed on the crest of the last thread. The number of crests are counted to the next inch graduation on the rule. This number of crests (threads) represents the pitch (Figure 10–2).

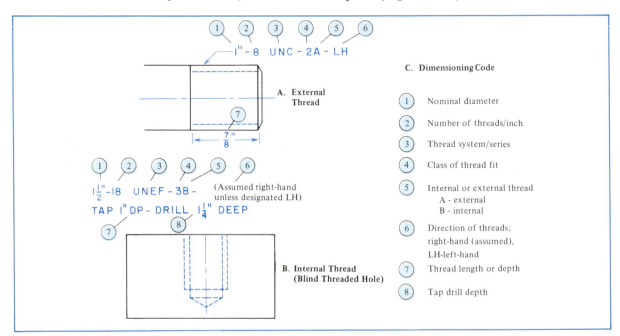

Figure 10–1 Simplified Drawings of Threads, with Dimensions and Dimensioning Codes

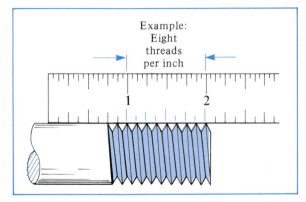

Figure 10–2 Using a Steel Rule to Measure the Number of Threads Per Inch

EXTERNAL THREAD CUTTING HAND TOOLS

External threads are hand cut with cutting tools known as *dies*. Dies are made in a variety of sizes, shapes, and types. Selection of dies depends on the thread form and size and the material to be cut. Dies are made of tool or high-speed steel and other alloys. Dies are hardened and tempered. Some dies are *solid* and have a fixed size. Other dies are *adjustable*.

SOLID DIES

The solid square-or hexagon-shaped die has a fixed size. Rarely used in the shop, it is applied principally to chase threads that have been poorly formed or have been damaged. Such threads require that a sizing die be run over them to remove any burrs or nicks or crossed threads. The damaged threads are reformed to permit a bolt or other threaded area to turn in a mating part. The hexagon shape may be turned with a socket, ratchet, or other adjustable wrench.

ADJUSTABLE SPLIT DIES

The adjustable split die is available in three common forms. The first form must be adjusted each time the die is changed in a *die holder*. In some cases the adjustment is made by turning the adjusting screws in the die holder (sometimes called a *die stock*). The split die is placed in the die holder and tried on a correctly sized threaded part. The screws are adjusted until there is a

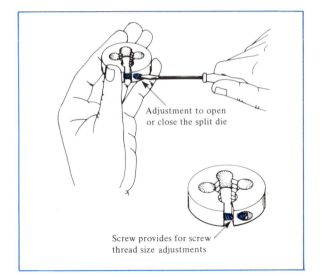

Figure 10–3 A Split Threading Die with a Screw Adjustment

slight drag between the die and the threaded part. In other cases the die is adjusted by a cut-and-try method. The die is opened. A few full threads are cut. The workpiece is tried on the mating part. The die is adjusted until the correct size is reached. The die is then locked at the setting for this size.

The second form of adjustable split die has a screw within the die head. This *screw adjustment* opens or closes the die thread teeth. Thus a range of adjustments for cutting oversized or undersized threads is provided (Figure 10–3).

The cutting teeth of the adjustable split die are formed by removing metal from portions of the thread form. Instead of flutes the area in back of the threaded section is removed to provide both a cutting edge and ample room for chips to clear the die. Since dies have more body than taps, they are considerably stronger and their chip channels may be cut deeper. The teeth are relieved in back of the cutting edge to make cutting possible and to avoid binding.

The third basic form of adjustable split die has three essential parts: a *cap*, a *guide*, and the *die halves* (which are the threading die). All the parts and their complete assembly are shown in Figure 10–4. The sides of the two die halves are cut at an angle. The die halves are held securely against the machined, tapered surfaces in the cap. The threads on the die halves are tapered at the front end. The taper permits easier starting

Figure 10-4 Parts of an Adjustable Split Die

and relieves excessive pressure on the first few cutting teeth. The cap has two adjusting screws. These screws move and position the die halves. The cap also has a lock screw recess by which the die is secured in the holder.

The guide serves two purposes:

- It forces the tapered sides of the die halves against similarly tapered sides in the cap. The force holds the two halves tightly in position so that the size cannot change.
- It pilots the die (halves) so that it is centered over a workpiece. Once the die halves are set to the required size, they are locked in position with the cap. The assembled die is then locked in the die holder.

PIPE THREAD DIES

Pipe threads are tapered in order to make tight joints in air and liquid lines. Pipes are measured (sized) according to their inside diameter. Because of this inside measurement, pipe threads are larger in diameter than regular screw threads. Pipe threading taps and dies are, therefore, larger for a specific diameter than regular taps and dies.

A 1/2″ standard taper pipe thread (NPT) requires a tap drill of 23/32″. This drill size permits a hole to be tapered with a thread taper of 3/4″ per foot. The actual outside diameter of the 1/2″ threaded pipe is 0.840″. Other com-

Table 10-1 National Standard Taper (NPT) Pipe Threads

Nominal Pipe Size	Threads per Inch	Outside Diameter	Tap Drill Size
1/8	27	0.405	11/32
1/4	18	0.504	7/16
3/8	18	0.675	37/64
1/2	14	0.840	23/32
3/4	14	1.050	59/64
1	11 1/2	1.315	1 5/32
1 1/4	11 1/2	1.660	1 1/2
1 1/2	11 1/2	1.990	1 47/64
2	11 1/2	2.375	2 7/32

mon National taper pipe thread sizes are given in Table 10-1.

Dies for pipe threads may be solid or adjustable. The adjustable die halves are held in an *adjustable stock* that serves as the die holder. There are indicating marks on both the die and the stock. When the die halves are aligned with these marks, the die is set to cut a standard pipe thread. The adjustable stock is provided with a positioning cap. This cap centers the work with the die halves.

How to Cut Threads with a Hand Threading Die

Selecting the Threading Die

STEP 1   Select an adjustable threading die that meets the job requirements for thread form and size. Select an appropriate type and size of die holder.

STEP 2   Check the outside diameter of the workpiece. Chamfer the sharp edge to a depth of one thread by turning or hand filing. Chamfering to this depth is necessary to start the die squarely so that the thread will be cut straight.

Adjusting the Threading Die

STEP 1   Test the adjustable threading die. Thread (turn) the mating part into the die by hand. Adjust the screw on a screw-adjusting die until a slight force is needed to turn the die.

Note: On some split dies the die may be positioned in a holder. The adjusting screws are then tightened to bring the die to size and to lock it in a fixed position.

Note: The lead side (tapered threads) of the die should be placed opposite the shoulder of the die stock.

STEP Tighten the lock screw in the die holder.
2

Starting the Thread

STEP Mount the work securely in a vise. Use
1 protecting soft jaws where required. Place the die on the work. The tapered starting cutting area is on the bottom facing the workpiece.

STEP Apply a cutting lubricant to the work
2 surface and the die cutting edges. Press down firmly. At the same time move the die on the top surface of the work.

STEP Continue to apply a downward and cir-
3 cular force (Figure 10–5). Make two or three revolutions to start the first few threads. Back off the die. Check the threads for squareness and to see that they are started correctly.

STEP Continue to turn the die until a few full
4 threads are produced. Remove the die. Clean the part. Check the size against the mating part, a ring gage, or other measuring tool. If necessary, adjust the die jaws.

Note: A finer finished thread is produced when the die is opened as much as possible. The thread is cut to length. This step is then followed by readjusting the die and taking a fine cut to produce the required fit.

Note: As in tapping, the cutting direction should be reversed every few turns (Figure 10–5). A reverse cutting action breaks the chips and clears them away from the work surface and die.

Note: A cutting lubricant should be applied throughout the cutting process.

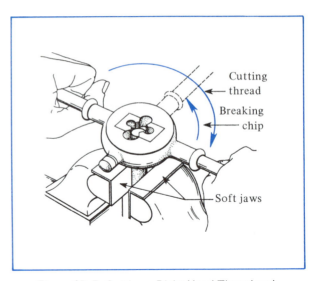

Figure 10–5 Cutting a Right-Hand Thread and Reversing the Stroke to Break the Chips

STEP Thread to the required length. Remove
5 the die. File off the burrs. Clean the threads. Retest the threaded part for size by turning it into the mating part. If required, readjust the die and take a finishing cut over the threaded part.

Note: When a thread is to be cut to a shoulder or close to a head, the die is turned 180° (inverted). The last few threads are cut with the die inverted.

Caution: Eye protection devices must be used and safety precautions must be followed while removing chips from dies.

SCREW EXTRACTOR

Bolts, screws, studs, and other threaded parts may be sheared off, leaving a portion of the broken part in the tapped hole. This portion may be removed by using a tool called a *screw extractor.* This tool is a tapered metal spiral. A right-hand thread screw extractor is illustrated in Figure 10–6. The spiral is formed in the reverse direction from the threaded part that is to be removed. In other words, a left-hand spiral is needed for a right-hand thread.

Figure 10-6 Right-Hand Thread
Screw Extractor

A pilot hole is first drilled into the broken portion. The diameter of the hole must be smaller than the root diameter of the thread. The drill size to use is generally given on the shank of the screw extractor.

Then, as the point of the spiral taper is inserted and turned counterclockwise (for right-hand threads), the screw extractor feeds into the hole. The spiral surfaces are forced against the sides of the hole in the broken part. The extractor and part become bound as one piece. The threaded portion then may be removed by turning the extractor.

Safe Practices in Cutting External Threads by Hand

- Analyze and correct the causes of torn threads. Torn threads indicate any one or all of the following conditions:
  - The jaws may be set too deep,
  - The threads are being cut at an angle,
  - The diameter of the workpiece is too large,
  - The cutting teeth are dull,
  - A proper lubricant (if required) is not being used.
- Remove burrs produced by threading.
- Exercise care in threading long sections. Turn a die holder from a position in which the hands do not pass across the top of a threaded part.
- Use an eye protective device when removing chips from the die or workpiece.

TERMS USED IN SCREW THREAD DRAWINGS AND HAND THREADING

Thread representation	Graphic techniques of drawing threads in order to describe them accurately and provide full thread specifications. Pictorial, schematic, and simplified techniques of drawing and sketching a thread. A drawing with all features and dimensions.
Standard dimensioning of a screw thread	A set of symbols, forms, and technical data. A universally accepted dimensioning system that provides the full specifications of a screw thread.
Screw pitch measurement	Establishing the number of threads per inch in the Unified and National Form thread systems. Measuring the distance between teeth in millimeters in the ISO metric system. Gaging the pitch of a screw thread with a screw pitch gage, a steel rule, or by other measuring techniques.
Dies (screw thread)	A screw thread forming tool. (The cutting edges conform to a required screw thread.) Solid, split adjustable, or two-part adjustable screw thread cutting tools.
Adjustable die (thread-size adjusting)	A thread cutting tool that provides for variations in thread size. (Die segments may be expanded or reduced with adjusting screws.)
Adjustable die guide	A threaded part that screws into the cap against the two die sections (halves) to lock them in position. (The hole in the guide centers the work surface in relation to the die. The guide helps to start the threads squarely.)
Nominal pipe size	The inside diameter of a standard pipe.
Screw extractor	A hardened, spiral-fluted tool. (When turned into a drilled hole in the direction in which the screw thread can be backed off, the extractor turns the broken portion out of the mating part.)

SUMMARY

- Screw threads are drawn pictorially, schematically, or by simplified representation techniques to represent particular features. Dimensions provide further specifications about size, degree of accuracy (class of thread fit), depth or length, and so on.

  - The pitch of a screw thread in the Unified, American National, and ISO Metric thread systems may be measured directly with a steel rule or gaged with a screw pitch gage.

- Three common forms of dies for regular straight threads and pipe threads include the solid die, split die, and adjustable split die.

  - Split dies and adjustable split dies are adjusted to the specific size of a workpiece. They are then locked in position. Adjusting screws in the die head or the die holder provide for a variation in pitch diameter.

- The pitch diameter of a thread controls the fit of the thread. The die head is locked in a fixed position (pitch diameter) by a lock screw in the die holder.

  - The die is positioned over the workpiece and centered by a guide. The guide aligns the die in relation to the work surface.

- Pipe thread diameters are expressed in terms of the inside diameter of the pipe. The tap drill size permits threads to be cut at full thread depth and at a taper of 3/4″ per foot.

  - The threading die halves of a pipe thread die are aligned with an index line on the die stock. The die halves are adjusted to produce a thread that meets specific job requirements. Once set, the threading dies may be used to thread a number of pieces.

UNIT 10 REVIEW AND SELF-TEST

1. Explain the statement: A drawing provides full specifications for cutting and measuring a required thread.

2. Give three reasons why adjustable split dies are preferred over solid dies.

3. State three major differences between pipe thread dies and regular screw thread dies.

4. List the steps required to thread a workpiece on the bench using an appropriate size adjustable threading die.

5. Identify how a broken threaded part may be removed.

6. State three conditions that result in cutting torn external threads when threading by hand.

PART 4 Basic Cut-off and Grinding Machines: Technology and Processes

SECTION ONE

Metal-Cutting Saws

This section describes the types, features, and operation of power hacksaws and horizontal band machines. Basic cutoff operations performed with these machines are considered in terms of everyday applications. Cutting speed, feeds, cutting fluids, machine adjustments, and cutting blades or saw bands are described in depth because the craftsperson must make correct selections. Also examined are step-by-step procedures for mounting workpieces, setting up the machines, and performing other processes essential to accurate and efficient cutting off.

UNIT 11

Power Hacksaws: Types and Processes

OBJECTIVES

After satisfactorily completing this unit, you will be able to:

- Identify design features of power metal sawing machines.
- Determine applications of utility, heavy-duty, and heavy-duty production power hacksaws for dry- and wet-cutting operations.
- Explain the function of power drives, speed range mechanisms, and other major machine components.
- Select appropriate hacksaw blades and set patterns of saw teeth.
- Determine work positioning and stacking techniques.
- Trace step-by-step procedures on how to operate a utility-type power hacksaw for square and angle cuts.
- Follow the recommended *Safe Practices* for power hack sawing.
- Use trade *Terms* whenever dealing with power hacksaws and sawing.

METAL SAWING/CUT-OFF SAWING MACHINES

The cutting apart of metals and other hard materials is referred to as *metal sawing* or *cutting off*. There are three general designs of metal-sawing machines.

- The *power hacksaw*, which uses a saw-toothed blade,

- The *horizontal (cutoff sawing) band machine*, which uses a saw band and the *vertical band machine* (which is more universal and advanced than the horizontal machine). The vertical machine has many accessories and is adaptable to contour cutting, three-dimensional cutting, and rough and finish filing of regular- and irregular-shaped surfaces,

- The *cold saw*, which uses a circular metal-cutting saw. This saw is designed for heavy-duty operations such as cutting off large-sized stock.

TYPES OF POWER HACKSAWS

Power hacksaws may be classified as *utility*, *heavy-duty*, or *heavy-duty production* machines. The first two types may be either *dry cutting* or *wet cutting*.

Wet-cutting machines normally operate at higher speeds than dry-cutting machines. Thus, they may be used on many different materials. The base of a wet-cutting machine is enclosed. The enclosed area serves as a reservoir for a coolant and houses the circulating pump.

Power hacksaws have a *reciprocating stroke*. On the cutting stroke the blade is forced into the material. Each tooth takes a cut. At the end of this stroke the blade is raised automatically. The teeth are not in contact with the workpiece as the blade is returned to its starting position. This cycle is repeated until the workpiece is cut and the machine stops.

On utility, heavy-duty, and production hacksaws of the wet-cutting type, the cutting fluid is recycled through screens into a reservoir. The large chip particles settle in a container. These chips must be disposed of regularly. The fine chip particles settle as sediment at the bottom and must be scraped out. For hygienic reasons and to retain the specific strength qualities of the cutting fluid, it is necessary to completely clean and wash out the reservoir and fluid-circulating parts.

POWER HACKSAWING PROCESSES

Power hacksaws are used in metal sawing to cut sections of materials such as rods, bars, tubing, and pipes to length. They can also be used to cut castings, forgings, and other parts. The hardness range of these parts and materials is from comparatively soft nonferrous metals (like aluminum, brass, and bronze) to mild steels, tool steels, and harder alloys. The sections of stock may be cut square (straight) with the work axis or at an angle up to 45°.

The power hacksawing process is similar to the hand hacksawing process. In both processes the cutting action takes place by applying force on the teeth of a saw blade as it moves across a work surface. A series of chips are cut by the shearing action of the teeth. Cutting action continues until a section is cut off. Materials that are smaller than 1/2″ are usually cut by hand. Larger sizes are cut by machine.

POWER DRIVES AND MACHINE OPERATION

Most utility power hacksaws, whether wet or dry cutting, are equipped with an oil *hydraulic system*. This system provides smooth uniform control for the speed and cutting force. The hydraulic system actuates the saw frame. The system controls the downward feed (force) during the cutting stroke. It also automatically raises the frame and blade on the return stroke. At the end of the cut, the blade and frame are brought up to the highest position to clear the work.

Usually, dry-cutting power hacksaws have two cutting speeds: 70 and 100 strokes per minute. Where a greater range of materials and sizes is to be cut, wet-cutting machines are made with a four-speed drive motor. Four-speed machines may cut at 45, 70, 100, or 140 strokes per minute. The general range for six-speed machines is 45, 60, 80, 85, 110, and 150 strokes per minute. These machines are provided with a speed/feed chart for different materials. The cutting speeds are obtained by positioning the speed-change lever at the required speed setting.

MAJOR MACHINE PARTS

The power hacksaw is a comparatively simple machine. A utility-type, wet-cutting power saw is shown in Figure 11–1 to illustrate the major parts of a power hacksaw.

The workpiece is held in a vise. The vise has a fixed jaw and a movable jaw (Figure 11–2). Some vises are designed for making square cuts. Other vises have elongated slots in the base. These slots permit the fixed (solid) jaw to be positioned at right angles to the blade. The

fixed jaw may also be positioned at any angle up to 45°. The movable jaw may be swiveled. Force is applied by the movable jaw to hold the workpiece securely against the face of the fixed jaw.

An adjustable work stop is attached to the table. The stop is set at a particular linear dimension and locked in place. The workpiece then is positioned against the stop. Additional pieces of the same linear dimension may be cut without further measurement. Long, overhanging bars of stock are usually supported by a floor stand.

HEAVY-DUTY AND PRODUCTION HACKSAWS

Heavy-duty and production power hacksaws are built heavier than the utility type. These hacksaws have additional accessories.

Production power hacksaws have hydraulically operated work tables (carriage). When the movable carriage is fully loaded and the cutoff length is set, the cutoff process is a continuous one.

Square and round bars and structural and other shapes may be stacked for production cutting. *Stacking* refers to the grouping of many

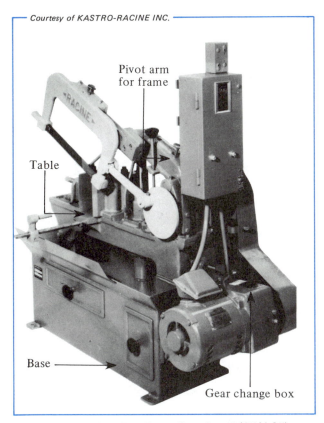

Pivot arm
for frame

Table

Base

Gear change box

Figure 11–1 Utility-Type, Four-Speed (6″ X 6″),
Wet-Cutting Power Hacksaw

Movable
vise
jaw

Stationary jaw angle setting range 0° to 45°

Figure 11–2 Vise Jaws Positioned to
Permit Angle Settings to 45°

Table 11–1 Recommended Pitches, Cutting Speeds, and Feeds for Power Hacksawing Ferrous and Nonferrous Metals (Partial Table)

Material	Pitch (teeth per inch)	Cutting Speed (feet per minute)	Feed (force in pounds)
Iron			
cast	6 to 10	120	125
malleable	6 to 10	90	125
Steel			
carbon tool	6 to 10	75	125
machine	6 to 10	120	125
Aluminum			
alloy	4 to 6	150	60
Brass			
free machining	6 to 10	150	60

pieces together. They are clamped at one time, positioned, and cut to the required length. The operation of the vise jaws, the resetting of the stacked bars to the required length, the raising of the frame, the application of the cutting force, and the circulating of the cutting fluid are all controlled hydraulically.

POWER HACKSAW BLADES

BLADE TERMINOLOGY

Each numbered *term* that follows is commonly used to identify and describe a corresponding numbered design feature as illustrated in Figure 11–3.

①	Blade Thickness	Measurement (gage) of the body.
②	Blade Width	Overall width from the tooth cutting edge to the back (top edge) of the blade.
③	Tooth Face	Surface on which chips form.
④	Saw Tooth Set	Right and left tooth offsets to produce clearance (kerf) for the saw blade.
⑤	Tooth	Cutting area of the saw blade.
⑥	Tooth Pitch	Distance from the tip of one tooth to the corresponding tip of the next tooth.
⑦	Teeth Per Inch	Number of teeth per inch (TPI).

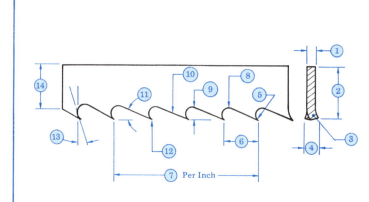

Figure 11–3 Power Saw Blade Terminology

⑧	Gullet	Curved shape at the bottom of the tooth.
⑨	Gullet Depth	Distance from the tooth tip to the bottom of the tooth gullet.
⑩	Tooth Back	Surface adjacent to the tooth face.
⑪	Tooth Back Clearance Angle	Angle formed by the cutting plane and the tooth back surface.
⑫	Tooth Tip	Cutting edge of a saw tooth.
⑬	Tooth Rake Angle	Angle formed by the tooth face and a line perpendicular to the cutting plane.
⑭	Blade Back Width	Measurement from the top of the gullet to the top edge of the blade.

Tables of machinability ratings of metals indicate the range of ease or difficulty in cutting a given metal. The tables provide information for establishing whether a cutting process should be carried on wet or dry and what kind of cutting fluid should be applied.

Other tables are used to determine the correct pitch, cutting speeds, and feeds for power hacksawing. The particular material to be cut and the size and shape of its sections must be considered. Table A–4 in the Appendix lists the recommended pitches, cutting speeds, and feeds for selected ferrous and nonferrous metals. A portion of this table is illustrated in Table 11–1.

The pitch recommended represents an average for material sizes of 2″ or smaller. Coarser pitches may be used for thicker or larger sizes. Finer pitches are used for thinner sections.

SELECTION OF POWER HACKSAW BLADES

Some of the principles governing the selection of hand hacksaw blades apply equally as well to power hacksaw blades:

- There must be at least two teeth in contact with the work surface at all times;
- The greater the cross-sectional area is, the coarser the required pitch (the coarser pitch provides greater chip clearance);
- Easily machined and soft materials require a coarse pitch and large chip clearance;
- Hard materials and small cross-sectional areas require finer pitches.

While some power hacksaw blades are made of carbon alloy steel, high-speed and other alloy steel blades are more durable. High-speed tungsten, high-speed molybdenum, and molybdenum steels are excellent cutting-blade metals.

Standard blade lengths range from 12″ to 14″, 16″, 18″, and 24″. Length depends on machine size and the nature and size of the material to be cut. The general range of pitch is from 4 to 14 teeth per inch. General metric size blade lengths are 300mm, 350mm, 400mm, 450mm, and 600mm. The common metric pitch range is from 6mm to 2mm.

SET PATTERNS FOR SAW BLADE TEETH

A common set pattern for the teeth on power saw blades is called *raker set*. The raker set pattern consists of a repeat design for every three teeth. Figure 11–4 shows the raker set. One tooth is unset. This tooth is followed by two teeth, one of which is offset to the right. The other tooth is offset to the left. This pattern continues for the length of the blade. The raker set blade is recommended for heavy work on bar stock, forgings, die blocks, and parts that have a constant cross section.

A second common set pattern is called *wave set* (Figure 11–4). Wave-set teeth are offset in groups. These groups alternate from right to left to form a wave pattern. Wave-set teeth are used where there is a considerable range of material sizes to be cut. Blades with wave-set teeth are

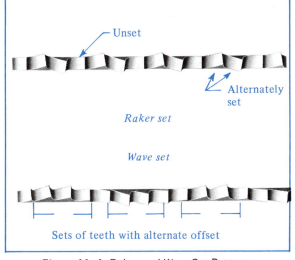

Figure 11–4 Raker and Wave Set Patterns for Saw Blade Teeth

used for cutting materials with cross-sectional areas that change. Structural forms and pipes are examples of parts that have changing area sizes.

A third set pattern, which has limited use in metal working, is called *straight set*. In this older pattern the teeth are offset alternately to the right and left of the blade.

How to Operate a Utility-Type Power Hacksaw

Cutting Off a Square Section

STEP 1 Determine the pitch of the blade. Select a cutting speed according to the shape, size and kind of material to be cut.

STEP 2 Set the speed-change lever at the required speed.

STEP 3 Check the direction of the teeth on the blade.

Note: The blade may be changed by placing the new blade on the pins in the frame. The teeth are pointed so that they cut on the *cutting* stroke. Turn the clamping screw until the blade is held tautly in the frame.

STEP 4 Check the position of the stationary vise jaw for squareness.

STEP 5 Position the adjustable work stop at the required cutoff distance from the saw blade.

Note: Sometimes the length is marked out directly on the workpiece.

STEP 6 Move the workpiece into the vise until it touches the work stop.

Note: The end of the part is checked first to see that the surface is square and flat. It may be necessary to square the end by cutting off a thin wafer-like section.

STEP 7 Support the end of any long bar on a floor stand.

Caution: Place either a warning flag, cloth, or a protecting screen around the end of the bar. Care must be taken so that the length being cut does not project into an aisle or walking area.

STEP 8 Position the cutting fluid nozzle. The fluid must be directed to the cutting area of the blade and workpiece.

STEP 9 Recheck the vise to see that the work is held securely. The blade must be taut. The speed and blade pressure also require checking.

STEP 10 Turn the power switch to ON. Position the frame with the handle. Bring the blade down almost to the work.

STEP 11 Move the clutch handle to the engaged position to automatically start the cutting action.

STEP 12 Remove burrs from the cutoff section.

Caution: Although the cutting fluid serves as a coolant, a part nevertheless may be too hot to handle safely. Allow it to cool before filing the burrs.

Cutting Off an Angular Section

STEP 1 Select the proper power hacksaw blade.

STEP 2 Set the machine at the recommended speed and feed.

STEP 3 Loosen the stationary vise jaw. Position it at the required angle.

Note: The vise base is graduated in degrees. Tighten the stationary jaw at the correct angle.

STEP 4 Mount the workpiece in the vise. The movable jaw is moved to the required angle.

STEP 5 Measure the cutoff length of the workpiece.

Note: The end of a steel rule is usually placed against the teeth of the blade. The cutoff length is represented by the measurement to the end of the workpiece.

Caution: *Before* making this measurement, be sure the machine power switch is in the OFF position and the blade is *not* moving.

STEP 6 Tighten the movable jaw. Check to see that the workpiece is held securely at the correct angle.

STEP 7 Proceed with the cut by following steps 7–12 under *Cutting Off a Square Section.*

Safe Practices in Power Hacksawing

- Position the cutting teeth and the blade to cut on the cutting stroke.
- Tighten the blade tension until it is adequate to hold the blade taut during the cutting operation.
- Check the blade pins regularly to see that they are not being sheared.
- Check the workpiece to be sure that it is tightened securely before starting the cut.
- Make sure the blade is moved away from the work before starting the power hacksaw.
- Start a new blade (after a cut has been started) in a new location. Otherwise the teeth may bind in the old kerf and break the blade.
- Direct the flow of the cutting fluid (when fluid is required) over the cutting area. The flow must be as close to the cutting saw teeth as possible.
- Support the ends of long pieces that project from the power hacksaw by using a roller stand.

- Place a protecting screen or a danger flag at the end of parts that extend any distance from the saw frame.
- Cool the cutoff section before handling. Cooling helps to avoid burns and cuts resulting from hot, burred pieces.
- Remove cutting fluids and clean the reservoir regularly.

POWER HACKSAW AND SAWING TERMS

Cutting off (metal sawing)	The sawing apart by the shearing action of a series of cutting teeth. The removal of a section of metal or other material by a metal-sawing process.
Utility power hacksaw	An all-purpose wet- or dry-cutting machine for cutting off square or angular sections of workpieces. A sawing machine. Workpieces are cut off by a hacksaw blade that moves in a reciprocating motion.
Cutting stroke	A stroke of a power hacksaw. A downward force is applied on the cutting teeth of the blade as it is drawn through the workpiece. The cutting of a saw kerf by the shearing action of the saw teeth on the cutting stroke.
Machinability	Properties of a material that relate to the ease or difficulty with which it may be cut or formed. Qualities of a material that influence the selection of a blade, its pitch, machining speed and feed, and an appropriate cutting fluid.
Raker set	A pattern of setting teeth to cut a kerf larger than the width of the hacksaw blade. A pattern of three teeth. An unset tooth is followed by one tooth offset to the right. Another tooth is offset to the left of the blade body.
Cutting speed (power hacksaw)	The number of full strokes the saw frame makes per minute.
Four-speed (power hacksaw)	A range of four speeds. A speed range that permits the hacksaw blade to make 35, 70, 100, or 140 strokes per minute.
Elongated slots	Two parallel slots in a vise base. Long slots that permit the stationary jaw to be positioned square or at an angle to the cutting blade.
Production power hacksaw	A power hacksaw with production capabilities. Multiple pieces of stock are stacked and clamped in position. The cutting fluid is controlled, the parts are cut off, and the hacksaw is stopped automatically.

SUMMARY

- Two common power metal-cutting saws are the power hacksaw and the horizontal band (cutoff sawing) machine.
  - Mild to hard ferrous metals and soft to hard nonferrous metals and other materials may be cut off by power hacksawing. The section may be a square, round, structural, or other preformed shape.
- Power hacksawing is the process of shearing a narrow width (kerf). The cutting teeth on the blade are forced to cut into the workpiece as the blade is drawn through.
  - Dry-cutting power hacksaws operate at slower speeds than wet-cutting power hacksaws. These hacksaws cut materials that do not require a cutting fluid or where there is a limited amount of heat generated. Any

small amount of heat that is generated does not affect the hardness or efficiency of the saw teeth.

- Wet-cutting power hacksaws of the utility and heavy-duty types are used when a great deal of heat is generated. A cutting fluid serves as a coolant to control the heat. The lubricating qualities of the cutting fluid protect the blade surfaces from being scored. Cutting efficiency is thus improved.
  - The pitch selection of power hacksaw blades is governed by the same principles that apply to hand hacksawing.
- Common lengths of power hacksaw blades range from 12″ to 24″, and pitches range from 4 to 14. Metric lengths range from 300mm to 600mm, with pitches from 6mm to 2mm.
  - A raker set tooth pattern consists of sets of three teeth. An unset tooth is followed by two other teeth that are alternately offset.
- The hydraulic system of a power hacksaw:
  — Actuates the saw frame,
  — Controls the down feed during cutting,
  — Raises the frame and blade on the return stroke,
  — Maintains a constant flow of cutting fluid,
  — Automatically raises the frame to clear the work at the end of the cut,
  — Stops all motion.
  - Two-speed power hacksaws generally operate at 70 to 100 strokes per minute. Four-speed machines have a range of 35, 70, 100, and 140 strokes per minute.
- Work may be held in a vise and positioned square or at an angle to the cutting blade. The stationary vise jaw may be set to a required angle.
  - The cutting fluid and reservoir must be kept clean. The fluid must be tested regularly. Testing and correcting ensures that the evaporation of liquid does not change the efficiency of the cutting fluid.
- Personal and machine tool safety-related precautions must be observed. These precautions cover all steps in handling the workpiece, cutting tools, and the power hacksaw.

UNIT 11 REVIEW AND SELF-TEST

1. Describe one cutting cycle of a common 6″ × 6″ (150mm × 150mm) utility power hacksaw.

2. Cite three guidelines that especially apply to selecting power hacksaw blades.

3. Give a general application of the following set patterns for power hacksaw blade teeth: (a) raker set, (b) wave set, and (c) straight set.

4. Explain two design features of a six-speed power hacksaw that make this machine more versatile than a two-speed machine.

5. Describe how stacked bars are automatically cut on a production power hacksaw.

6. State three safe practices that relate particularly to power hacksawing.

Horizontal Band Machines and Cutoff Sawing Processes

OBJECTIVES

After satisfactorily completing this unit, you will be able to:

- Describe utility and heavy-duty dry- and wet-cutting band machines, processes, and operating principles.
- Know the effect of heat generation and control upon saw teeth, performance, and tool life.
- Apply information about the properties of straight oils, soluble oils, and synthetic cutting fluids.
- Assess the features and properties of carbon alloy, high-speed steel, and tungsten carbide blades.
- Differentiate and recommend uses of precision, buttress, claw, and carbide tooth forms for cutoff sawing.
- Perform the following processes.
  - Select the Saw Blade and Determine Cutoff Sawing Requirements.
  - Operate a Horizontal Cutoff Band Machine.
- Follow recommended *Safe Practices* for machine setups and operation.
- Interpret each new band machine *Term*.

The horizontal (cutoff sawing) band machine is a second type of power saw. This machine is often referred to as a *metal-cutting band saw*, a *cutoff band saw*, or a *cutoff band machine*. The term *band* indicates that the cutoff sawing process requires a closed saw blade.

There are also vertical band machines with saw bands and file bands. These machines are used for intricate and precise sawing and filing processes.

BASIC TYPES AND FEATURES OF HORIZONTAL BAND MACHINES

Metal-cutting horizontal band machines for cutoff sawing may be either *dry cutting* or *wet cutting*. They may also be of a *utility* or *heavy-duty* design. The utility design is illustrated in Figure 12–1.

The saw frame of the band machine has two wheels. These wheels hold and drive a continu-

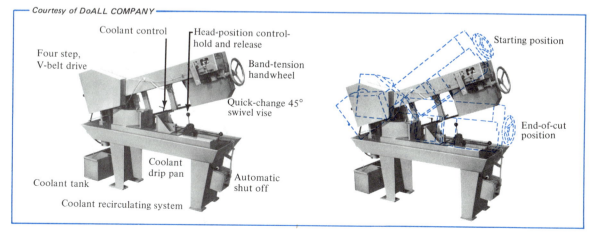

Courtesy of DoALL COMPANY

Four step, V-belt drive — Coolant control — Head-position control-hold and release — Band-tension handwheel — Quick-change 45° swivel vise — Coolant drip pan — Automatic shut off — Coolant tank — Coolant recirculating system — Starting position — End-of-cut position

Figure 12–1 Utility-Type Horizontal Band Machine

ous (closed) *saw blade* (band). A tension control adjusts the blade to track properly at the correct tension. The teeth are thus forced through the work. Blade guide inserts are provided for positioning the blade vertically at the cutting area. The inserts guide the blade to cut squarely. The saw frame is hinged to permit raising the saw band (blade) to clear the work and lowering it to take a cut. A *pneumatic system* controls the circulation of the cutting fluid and many of the machine mechanisms.

OPERATING PRINCIPLES

The continuous saw band revolves around the *driver* and *idler wheels*. Attached to the frame are two adjustable *(blade) supports* with *guide inserts*. These supports and guides serve two functions:

- Guide the blade in a vertical position so that it does not bend from the work.
- Support the blade so that a cutting force can be applied.

DRY CUTTING AND WET CUTTING

Materials are dry cut in both power hacksawing and band machine sawing when:

- A slow cutting speed is used and the frictional heat that is generated during the cutting process is minimal;
- A metal is comparatively soft and easy to cut;

- A material, such as gray and malleable cast iron, produces loose graphite that acts as a lubricant;
- Wet cutting hard materials has a tendency to produce a work-hardened surface.

Wet cutting is recommended to:

- Dissipate the heat generated over the small surface area of each saw tooth,
- Remove heat to prevent softening the cutting edges of the saw teeth,
- Reduce the friction between the chips and the saw teeth,
- Prevent the depositing of metal (caused by the cutting action) at or near the edges of the teeth,
- Clear chips away from the workpiece,
- Keep the sides of the saw blade from being scored,
- Increase productivity and tool wear life.

CUTTING ACTION AND CUTTING FLUIDS

The action of the saw teeth cutting through a workpiece usually produces a tremendous amount of heat. Heat is generated in three main places (Figure 12-2):

- At the point of contact and along the cutting face,
- On the cutting edge,
- At the shear plane in the forming of a chip.

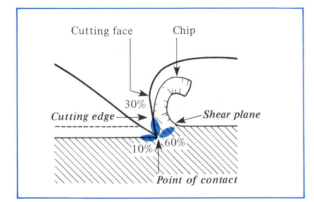

Figure 12-2 Three Main Places Where Heat Is Generated by the Cutting Action of Saw Teeth

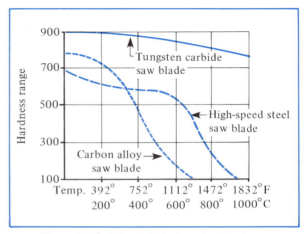

Figure 12-3 Effect of Heat on the Hardness of Saw Band Teeth

The heat must be carried away, and the temperature must be held within a specific range. Excess heat can cause the teeth to soften and dull.

The chart in Figure 12–3 shows the effect of heat on the hardness of three major kinds of saw blades. Note that the teeth of the carbon alloy saw blade soften at a comparatively lower temperature than do teeth of the high-speed steel saw blade. As the saw teeth dull and wear, their cutting efficiency is reduced drastically. Over the same temperature range, tungsten carbide saw blade teeth maintain their cutting qualities.

Cutting fluids are used to dissipate the heat produced by sawing and cutting. The fluids also wash chips away from the cut and increase the saw life.

There are three basic groups of cutting fluids that are widely applied in wet cutting: straight oils, soluble oils, and synthetic (chemical /water) cutting fluids.

STRAIGHT OILS

Straight oils are mineral oils. They are used on very tough materials that must be cut at slow speeds. Straight oil has *lubricity*—that is, high lubricating properties in contrast with its heat removing capability. Straight oils may have sulphur or other chemical compounds added (additives).

SOLUBLE OILS

Soluble oils are used in the cutting off of a wide variety of materials. Since band machine sawing is done at high speeds, a great deal of heat is generated. It is necessary to remove the heat rapidly to maintain efficient cutting and to protect the life of the saw teeth and saw band.

Soluble oils are mineral oils that are readily *emulsified* (suspended) into fine particles, or globules, when mixed with water. The resulting mixture combines the properties of straight oil for lubricity with the high cooling, or heat dissipating, rate of water.

The more concentrated the mixture- for example, one part of soluble oil to three parts of water (1:3)—the greater the lubricity. The more dilute the mixture—for example, 1:7—the greater the heat removal capability.

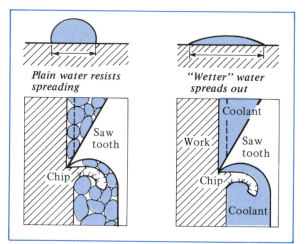

Figure 12–4 Heat Dissipation Properties of Cutting Fluids Improved by Chemical Additives

SYNTHETIC (CHEMICAL/WATER) CUTTING FLUIDS

Synthetic cutting fluids do not contain mineral oils. They are a mixture of chemicals added to water to produce "wetter" water. Water tends to resist spreading because of its surface tension. This surface tension may be reduced by chemical additives, as shown in Figure 12–4. Synthetic cutting fluids then have better properties of spreading and flowing easily and quickly. These properties provide for a high rate of heat removal. Another desirable property is that many synthetic cutting fluids are transparent. This transparency makes the cutting area and action more visible.

KINDS AND FORMS OF SAW BLADES (BANDS)

There are three general kinds of cutoff saw blades: *carbon alloy, high-speed steel,* and *tungsten carbide.*

Carbon alloy saw blades are generally used in the toolroom and maintenance shop and in light manufacturing where accuracy is required. High-speed steel blades are used in heavy-duty and full-time production work. Tungsten carbide blades are suited for heavy production and for rough cutting through tough materials.

Carbon alloy and high-speed steel saw bands have three forms of saw blade teeth: *precision*, *buttress*, and *claw*. Tungsten carbide saw bands have a special tooth form. The general design features of saw tooth forms are illustrated in Figure 12–5.

PRECISION TOOTH FORM

The precision form is the most widely used tooth form. The tooth has the following features: a rake angle of 0°, a back clearance angle of 30°, a deep gullet, and a radius at the bottom. Each of these features is shown in Figure 12–6A. The precision form produces accurate cuts and a fine finished surface. The clearance angle and gullet provide ample chip capacity for most cut-off sawing operations.

BUTTRESS TOOTH FORM

The buttress form is also known as a *skip tooth* (Figure 12–6B). The 0° rake angle and the 30° back clearance angle are similar to the precision form. However, the teeth are spaced wider apart. Wider spacing provides greater chip clearance. Buttress-form teeth cut smoothly and accurately. The buttress tooth form is recommended for thick work sections, deep cuts, and soft material.

CLAW TOOTH FORM

The claw tooth is also called a *hook tooth* (Figure 12–6C). The tooth has a positive rake angle. The clearance angle is smaller than the clearance angle of the precision and buttress forms. The gullet is specially designed to be stressproof. The claw tooth form makes it possible to cut at a faster rate and at reduced feed pressures—features that provide longer tool life.

TUNGSTEN CARBIDE TOOTH FORM

The tungsten carbide tooth form has a positive rake angle and a smaller clearance angle than any of the other tooth forms (Figure 12–7). The tungsten carbide teeth are fused into a fatigue-resistant blade that is necessary for heavy, tough cutoff sawing operations.

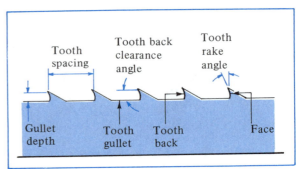

Figure 12–5 General Features of Saw Tooth Forms

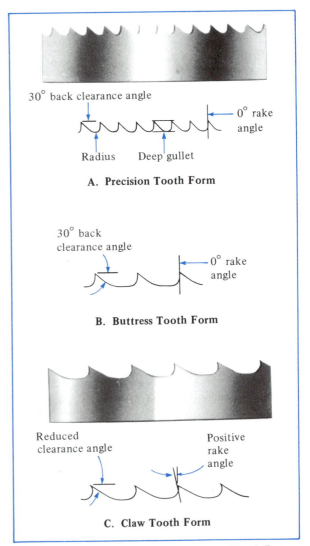

A. Precision Tooth Form

B. Buttress Tooth Form

C. Claw Tooth Form

Figure 12–6 Precision, Buttress, and Claw Tooth Forms

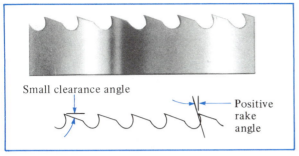

Figure 12-7 Tungsten Carbide Tooth Form

CUTOFF SAWING REQUIREMENTS AND RECOMMENDATIONS

More cutoff sawing operations are required in the manufacture of parts than any other single machining process. The craftsperson must make a number of decisions that center around the saw band alone. Some of the major considerations are as follows:

- Kind of saw blade,
- Form of tooth,
- Cutting fluid (if required) and its rate of flow,
- Pitch of the teeth,
- Velocity of the saw band in feet per minute.

The decisions must be based on certain information:

- Size and shape of the workpiece,
- Properties of the material to be cut,
- Required quality of the finish of the sawed surface,
- Quantity to be cut off,
- Overall cross-sectional area if multiple pieces are cut at one setting.

Band machine and saw band manufacturers provide tables to aid the worker in selecting the correct saw band for maximum cutting efficiency. Cutoff sawing requirements and recommendations are further simplified and combined in a *job selector*. A section of job selector is illustrated in Figure 12-8.

How to Select the Saw Blade and Determine Cutoff Sawing Requirements

Sample Job Requirement: Select the correct saw blade for cutting off a quantity of workpieces. The required lengths are to be cut from a 2″ diameter bar of low carbon steel in the 1015 to 1030 range. A band machine with a welded and dressed blade is to be used.

STEP 1 Turn the job selector to the kind of material to be cut. Figure 12-9 at ①.

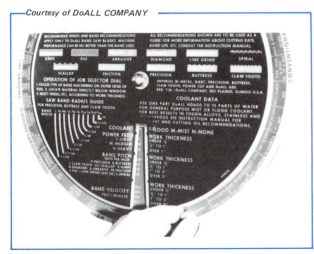

Courtesy of DoALL COMPANY

Figure 12-8 Band Machine Job Selector and Functions

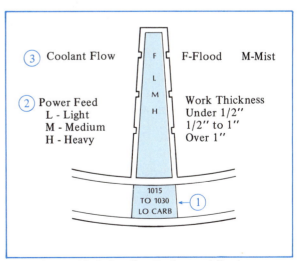

Figure 12-9 Reading Coolant and Feed Requirements in the Job Selector

Note: The band machine settings and operating conditions are read in the window above the material setting. The feed, band pitch, kind of saw blade, and band velocity are given in the column to the left of this window. The thickness of the workpiece and the rate of coolant flow are given in the column to its right. In this example, a workpiece thickness of 1 1/2″ is assumed.

STEP 2 Read the setting for the power feed (H, or heavy, in Figure 12–9). Set the machine for a heavy power feed (H), as shown at ②.

STEP 3 Read the rate of coolant flow. (The selector shows F [flood] at ③.)

STEP 4 Refer to a machinability table on cutting fluids for ferrous metals. (Appendix A–18 is an example.) Select the recommended fluid in the low-carbon steel column (machinability group II) under sawing processes. Either a sulphurized

oil (Sul), a mineral-lard oil (ML), or a soluble or emulsified oil or compound (Em Sul ML) is recommended.

Note: Some job selectors provide information on cutting fluids.

STEP 5 Read on the job selector the recommended kind of saw blade and pitch. (The 6P at ⑤, Figure 12–10 indicates 6 pitch, precision form, carbon alloy blade.)

STEP 6 Select and mount the appropriate saw band.

STEP 7 Read on the job selector the recommended band velocity. (The setting recommended at ④ is 155 feet per minute.) Set the band velocity.

Safe Practices in Band Machining (Cutoff Sawing)

- Check the blade tension. It must be adequate to permit proper tracking on the driver and idler wheels and to transmit sufficient force to cut off the workpiece.

- Determine the speed, feed, kind, and form of saw blade teeth, and the cutting fluid. Cutoff sawing specifications are obtained from machine and blade manufacturers' tables.

- Lock the frame guards in the locked position before any control switches are turned on.

- Direct the cutting fluid to provide maximum lubrication between the blade, the guides, and the workpiece. The cutting fluid must also provide for dissipating the heat generated at the cutting edges.

- Position the power controls in the OFF position when making blade or work adjustments.

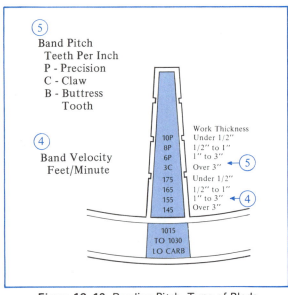

Figure 12–10 Reading Pitch, Type of Blade, and Velocity on the Job Selector

TERMS USED IN BAND MACHINING (CUTOFF SAWING)

Band machine (cutoff sawing)	A rotating machine with driver and idler wheels. (The wheels drive and force a metal-cutting sawtoothed band to shear and cut off materials.)
Blade supports	Two adjustable supports on the band machine frame. Arms that are positioned to provide support for the saw band.
Precision form	A saw tooth form having a 0° rake angle, a 30° back clearance angle, a deep gullet, and a bottom radius. An accurate-cutting saw tooth form that produces a fine finished surface.
Buttress (skip tooth) form	A saw tooth form having a 0° rake angle and 30° back clearance angle. (The teeth are spaced farther apart than are the teeth of the precision form.)
Claw (hook tooth) form	A saw tooth form that has a positive rake angle and a smaller clearance angle and wider-spaced teeth than the precision form.
Tungsten carbide form	Tungsten carbide teeth fused on a fatigue-resistant blade back. A specially shaped, positive-rake-angle tooth form. A tooth form that is particularly adapted for heavy sawing operations through tough materials.
Job selector	An information dial guide on band machines. Manufacturers' recommendations on saw blade characteristics, speeds, feeds, cutting fluids and flow rate, and other machine operating conditions.
Blade guides	Inserts in the blade supports that nest the saw blade (above the teeth) and guide the saw blade in a vertical direction through the workpiece.
Wet cutting	The use of a cutting fluid to dissipate heat, flow chips away, reduce friction, and improve cutting efficiency.
Cutting fluids	Fluids that have two basic properties: (1) rapid heat removal and (2) lubricity. (Lubricity helps to produce a finer cutting action and reduce friction.)
Straight oils	Mineral oils used on tough materials at low cutting speeds where a high degree of lubrication is required.
Soluble oils	A mixture of an emulsified mineral oil and a water soluble oil.
Synthetic cutting fluids	A mixture of chemical additives that improve the wetting and cooling properties of water.

SUMMARY

- Accuracy of the cut depends in part on the close positioning of the band supports in relation to the overall width of the material to be cut. Careful adjustment of the guides that steady the saw band also influences this accuracy.

    - The saw band must track evenly on the idler and driver wheels. The tension on the saw band must be adequate to drive the saw teeth to cut uniformly.

- Horizontal band machines may be operated dry at slow speeds. Dry cutting is used when a limited amount of heat is generated or the material does not require a cutting fluid.

■ Wet cutting is employed when (1) there is considerable heat to be dissipated, (2) friction is to be reduced to a minimum, (3) chip deposits are to be prevented from forming on the teeth, and (4) the chips are to be carried away.

■ Slow-speed cutoff sawing operations require a high degree of lubricity. A straight mineral oil is satisfactory.

■ Soluble oils are used when a combination of a coolant and a lubricant is required.

■ Synthetic cutting fluids with a chemical additive make a water base a more efficient cooling agent. These fluids are recommended for cutoff sawing operations that require a high level of heat removal.

■ Carbon alloy, high-speed steel, and tungsten carbide are three basic kinds of saw blades. They are used for general precision sawing, production, and heavy-duty cutting through tough materials, respectively.

■ The precision and buttress tooth forms have a 0° rake angle and a 30° back clearance angle. They differ in the distance between teeth, the depth of the gullet and bottom form, the quality of the finished surface produced, and the cutting time.

■ A job selector chart provides saw blade and machining information. Recommendations include the kind of saw blade and tooth form, cutting speeds, feeds, cutting fluids, and other operating conditions. The recommendations are read easily on the chart. The technical information is related to the type of material to be cut off, its size, and the surface finish required.

UNIT 12 REVIEW AND SELF-TEST

1. List three distinct design features of the band machine and blade.

2. Explain why the horizontal band machine cuts faster than a reciprocating power hacksaw.

3. State three conditions under which dry band machine sawing is recommended over wet cutting.

4. Identify one saw blade material of which the cutting quality of the saw blade teeth is maintained longer (over the same temperature range) than the cutting quality of a high-speed steel blade.

5. State why synthetic cutting fluids are used in cutoff sawing processes that require a high heat-removal rate.

6. Distinguish between (a) the precision form and (b) the tungsten carbide form of saw band teeth.

7. Indicate the kind of information that is provided by a job selector on a horizontal cutoff band machine.

8. List the preliminary steps that must be taken before mounting and cutting off parts on a horizontal cutoff band machine.

9. Identify three safe practices to observe when band machining (cutoff sawing).

Bench and Floor Grinders

UNIT 13

Hand Grinding Machines, Accessories, and Basic Processes

OBJECTIVES

After satisfactorily completing this unit, you will be able to:

- Relate construction and other features of bench and floor grinders to offhand grinding processes.
- Understand the reasons for truing and dressing grinding wheels using an abrasive stick or a mechanical dresser.
- Analyze wheel conditions to determine dressing needs.
- Perform the following processes.
  - Dressing a Bench or Floor Grinder Wheel.
  - Offhand Grinding Flat Surfaces.
  - Offhand Grinding Angular Surfaces, Round Points, and a Mushroomed Edge.
- Follow recommended *Safe Practices* for truing, dressing, and using grinders and grinding wheels.
- Use bench and floor grinder and grinding wheel *Terms.*

BENCH AND FLOOR (PEDESTAL) GRINDERS

The grinding of the cutting edges of tool bits, drill points, and chisels; the forming of blades and other tools; and various other metal-removing processes all may be performed on a simple grinding machine. This machine is called a *bench grinder* if it is mounted on a bench as shown in Figure 13–1. The terms *floor* or *pedestal grinder* are used interchangeably when the grinding head is mounted on a pedestal for floor work. Pedestal grinders range in size from grinders used for small tool grinding operations to large grinders used for heavy-duty, rough snagging processes on large forgings and castings.

CONSTRUCTION AND FEATURES OF GRINDERS

Grinding operations on the bench or floor grinder are referred to as *offhand grinding*. In offhand

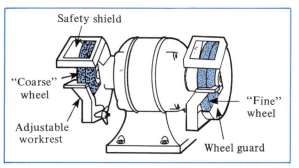

Figure 13–1 Bench Grinder

grinding the object is held and guided across the revolving face of the grinding wheel by hand.

Regardless of the type, bench and pedestal (floor) grinders have a direct power driven *spindle*. The spindle is flanged and threaded at both ends to receive two *grinding wheels*. The grinding wheel at one end is coarse, for rough cutting operations. The wheel on the opposite end is fine, for finish grinding.

Each grinding wheel is held between two *flanged collars*. There is a simple *adjustable tool rest* for each wheel. Each rest provides a solid surface upon which the part may be placed and steadied while being ground. The tool rest may be positioned horizontally or at an angle to the wheel face.

Wheel guards (Figure 13–1) are an important part of the grinder. Safety glass shields permit the work surface to be observed and provide additional protection against flying abrasives and other particles. However, because of the hazards of machine grinding it is also necessary for the worker to use personal eye protection devices.

GRINDING WHEELS

MOUNTING ON A SPINDLE

Grinding wheels are mounted directly on the spindles of bench and floor grinders. Figure 13–2 shows a mounted grinding wheel assembly. One end of the spindle has a right-hand thread; the other end, a left-hand thread. The thread direction safeguards the wheel from loosening during starting and stopping and during cutting action.

Grinding wheels have a soft metal core that is bored accurately to the diameter of the spindle. The wheel slides on the shaft with a snug fit. The wheel should never be forced on the spindle or tightened with too great a force against the flanged collars. A soft compressible material (washer) is placed between the sides of the wheel and the flanged collars. The washer, which is sometimes called a *blotter*, is often used as the label for the manufacturer's specifications. This washer helps to avoid setting up strains in the wheel. The clamping nut is drawn against the flanged collar only tight enough to prevent the grinding wheel from turning on the spindle during cutting.

TRUING AND BALANCING GRINDING WHEELS

A wheel is tested for trueness and balance before grinding. *Trueness* refers to the concentricity of the outside face and sides of a grinding wheel as it revolves at operating speed. *Balance* relates to whether the mass of a grinding wheel is evenly distributed about its axis of rotation. Excessive force may cause a wheel to fracture and fly apart. A fractured wheel is a safety hazard and should be replaced. Grinding wheel manufacturers true and balance wheels so that they are ready to be mounted and used directly. Wheels over 12″ (304.8mm) diameter should always be balanced before mounting.

Truing is a process of shaping a grinding wheel. A wheel is *trued* to restore its concentricity. The

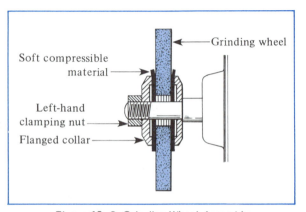

Figure 13–2 Grinding Wheel Assembly

A. Glazed (Loaded) Wheel Before Dressing   B. Open, Dressed Wheel

Figure 13–3 Glazed and Dressed Grinding Wheels

wheel may be dressed so that it has a straight, angular, or special contoured face. However, (after truing), the grinding face (periphery of the grinding wheel) must run concentric with the spindle axis.

DRESSING AND HAND DRESSERS

DRESSING

During the grinding process some of the sharp-edged abrasive grains become dull (glazed). The grains may not fracture as they should to provide new, sharp cutting edges. Also, metal particles from the workpiece or tool may become imbedded in the wheel. Such a wheel is said to be *loaded* and must be *dressed*. A loaded, glazed wheel is shown in Figure 13–3A.

Dressing is the process of reconditioning a grinding wheel. Reconditioning a wheel is necessary for efficient grinding and to produce a good quality of surface finish. Figure 13–3B illustrates an open, dressed wheel. The grains are sharp. Fast cutting takes place.

Hand dressers are used for both dressing and truing grinding wheels. The two common hand types to be considered next are the *abrasive stick* and the *mechanical dresser*.

DRESSERS

Abrasive Stick. The abrasive dressing stick is used to remove particles that produce a glazed surface and to sharpen the cutting face. The abrasive stick shown in Figure 13–4 is held by hand. It is brought carefully into contact with the grinding wheel. Usually a few passes across the face are sufficient to correct the glazed condition and produce sharp-cutting abrasive grains.

Mechanical Dressers. Mechanical dressers are usually used for heavier dressing and truing operations than the abrasive stick. Mechanical dressers are designed with many different forms of metal wheel discs (Figure 13–5). These discs are mounted on a shaft at one end of a hand holder. The discs have star or wavy patterns. They are made of hardened steel and are replaceable. There is a circular disc separator between each pair of star discs that allows for a great deal of side play between the discs.

As the disc faces are brought into contact with the face of the grinding wheel, they turn at considerable velocity in comparison to the speed of the grinding wheel. The action between the dresser discs and the grinding wheel causes the abrasive grains to fracture. As new, sharp cutting edges are exposed, the imbedded particles are torn out. The wheel is also trued during the process. A diamond dresser, guided under more controlled conditions, is used to true a grinding wheel to a finer degree of accuracy.

Caution: Extreme care must be taken when hand dressing. The grinding wheel guard and shield must be in place, the tool rest must be correctly positioned to support and serve as a guide for the dresser, and personal goggles must be worn.

Mechanical dressers are practical where speed in reconditioning a grinding wheel is more essential than accuracy. The heavy-duty dresser in Figure 13–5 has dust-protected ball bearings

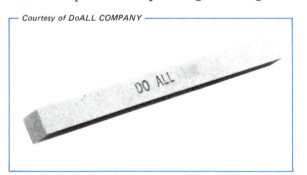

Courtesy of DoALL COMPANY

Figure 13–4 Abrasive Dressing Stick

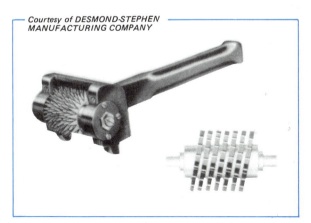

Courtesy of DESMOND-STEPHEN MANUFACTURING COMPANY

Figure 13–5 Heavy-Duty Ball Bearing Dresser

for the cutter spindle. This dresser is practical on large, coarse grinding wheels, particularly wheels that are bakelite and rubber bonded.

WHEEL CONDITIONS REQUIRING DRESSING

The craftsperson must make decisions as to when a grinding wheel must be dressed. Dressing is needed under the following conditions:

- An uneven chatter surface is produced,
- The workpiece is heated excessively during grinding,
- Colored burn marks appear on the work surface,
- There are load lines on the workpiece that indicate the wheel is loaded with metal particles,
- There is considerable vibration when the wheel is in contact with the workpiece.

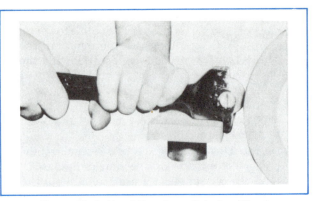

Figure 13–6 Hand Dressing a Grinding Wheel with a Mechanical Dresser

---

How to Use and Dress a Grinding Wheel (Bench and Floor Grinders)

Using a Mechanical Dresser

STEP 1 Adjust the work rest (tool rest). The front face may be used as a guide for the mechanical dresser. The distance between the work rest and the wheel should permit the discs to almost touch the face of the grinding wheel.

STEP 2 Remove the dresser. Check to see that the work rest and wheel guards are secure. Check to see that the wheel is tight on the spindle.

Caution: The safety shield should be positioned to provide good visibility. Safety goggles *must be worn*.

STEP 3 Start the grinder.

Caution: In all grinding operations the operator's position at start-up and during the grinding operation is to stand to one side and out of the path of the grinding wheel.

STEP 4 Place the guiding lugs of the dresser against the front edge of the work rest.

Note: There should be no contact at this time between the cutter and the revolving grinding wheel.

STEP 5 Hold the dresser down on the work rest and back against the edge with one hand. Move the dresser across the wheel face and guide it with the other hand. (Figure 13–6 shows the position of the hands and dresser.)

STEP 6 Tilt the dresser upward a small amount. Move it across the wheel face.

Note: The dresser should be raised slightly after each pass. Otherwise there may be a tendency to hog in (dig in). Excessive amounts are cut out in hogging, and undesirable results are produced.

STEP 7 Continue to raise the dresser for each pass across the face of the wheel. After a few passes the wheel should run true, the face should be straight, and the glazed condition and dull grains should be eliminated. The correct conditions produce an efficient cutting action.

Caution: There are a great number of abrasive and metal particles that are released at high speeds during the dressing process. Care must be taken to avoid inhaling these particles.

STEP 8 Stop the machine. Reset the work rest to within no more than 1/8″ of the wheel face. Reset the spark arrester to within 1/16″.

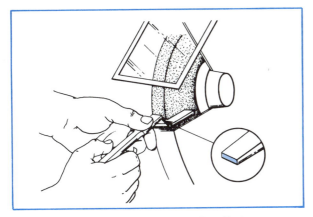

Figure 13–7 Grinding a Flat End

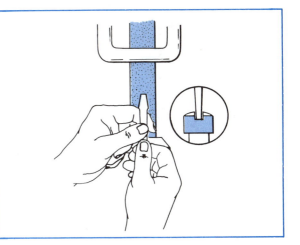

Figure 13–8 Grinding a Flat Side

How to Grind a Flat Surface

Grinding a Point

STEP 1 Check the grinding wheel for cutting ability and trueness. See that the machine guards, safety shields, and the work rest are in position and are secure.

STEP 2 Start the grinder. Hold the flat face of the tool blade down against the work rest with one hand. The tool is held at right angles to the wheel face. With the other hand ease the edge of the tool into contact with the grinding wheel face.

Caution: The tendency with narrow width tools is to jam them into the wheel face. Jamming may cause unsafe conditions when abrasive grains are dug out or the edge of the wheel is chipped.

STEP 3 Move the tool edge across the whole face of the wheel (Figure 13–7). Use one hand to keep a steady downward force. With the other hand move the tool edge across the wheel face. Continue the grinding process until the tool face is ground square. First rough grind with the coarse wheel. Then finish grind.

Note: A hardened tip becomes heated. It must be frequently dipped in water to prevent drawing the temper and softening the part.

Grinding Sides and Flat Faces

STEP 1 Repeat steps 2 and 3 to grind the sides of the tool (or part) flat.

STEP 2 Hold one face of the tool against the wheel at an angle (tangentially). Move it across the face of the wheel (Figure 13–8). The ground surface should be parallel with the tool axis.

STEP 3 Turn the tool 180° (Figure 13–8 enlargement) and repeat step 2 on the second face. Continue to grind. Check the two faces for parallelism and thickness.

STEP 4 Stone the edges to remove any burrs (Figure 13–9). Clean the machine.

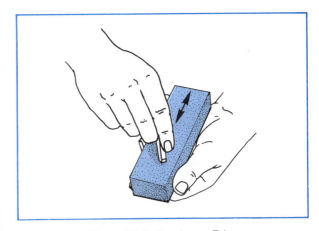

Figure 13–9 Stoning an Edge

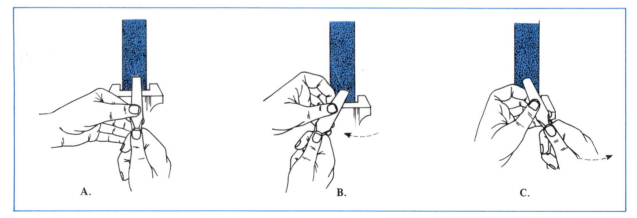

Figure 13–10 Grinding an Angular Convex Cutting Edge

How to Grind an Angular Surface and a Round Point

Grinding an Angular Surface

STEP 1 Set the work rest at or near the required angle. Move the work rest to within 1/8″ of the grinding wheel.

STEP 2 Hold the tool or workpiece firmly on the work rest with one hand.

STEP 3 Bring the tool into contact with the revolving grinding wheel face and guide the tool across the entire wheel face with the other hand.

STEP 4 Continue grinding until the required width of the angular surface is reached.

Note: A slightly different procedure may be followed in grinding the convex angular face on a cold chisel. An angular and convex surface is produced by holding the flat surface of a chisel against the work rest at an angle (Figure 13–10A) while the chisel point is swung through a slight arc (Figures 13–10B and C). This cutting procedure produces the convex cutting face.

Caution: Quench the tool point frequently in water so that the tool retains its hardness. Nonhardened workpieces should also be quenched in water to keep the temperature within the range in which the part may be handled safely.

Grinding a Point

STEP 1 With one hand hold the tool against the work rest at the angle to which the point is to be ground.

Note: For a center punch the included angle is 90°. A prick punch is ground at 30°.

STEP 2 Bring the tool point carefully against the revolving grinding wheel and simultaneously turn the tool with the fingers of the other hand.

Note: While the tool or part is being rotated and ground, it is also moved across the face of the grinding wheel.

STEP 3 Check the ground point angle against a gage or protractor. Change the angle position if necessary. Then continue to grind to the required angle.

How to Grind a Mushroomed Edge

STEP 1 Steady the tool or part against the work rest. Hold it at an angle with one hand.

STEP 2 With the other hand bring the mushroomed edges into contact with the grinding wheel.

STEP 3 Rotate the part continuously as it is being ground. Continue to grind around the edges of the tool. Grind at a slight angle until all mushroomed edges are ground away.

Safe Practices in Hand Grinding

- Secure the collars against the sides of a grinding wheel by applying a slight tightening force to the spindle nuts. Apply only enough force to prevent the wheels from turning on the spindle during the grinding process. Excessive force can strain and fracture the wheels and thus create an unsafe condition.
- Store grinding wheels carefully in a special rack. Damage may result if the wheel is laid flat or placed where it can be hit.
- Dress glazed and dull grinding wheels that are inefficient and produce surface imperfections in the workpiece. After a grinding wheel is trued or dressed, the tool rest must be readjusted to within no more than 1/8″ of the wheel face.

- Provide a small space between the tool rest, the dresser, and the grinding wheel. The tool rest supports and guides the dresser and permits it to be positioned at a cutting angle.
- Keep machine guards and shields in position to provide good visibility. Put on safety glasses before the grinder is started. Keep them on throughout the grinding process.
- Maintain a grinding wheel speed below the maximum revolutions per minute specified by the manufacturer. The manufacturer's recommended speed is marked on the grinding wheel.
- Perform grinding operations from a position that is out of direct line with the revolving wheel.

TERMS USED IN BENCH AND FLOOR (PEDESTAL) GRINDING

Offhand grinding	The shaping and reducing to size of a workpiece or tool on a bench or floor grinder. Positioning, holding, and guiding a workpiece or tool against a revolving grinding wheel. Producing a cutting edge, a correct working surface, or forming a hardened or other part to a specific size and shape by hand grinding.
Stoning	A hand finishing process for removing burrs from a hardened metal surface. Finish grinding a fine cutting edge. Forcing a metal part against an abrasive sharpening stone to remove a limited quantity of metal.
Bench and floor (pedestal) grinder	A simple, basic grinding machine mounted on a bench or on a pedestal for floor grinding. A flanged, threaded spindle with separate collars against which grinding wheels are secured.
Bench and floor (pedestal) grinding	The process of cutting away material with grinding wheels. The grinding of cutting, fastening, forming, and other small tools (such as tool bits, drills, chisels, and punches). Sharpening or forming a part to a desired shape or size. Grinding by hand with the aid of a work rest on a bench or floor grinder.
Trueness	The concentricity of the outer surface (periphery) of a grinding wheel in relation to its axis.
Truing	The process of cutting away the outer surface (periphery) of a grinding wheel. Producing a grinding wheel that runs concentrically and true.
Loaded wheel	A condition in which a great number of particles of the material being ground become lodged in the grinding wheel. A grinding wheel condition that (1) limits the quality of the grinding process, (2) produces additional heat, and (3) reduces the cutting efficiency.
Mechanical dresser (star dresser)	A series of hardened star- or corrugated-shaped discs with a circular separator between each pair. Discs that rotate freely in a holder when brought into contact with a grinding wheel face. A common hand dressing tool used to true and recondition a wheel.

SUMMARY

- Hardened cutting, forming, fastening, layout, and other tools and parts require grinding. Grinding reduces a tool to a correct form, size, and efficient cutting or operating condition.
  - Bench and floor grinders are made to accommodate a wide variety of grinding wheel sizes. Grinding processes range from finish grinding small surfaces to rough snagging on welded, forged, cast, and other manufactured parts.

- Bench and floor (pedestal) grinding is a cutting process. The cutting face of a grinding tool consists of a series of sharp-cutting abrasive crystals (grains). Each grain removes chips of material.
  - Workpieces and tools are ground by holding the part against a work rest. The surface may be guided straight across the face of the grinding wheel. The part may also be held at an angle or turned. A desired angle, point, or flat or convex surface is produced by manipulating the tool or part by hand.

- Bench and floor (pedestal) grinders usually have one coarse and one fine wheel. The coarse wheel is used to remove metal as quickly as possible for rough grinding. The fine wheel produces a fine-grained surface finish and a good-quality cutting edge.
  - Grinder spindles are threaded right- and left-hand. These threads prevent the clamping nut and grinding wheel from loosening during the grinding process.

- Grinding wheels must be true and in balance to grind properly and produce quality finished surfaces.
  - Care must be taken in mounting and securing a grinding wheel. The bore should be the exact size of the spindle. Blotters should be placed between the faces of the flanges and the wheels. The clamping nut should be tightened carefully.

- Wheel dressers of the abrasive stick or mechanical type are used to dress loaded wheels. Dressers are used to true the face and produce new, sharp cutting faces on the abrasive grains.
  - Chatter marks and vibration are signs that a wheel must be trued. Excessive heating, burn marks, and load lines indicate that a wheel must be dressed.

- Flying abrasive and metal particles and the possible fracturing of an abrasive wheel require that grinders be adequately guarded. Safety shields must be in place. Protective goggles should be worn at all times.

UNIT 13 REVIEW AND SELF-TEST

1. Compare the stoning of burrs with the removal of burrs by filing.
2. Describe the mounting of grinding wheels on bench and floor grinders.
3. State how a heavy-duty grinding wheel may be hand dressed.
4. Indicate three unsafe conditions that require the worker to dress a grinding wheel.
5. List the main steps for offhand grinding the flat face on a tool.
6. Give three safe practices to follow before starting a bench or floor grinder.

PART 5 Drilling Machines: Technology and Processes

Drilling Machines and Accessories

This section identifies the design features and functioning of conventional hole-making machines and processes. These include *drilling machines* and such *conventional cutting tools* as: drills, reamers, taps, countersinks, and counterbores. Workholding devices and accessories generally used in hole-making processes are also described. High production, numerically controlled, and unconventional hole producing machines and tooling (including electrasonic, laser beam, electrochemical, and other processes) are covered in the *Advanced Machine Technology* textbook.

UNIT 14

Drilling Machines and Processes

OBJECTIVES

After satisfactorily completing this unit, you will be able to:

- Describe the functions of major components of light model drilling machines for hole-producing operations.
- Identify applications of multiple-spindle, heavy-duty and special hole-forming machines.
- Apply technical information about the main features and operations that are basic to all drilling machines.
- Determine which workholding and positioning device to use in drilling.
- Follow recommended *Safe Practices* in drilling machine operation and maintenance.
- Apply drilling machine/accessory *Terms*.

Holes are an essential feature in design and manufacture. Holes are produced in simple parts as well as in complex machines. *Hole making* relates to machining processes that:

— Produce a hole of any size or shape where no hole previously existed;
— Alter the size of a hole, as is done in reaming or boring;
— Enlarge part of a regular hole, as is done in countersinking (at an angle) or counterboring (to produce a recessed surface), or an irregular-shaped hole;
— Tap and form internal threads;
— Grind and produce a hole to precise dimensions.

These hole-making processes are general processes performed on both nonproduction and production drilling machines. The work must be positioned and held securely for all processes. Various types of cutting tools and drilling machines are required.

CONVENTIONAL DRILLING MACHINES

Hole making refers to many different processes. Some of these processes are drilling, reaming, counterboring, countersinking, spotfacing, and tapping (Figure 14–1). These processes are performed on a group of machines that are identified as *drilling machines.* Some drilling machines are light and simple. Operations are controlled by the operator, who feeds a cutting tool into the work. Simple drilling machines are usually of the bench or floor type. Other drilling machines are large, for heavy-duty operations. In mass manufacturing there are multiple-spindle drilling machines that are used for the rapid production of many holes.

Some work may be brought conveniently to a machine and positioned for drilling and other operations. Large, heavy, cumbersome parts that are cast, forged, or produced by other methods are secured on a machine table. Hole-making operations are then performed by positioning the movable tool head and the cutting

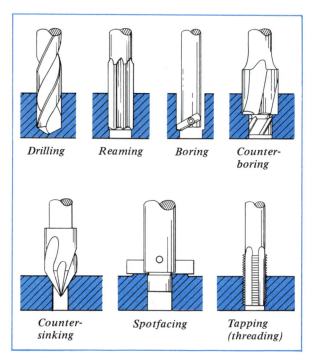

Figure 14–1 Basic Drilling Machine Processes

tool at a desired location. The radial drill press is used for such operations.

Most holes are machined straight, at a right angle to a surface. Some holes are drilled at various angles. Other holes require deep-hole drilling, in which the workpiece is rotated instead of the drill. Drilling machine tables may be stationary or adjustable for angle operations. Whether holes are machined straight, angled, or deep, conventional hole-making machines are used.

Six different groups of conventional machines are described in this unit:

• Sensitive and heavy-duty drilling machines (bench and floor types);

• Multiple-spindle types, for production;

• The deep-hole drilling machine;

• The radial drill press, for operations requiring a long overhanging arm and a machine head that may be swung through a large radius;

• The multi-operation turret drilling machine;

• A numerically controlled unit, the operations of which are programmed.

SENSITIVE DRILL PRESS
(BENCH AND FLOOR TYPES)

Of all conventional drilling machines, the simplest drilling machine design and the easiest machine to operate is the *sensitive drill press* (Figure 14–2). When the length of the drill press column permits the machine to be mounted on a bench, it is referred to as a sensitive *bench* drill press. The same machine is also produced in a floor model. This model is referred to either as a sensitive *pedestal* or sensitive *floor* drill press.

The machine size indicates the maximum work diameter that can be accommodated. For instance, a 24″ drill press will accommodate a workpiece with up to a 12″ radius (24″ diameter). The speed range of the sensitive drill press permits drilling holes from 1/2″ down through the smaller fractional, number, letter, and metric drill sizes. The smaller the drill size, the higher the speed required. Bench and floor drill presses are widely used for drilling, reaming, boring, countersinking, counterboring, spotfacing, and other operations.

MAJOR COMPONENTS OF THE
SENSITIVE DRILL PRESS

Regardless of the manufacturer, there are four major components of the sensitive drill press: base, work table, head, and drive mechanism.

Machine Base. The base of most drilling machines has a stationary upright column. The axis of the column is at a right angle to the base. The upright column and base are rigid. The workpiece may rest on the machined surface of the base. Long parts may be accurately positioned and held on the machined base.

Work Table. Mounted on the upright column is a work table. The work table may be round, square, or rectangular. The table may be moved vertically and then locked at a required height to accommodate different work lengths. The work table may be held square in relation to the spindle, or it may be adjustable for angle work. Tables have either elongated slots or grooved slots. Grooved slots provide a surface against which T-head bolts or nuts may be tightened. The slots permit locating the work in relation to the spindle.

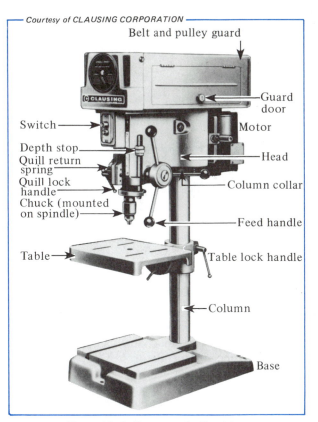

Belt and pulley guard

Switch

Depth stop

Quill return spring

Quill lock handle

Chuck (mounted on spindle)

Table

Guard door

Motor

Head

Column collar

Feed handle

Table lock handle

Column

Base

Figure 14–2 Features of a Sensitive Drill Press (Bench Model)

Drill Press Head. The head includes a sleeve through which a spindle moves vertically. Vertical movement of the spindle is controlled by a feed lever that is hand operated. Where there is a power feed, the spindle may be moved up and down both by hand and by power.

The end of the spindle has a taper bore. Some drilling attachments, chucks, and cutting tools have a corresponding taper shank and tang. There are adapters to fit the taper bore. The adapters accommodate tool shanks or other parts that have a taper smaller or larger than the taper bore.

Drill chucks are a practical and widely used device for holding straight-shank tools and pieces of round stock. The chuck jaws are adjustable over a wide range of sizes. For example, one drill chuck may hold all diameters from a #60 drill to 3/8″ (or 10mm).

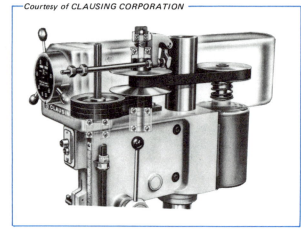

Figure 14–3 A Quick-Setting Variable Speed Drive

Drive Mechanism. The spindle may be turned at different speeds (revolutions per minute) by a V-belt drive mechanism. The required operating speed depends on the material, nature and size of the operation, type of cutting tool, and desired surface finish. Speed adjustments are made by changing the step-cone pulley over which the V-belt rides. The range of bench and floor drill press speeds is from 30 to 5,000 revolutions per minute.

THE STANDARD UPRIGHT DRILL PRESS

The *standard upright drill press* is of heavier construction than the sensitive drill press. In the shop the standard upright drill press (drilling machine) is usually called just a *drill press*. The column may be round, cast, or fabricated in a rectangular form. Speed variations on the lighter upright drill presses are obtained by a variable speed drive like the speed drive illustrated in Figure 14–3. The operator sets the variable speed control at the desired speed.

The floor-type standard drill press has both a manual (hand) feed and an automatic mechanical feed. A variable feed range from 0.002″ to 0.025″ is standard. The feed control lever is set at the required feed per revolution. The cutting tool automatically advances the feed distance at each complete turn.

HEAVY-DUTY, GEARED-HEAD DRILL PRESS

The *heavy-duty upright drilling machine* uses a gear drive system. The selected speed is obtained by shifting gear selector handles while the machine is stopped. Different combinations of gears are thus engaged. The range of speeds is increased further by using a two-speed motor. This motor provides speed versatility for a low-speed range and a high-speed range.

The cutaway section of the heavy-duty, geared drill press head (Figure 14–6) shows the following design features.

- The *gear train* from the two-speed motor to the *spindle*.
- The rack and pinion for raising and lowering the drill press *head* on the *machine column*.
- The rack and pinion for moving and feeding the *quill assembly* (that houses the machine spindle).
- *Adjusting levers* for spindle speeds.

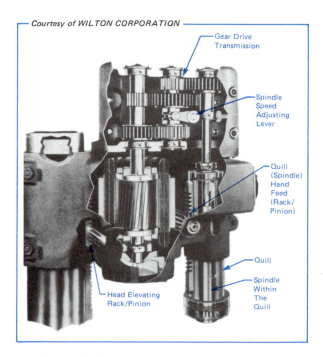

Gear Drive Transmission

Spindle Speed Adjusting Lever

Quill (Spindle) Hand Feed (Rack/Pinion)

Quill

Spindle Within The Quill

Head Elevating Rack/Pinion

Figure 14–4 Cutaway Section of a Geared Head Drill Press Transmission System

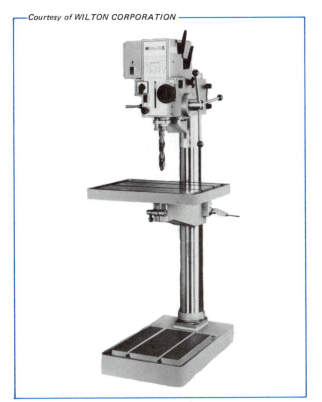

Figure 14–5 Heavy-Duty, Geared-Head,
Power Feed Drill Press

The drill press displayed in Figure 14–5 is available for manual feed or power feed. There are four power feeds on this machine: .003", .006", .008", and .012" (.08 to .30mm) per revolution. Once the drilling depth is preset, the power feed is engaged by turning the handwheel (or, on other machine designs, by engaging power feed levers). The power feed automatically disengages when the required depth is reached. The quill, spindle, and cutting tools are then returned to the starting position.

The drill press, as illustrated, has eight spindle speeds. The four spindle speeds in the *low range* are: 85, 335, 410, and 1620 revolutions per minute. The four spindle speeds in the *high range* are: 170, 670, 820, and 3240 RPM. Drill presses are also available with 12 spindle speeds.

The *movable* table (that may be rotated around the column and adjusted for height) and the *base table* are precision machined. The tables are designed with T-slots to permit flexibility in positioning and securing workpieces and fixtures.

The spindle is designed with a standard *Morse #3 taper*.

THE GANG DRILL

The *gang drill* is a type of multiple-spindle drilling machine. A number (gang) of drilling heads are lined up in a row. They may be driven individually or in the gang.

The number of spindles is determined by the number of operations to be performed. Holes in a workpiece may be drilled, reamed, bored, tapered, and so on. Each hole may be of a different diameter.

Each cutting tool is usually positioned in relation to a specific dimension by using drill jigs or special work-holding fixtures. On especially large drilling machines, the table on which the work is held may be positioned by hydraulic feed mechanisms that are numerically controlled.

THE TURRET DRILLING MACHINE

The *turret drilling machine* has a number of drill head units. The turret of the drilling machine may have six, eight, or ten separate heads (stations). A different cutting tool or size may be inserted in each head. For example, the hexagonal turret may be set up with a combination of center drill, pilot drill, large-sized drill, spotfacer, counterbore, and reamer. This combination represents the sequence in which the hole is started with the center drill, drilled with the pilot drill at a high speed, followed at slower speeds with the reamer drill, spotfaced, counterbored to a particular depth, and finally reamed.

The turret drilling machine eliminates tool changing time. Also, a single machine may be used for all processes. Small-sized machines may be hand fed. Some large machines, on which operations like face milling are performed, may also be manually operated. Other turret drilling machines may be operated by a numerical control unit.

THE MULTIPLE-SPINDLE DRILLING MACHINE

The *multiple-spindle*, or *multispindle drilling machine* is a mass-production hole-making machine (Figure 14–6). It consists of a number of

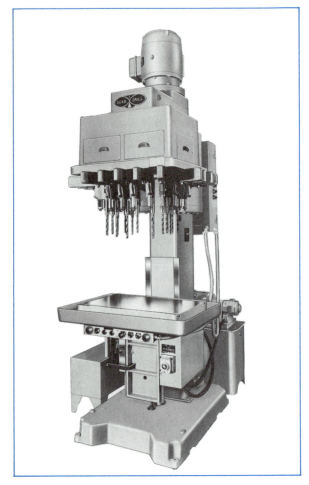

Figure 14–6 A Multiple-Spindle Drilling Machine

Courtesy of ZAGAR INCORPORATED

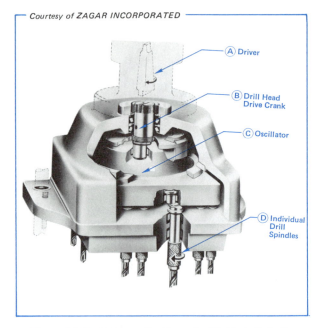

Ⓐ Driver

Ⓑ Drill Head Drive Crank

Ⓒ Oscillator

Ⓓ Individual Drill Spindles

Figure 14–7 Cutaway Section of a Multiple Spindle Gearless Drilling and Tapping Head

drilling heads that are positioned at desired locations. Each spindle has at least two universal joints. The drivers may be expanded or closed to bring a spindle to a new location.

It is important that a guide plate (mounted with the drill head above the work) or a drill jig be used in multiple-spindle drilling operations. Also, the drill lengths must be staggered to distribute the starting load.

MULTIPLE SPINDLE HEADS

Multiple-spindle drilling and tapping heads (that are geared head or gearless) may also be used for multiple hole-producing processes. The cutaway model (Figure 14–7) shows a gear-

less drilling and tapping unit that is driven by the machine spindle.

The *tang adapter* secured in the *machine spindle* Ⓐ drives *crank* Ⓑ in the *drill head*. The drive crank in turn moves the *oscillator* Ⓒ in a small oscillating motion. This motion drives (turns) the *individual drill spindles* Ⓓ at the same speed and in the same direction as the drive crank B. A *jig plate* is used with a multiple spindle head to accurately position each cutting tool. Workpieces are usually secured in a part fixture mounted on the machine table.

THE WAY-DRILLING MACHINE

The *way-drilling machine* is an adaptation of the multiple-spindle drilling machine. The difference is that the spindles of the way-drilling machine operate from more than one direction. The production of automotive engine blocks is an application of way-drilling. A series of holes is produced simultaneously on several sides of these blocks.

THE DEEP-HOLE DRILLING MACHINE

One problem in drilling a deep hole is producing a hole that continues to follow a required

Courtesy of DoALL COMPANY

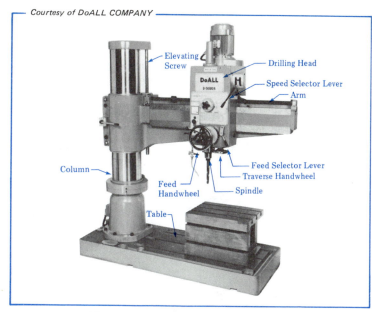

Figure 14–8 Features of a Radial Drill

axis. The longer the hole, the greater the tendency for the drill to be deflected. A true, straight hole may be drilled by holding the tool stationary and rotating the work. The cutting tool then follows the axis of rotation. A straighter and more precisely drilled hole is produced.

In deep-hole drilling, the drill is supported and guided by a guide bushing. It is always necessary to keep a coolant on the drill point. The chips must be removed continuously.

Special drills are available that permit a cutting fluid to flow through the drill body to the cutting edges. The fluid cools the cutting edges, drill, and workpiece while chips are forced out of the drilled hole.

Deep-hole gundrilling machines are designed for single-spindle, dual-spindle, and multiple-spindle operation, depending on hole-producing production requirements. Manual or automatic fixtures, indexing tables, and self-centering components to meet varying deep-hole drilling specifications are used with the appropriate spindle head module(s).

THE RADIAL DRILL

The *radial drill* derives its name from its design. A radial drill, with its major components identified, appears in Figure 14–8.

A heavy workpiece is positioned and clamped to the base. The tool head may be moved with a handwheel. It may also be motor driven rapidly close to the desired hole location. The cutting tool is usually brought by hand to the exact point where an operation is to be performed.

The spindle is moved to the work until the cutting tool is almost in contact with the work. The automatic feed is then engaged. Some machines have automatic trips to regulate the depth. A predetermined depth may be set. As the spindle and cutting tool reach the required depth, the feeding stops.

The radial drill may produce straight holes or angular holes. The holes may be located from a direct layout on the workpiece or with a template or a drill jig.

SPECIAL HOLE-PRODUCING MACHINES

The drilling head is versatile. A single operation like drilling may be performed. Other operations may be combined to include drilling, reaming, boring, and so on. One head may be used, or any number of heads may be combined. They may be positioned vertically, horizontally, or at a simple or compound angle.

A workpiece may be positioned and held in a drill fixture. A part may also be held in a drill jig with a drill bushing to guide the drill. The table may be round, rectangular, or rotary. The combination of machine and work-holding devices and cutting tools depends on the special quantity production requirements of the job. *Special hole-making machines* incorporate many of these requirements and design features.

THE NUMERICALLY CONTROLLED DRILLING MACHINE

A turret drilling machine or any other production machine may be *tooled up* to perform a combination of hole-making and additional machining operations. These operations may be performed automatically according to a sequence of directions that are *programmed*. In *programming*, most of the mental and physical processes that the craftsperson follows in setting up and machining a part must be recorded. Information such as the type and sequence of each operation, the dimensional position of each operation, the tool number, the depth of cut, the cutting speeds, the rate of feed, and the speed of the spindle must all be coded in a series of letters and numbers. The coded letters and numbers, (when transferred to a tape or disc) provide directions for the *numerical control system*.

The drilling machine or other machine tool is controlled by the numerical control unit. The workpiece and cutting tool are positioned in proper sequence. The operations are performed to the required depth and required degree of dimensional accuracy. The table, spindle, and other movements and the various feeds and speeds are all numerically controlled. The system uses electronic, hydraulic, pneumatic, and other automatic mechanical devices. Each device is activated by the numerical control unit.

A craftsperson must initially set up the tools and the work-holding fixtures and devices. During the processes the cutting qualities of each tool, the surface finish, and the accuracy of each operation must be observed by the operator. The setup person, programmer, and operator must know the technology and drilling processes that are essential to program effectively and to actually operate a numerically controlled machine tool.

WORK-HOLDING DEVICES

The accessory in which the work is positioned and held may be a *work-holding*, or *clamping*, *device*. Some of the common methods of positioning and holding a workpiece include:

- Clamping directly in a vise.
- Mounting on parallels or V-blocks. The part is clamped in position with hold downs, straps, clamps, and T-bolts.
- Nesting and holding in a specially designed form. The drill or cutting tool may be guided for location by a drill bushing. The drill bushing also helps to produce a concentric hole. Another holding device is called a drill jig or fixture.

WORK-HOLDING VISES

There are three basic designs of vises and work holders that are generally used for drill press work. These are identified as: the *multipurpose* (or *standard*) *vise*, the *utility angle base vise*, and the *column-mounted safety work holder*.

Workpieces of regular shape (square, rectangular, and round) may be held in a *standard drill press vise* (Figure 14–9). After the hole location is layed out, the work is placed on parallels. The work then is securely tightened in the vise and positioned on the drill press table for the hole-making operation. The drill press vise should be clamped to the work table for heavy drilling operations.

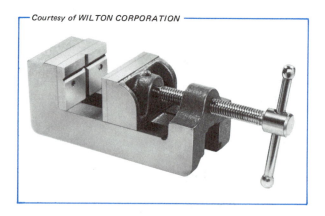

Courtesy of WILTON CORPORATION

Figure 14–9 A Multipurpose Drill Press Vise

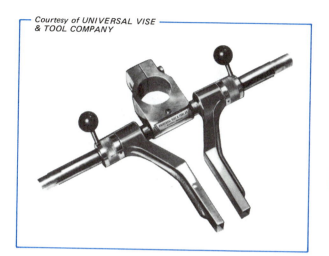

Figure 14-10 A Safety Work Holder for Column-Mounting on a Drill Press

Figure 14-11 Precision Swivel Base Vise

An *angle vise* is used for some angular machine operations. This vise holds the work at the required angle and eliminates the need for tilting the work table. The vise on this holding device is mounted on an angle-graduated base. Adjustments are easily made to position the vise and workpiece at any angle from 0° to 90° vertical, in relation to the spindle.

The *safety work holder* (Figure 14-10) is another holding device. It fits around the machine column and has two easily adjusted arms. These arms hold the work in position on the work table.

Two additional basic types of vises are used for machining operations on machine tools other than drill presses. These vises are called the *swivel base vise* and the *universal* (three-way) *compound angle vise*.

The *swivel base vise* is used to position and to secure workpieces at a required angle in a horizontal plane. The swivel base vise shown in Figure 14-11 incorporates the design features of a precision standard machine vise mounted on a base that is graduated in degrees.

Surfaces and holes that require machining at compound angles are positioned and held in a *universal* (three-way) *vise*. This vise permits three separate angular adjustments: 360° in a horizontal plane, within 90° from a horizontal to a vertical plane, and through 90° in the third plane.

V-BLOCKS

V-blocks are widely used to position and hold cylindrical parts for layout, machining, and assembly operations. A V-block is a rectangular metal solid. It has V-shaped angular surfaces on at least two opposite sides.

Round workpieces of different diameters may be centered and nested in V-blocks. A U-shaped clamp is provided to hold the workpiece in position.

When a hole is to be drilled through the axis of a cylindrical workpiece, a center line is usually scribed on the end. The workpiece is mounted in V-blocks. The center line (axis) is then aligned with a steel rule or other square head. The square is held against the surface of the

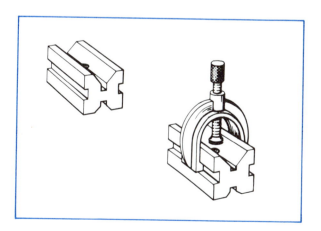

Figure 14-12 A V-Block Set and a U-Clamp

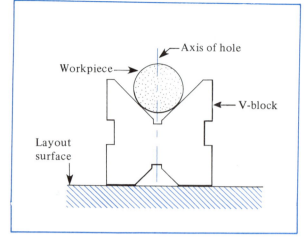

Figure 14-13 Aligning the Center Line
of a Hole to Be Drilled

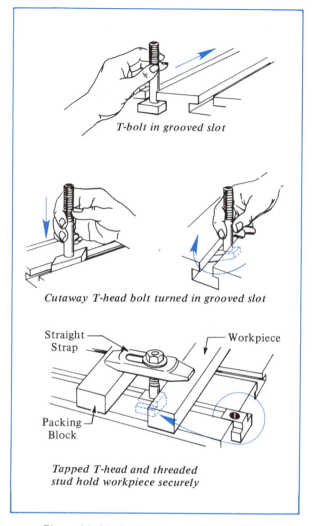

T-bolt in grooved slot

Cutaway T-head bolt turned in grooved slot

*Tapped T-head and threaded
stud hold workpiece securely*

Figure 14-14 Applications of T-Head Bolts,
Straight Strap, and a Packing Block

machine table or other layout plate. The alignment for the workpiece is shown in Figure 14-13.

STRAP CLAMPS AND T-BOLTS

Irregular-shaped workpieces are sometimes held directly on machine tables with a number of differently shaped *strap clamps*. (The terms *strap*, *clamp*, and *strap clamp* are used interchangeably.) Each clamp has an elongated slot. A *T-bolt* fits in the slot or in a machined groove in a table and passes through the clamp. The opposite end of the clamp is raised by *blocking*. It is then positioned to apply a maximum force on the workpiece.

The flat surface of the clamp should be parallel to the table to permit the force applied with the T-bolt to hold the workpiece securely and in the position required for the machining operation. Examples of square head, cutaway head, and tapped T-head bolts are shown in Figure 14-14.

The judgment of the craftsperson is important. The force exerted in the clamping process should not cause conditions in which the part is bent or fractured. Applying excess force causes such conditions.

Blocking to raise the clamp to a correct height may be done in several ways. A step block with a number of steps of different heights may be used. Sometimes sets of blocks are used. These blocks may be added together to reach a particular height. Use of an adjustable jack permits even greater variation in height and makes adjustment easier.

The four common shapes of strap clamps, each of which has a number of variations, are the *straight*, *finger*, *bent*, and *U*-strap clamps.

Straight Strap Clamp. The straight strap clamp is a rectangular bar with an elongated slot running lengthwise. The two ends are flat. One end rests on the workpiece; the other, on the blocking. The necessary force is produced by tightening the T-bolt, washer, and nut assembly (Figure 14–14). Some straight strap clamps have an adjusting screw on one end.

Other examples of clamps include the *finger strap*, *bent-tail strap (gooseneck clamp)*, and *U-strap* with a finger.

THE ANGLE PLATE

The *angle plate* that is used on a drilling machine usually has a number of holes or elongated slots. These holes or slots permit strapping the angle plate to the table and securing a workpiece to the right-angle face. Figure 14–15 shows a typical angle plate setup.

DRILL JIGS AND FIXTURES

Mentioned earlier as devices for positioning and holding workpieces, drill jigs and fixtures are widely used with workpieces of irregular shape when a quantity must be machined accurately.

A *drill fixture* is a specially designed holding device. The fixture contains a nesting area into which the workpiece fits. This area positions and holds the workpiece. The relationship is fixed between the workpiece, the spindle of the machine, and the cutting tool.

A *drill jig* is an extension of the drill fixture. It serves the same functions and, in addition, has guide bushings. These bushings guide

cutting tools such as drills and reamers to a precise dimensional location. The guide bushings help to produce accurate concentric holes. Figure 14–16 shows a simple drill jig for accurately drilling a hole in a cylindrical part.

MACHINE ACCESSORIES

The term *accessories* refers to the tools and the mechanical parts and machines that are used in conjunction with a machine tool. These accessories aid in setting up and holding a workpiece so that a particular process may be performed. Five accessories that are widely used in hole-producing operations are: *parallels*, *jack screws*, *step blocks* and *packing blocks*, *shims*, and *universal swivel blocks*. Clamps (straps), bolts, and V-blocks, which are also accessories, were described earlier.

JACK SCREW

The jack screw, or jack, is a simple, easily adjusted support device. The jack is designed to provide for a wide range of adjustment. With a jack it is possible to make up whatever length is required between blocks, a machine table, and a work surface. All jacks consist of a base, an elevating screw, and a swivel head. Jacks are placed under a workpiece and brought into contact with the surface that is to be supported. A jack may also be used under a clamp to replace a number of blocks.

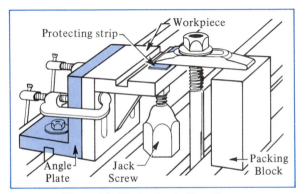

Figure 14–15 An Angle Plate Setup

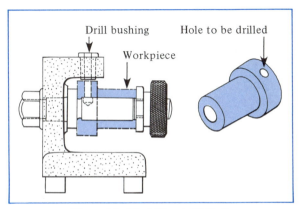

Figure 14–16 A Simple Drill Jig

STEP BLOCKS AND PACKING BLOCKS

The step block derives its name from the fact that each block consists of a series of steps. These steps permit the end of a clamp to be located at whatever height (step) is needed for a job. Each step also provides a surface against which a clamping force may be applied.

Packing blocks are individual blocks of different sizes. The blocks may be used alone or in any combination. When a desired height is reached, a force may be applied correctly on a workpiece.

SHIMS AND PROTECTING STRIPS

Shims are thin metal strips. They are used to compensate for surface variations in measurement. Since it is necessary to have a good bearing surface so that a workpiece will not move during a cutting process, shims may be inserted between the workpiece and a parallel. Shims level an uneven surface.

Protecting strips are thin metal pieces. These strips are placed under any metal part, such as a clamp. Protecting strips are used to prevent damage to a finished work surface or machine table when a force is applied in a work-holding setup.

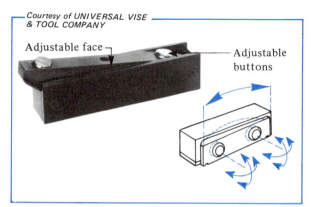

Courtesy of UNIVERSAL VISE & TOOL COMPANY

Adjustable face
Adjustable buttons

Figure 14–17 Universal Movement of the Adjustable Face and Buttons of a Universal Swivel Block

UNIVERSAL SWIVEL BLOCK

Many surfaces that are tapered or irregularly shaped must be held securely. A universal swivel block is used with a vise for such purposes (Figure 14–17). The universal swivel block consists of two major faces. One face on the body is flat. This face rests against the stationary jaw of the vise. The other face has two buttons. This face is adjustable. It may be positioned against an irregular or tapered surface. The universal motion of the swivel block permits it to accommodate the shape of a workpiece.

Safe Practices in Drilling Machine Operation and Maintenance

- Check the table surface, work-holding device, and the workpiece. Each must be free of burrs, chips and foreign particles. Cleaning should be done with a brush or a wiping cloth with the machine turned off.
- Fasten each workpiece securely to withstand all cutting tool forces, to ensure personal safety, and to avoid damage to the machine, work, cutting tool, and accessories.
- Stop the machine when a feed change requires the shifting of a gear lever.
- Check to see that covers and guards are in place and secured. All extra tools and parts should be removed before starting the machine.
- Check personal clothing. Loose clothing may be caught by a revolving spindle, chuck, or cutting tool.
- Use eye protection goggles or a protective shield.
- Use extreme caution if an air blast is used for cleaning the work, machine, or tools. The blast (1) must be downward and away from any other operators and (2) the work and the area around it must be covered to restrict chips and other particles from flying.

DRILLING MACHINE TERMS

Hole making	A series of machine processes for producing, altering, and enlarging a hole in part or total. Producing regular through and blind holes, threaded holes, and irregularly shaped holes.
Conventional drilling machine	A classification given to upright bench and floor drill presses. Drill presses of the sensitive and heavy-duty types; multiple-spindle, radial, and turret drilling machines; and adaptations of these machines.
Multiple-spindle drilling machine	A drilling machine with many spindles that may be positioned to produce a number of holes at one time. A mass-production hole-producing machine.
Deep-hole drilling	A process of rotating the workpiece and feeding a guided stationary drill.
Radial drill	A machine tool for producing holes. A movable tool head extending radially from a machine column.
Turret drilling machine	A series of cutting tool stations on a drilling machine. A six-, eight-, or ten-sided turret that has a cutting tool positioned in each turret station.
Numerically controlled machine tool	A system of machining in which each operation is programmed and controlled by using a system of letters and numerals to indicate the nature, sequence, and timing of machining operations.
Nontraditional machine drilling	The making of irregular- and intricate-shaped holes by using unconventional tools. (Properties of sound, light, heat, electricity, chemical reactions, and velocity are used to remove material.)
Work-holding device	A vise, clamp, fixture, or drill jig that is used to position and hold a workpiece.
Accessories	Tools, parts, and mechanical devices used to position, hold, or safely perform certain operations or to extend the range of operations.

SUMMARY

- Standard drills, reamers, boring tools, spotfacers, counterbores, countersinks, machine taps, and other cutting tools are used in conventional hole-producing methods.
    - Drilling machines may be grouped according to size, number of spindles, and other design features. The groups include upright bench and pedestal drill presses of the sensitive and heavy-duty types and multiple-spindle, deep-hole, radial, and turret drilling machines.
- The nature, sequence, and processes performed on drilling machines may be programmed and numerically controlled.
    - Heavy-duty drilling machines have variable speed controls and a great number of feeds and are equipped for both manual and automatic feeds. Coolant systems are required to lubricate and dissipate the heat developed during cutting processes.
- The gang drill, multiple-spindle, and way-drilling machines are production types of drilling machines.

- When the depth of a hole exceeds five or six times the drill diameter, a deep-hole drilling technique is used.
- Heavy workpieces that cannot be positioned and secured on a standard drilling machine may be secured on the work table of a radial drill.
  - A series of machining processes may be performed using a turret drilling head. Each cutting tool is positioned in sequence.
- Nontraditional drilling machines require specially designed tools. The cutting or disintegrating action is produced through: electrical discharge machining (EDM), electrochemical machining (ECM), ultrasonic machining (USM), or laser beam machining (LBM), plasma arc, and other machine tools.
  - Standard, angle, swivel base, and universal drill press vises hold and position workpieces. These vises are used for straight, single-angle, and compound-angle machining operations.
- The four most widely used strap clamp shapes are the straight, finger, bent-tail, (gooseneck and arc styles), and the U-clamp. T-bolts are used as fasteners for holding parts to the machine table.
  - Parallels and V-blocks provide precision surfaces. Workpieces may be positioned and held on these surfaces.
- Jacks, step blocks, regular blocks, and universal swivel blocks are common accessories. They are used for positioning and securing workpieces.
  - Safety precautions must be observed in tool and machine setting, cutting, and cleaning and burring of the workpiece.

UNIT 14 REVIEW AND SELF-TEST

1. State three main differences between the operation of (a) a sensitive bench drill press and (b) a multiple-spindle drilling machine.

2. Distinguish between gang drilling and turret drilling machines.

3. Indicate the kinds of specifications the craftsperson must code in a series of letters and numbers to transfer to a tape for a numerically controlled drilling machine.

4. Identify (a) three unconventional drilling machines, (b) the form of energy used by each machine, and (c) the type of hole-producing tool or process required for each machine.

5. Give three examples of the unique hole-producing functions of unconventional drilling machines.

6. Cite the advantages of using a safety work-holder attachment on a drill press to position and hold a workpiece.

7. Describe briefly the function served by each of the following work-holding devices or accessories as related to drilling machines: (a) step blocks, (b) adjustable jack, (c) gooseneck clamp, (d) drill fixture, and (e) universal swivel blocks.

8. List three safety precautions to follow in setting up for drilling to avoid personal injury.

SECTION TWO

Cutting Tools and Drilling Machine Operations

This section presents the technologies involved in and procedures for hole-making processes performed on drilling machines. These processes include drilling, reaming, threading, countersinking, counterboring, spotfacing, and boring. Emphasis is placed on drill point geometry for general drilling processes. Features of the blunt 140° point angle, multi-facet point grinding and the latest drill point geometry for production and numerical control applications are introduced.

UNIT 15

Drilling Technology and Processes

OBJECTIVES

After satisfactorily completing this unit, you will be able to:

- Understand the functions of drill gages, adapters, and other related drilling accessories and tools.
- Translate information about inch-standard and metric-standard drill size systems into the selection of correct size drills.
- Work through problems requiring the application of simple formulas relating to cutting speeds (CS) and feeds (SFPM).
- Use reference tables of CS and SFPM and others on drill specifications.
- Perform the following processes.
  - Grind a Drill.
  - Drill Holes in Flat and Round Surfaces.
  - Insert and Remove a Drill.
- Follow recommended *Safe Practices* in drilling.
- Communicate in shop language using drilling machine *Terms.*

The tool most commonly used with a drill press is the *twist drill*, usually called just a drill. In principle a drill has:

- An angular point where the cutting action takes place;

- A series of flutes cut along the body that permit forming of the cutting edges and removing of chips;

- A channel through which a cutting fluid may reach the drill point.

Drilling processes are performed on metals and other materials. These materials vary in hardness. Holes are drilled to different depths and degrees of accuracy. Some drills are started in center-punched locations. Other drills are positioned and guided by drill jigs or fixtures.

The actual process of drilling is largely influenced by a thorough understanding of twist drills. The operator must know about the materials used in drills, drill parts and their functions, grinding and checking of drill points, relationships of drill sizes in the inch and metric systems, and speeds and feeds. This unit deals with the background technology related to twist drills, drill grinding, and drilling processes.

DESIGN FEATURES OF TWIST DRILLS

MATERIALS

There are four common materials used in the manufacture of drills: *carbon steel*, *high-speed steel*, *cobalt high-speed steel*, and *tungsten carbide*. Although some drill design features are similar, each of these materials is adapted to special drilling requirements.

Carbon Steel Drills. Drills of carbon steel are used in comparatively slow-speed drilling operations. The limited amount of heat that is generated by the cutting action must be dissipated quickly, because the hardness and cutting ability of a carbon steel drill is lost once the drill point is overheated and the temper is drawn. *Temper* refers to the degree of hardness and the relieving of internal stresses within a tool. Overheating causes the drill to soften and lose its cutting edges. A carbon steel drill may be identified by white sparks in a spangled pattern that are evident during drill grinding.

High-Speed Steel Drills. High-speed drills have largely replaced carbon steel drills. As its name implies, a high-speed steel drill retains its cutting edges at high cutting speeds and at temperatures up to $1,000°F$. The cutting speeds are twice as fast as for carbon steel drills.

High-speed steel drills are used in production. Such drills are versatile cutting tools for ferrous and nonferrous metals and other hard

Courtesy of DoALL COMPANY

Figure 15–1 A Heavy-Duty, Cobalt High-Speed Steel Drill

materials. The shank end of a high-speed steel drill is stamped HS or HSS. The mark identifies the metal used in the drill. When a high-speed steel drill is ground, it gives off a dull red spark.

Cobalt High-Speed Steel Drills. The addition of cobalt to high-speed steel produces properties that are necessary for tough drilling operations. Cobalt high-speed drills are used for drilling tough castings and forgings, armor plate, work-hardened stainless steel, and other hard steels. Cobalt high-speed steel drills are run at high cutting speeds. A steady, uninterrupted feed is used. Such drills may be operated at higher temperatures than high-speed steel drills. Figure 15–1 illustrates a heavy-duty, cobalt high-speed steel drill.

Tungsten Carbide Drills. Carbide-tipped and solid carbide drills are used for drilling abrasive materials such as glass-bonded plastics; brass, aluminum, bronze, and other nonferrous metals; and hard-scaled castings.

A carbide-tipped drill is produced by altering the spiral flute of a high-speed steel body. A tungsten carbide cutting tip is inserted and the clearance angle is changed. A common two-flute, carbide-tipped drill is shown in Figure 15–2. Carbide-tipped drills must be sharpened on special grinding wheels. A silicon carbide or diamond-impregnated grinding wheel is used.

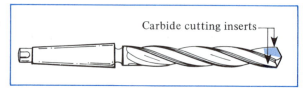

Carbide cutting inserts

Figure 15–2 A Two-Flute, Carbide-Tipped Drill

TYPES OF TWIST DRILLS

There are several types of twist drills. These types include *two-flute*, *three-* and *four-flute*, *sub-land*, and *coolant-feeding* drills.

Two-Flute Twist Drill. The most commonly used twist drill for all-purpose and production drilling is the two-flute, high-speed steel drill. As the name indicates, there are two flutes. These flutes may be either straight or spiral. The depth of the flutes varies. The depth depends on the degree of strength and rigidity required. In addition to the standard form, heavy-duty two-flute drills are manufactured.

Three- and Four-Flute Drills. Three- and four-flute drills are used where holes are to be enlarged and close tolerance control is necessary. These drills are referred to as *core drills*. The enlarging is done on holes that may be formed by casting, drilling, punching, and other production methods.

Sub-Land Drills. A sub-land, multidiameter drill is an adaptation of a two-flute drill. The cutting end is reduced in diameter for a distance along the body. The cutting edges between the first and second diameters may be ground to produce a square, round, or angular connecting area.

The advantage of using a sub-land drill is that it is possible to drill two or more concentric holes of different diameters in one operation. Only one drill and setup are required to drill two holes, drill and counterbore, or drill and recess holes.

Oil-Hole (Coolant-Feeding) Drills. Many machining processes require a coolant and a lubricant. The use of a cutting fluid increases cutting tool life, production, dimensional accuracy, and quality of surface finish. An adequate supply of cutting fluid may be provided from a coolant nozzle. The coolant is flushed into a drilled hole under a slight pressure.

However, for deep-hole and heavy-duty drilling, it may be necessary to use a *coolant-feeding drill*. A phantom view of this drill is shown in Figure 15–3. A coolant-feeding drill has a hole drilled in the core or through the

Holes for coolant

Figure 15–3 A Phantom View of a Coolant-Feeding Drill

clearance area. The hole extends to the point of the drill. The cutting fluid is forced under pressure through the drill. The fluid thus reaches the area where the cutting action takes place. The cutting fluid keeps the cutting lips flooded and cooled. The cutting fluid also helps to remove the drill chips from the workpiece.

FUNCTIONS OF TWIST DRILL PARTS

There are two main parts of a twist drill: the *shank* and the *body*. The shank provides a gripping surface. A force may be exerted through the shank to turn and feed the drill. The body is designed to provide efficient cutting action, chip removal, and heat control. Other parts of a twist drill are the *flutes*, the *web*, the *land*, and the *point*. The principal design features and part names of a twist drill are given in Figure 15–4.

The Shank. Twist drill shanks are either *straight* or *tapered*. The standard straight-shank drill usually fits into and is driven by a *drill chuck*. The straight shank is found mainly on letter size, number size, metric, and fractional size drills. These drills are 1/2″ (12mm) or smaller in diameter. The straight shank of a drill larger than 1/2″ in diameter is sometimes reduced in diameter to fit a regular drill chuck.

The end of a taper-shank drill is machined flat for a short distance to form a *tang*. The tapered area (taper shank) fits into the tapered drill press spindle. The taper shank permits the drill to be centrally aligned. The tang serves three functions:

- Together with the holding power of the taper shank, to transmit the spindle force to the drill;
- To prevent the drill from turning in the spindle;
- To provide a way of removing a drill.

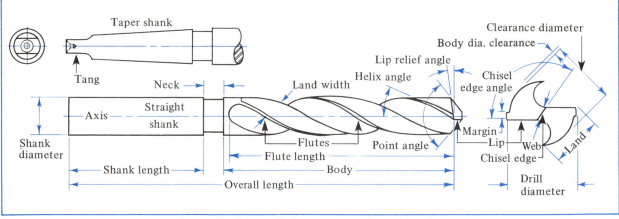

Figure 15–4 Main Design Features of a Twist Drill

The Body. The body extends from the drill point to the shank, including the flutes. *Flutes* are specially formed spiral grooves that are cut around the body. The flutes cause chips to curl and flow out of the work and provide a passage for a cutting fluid to reach the cutting edges.

The Point. The cutting action of a twist drill results from the design of the point. The point includes the cone-shaped area formed with the flutes (Figure 15–5A). The angle of the cone depends on the material to be drilled, the drill size, and the nature of the operation. Cutting edges, or *lips*, are formed by grinding a lip clearance on the angular surface of each flute

(Figure 15–5B). In other words, a lip is ground at a clearance angle. The area behind the cutting edge slopes away to produce a chisel-edged shape that provides the shearing, cutting action. Since the clearance angles are on opposite sides of a two-flute drill, a flat surface is formed across the drill axis (dead center). A chisel point angle of 120° to 135° is used for general drilling.

DRILL POINT SHAPES

Cutting conditions vary for different materials. To compensate for these conditions, it is necessary to change the shape of the cutting edge and the cutting angle. The *standard drill angle* for general-purpose drilling operations is

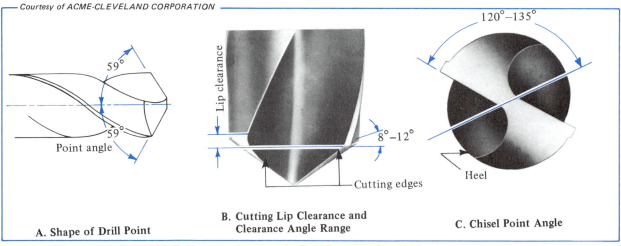

Figure 15–5 Shape and Angles of a Drill Point for General-Purpose Drilling

Table 15–1 Recommended Lip Clearance Angles
(Inch-Standard Drill Ranges)

Drill Sizes	Drill Diameters	Lip Clearance Angle
#80 to #61	0.0135 to 0.0390	24°
#60 to #41	0.0400 to 0.0960	21°
#40 to #31	0.0980 to 0.1200	18°
1/8 to 1/4	0.1250 to 0.2500	16°
F to 11/32	0.2570 to 0.3438	14°
S to 1/2	0.3480 to 0.5000	12°
33/64 to 3/4	0.5156 to 0.7500	10°
49/64 and up	0.7656 to	8°

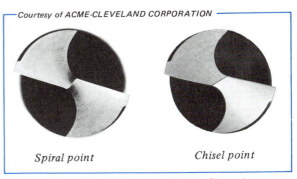

Spiral point   *Chisel point*

Figure 15–6 Features of Drills Ground
to Spiral and Chisel Points

118°. There is a 59° angle on each side of the drill axis. The angles are shown in Figure 15–5A, B, and C.

The cutting edge is formed by relieving the flute. A general *clearance angle* of from 8° to 12° is illustrated. Too small a clearance angle causes the drill to rub instead of cut. At too steep an angle, the drill tears into the work. Thus, the cutting edges may dull rapidly or parts of the drill may break away. After extensive research drill manufacturers have prepared tables of recommended lip clearance angles. Table 15–1 contains recommended lip clearance angles for drill sizes #80 to 1″ and larger. The same lip clearance angles apply to equivalent metric drill sizes.

The precision to which a drill produces an accurate size hole depends on the grinding of the cutting lips. They must be equal in length and at the same angle.

Thus far, only the *chisel point* of a general-purpose drill has been described. The flat shape of this drill point helps to produce a hole that meets most work requirements. These requirements relate to conditions of roundness and size within rough tolerance limits. Roundness may be more accurately produced when the chisel point is guided by a drill bushing in a drill jig.

Another shape of drill point is called the *spiral point*. The spiral-point drill, in contrast to the chisel-point drill, has a sharply pointed end. With this sharp end, the drill may be started accurately without center punching, center drilling, or using a drill jig. The hole produced

with the spiral point is concentric with the axis and is round. The photos in Figure 15–6 provide a comparison of a spiral- and a chisel-point drill.

Spiral points may be ground on standard drills. A special spiral-point drill-grinding machine is used for the grinding.

The self-centering feature of the spiral-point drill eliminates the need for a drill-positioning device. The spiral-point drill is particularly adaptable to numerically controlled drilling operations. Spiral-point drills have longer tool life than drills ground with a chisel point. Test records of spiral-point drills show that a greater number of workpieces may be produced between each spiral-point grinding in contrast to the number of workpieces produced between drill-point conventional grindings.

SPLIT-POINT DRILL

Another point, the *split point*, reduces the web to a minimum shape. This point forms two additional cutting edges. Figure 15–7 shows that the clearance angle of the added cutting edges is formed by grinding. Part of the web section is ground parallel with the drill axis. In contrast to the chisel and spiral points, the design of the split point permits:

- Easier and faster cutting,
- Easier access for the chips to clear the cutting edges and workpiece,
- Less force to be used in the drilling process,
- Cutting with less power,
- Less heat to be generated from cutting.

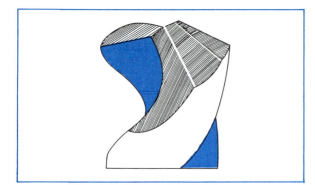

Figure 15–7 A Split-Point Drill Grind

The split point reduces tool wear. The grinding does not disturb the cutting lips. There is also less tool breakage when drilling into hard outer surfaces (scale) of castings and forgings.

THE DRILL POINT GAGE

The accuracy of the drill point angle and the dead center may be measured with a *drill point gage*. A common drill point gage has a graduated blade shaped to the required drill angle. This blade is attached to a steel rule.

The body of the drill is held against the edge of the steel rule. The point end is brought up to the blade. Any variation between the cutting edge and the correct angle may be sighted. The location of dead center is read on the graduated blade. The drill is then turned 180°. A fractional dimension reading is again taken. The drill point angle is correctly ground when there is no light between the cutting edge and the angular blade. The lips are correct when the dimensional reading to the chisel point of each lip is the same.

TAPER DRILL SHANK ADAPTERS

Adapters are used to accommodate the difference between a taper-shank drill and the drill press spindle taper. Two common types of adapters are called a *drill sleeve* and a *taper socket*.

DRILL SLEEVE

When the taper shank of a drill is smaller than the spindle taper, the shank may be brought up to size with a drill sleeve (Figure 15–8). The

Figure 15–8 A Drill Sleeve

standard taper that is used on taper-shank drills is known as a *Morse taper*. The same standard is used on many other cutting tools like reamers, boring tools, and counterbores. Morse tapers are numbered from 0 through 7. The number 2, 3, and 4 tapers are used on drills that range from 3/8″ to 1 1/2″ diameter (10mm to 38mm). For example, a drill sleeve is required for a 1/2″ drill that has a #2 Morse-taper shank when it is to fit a spindle with a #3 or a #4 Morse taper.

FITTED (TAPER) SOCKET

The taper socket is also known as a *fitted socket* (Figure 15–9). It has an external Morse-taper shank. The body has an internal taper. There is an elongated slot at the small end of the taper. This slot is where the tang of the drill or cutting tool fits. It also serves as a way of removing a tool.

A fitted (taper) socket often provides a method of extending the length of a cutting tool. Different combinations of taper sizes are available. For example, a fitted (taper) socket with a #3 Morse-taper shank may have a #2 taper in the body. The #3 taper fits a #3 taper spindle. The #2 taper accommodates a drill or reamer that has the same size taper shank.

THE DRILL DRIFT

The *drill drift* is a flat, metal, tapered bar. It has a round edge and a flat edge. The round edge fits the round portion (end) of the elongated

Figure 15–9 A Fitted (Taper) Socket

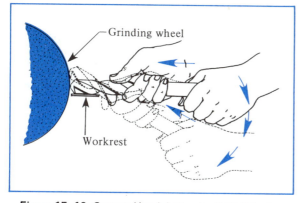

Figure 15–10 Correct Hand Action in Drill Grinding

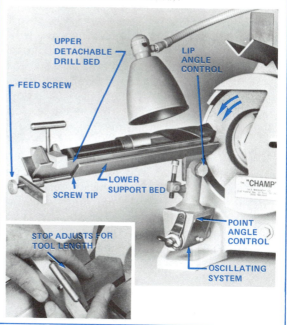

Figure 15–11 Drill Point and Tool Grinder

slot of a drill sleeve or a fitted socket. The flat edge rests and drives against the tang end of a cutting tool. The joined tapered surfaces are freed by a sharp, slight tap on the drift.

The *safety drill drift* is a type of drift. A floating handle is attached to the drift. The body of a cutting tool may be held in one hand. The other hand is used to quickly slide the handle so that it delivers a sharp tap. The force disengages the tapers. It is good practice to protect the surface under a drill so that it will not be hit by a cutting tool when it is freed.

DRILL POINT GRINDING

Drill points may be ground by several methods. They may be ground off-hand (Figure 15–10), on a *drill point grinding machine* or by using a drill point attachment fitted to a bench or floor grinder. Regardless of the method, the drill must be positioned, held, and fed into the grinding wheel to produce a cutting edge and clearance in correct relation to the flute.

There are two main differences between off-hand grinding and grinding with a drill point machine or attachment. First, the skill and judgment of the operator controls the accuracy of the drill point when grinding by hand. Second, the face of a straight grinding wheel is used in hand grinding. The cutting is done in machine grinding on the side face of a *cup wheel*.

In drill point grinding, particular attention must be paid to the following factors: clearance angle, heat generated, lips, and web.

DESIGNATION OF DRILL SIZES

There are three standard designations of drill sizes in the inch-standard system and one designation in the metric system. In the inch-standard system, a drill may be given a *wire number size*, *letter size*, or *fractional inch size*. Each drill size has its equivalent decimal value. Metric sizes are given in terms of millimeters. The common sizes of straight-shank drills in the metric system, for example, range from 0.20mm to 16.00mm.

NUMBER SIZE DRILLS (WIRE GAGE DRILLS)

The sizes of the number size series of drills range from #80 (0.0135″) to #1 (0.228″). The most widely used set of drills ranges from #1 to #60 (0.040″).

LETTER SIZE DRILLS

The sizes of the letter size series of drills are designated by letters from A (0.234″) to Z (0.413″). Letter size drills are larger than the largest diameter in the number drill series, which is #1 (0.228″).

FRACTIONAL SIZE DRILLS

The sizes of the fractional size series of drills begin at 1/64″ diameter. Each successive drill is 1/64″ larger. The general range of diameters extends to 3″. The term *set* is used for a series of fractional size drills in a range, such as the 1/64″ to 1/2″ range. When a series of fractional size drills is part of a set—for example, from 1/64″ to 1/2″—it is sometimes called a *jobber's set*.

Part of a drill size table with selected number, letter, fractional, and metric sizes is represented in Table 15–2. Note that there are variations in diameter among all four drill size designations. These variations provide an adequate range for most hole drilling from a #80 drill (0.0135″) to 3/4″ or 19.0mm (0.748″).

DRILL SIZE MEASUREMENT

Drill sizes may be measured most accurately with a micrometer. A micrometer measurement is taken across the margins of a two-flute drill.

A quick, common practice in the shop is to use a *drill gage*. A drill gage is a flat plate with the same number of holes as there are drills in the series. There are number size, letter size, fractional size, and metric size drill gages. The decimal equivalent appears for each drill size.

Table 15–2 Selected Drill Sizes and Decimal Equivalents

Number (wire)	Letter	Fractional (inch)	Metric (mm)	Decimal Equivalent
1				0.2280
			5.8	0.2283
			5.9	0.2323
	A			0.2340
		15/64		0.2344
			6.0	0.2362
	B			0.2380
			6.1	0.2402
	C			0.2420

CAUSES AND CORRECTION OF DRILLING PROBLEMS

A number of problems can arise in drilling if correct practices are not followed. These problems affect dimensional accuracy, surface finish, and drilling efficiency. Tool life is increased and the accuracy and quality of drilling are improved when correct practices are followed.

Table 15–3 lists a number of common problems that relate to drills and drilling procedures. Possible causes of each problem are given. These causes are followed by a series of check points for correcting each problem. The craftsperson must be able to recognize a problem, establish its cause, and take corrective steps.

CUTTING SPEEDS FOR DRILLING

Every cutting tool performs most efficiently and with greater tool life when it is cutting at the correct cutting speed. *Cutting speed* is expressed as the *surface feet the circumference of a drill travels in one minute*. This speed is stated as surface feet per minute (sfpm).

Manufacturers' tables contain recommended cutting speeds for different types of cutting tools and materials. Table 15–4 shows part of a table that lists the high-speed steel drill speeds for various steels. These speeds are reduced by 40–50% when a carbon steel drill is used instead of a high-speed drill.

The craftsperson must determine the spindle *revolutions per minute* (RPM) from the information provided in a cutting speed table. The RPM must produce the required cutting speed for a particular drilling operation.

CALCULATING RPM FOR A CUTTING SPEED

The rpm required for a particular surface feet per minute cutting speed (or circumferential speed) may be calculated with the simple formula:

$$RPM = \frac{Cutting\ Speed \times 12}{Drill\ Circumference}$$

Values are rounded off to produce a more simplified formula that is used in the shop:

$$RPM = \frac{CS \times 4}{d}$$

Table 15–3 Drilling Problems: Causes and Corrective Action

Problem	Possible Cause	Corrective Action
Drill breaking	—Dull drill point	—Regrind drill point
	—Insufficient lip clearance	—Check angle of lip clearance
	—Spring in the workpiece	—Reclamp workpiece securely
	—Feed too great in relation to cutting speed	—Decrease feed or step up the cutting speed
Drill breaking (nonferrous and other soft materials)	—Flutes clogged with chips	—Use appropriate drill for the material
	—Cutting point digging into the material	—Remove the drill a number of times so the chips clear the hole
		—Increase the cutting speed
		—Grind a zero rake angle
Rapid wearing of cutting edge corners	—Cutting speed too fast	—Reduce cutting speed
	—Work hardness; hard spots, casting scale, or sand particles	—Sand blast or tumble castings, or heat treat
		—Change method of flowing the coolant and check for correct cutting fluid.
Tang breaking	—Taper shank improperly fitted in sleeve, socket, or spindle	—Check condition of adapter or spindle
		—Clean taper hole and remove nicks or burrs by reaming carefully
Margin chipping (using drill jig) Cutting edge and lip chipping	—Jig bushing too large	—Replace drill bushing with same size as drill
	—Lip clearance too great	—Regrind to proper lip clearance
	—Rate of feet too high	—Reduce the feed rate
Cracking or chipping of high-speed steel drill	—Excessive feed	—Reduce the feed rate
	—Improper quenching during grinding or drilling	—Quench gradually with warmer coolant when grinding
Change in cutting action	—Drill dulling; cutting edge damaged	—Regrind the drill point correctly
	—Variations in the hardness of the workpiece	—Anneal the workpiece
	—Change in the amount of cutting fluid reaching the drill point	—Increase the direction or rate of flow of the cutting fluid
Oversize drilled hole	—Improper grinding of point with equal angles and length of cutting edges	—Check reground cutting angle and length of cutting edges
	—Excessive *play* in spindle	—Adjust or correct amount of free movement in spindle
	—Slightly bent drill	—Straighten drill carefully
Cutting on one lip	—Angle of cutting lips varies —Unequal length of cutting lips	—Regrind and check accuracy of cutting angle and length of cutting edges
Poor quality of surface finish (rough drilled hole)	—Improperly ground drill point	—Regrind and check accuracy of drill point
	—Excessive drill feed	—Decrease drill feed
	—Wrong cutting fluid or limited rate of flow	—Change cutting fluid for the kind of material
	—Workpiece not correctly secured	—Reposition and secure the workpiece securely
Drill splitting (along the center)	—Excessive feed	—Decrease drill feed
	—Insufficient lip clearance	—Increase the lip clearance to the correct angle

Table 15–4 Partial Table of Cutting Speeds for High-Speed Steel Drills

Material	Cutting Speed Range (surface feet per minute)
Steel	
Low-carbon	80 to 150
Medium-carbon	60 to 100
High-carbon	50 to 60
Tool and die	40 to 80
Alloy	50 to 70

Table 15–5 Feeds for Selected Drill Sizes

Drill Size (inches)	Feed (inches per revolution)
Small # sizes to 1/8	0.001 to 0.002
1/8 to 1/4	0.002 to 0.004
1/4 to 1/2	0.004 to 0.007
1/2 to 1	0.007 to 0.015
Larger than one inch	0.015 to 0.025

For example, assume that a series of holes is to be drilled in a low-carbon steel. The cutting speed of 150 sfpm is to be used with a 1/2″ drill. The known values are inserted in the simplified RPM formula as follows:

$$RPM = \frac{150 \times 4}{1/2} = 1,200$$

The drill must turn 1,200 RPM to produce a cutting speed of 150 sfpm.

FACTORS AFFECTING CUTTING SPEED

The cutting speed that the worker finally selects is influenced by seven different factors:

- Material of the workpiece,
- Material of which the cutting tool is made,
- Size of the hole,
- Quality of surface finish that is required,
- Nature and flow rate of the cutting fluid,
- Type and condition of the drilling machine,
- Manner in which the workpiece is mounted and held.

CUTTING FEEDS FOR DRILLING OPERATIONS

Cutting feed indicates the distance a tool cuts into the workpiece for each revolution of a cutting tool. Tables of cutting feeds are provided by cutting tool manufacturers. The tables provide a recommended range of feeds according to the size of the tool. Table 15–5 gives feeds for selected drill sizes. These sizes range from the smallest number size drill to drills of 1″ diameter and larger.

The *feed indicator* on a machine that is equipped with an automatic feed device is set within the recommended feed range.

DRILL POINT GEOMETRY: NEW MANUFACTURING PROCESSES AND NUMERICAL CONTROL APPLICATIONS

The revival and acceptance of multi-faceted, self-centering drill points is due to the following industrial developments.

- The advent of numerical control machinery and changing manufacturing technology and processes.
- The availability of precise, versatile, drill point grinding, splitting, and web-thinning drill sharpeners, and tool geometry inspection/measuring machines.
- The widespread use of a greater variety of standard and new materials.

MULTI-FACETED DRILL POINTS

Multi-faceted drill points require less torque in drilling, generate less heat, require less frequent regrinding, are self-centering, and have longer tool life.

The *four-faceted drill point* shown in Figure 15–12(A) has *primary clearance* and *secondary clearance facets*. The secondary facet (generally ground at an angle of 20°) produces a self-centering *chisel point apex* instead of the sharp edge, chisel point of the conventional drill.

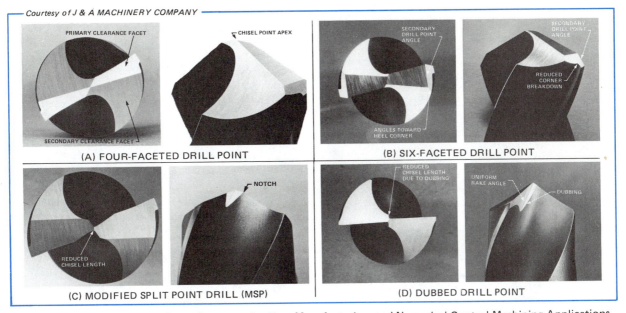

Courtesy of J & A MACHINERY COMPANY

(A) FOUR-FACETED DRILL POINT

(B) SIX-FACETED DRILL POINT

(C) MODIFIED SPLIT POINT DRILL (MSP)

(D) DUBBED DRILL POINT

Figure 15–12 Modified Drill Point Geometry for New Manufacturing and Numerical Control Machining Applications

The *six-faceted drill point* has a *secondary drill point angle* (Figure 15–12B). For general drilling purposes, the secondary drill point angle is ground at one-half of one side of the **primary point angle**. For example, for a 140° point angle, one side of the point angle is 70°. The secondary drill point angle (1/2 of 70°) is 35°.

Generally, the *secondary lip clearance angle* and the primary lip clearance angles are equal. The *secondary drill point angle* on a six-faceted drill point overcomes a great deal of cutter stress, wear, and breakdown on the outside cutting lip corner.

MODIFIED SPLIT POINT DRILL

The *modified split point* (Figure 15–12C) drill has a *notch* ground into the web parallel to the flute angle. The notch produces a positive secondary cutting edge up to the chisel edge. The notch allows chips to flow up the flute freely.

The modified split point is self-centering and can withstand the increased thrust required for drilling difficult-to-machine metals such as stainless steels and titanium.

The modified drill point overcomes the difficulties associated with the conventional split point (as described earlier in this unit). For instance, (1) the conventional split point is limited to drill

diameters of 1/2″ (12.7mm) or smaller, and (2) the cutting action of the conventional split drill creates problems at the drill corners where the flow of chips is impeded and results in the forming of a weld bead.

DUBBED DRILL POINT

Dubbing relates to the grinding process for producing a uniform positive or negative rake angle at the drill cutting edge. *Lip blunting* or *dubbing* is necessary to prevent hogging-in, drill chatter, and excessive force on the cutting point when drilling brass, bronze, copper, fiberglass, and other similar materials.

Dubbing the cutting edges (Figure 15–12D) reduces the chisel length and cuts down on the tremendous stress on the lip corners as is the case with standard drill points.

BLUNT DRILL POINT ANGLES

Greater productivity is often achieved by changing the drill point angle from the basic 118° to a 140° (or another included angle). The blunt 140° angle produces overall cutting lip lengths that are shorter than the 118° angle for the same diameter drill. This means the drill reaches maximum cutting action faster with less cutting torque.

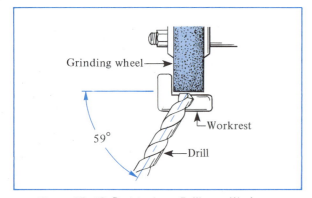

Figure 15–13 Positioning a Drill on a Workrest

How to Grind a Conventional Drill Point

Offhand Grinding

STEP 1 Check the trueness of the coarse and the fine bench grinder wheels. The workrest should be horizontal and within 1/8″ of the wheel face.

STEP 2 Examine the point of the drill. Be sure it is not burned. Check the condition of the margins and the dead center.

Note: The drill point may need to be ground back beyond the point of damage.

STEP 3 Hold the drill on the workrest with one hand. The drill should be positioned at a 59° angle to the wheel face (Figure 15–13). The forefinger of the hand serves as a positioner. It also applies a slight force to hold one cutting edge steady against the coarse wheel face.

STEP 4 Hold the shank end of the drill in the other hand. Move the shank end down while applying force against the cutting face.

Note: There is a natural tendency to move the drill in an arc. An arc movement is the pattern to which the drill is to be ground.

Caution: Quench the drill point frequently. Cooling avoids burning the cutting edge. Cooling also makes it possible to handle the drill safely.

Note: Avoid overheating the point followed by rapid quenching. Extremes of heat and cold may cause microscopic cracks along the cutting edges. Such cracks cause the lips to break down rapidly.

STEP 5 Check the 59° angle against a drill gage.

STEP 6 Repeat steps 3 and 4 until the cutting edge and lip clearance are correct. A sharp cutting edge is required.

Note: It is good shop practice to alternate the grinding process between the opposite cutting edges.

STEP 7 Check the drill point for (a) the correct cutting angle of 59°, (b) the length of the cutting edges, and (c) the required clearance angle (usually 8° to 12°).

Note: Finish grind the cutting edges on the fine grinding wheel. Finish grinding produces a fine cutting edge and increases tool wear life.

Thinning the Web

STEP 1 Hold the cutting edge of the drill parallel to the side of a straight grinding wheel. The wheel must have a square corner with a very slight radius.

STEP 2 Grind the front end of each flute. The web thickness must be ground equal to the web thickness of a new drill of the same size.

Using a Drill Grinding Attachment

STEP 1 Clamp the drill in the V-block. The point must overhang.

STEP 2 Set the end guide (tailstock center) so that the tool point may be brought into contact with the front face of a cup wheel.

STEP 3 Move the setting finger on the grinding attachment to contact and position the flute to be ground. The contact point should be close to the wheel.

STEP 4 Set the base angle of the attachment at 59°. The body scale is set to the clearance angle of from 8° to 12°.

STEP 5 Feed the drill until contact is made with the wheel face. The movement of one lever produces the 59° angle. The other lever is moved to produce the clearance angle.

STEP 6 Turn the drill 180°. Align the second flute. Grind to the same depth as the first cutting edge.

Note: The same setup and grinding principles are applied when a grinding fixture is used. The tool is fed with a vertical movement of the cutter grinder head.

How to Drill Holes

Flat Surfaces Held in a Drill Vise

STEP 1 Select the correct size drill. Check the shank to see that it is free of burrs. Check the drill point for the accuracy of the cutting and clearance angles.

STEP 2 Determine the required speed and the feed if a power feed is to be used. Set the speed selector and the rate of feed.

STEP 3 Clamp the work-holding device securely to the table. On small-sized drills the vise may be held by hand.

Note: If an angle plate setup is required, the overhanging leg should be supported by a jack (Figure 15-14).

STEP 4 Chuck a straight-shank drill or center drill.

STEP 5 Move the workpiece so that the center of the required hole aligns with the axis of the spindle and drill.

STEP 6 Start the machine. Bring the drill point to the work. Apply a slight force to start the drill.

STEP 7 Set the coolant nozzle if considerable heat might be generated. The cutting fluid should flow freely to the point of the drill.

Note: In general practice, where there is limited heat, a cutting fluid is brushed on.

Figure 15-14 An Angle Plate Setup for Drilling

STEP 8 Continue to apply a force to feed the drill uniformly through the workpiece.

Note: On deep holes (depth in excess of five to six times the diameter of the drill), feed the drill until it is the equivalent of one diameter. Then remove the drill to help clear the chips and permit the cutting fluid to flow freely into the holes.

The power feed is set so that it trips off after the drill clears the workpiece on a through hole. The depth of a blind hole may be read directly on a graduated scale on the drill head.

STEP 9 Stop the machine. Remove the chips with a brush. The burrs may be cut away with a triangular scraper or a countersink.

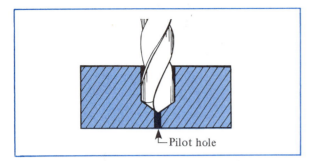

Pilot hole

Figure 15-15 Drilling a Pilot Hole for a Large-Sized Drill

Drilling a Hole in a Cylindrical Surface

STEP 1 Position the workpiece in a set of V-blocks. Square the centerline on the end surface with the machine table. Squaring also aligns the hole in relation to any other hole or surface.

STEP 2 Select a center drill. The pilot hole and its outside diameter should be in proportion to the size of the hole to be drilled. Chuck the center drill.

STEP 3 Set the center-punched hole central with the point of the center drill. Clamp the workpiece and setup if the cutting force requires it.

STEP 4 Determine and set the speed and feed controls. Start the machine.

STEP 5 Bring the center drill carefully to the workpiece. Start drilling. Feed slowly until the point end and the angle portion provide a large enough area in which to center a drill.

Note: A cutting oil is usually applied to the drill point with a brush. This oil produces a finer cutting action. The machinability of the material and tool life and productivity are also increased.

STEP 6 Replace the center drill with the required drill.

Note: A small pilot hole is often drilled first. This hole is followed by a larger diameter drill, for more efficient drilling (Figure 15–15).

STEP 7 Change the cutting speed if necessary. Start the drill and feed it by hand. En-

gage the power feed for large-diameter holes.

Note: The lips of a drill are ground to a 0° rake angle for drilling soft materials such as aluminum and brass.

STEP 8 Drill and burr, following the same steps used in drilling holes in flat surfaces.

STEP 9 Stop the machine. Clean the workpiece, machine surfaces, and cutting tools.

How to Remove a Drill

Removing a Taper-Shank Drill

STEP 1 Move the spindle down so that the drill moves only a short distance when the tapered surfaces are freed.

STEP 2 Place a protective block of soft material under the drill. Small-sized drills may be held by hand.

Caution: The drill must be checked carefully to be sure it is cool enough to be handled. A wiping cloth may be placed around the drill for handling large sizes.

STEP 3 Slide the drift handle quickly to deliver a sharp, light blow on the drift. The force causes the taper-shank surfaces to become free.

Note: A slightly different procedure is used when the drift must be tapped firmly with a hammer. In such a procedure the drift is held in one hand. The end of the drift is tapped firmly. The drill is released against a protecting block.

Removing a Straight-Shank Drill

STEP 1 Insert the pin of the chuck wrench into the chuck body. The wrench teeth must mesh properly with the teeth in the releasing collar.

STEP 2 Turn the wrench handle counterclockwise to release the drill.

STEP 3 Examine the shank of the drill. Remove any burrs or cut edges. Burrs and cut edges are produced if the drill turns in the chuck during the drilling process.

Safe Practices in Drilling

- Check the straight and taper shanks and tangs of drills, sleeves, and drill sockets. They must be free of burrs and nicks. The tapers and tangs are checked for correct taper fit.
- Set the tapers on drills, adapters, chucks, and so on, by a firm tap with a soft-face hammer.
- Grind the cutting edges of a drill to the same length and angle. Otherwise oversized, roughly drilled, and uneven holes will be produced.
- Move (flow) chips out of the drilled hole to prevent clogging, the generation of excess heat, and drill breakage.
- Quench carbon steel drills regularly during grinding to prevent softening (drawing the temper). A burned drill must be reground beyond the softened portion.
- Avoid excessive speed, feed, heat, and force, all of which can cause rapid wearing of a drill point and drill margin or drill breakage.
- Check the lip clearance angle. Too great a lip clearance angle causes chipping of the cutting edge. Too small a clearance angle and excessive speed causes drill splitting.
- Wear a protective shield to prevent eye injury and to keep hair away from a revolving spindle and cutting tools.
- Avoid wearing loose clothing near rotating parts. Remove all wiping cloths from the drilling area before starting a machine.

TERMS USED IN DRILLING TECHNOLOGY AND PROCESSES

High-speed steel and cobalt high-speed steel drills	A twist drill capable of cutting at high speed without changing the temper, softening the cutting edges, or reducing the cutting efficiency. (The addition of cobalt to high-speed steel produces a cutting material that is capable of drilling through tough metals and other compositions.)
Carbide-tipped drill	Inserts of tungsten carbide on the drill point and point end of a regularly formed twist drill. Cutting edge and body inserts of tungsten carbide.
Drill point angle	The included angle of the drill point. Equal angles on each side of a drill axis (centerline). A 118° included angle of a drill point for general, everyday applications.
Drill clearance angle	The angle at which the cutting edge of a drill is ground away in relation to a plane.
Multi-faceted drill point	Changed drill point geometry produced by grinding a secondary clearance facet that forms a self-centering drill point apex (four-faceted point) and (in the case of a six-faceted point) a secondary drill point angle and lip clearance.
Modified split point drill (MSP)	A split point drill with a notch ground into the drill web parallel to the flute angle.
Dubbed drill point	A ground positive or negative rake angle at the drill cutting edge. Lip blunting to prevent a drill from hogging-in.
Drill sleeve	An adapter. (The outside taper fits the drill press spindle. The inside taper accommodates the taper shank of a drill.)
Taper (fitted) socket	An adapter for taper-shank drill sizes that are larger or smaller than the spindle taper.
Letter, number, fractional, and metric drill sizes	Separate systems of designating drill sizes. A combination of systems having different drill sizes among the drills in each series. Four series of small drill sizes that permit drilling a wide range of holes.

Drill point gage	A flat, angular, graduated blade that may be attached to a steel rule. A gage for comparing the drill point angle with a standard. A gage for measuring the cutting lips of a drill and for checking the angle.
Cutting speed (drilling)	The number of surface feet the circumference of a drill travels in one minute (sfpm).
Cutting feed	The distance a drill advances for each revolution. The feeds recommended by manufacturers for drills of different sizes and materials.
V-block (drilling)	A layout, positioning, and holding device. A squared metal block with V-slotted sides that nest cylindrical workpieces.

SUMMARY

- Drills are manufactured from carbon steel, high-speed steel (HSS), and cobalt high-speed steel. The HSS bodies may also be tipped with tungsten carbide cutting edge inserts.

  - The two-flute drill is an all-purpose drill. Design features of two-flute drills are:
    - An angular point on which the cutting edges are formed,
    - Flutes that are connected with a web,
    - Lands that are relieved to leave a thin margin,
    - A channel to remove chips and to feed cutting fluid,
    - A shank.

- Special sub-land (multidiameter) drills produce two or more holes of different diameters in one operation. The connecting shoulder is also cut during the operation.

  - Oil-hole drills feed a cutting fluid through the drill shank and body to the cutting edges.

- The most widely used drill point angle is 118°, with an 8° to 12° clearance angle. The cutting point angle and the length of the cutting edge may be checked with a drill point gage.

  - Four- and six-faceted drill points and modified split point drills are required in numerical control machining.

- Multi-faceted drill point, spiral point, and modified split point drills are self-centering, require less torque in drilling, provide greater tool wear life, and have higher productivity rates than conventional drill points.

  - The four main drill size designations are:
    - Number sizes from #1 to #80 (0.228″ to 0.0135″);
    - Letter sizes from A to Z (0.234″ to 0.413″);
    - Fractional sizes from 1/64″ to a number of inches (in increments of 1/64″);
    - Metric sizes in millimeters.

- The drill sleeve and taper socket are two common adapters for taper-shank tools. The Morse taper series is used on drilling machine spindles and taper-shank drills, reamers, and other tools.

■ Cutting speed depends on the material in the workpiece and cutting tool, hole size, required surface finish, use of a cutting fluid or coolant, condition of the machine, and how the work is secured.

■ Drill breakage, rapid dulling and wearing, oversized holes, rough hole surfaces, and broken tangs result from one or more of the following causes:
  • Improper fitting of tapers and tang;
  • Too great or too small a clearance angle, cutting angle, rate of feed or speed;
  • Unequal cutting angles and/or length of cutting edge;
  • Too thick a web;
  • A burned drill point.

  ■ Tools and workpieces must be protected from the effects of excessive heat and force, improper fitting of cutting tools in holders, and improper grinding and machine processes. The work must be held securely either in a vise or by clamping to the table.

UNIT 15 REVIEW AND SELF-TEST

1. a. Cite the effect the addition of cobalt has on the cutting ability of high-speed drills.
    b. Identify the kinds of materials that may be efficiently drilled with cobalt high-speed steel drills.

2. Explain the principal design feature of a coolant-feeding drill.

3. Tell what purpose a drill bushing in a drill jig serves in drilling.

4. Give three advantages of using a spiral-point drill over a chisel-point drill.

5. State two conditions relating to numerical control drilling processes that require different drill point design features than those of conventional twist drills.

6. List two of the newer drill point designs that are used in manufacturing and numerical control applications.

7. a. Describe the drill point geometry of a dubbed drill.
    b. Identify two functions of a dubbed drill point.

8. Differentiate between two common types of adapters for drills.

9. Indicate three checkpoints to observe in grinding drill points.

10. Name four common drill size gages.

11. List five factors a drilling machine operator must consider when selecting the cutting speed for a particular drilling operation.

12. Give two cautions to observe in grinding a drill offhand.

13. List the steps to follow in setting up a part and drilling a hole at a right angle to the axis.

Machine Reaming: Technology and Processes

OBJECTIVES

After satisfactorily completing this unit, you will be able to:

- Understand features, characteristics, and applications of jobber's, shell, fluted chucking, rose, expansion, adjustable, step, and taper machine reamers and reaming.
- Determine undersize hole allowances for machine reaming.
- Apply factors affecting machine reaming cutting speeds and feeds.
- Deal with reamer alignment and conditions that produce chatter.
- Select cutting fluids for various reaming processes.
- Take corrective steps to remedy reamer breakage or excessive wear.
- Perform the following machine reaming processes.
  - Ream Through and Aligned Holes.
  - Produce a Duplicate Reamed Hole.
  - Hand Ream on the Drill Press.
- Work cutting speed and spindle RPM formulas to solve general shop problems on reaming processes.
- Follow the recommended *Safe Practices* for machine reaming.
- Apply new *Terms* defining machine reaming technology and processes.

Holes may be produced to meet accurate dimensional tolerances, standards of roundness, and quality surface finishes by reaming. Unit 8 treated reamers and the reaming process in terms of *hand reaming.* Industry also requires that reaming be done by machine. The drilling machine, lathe, milling machine, jig borer, and other machine tools are used for *machine reaming.*

There are common design and cutting principles that apply to both hand and machine reamers and reaming processes. However, the applications are so varied that many different reamer materials, types, shapes, and cutting angles are required. This unit provides additional information about the functions of machine reamers, general design features, a few machine setups, and the reaming process using a drill press.

MAIN PURPOSES OF REAMING

Reaming serves four main purposes:

- To enlarge a hole to a precise dimension,
- To form a straight (concentric) hole when the hole first is bored concentric and then reamed to size,
- To produce a smooth hole with a high-quality surface finish,
- To correct minute out-of-round conditions of a drilled hole.

VARIABLES IN REAMING

The cutting action of reamers depends on a number of variables. The judgment of the craftsperson is based on considering these variables:

- Speed and feed,
- Material of the workpiece,
- Condition of the machine,
- Rigidity of the setup,
- Kind of reamer used,
- Rake of the cutting edge,
- Clearance angle,
- Required degrees of dimensional accuracy and surface finish,
- Amount of stock to be removed,
- Proper cutting fluid (where required) and its application,
- Accuracy and type of reamer guide bushings (when used),
- Condition of the drilled hole.

UNDERSIZED HOLE ALLOWANCE FOR MACHINE REAMING

A sufficient amount of material must be left in a predrilled or machined hole to permit accurate reaming. The amounts and conditions for hand reaming to close tolerances were covered as a bench work process. The amount stated for hand reaming ranges from 0.002" to 0.005" (0.05mm to 0.12mm).

By contrast, larger amounts of material must be left for machine reaming. A common practice is to drill a hole 1/64" smaller than required, or *undersized*. This practice is followed in a *one-step reaming process* with a fluted reamer. The reamer is able to correct minor surface and shape irregularities of the hole.

Table 16–1 gives maximum material allowances for five hole sizes. The sizes range from 1/4" to 3", and metric equivalents from 6mm to 72mm. Holes that require machining to finer dimensional tolerances and a higher quality surface finish are often produced by a *two-step reaming process*. An undersized hole is first rough reamed to within 0.002" to 0.005". A second fluted machine reamer is used for finish reaming.

Table 16–1 Material Allowances for Reaming (General)

Inch Standard		Metric Standard	
Reamer Diameter	Maximum Allowance	Reamer Diameter	Maximum Allowance
1/4"	0.010"	6mm	0.2mm
1/2"	0.016"	12mm	0.4mm
1"	0.020"	24mm	0.5mm
2"	0.032"	48mm	0.8mm
3"	0.047"	72mm	1.2mm

FEATURES OF MACHINE REAMERS

The shank of a machine reamer is either straight or tapered. The taper shank has a tang. The flutes may be straight or helical (spiral) with either a right-hand or left-hand helix. The body and shank may be one piece, or the fluted body may be a separate cutting tool fitted to an arbor.

The cutting action is performed by the cutting edges. The cutting end features of a machine reamer are labeled in Figure 16–1. Each cutting edge is formed by a radial rake angle, a chamfer angle from 40° to 50°, and a chamfer relief angle. A machine reamer cuts to the outside diameter.

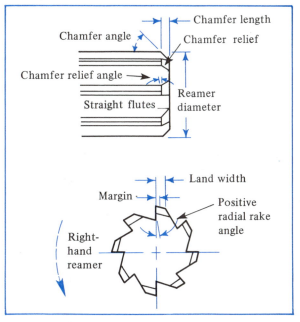

Figure 16–1 Cutting End Features of a Machine Reamer

TYPES OF MACHINE REAMERS

The following eight common types of machine reamers are described.

- General jobber's reamer,
- Shell reamer,
- Fluted chucking (machine) reamer,
- Rose chucking reamer,
- Expansion chucking reamer,
- Adjustable chucking reamer,
- Step reamer,
- Taper (machine) reamer.

Each of these types is available with a straight or taper shank and straight or spiral flutes.

JOBBER'S REAMER

The *jobber's (chucking) reamer* is a one-piece machine reamer. Jobber's reamers are general-purpose reamers.

SHELL REAMER

The *shell reamer* consists of two parts. An arbor and shank form one part. The second part is a reamer shell that includes the flutes and cutting teeth. The bore of the shell has a slight taper that fits the arbor taper. A pin (lug) in the arbor fits the two slots in the shell. Figure 16–2 illustrates straight- and spiral-fluted shell reamers and an arbor.

A shell reamer is used as a finishing reamer for dimensional accuracy. The shell is replaceable, making this reamer an economical type.

Also, the same arbor may be used with a number of differently sized shell reamers.

FLUTED CHUCKING REAMER

The *fluted chucking reamer* (Figure 16–3) is designed for cutting on the end and along the teeth. The teeth have a slight chamfer ground on the ends, for end cutting. A clearance angle is also ground along the entire length of each tooth. The fluted reamer is used for finish and difficult reaming operations.

A circular margin (outside diameter) of 0.005″ to 0.020″ (0.12mm to 0.5mm) runs the length of the flute. The lands are backed off to provide body clearance. The cutting end has a 45° bevel. Helical flutes are especially adapted to produce smooth, accurate holes and a free-cutting action. A helical fluted reamer must be used when there is an interruption in the roundness of a hole, such as a keyway or a cutaway section.

ROSE CHUCKING REAMER

A *rose chucking reamer* is a coarse-toothed reamer in comparison to a fluted machine reamer. The teeth are ground at a 45° end-cutting angle. The beveled teeth do the cutting. The rose chucking reamer is used when a considerable amount of material is to be removed. Under these conditions the rose reamer produces a rough reamed hole.

These reamers are used to rough ream undersized holes to within 0.003″ to 0.010″ (0.08mm to 0.3mm). The teeth have a back taper of 0.001″ per inch running the entire length of

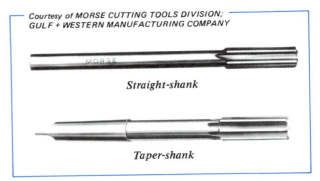

Courtesy of MORSE CUTTING TOOLS DIVISION; GULF + WESTERN MANUFACTURING COMPANY

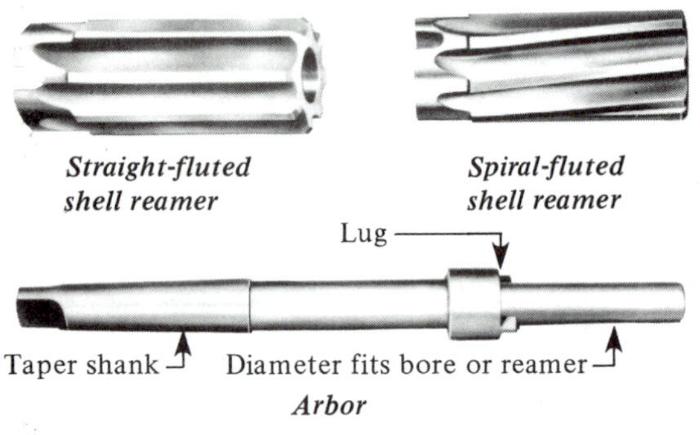

Straight-fluted shell reamer

Spiral-fluted shell reamer

Lug

Taper shank

Diameter fits bore or reamer

Arbor

Figure 16–2 Shell Reamers and an Arbor

Straight-shank

Taper-shank

Figure 16–3 Straight- and Taper-Shank Fluted Chucking Reamers

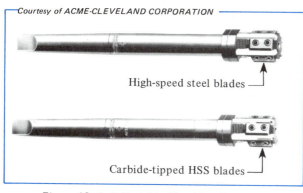

High-speed steel blades

Carbide-tipped HSS blades

Figure 16–4 Adjustable Chucking Reamers

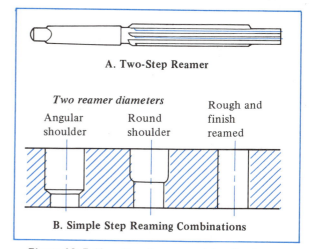

A. Two-Step Reamer

Two reamer diameters

Angular shoulder

Round shoulder

Rough and finish reamed

B. Simple Step Reaming Combinations

Figure 16–5 Two-Step Reamer and Applications

each flute. The back taper provides hole clearance so that the reamer does not bind in deep-hole reaming. The outside diameter is ground concentric. There is no land.

EXPANSION CHUCKING REAMER

As its name suggests, an *expansion chucking reamer* can be increased slightly in diameter. An adjustment feature on the end permits enlarging the reamer to produce a hole that is larger than a standard size. Compensation can also be made for wear. This feature increases the tool life.

ADJUSTABLE CHUCKING REAMER

Adjustable chucking reamers are produced with high-speed steel (HSS) blades or carbide-tipped, high-speed steel blades (Figure 16–4). The blades can be adjusted to increase the range of the reamer up to 1/32'' (0.8mm) in diameter. The blades are replaceable.

The carbide-tipped chucking reamer is especially adaptable to the reaming of ferrous and nonferrous castings and other materials. These materials require the cutting tool to have high abrasion-resistant qualities. The cutting tool must withstand high cutting speeds and temperatures.

STEP REAMERS

Step reamers produce two or more concentric diameters in one operation. A common step reamer has one portion of the fluted area ground at a small diameter for rough reaming, followed

by the standard diameter for final precision reaming. Other examples include the cutting of two concentric diameters which have angular or square shoulders (Figure 16–5).

TAPER REAMERS

Machine taper reamers are of similar design to hand taper reamers. The tapers conform to standard Morse, Brown & Sharpe, SI metric system, and other special tapers.

MACHINE REAMING SPEEDS AND FEEDS

MACHINE REAMING SPEEDS

The speed at which a hole is reamed depends on the following conditions:

- Required dimensional accuracy,
- Material to be reamed,
- Kind and flow of the cutting fluid on the cutting edges and workpiece,
- Desired quality of surface finish,
- Type and condition of the machine spindle,
- Size of the hole to be reamed.

The cutting speed for general machine reaming of a material is approximately 60% of the cutting speed required for drilling the same material. A table may be used to determine the correct cutting speed. Table 16–2 gives some examples of cutting speeds for HSS drills in surface feet per minute (sfpm) and meters per minute (m/min). The values are converted to cutting speeds for reaming by multiplying by 0.6 (60%).

MACHINE REAMER FEEDS

Like hand reamers, machine reamers should be fed two or three times deeper per revolution than in drilling. The amount of feed depends on the material and required quality of surface finish. As a general rule the highest rate of feed should be used to produce the desired finish. Operator judgment is important. A starting feed of from 0.0015" to 0.004" (0.04mm to 0.1mm) per flute per revolution may be increased by observing the cutting action, dimensional accuracy, and surface finish.

REAMER ALIGNMENT AND CHATTER

Reaming accuracy depends on alignment. The axis of the hole to be reamed, the reamer, and the machine spindle must be aligned. When a reamer bushing is used on a jig, this, too, must be aligned with the reamer axis. Any misalignment produces excessive wear on the bushing and reamer. The hole is also reamed inaccurately. A *floating reamer holder* reduces misalignment.

Chatter always affects dimensional accuracy and the quality of the machined hole. As mentioned earlier, chatter is caused by:

- Excessive cutting speed;
- Too light a feed;
- Improper grinding with excessive clearance;
- An incorrect setup in which the workpiece is not held securely;
- Play, or looseness, in the spindle or floating holder.

When chatter is produced, the operation must be stopped and the cause of the chatter corrected.

CUTTING FLUIDS FOR REAMING

The reaming of most materials requires the use of a cutting fluid. As in other machining operations, the cutting fluid is a coolant. Where necessary, the cutting fluid also serves as a lubricant. A cutting fluid helps to produce a finer cutting action and surface finish and prolongs tool life.

Table 16–2 Selected Cutting Speeds for High-Speed Steel (HSS) Drills

| Material | Cutting Speed Range | |
	Surface Feet per Minute	Meters per Minute
Steel Low carbon (.05–30%)	80 to 100	24 to 30
Cast iron Soft gray	100 to 150	30 to 45
Nonferrous Aluminum and alloys	200 to 300	60 to 90
Plastics Bakelite, etc.	100 to 150	30 to 45

Single workpieces of gray cast iron are reamed dry. A stream of compressed air is used as a coolant in production reaming of cast iron. Mineral-lard oils and sulphurized oils are recommended in general reaming practices. Tables of cutting fluids and manufacturers' data are referred to in selecting an appropriate cutting fluid.

COMMON CAUSES OF REAMER BREAKAGE OR EXCESSIVE WEAR

Excessive wear and breakage of reamers may be caused by any one or a combination of the following incorrect practices or conditions:

- Burrs or other foreign particles on the surfaces of the taper socket or machine spindle;
- Misalignment of one or more parts of the setup that causes excessive wear on the lands of the reamer and produces a bell-mouthed hole;
- Cutting speeds that are too fast or too slow;
- Inappropriate composition of the cutting fluid used on the material being reamed or lack of lubrication between a guide bushing and the reamer;
- Too much or insufficient stock allowance for reaming;
- A rough, tapered, or bell-mouthed drill hole that causes a wedging action;
- Faulty grinding such as:
  - Improper grinding of the clearance of the cutting edges,
  - Unbalanced grinding of the cutting edges

where only a few teeth cut and take the whole cutting load,

- Grinding a wrong end-cutting angle,
- Producing saw-toothed cutting edges that result from too coarse a grind,
- Grinding the cutting edges so fast and with so heavy a cut that grinding cracks are produced;
- An oversized or undersized reamer bushing;
- Use of an improperly designed reamer,
- Spindle play, lack of rigidity in the setup, and chattering;
- Incorrect rotational direction of the workpiece or reamer;
- Forcing of the reamer when entering the workpiece or when bottoming;
- Failure to cut below scale and hard spots in cast and forged materials;
- Nonremoval of abrasive particles before reaming.

How to Machine Ream Holes

Reaming Through Holes

STEP 1 Align the hole axis and spindle axis. Center drill. Drill the hole with the reamer drill.

STEP 2 Measure the reamer diameter. Check the taper shank, spindle, and adapter (if used) for burrs or nicks.

STEP 3 Determine the reamer speed and feed. Set the spindle speed at the required sfpm and feed per revolution.

STEP 4 Guide the reamer carefully to start in the drilled hole. Apply a cutting fluid with a brush or spout container so that it reaches the cutting lips.

Note: If the reamer starts to chatter, withdraw it. Reduce the surface speed.

STEP 5 Ream through. Withdraw the reamer while it is still revolving.

STEP 6 Clean the workpiece. Check the dimensional accuracy. Reream if necessary.

STEP 7 Remove burrs from the workpiece.

Caution: Replace the machine reamer in an appropriate holder.

Reaming Two Aligned Holes

STEP 1 Clamp the workpieces so that the axis of the holes to be reamed are aligned.

Note: Whenever possible, holes that are to be reamed together should be drilled during the same setup.

STEP 2 Strap the workpieces on parallels in a vise, on the table, or in another holding device.

STEP 3 Align the axis. Drill the reamer-size hole. Follow with a finish-size reamer.

STEP 4 Check the dimensional accuracy. Reream if necessary.

Producing a Duplicate Reamed Hole

STEP 1 Align the second workpiece in relation to a hole that is already reamed. Clamp securely.

STEP 2 Use a drill the size of the reamed hole. Align the drill with the axis of the previously reamed hole.

Note: A transfer punch is sometimes used to locate the position of a reamed hole in another workpiece.

STEP 3 Spot the center by drilling for a short distance to produce a cone-shaped center. Remove the spotting drill.

STEP 4 Use the reamer (undersized) drill and drill the hole.

STEP 5 Select an appropriate type of machine reamer. Machine ream.

STEP 6 Test for dimensional accuracy.

STEP 7 Insert a pin (plug) that fits snugly in the reamed hole.

STEP 8 Proceed to spot, drill, and ream the second hole.

STEP 9 Insert a pin or plug in the second reamed hole if additional holes are to be reamed to keep the holes in the correct position.

STEP 10 Continue to spot, drill, and ream any remaining holes. The two locating pins are enough to maintain the hole locations accurately.

Note: Continue to check the dimensional accuracy and the quality of surface being produced.

STEP 11 Clean the workpieces, tools, and machine. Carefully remove the reamer. Check and stone off any burrs.

Caution: A sharp, fine burr may be produced in machine reaming. Avoid rubbing over a reamed hole.

How to Hand Ream on a Drill Press

Note: It is assumed that the workpiece has been aligned and the reamer hole has been drilled to within 0.002″ to 0.005″ (0.05mm to 0.12mm) of the required size.

STEP 1 Mount a center point in a drill chuck.

STEP 2 Select an appropriately sized tap wrench for the size of reamer to be used.

STEP 3 Bring the square shank end of the hand reamer to the center point.

STEP 4 Hold the cutting edge of the reamer close to the hole to be reamed.

STEP 5 Bring the spindle down. Align the center point with the centered end of the reamer in order to guide it.

STEP 6 Turn the reamer so that the cutting edges touch the workpiece.

Note: An equal force must be applied to both handles. Brush a small quantity of cutting fluid on the reamer and cutting area.

STEP 7 Continue to turn and feed the reamer. Remove the reamer when the hole has been reamed a short distance.

Note: The reamer is turned clockwise for both reaming and withdrawing.

STEP 8 Check and correct any condition causing chatter, improper cutting, an inaccurately reamed hole, a poor quality of surface finish, or misalignment.

STEP 9 Ream through (or to the required depth).

STEP 10 Deburr the reamed hole. Check for dimensional accuracy. Ream if an undersized hole is produced.

Note: Whenever possible, it is good practice to countersink the edge of a hole before reaming.

STEP 11 Clean the workpiece, machine, and tools. Examine and remove any burrs. Carefully store all tools.

Safe Practices in Machine Reaming

• Examine the shank, tang, and margins of the reamer for nicks and burrs. Nicks and burrs impair accuracy and produce scratches in the work surface. Remove burrs with an oilstone.

• Store reamers in separate storage sections to prevent the reamers from rubbing against or touching any tool, instrument, or hardened part.

• Set a taper-shank reamer in a sleeve, socket, or taper spindle by tapping firmly and gently with a soft-face hammer.

• Turn the reamer in a clockwise direction only to prevent damage to the margins, cutting edges, and work surface.

• Align the axis of the reamer, workpiece, and spindle to avoid drifting of the reamer and assure reaming to specifications.

• Strap workpieces securely. Use adequate blocking and protecting strips.

• Stop the operation whenever chatter occurs. Diagnose the cause. Take steps to remedy the condition.

• Withdraw the machine reamer while it is still turning.

• Wear a shield to keep hair away from moving parts. Wear an eye-protection device to protect the eyes from chips and other particles. Loose clothing is dangerous and should be secured.

MACHINE REAMER AND REAMING TERMS

Undersized hole allowance (stock removal allowance)	The material left to produce a dimensionally accurate hole that meets specific surface finish and other hole geometry requirements.
Two-step reaming	Producing two concentric reamed holes simultaneously with one reamer. Using a single two-step reamer to ream two different size holes and the shoulder in one operation.
Chamfer angle	An angle ground uniformly on each flute to produce a cutting edge.
Reamer body	That portion of a reamer consisting of flutes, the area between them (lands), and the margin.
Relief, or body clearance	A relief machined in back of the margin and sloping toward the flutes.
Rake angle	The angle of a cutting face in relation to the centerline of the reamer. A line extending from the front edge of a margin through the center.
Cutting lip clearance	The relief for the cutting lip of each tooth.
Machine (chucking) reamers	Two types of machine reamers: rose reamers and fluted reamers.
Rose reamer	A reamer with teeth beveled on the end and provided with a clearance. A machine reamer on which the cutting takes place on the beveled end-cutting teeth instead of the sides. A reamer with teeth that have a slight back taper (0.001″ per inch) along the full length of each flute.
Fluted reamer	A machine reamer with narrow lands backed off the entire length. A rough finishing reamer with a greater number of cutting teeth than a rose reamer.
Shell reamer	A replaceable reamer section having a slight taper bore. A shell cutter with a taper bore that aligns on a tapered arbor.
Cutting speed (reaming)	The sfpm recommended for a particular size of reamer to cut to a desired degree of accuracy.
Reamer alignment	A condition in which the drilling machine spindle is positioned with the axis of the reamer and hole to be reamed.

SUMMARY

- Precision hand and machine reaming allowances range from 0.002″ to 0.005″ (0.05mm to 0.12mm). The general rule for rough machine reaming is to allow 0.010″ for 1/4″ diameter reamers to 0.020″ (1″) to 0.032″ (3″). Similar amounts in fractional millimeters are allowed on metric-sized holes.
  - The fluted chucking reamer cuts at the front cutting lips and along the cutting edge of each margin. The lands are backed off, leaving a narrow margin.
- Machine reamers have either a straight or taper shank. The teeth may be straight or have a right-hand or left-hand helix.
  - The cutting action of machine reamers is produced by cutting teeth that are beveled at a 40° to 50° angle and relieved to form a cutting lip.
- A jobber's reamer is a general-purpose reamer.

■ A shell reamer includes a replaceable reamer shell with a slight taper bore that fits an arbor. The shell reamer is a finishing reamer for dimensionally accurate holes.

■ Machine reaming speeds are slower than speeds used for drilling. The feeds are considerably greater.

■ Manufacturers' recommended cutting speeds, in sfpm, should be followed for the reamer size and the kind of material to be reamed.

■ The rose reamer cuts on the beveled-angle cutting teeth only. This reamer is adaptable for rough reaming undersized holes. The teeth taper slightly along their length to prevent rubbing. The outside diameter of the teeth is concentric and not relieved.

■ Expansion chucking reamers have the added feature of being adjustable. With these reamers, holes may be reamed slightly larger or smaller than a standard size.

■ Step reamers produce multiple, concentric reamed holes.

■ Adjustable blade reamers permit adjustments up to 1/32″ (0.8mm). The blades may be reground and replaced.

■ A coolant and a lubricant, where applicable, should be used in machine reaming.

■ Helical (spiral)-fluted reamers produce a finer shearing action than straight-fluted reamers. They are most practical where an interrupted section in the workpiece is to be reamed.

■ Chatter may be overcome by:
• Reduced cutting speed,
• Increasing the feed,
• Reducing the clearance on the cutting lips,
• Correcting misalignment,
• Securing the workpiece properly,
• Removing spindle play.

■ Reamed holes are dimensioned on sketches with drill and reamer size, depth, and the number of holes of a similar diameter. The surface finish may also be included.

UNIT 16 REVIEW AND SELF-TEST

1. Tell how the cutting edges of a machine reamer are formed.

2. Distinguish between general applications of fluted chucking reamers and rose chucking reamers.

3. Make a rule-of-thumb statement about the cutting speeds for reaming compared to the cutting speeds for drilling the same material and hole size.

4. State three considerations and/or checks by the machine operator before increasing the cutting feed for a machine reaming process.

5. List five incorrect practices requiring corrective operator action to avoid excessive reamer wear and breakage.

6. Explain the function of a floating reamer holder.

7. Set up the series of steps to be followed in producing reamed holes.

8. Tell how to safely withdraw a machine reamer to prevent scoring a finish reamed hole.

Machine Threading on a Drill Press

This unit deals with the thread-cutting tools and accessories that are generally used on drilling machines. These thread-cutting tools either tap internal threads or cut external threads. The American Unified system and the SI metric system of dimensioning screw threads are illustrated. Common tapping troubles and their causes are diagnosed. Recommended corrective measures are presented.

OBJECTIVES

After satisfactorily completing this unit, you will be able to:

- Apply design features and functions of reversing and nonreversing torque-driven, micro-tapping, and heavy-duty tapping attachments, or tap drivers to the cutting of internal threads.
- Understand conditions that affect machine threading efficiency.
- Determine cutting speeds and cutting fluids for machine threading.
- Take corrective action for general machine tapping problems.
- Make simple checks of quality for threaded holes or parts.
- Refer to and use tables on cutting speeds and cutting fluids for machine threading.
- Follow the *Safe Practices* recommended for tool and personal safety.
- Apply new machine threading *Terms.*

Courtesy of TRW/GEOMETRIC TOOL DIVISION

Solid adjustable spring die

Die mounted in releasing die holder

Figure 17–1 An Adjustable Spring Die and Holder

EXTERNAL MACHINE THREADING ON THE DRILL PRESS

SOLID ADJUSTABLE SPRING DIE

The *solid adjustable spring die* is one type of external thread-cutting tool. The die is held in a *releasing die holder*. A solid adjustable spring die and a die that is mounted in a releasing holder are shown in Figure 17–1. The holder is usually mounted in a chuck. The die is adjusted for thread depth by adjusting the cap against the tapered portion of the spring die. As the die turns, it threads to the predetermined depth. When this depth is reached, the die releases and remains stationary. The die holder continues to revolve with the drill press spindle.

The die is removed by reversing the spindle direction. A reversing action causes the die holder to again engage the die and back off from the thread. Only a slight force should be applied when backing off, to avoid damage to the first thread.

RETRACTABLE DIE (CHASERS)

The *retractable die head* is another type of external thread-cutting tool. This die head includes die sections called *chasers*, a holder (carrier), and a shank. Whether the die head is stationary (hand-operated) or revolving, the chasers are closed to cut a thread. The chasers must then be *retracted* (opened) after the thread is cut.

Stationary die heads are not used on drill presses. They instead are adapted to automatic and hand screw machines, turret lathes, and other turning machines on which the die head does not rotate. The die head shown in Figure 17-2 is used for cutting right- or left-hand threads. Chasers and carriers are changed for the hand of thread and size. The carriers are rigidly supported in the die head. They are easily adjusted for size. This die head has a front end trip that ensures machining a thread to an accurate length and close to a shoulder.

REVOLVING DIE HEAD

A *revolving die head* is used for external threading on drill presses and other machines (Figure 17-3). The head is designed to rotate about the axis of the workpiece. The chasers in the revolving die head are actuated by the machine and close automatically. The chasers are replaceable and come in sets for a required diameter and pitch. The chasers are adjustable to accommodate different depth and tolerance requirements.

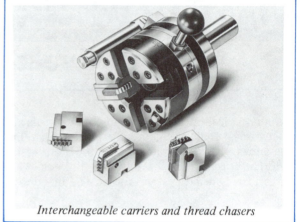

Interchangeable carriers and thread chasers

Figure 17-2 A Self-Opening Insert Chaser (Stationary-Type) Die Head

Revolving die heads are designed to be opened manually or automatically. They are available for fine and coarse threading on all kinds of materials. Revolving die heads may remain stationary when required.

MACHINE TAPPING ATTACHMENTS

The skilled craftsperson often uses power to cut a thread with a tap on a drill press. The machine must be equipped with a spindle-reversing mechanism.

For such an application the tap is secured in the drill chuck and a hole is tapped. The spindle is then reversed. The tap is backed out and removed. Care must be taken to apply only a

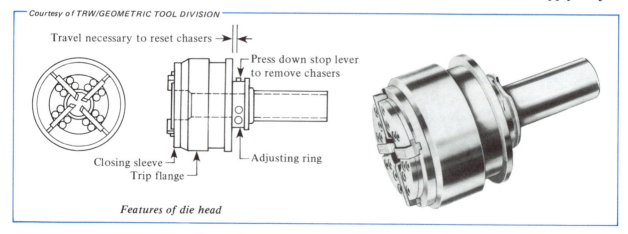

Features of die head

Figure 17-3 A Rotary, Self-Opening Die Head

limited force and to see that the tap starts to cut immediately. Further, the tap must thread completely out of the workpiece without rubbing on the first threads.

Tapping is also done with other machine attachments and accessories that provide for quality production and reduce tap breakage. There are a number of different types of machine tapping attachments:

- All-purpose drill press tapping attachment,
- Reversing-spindle tap driver,
- Nonreversing-spindle tap driver,
- Torque-driven tapping attachment,
- Micro-tapping attachment,
- Heavy-duty tapping attachment.

An all-purpose drill press tapping attachment is shown in Figure 17–4. This tapping attachment works equally well with mild steels, tool steels, and other nonferrous metals. Lubrication systems are provided to concentrate the cutting fluid within the cutting area.

Courtesy of BUCK SUPREME INC.

Figure 17–4 An All-Purpose Drill Press Tapping Attachment

REVERSING-SPINDLE TAP DRIVER

A *reversing-spindle tap driver* is used on a drilling machine that has a reversing spindle. The tap driver holds the tap securely while it is turned to cut the thread and while the tap is removed. Spindle direction is reversed to remove the tap.

NONREVERSING-SPINDLE TAP DRIVER

A *nonreversing-spindle tap driver* is adapted to a drilling machine where the spindle rotates in one direction only. The tapping attachment may be mounted in the drill chuck or in the tapered spindle. The tap is chucked in the attachment and fed to the workpiece. The downward force causes a *forward clutch* to engage and drive the tap. Raising the spindle lever releases the torque on the tap. A *reversing clutch* in the attachment is then engaged. The tap is backed out.

TORQUE-DRIVEN TAPPING ATTACHMENT

Conditions such as accumulation of chips in tap flutes, forcing of a tap at the bottom of a blind-hole tapping operation, and improper use of cutting fluids often require that excessive force be applied to a tap. These conditions and the resultant force may fracture the tap. Tap breakage is prevented by using a *torque-driven tapping attachment*. A *friction clutch mechanism* is provided for closely controlling the amount of torque (force) that may be applied. A *torque-setting device* is set according to the size and strength of a tap. When the torque setting is exceeded, the mechanism releases the tap to prevent tap breakage.

MICRO- AND HEAVY-DUTY TAPPING ATTACHMENTS

Microtapping attachments are designed for small taps in the #00 to #10 range. By contrast *heavy-duty self-contained tapping attachments* have speed reducers and are used for tapping operations that normally require a heavy-duty drill press. The speed-reduction feature increases the torque-driving capability of the attachment. A floating-driver feature compensates for any slight misalignment.

FACTORS AFFECTING MACHINE THREADING EFFICIENCY

CUTTING SPEED CONSIDERATIONS

There are many factors that affect the cutting speed required for machine tapping and other thread-cutting processes. Each of the following factors must be considered before the craftsperson makes a judgment about cutting speed:

- Kind of material to be tapped,
- Hardness and heat treatment of the workpiece,
- Pitch of the thread,
- Length of the threaded section,
- Chamfer in the hole or the beveled edge of the workpiece,
- Required percent of full thread depth,
- Quality and fit (class) of the thread,
- Type of cutting fluid (Figure 17-1),
- Design of the drilling machine and the machine tapping attachment.

The recommended cutting speeds for general machine tapping are given in Appendix Table A–11. These cutting speeds must be adjusted to compensate for the preceding factors. In addition the following conditions must be considered:

- As the length of a tapped hole increases, the cutting speed must be *decreased* (the chips tend to accumulate and thus prevent the full flow of the cutting fluid and produce additional heat on the cutting edges);
- Taps with long chamfers (taper taps) may be run faster than plug taps in holes that are tapped for a short distance (plug taps have just a few chamfered threads);
- Plug taps may be run faster than taper taps in holes that are to be tapped to a great depth;
- A 75% depth thread may be cut at a faster speed than a full-depth thread;
- The cutting speed for coarse-thread series taps that are larger than 1/2″ is slower than fine-thread series taps of the same diameter;
- The cutting speed for taper thread (pipe) taps is from one-half to three-quarters of the speed used for straight thread tapping;
- Greater tapping speeds are employed when tapping is automatically controlled (versus manual operation).

ANALYSIS OF COMMON TAPPING PROBLEMS

Table 17–2 analyzes the causes of eight common tapping problems and suggests corrective measures that may be taken for each problem.

Table 17–1 Recommended Cutting Fluids for Tapping Selected Metals (Partial Table)

Metal	*Recommended Cutting Fluid for Machine Tapping*
Ferrous	
Plain carbon and alloy steels	Sulfur-base oil with active sulfur
Malleable iron	
Monel metal	
Tool steel	Chlorinated sulfur-base oil
High-speed steel	
Stainless and alloy steels (heat treated to a higher degree of hardness)	
Nonferrous	
Aluminum, brass, copper, manganese bronze, naval brass, phosphor bronze, Tobin bronze	Mineral oil with a lard base
Aluminum and zinc die-casting metal	Lard oil diluted with up to 50% kerosene

Table 17–2 General Tapping Problems: Causes and Corrective Action

Problem	Probable Cause	Corrective Action
Tap breaking	—Tap drill size too small	—Use correct size tap drill
	—Dull tap	—Sharpen cutting edges
	—Misalignment	—Check holder and tap; align concentric with axis of tapped hole.
	—Excessive force in bottoming	—Drill tap hole deeper if possible —Reverse the tap direction earlier —Correct the torque (driving force) or replace the machine tapper
	—Tapping too deep	—Use spiral point or serial taps
Teeth chipping	—Chips loading on cutting edges	—Back the tap to break the chips —Check the cutting fluid
	—Jamming the tap at the bottom in blind hole tapping	—Correct the reversing stop or reverse sooner —Drill a deeper hole
	—Chips packed in the blind hole	—Remove and clean the tap during tapping
	—Work hardening	—Check prior hole-producing processes
Excessive tap wear	—Sand and abrasive particles	—Machine tumble or wire brush to remove foreign matter
	—Incorrect or inadequate quantity of cutting fluid	—Consult chart for appropriate cutting fluid —Position the lubricant nozzle so that the cutting fluid reaches the cutting edges.
	—Worn and dull tap	—Sharpen a dull tap —Replace a worn tap
Undersized threads	—Enlargement and shrinking of workpiece during tapping	—Use an oversized tap —Sharpen the cutting faces so that the tap cuts freely
Oversized threads	—Loading	—Check both the quality of the cutting fluid and the quantity that reaches the cutting edges.
	—Misalignment	—Align the tap and work axes before starting to tap
	—Worn tapping attachment or worn machine spindle	—Adjust the play in the spindle or tapping attachment
Bell-mouthed hole	—Excessive floating of machine spindle or tapping attachment	—Position the machine spindle and tapping attachment accurately
	—Misalignment	—Align the axes of the tap and workpiece
Torn, rough threads	—Incorrect chamfer or cutting angle	—Grind the starting teeth and cutting angle properly on a tool and cutter grinder
	—Dull tap teeth	—Grind the cutting faces of the tap
	—Loading	—Back the tap and remove the chips
	—Inadequate cutting fluid	—Consult the manufacturer's recommendations for the appropriate cutting fluid
Wavy threads	—Misalignment	—Align the axes of the spindle, tap, tap holder, and workpiece
	—Incorrect thread relief or chamfer of teeth	—Grind the cutting edges and chamfered teeth concentrically

GENERAL CUTTING FLUID RECOMMENDATIONS

Cutting fluids serve the same functions in machine tapping as they do in reaming, countersinking, counterboring, and other machine processes. Friction is reduced, tap life is increased, surface finish is improved, and thread dimensions may be better controlled when the correct cutting fluid is used. Table 17–1 lists a number of ferrous and nonferrous metals and their recommended cutting fluids.

Plastic and cast-iron parts are machine tapped dry or with compressed air. The air jet removes the chips and cools the tap.

FEATURES OF MACHINE TAPS

Machine taps are similar to hand taps except for two major differences. First, the shank end of a machine tap is shaped to fit the tapping attachment holder. Instead of the square head that provides a gripping surface for hand taps, the whole shank of a machine tap is round. The shank with a groove cut lengthwise fits a particular size and shape of chuck.

Second, machine tapping requires the use of a high-speed steel (HSS) tap. The HSS tap withstands higher operating temperatures and speeds than a carbon steel hand tap.

CHECKING AND MEASURING A THREADED PART

Internal and external threads may be measured in various ways. When a thread is machine tapped or cut with a die on a drilling machine, its size may be checked with a bolt, nut, or the mating part. Threads that are held to closer tolerances may be gaged with *thread plug* or *ring gages*. The pitch diameter of a thread may be measured with a thread micrometer or by using three wires and a standard micrometer.

How to Machine Tap (Drill Press)

STEP 1 Select the size and type of machine tap to use.

STEP 2 Check the drawing specifications for the percent of full depth of thread required. Check the size of the tap drill hole.

STEP 3 Select a tapping attachment that accommodates the tap size and is appropriate to the design of the drill press.

STEP 4 Mount the tapping attachment in a chuck or in the tapered spindle. Adjust the torque.

STEP 5 Set the spindle speed. Position the cutting fluid nozzle.

STEP 6 Align the axis of the tapping attachment and tap with the center line of the workpiece. Check to see that the work is held securely.

STEP 7 Start the drill press. Bring the tap carefully to the chamfered edge of the drilled hole. Apply a slight force so that the tapered threads of the tap begin to engage.

STEP 8 Continue to cut the thread. Continuously apply a cutting fluid.

STEP 9 Back the tap out of the hole. Brush away the chips in deep-hole or blind-hole threading. Chips tend to accumulate.

Note: Avoid using any force when withdrawing a tap, particularly when the last thread is reached.

STEP 10 Clean the workpiece and tap. Remove burrs. Check the threads with a bolt, stud, or other thread gage.

How to Cut an External Thread on a Drill Press

STEP 1 Select a solid adjustable spring die and holder or a retractable die and head. The die and holder should be appropriate to the size of thread to be cut.

STEP 2 Set the releasing die holder or the chasers to the required depth. Then release.

Note: A bolt, stud, or other threaded part is often used in the shop for setting an adjustable solid die or chasers.

STEP 3 Mount the threading attachment in a chuck or drill press spindle.

STEP 4 Check to see that the workpiece is held securely. The axes of the workpiece, die, and holder must be aligned. Position the cutting fluid nozzle.

STEP 5 Set the spindle feed.

STEP 6 Bring the die into contact with the chamfered surface of the workpiece.

STEP 7 Reverse the cutting process to break the chips and prevent the formation of an excessive amount of chips. Chips tend to clog the die.

STEP 8 Continue to cut the thread. Reverse the spindle direction if a releasing die holder is used.

Note: Run the die completely off the workpiece to clear the first few threads and to avoid damaging them.

STEP 9 Clean the workpiece, holder, and cutting surface. Remove burrs.

STEP 10 Check the quality of threads and size. Use a nut, mating part, or thread ring gage to measure the finished threads.

Safe Practices in Machine Threading and Tapping

• Align the axes of a thread-cutting attachment, tap or die, and workpiece. Alignment helps to prevent damage to a tap or die and produces an accurately threaded part.

• Direct an adequate supply of cutting fluid to reach the cutting edges of a threading tool. A cutting fluid improves tool life, increases cutting efficiency, and helps to produce a quality surface finish.

• Avoid applying excessive force or jamming of a tap. Excessive force and jamming cause a tap to fracture.

• Free a drill press spindle so that a tap, when once started, works freely into or out of a workpiece.

• Remove a broken tap safely with a tap extractor. Avoid brushing the hands over a broken tap.

• Clean and remove burrs before measuring or assembling a workpiece.

• Observe general safety precautions related to rotating machines and the safe handling of cutting tools.

TERMS USED IN MACHINE (DRILL PRESS) THREADING

Releasing type (external threading) holder A chucking device with a tapered nose in which a solid adjustable spring die is held. A die holder that permits a die to release at a preset depth and torque.

Retractable die (chasers) A series of chasers mounted in a die head. (The die head may be closed for threading or opened after a thread is cut.)

Revolving die head A die head with replaceable thread chasers. A die head that rotates about the axis of a workpiece.

Spindle reversing mechanism (drilling machine) A design feature of a drilling machine. A device for reversing the direction of a spindle to back out a machine tap or die.

Reversing tap driver A tapping device used on a reversable spindle drill press. A mechanism for both driving and removing a machine tap.

Nonreversing tap driver A machine tap holding and driving device that has an internal reversing clutch.

Torque setting A design feature of machine threading attachments. Setting the force needed to turn a tap or die during machine threading.

Threading efficiency factors A series of variable conditions that the craftsperson must consider when producing threaded holes and parts. (These holes and parts must meet standards that conform to quality and quantity specifications.)

Cutting speed considerations	Background information that must be weighed when selecting the most appropriate cutting speed in machine tapping processes.
Thread plug or ring gages	A measuring and gaging tool with precisely ground internal or external threads. Used for gaging the size of internal (plug gage) or external (ring gage) threads.

SUMMARY

- Machine dies are used for cutting external threads with the drill press.
    - The solid-type adjustable spring die permits adjusting to produce a thread of a specified depth. Spindle direction is reversed to remove the die when a releasing-type die holder is used.
- A retractable die (chasers) is mounted in either a stationary or revolving type of holder. The chasers are held in a closed position for threading. They are opened after a thread is cut.
    - A revolving die head rotates about the axis of a workpiece to cut an external thread. The head may be opened or closed manually or automatically.
- A reversing tap driver is used on a reversable spindle drill press for machine tapping.
    - A nonreversing tap driver applies the required clockwise force on a tap for thread cutting. Raising the spindle lever causes a reverse clutch to engage. The tap then turns counterclockwise out of the threaded hole.
- A friction clutch permits adjusting to the torque required to turn a thread-cutting tool. The amount of torque is affected by the thread size (pitch and diameter), percent of full depth, kind of material, and cutting fluid.
    - Heavy-duty tapping attachments contain speed-reducing units. These units increase the capability (force) of a drilling machine to produce threads. In contrast, microtapping attachments are used for small taps and fine threads.

UNIT 17 REVIEW AND SELF TEST

1. Describe how a revolving die head works.

2. Compare the operation of torque-driven and heavy-duty tapping attachments.

3. a. List three factors the craftsperson must consider for machine tapping.
   b. List three conditions that govern the selection of cutting speeds for machine tapping.

4. State what effect the use of a correct cutting fluid has on machine threading.

5. Suggest three general safety precautions that must be followed to cut quality external threads and to prevent damage to a tap, die, or workpiece

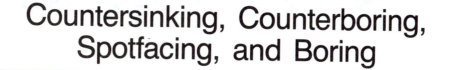

Countersinking, Counterboring, Spotfacing, and Boring

The outer surface of a hole may be cut away at an angle, recessed with a shoulder, or machined with a flat circular area. The general machining processes that produce such shapes on the outer surface are called countersinking, counterboring, and spotfacing. A hole may also be enlarged by boring. These processes and the cutting tools and accessories that are used in each process are treated in this unit.

OBJECTIVES

After satisfactorily completing this unit, you will be able to:

- Identify design features and applications of countersinks.
- Know when to use a combination drill and countersink and center drill tips and adapters.
- Select and use counterbores, drill and counterbore combination cutting tools, and spotfacers.
- Relate single-point boring tools and a boring head to precision boring.
- Perform each of the following processes.
  - Countersinking a Hole.
  - Drilling a Center Hole.
  - Counterbore a Hole.
  - Bore a Hole Using a Drill Press.
- Interpret drawings containing drilled and countersunk holes, counterbored holes, spotfaced areas, and bored holes.
- Follow *Safe Practices* in using hole forming tools.
- Apply new *Terms* related to hole-forming tools and processes.

COUNTERSINK AND COUNTERSINKING

Countersinking refers to the machining of a cone-shaped opening or a recess in the outer surface of a hole. The angle of the recessed indentation corresponds with the angle of a screw, a rivet head, or other tapered object that seats against the angular surface. A hole that is machined at an angle is called a *countersunk hole*. The tool that produces this cone shape is a *countersink* (Figure 18–1).

Standard countersinks are manufactured with an included angle of either 60° or 82°.

Courtesy of DoALL COMPANY

Figure 18–1 Standard Countersink

Countersinks may be made of carbon tool steel, high-speed steel, or carbides. Holes may be countersunk by chucking and revolving a countersink. The depth of cut is controlled with a stop gage or by measuring across the outside (large) diameter. The mating part may also be tried in the countersunk hole. Countersinking should be performed at about one-half the speed used in drilling the hole.

Chatter is produced when the speed is too great, the workpiece is insecurely held, or the countersink is forced.

COMBINATION DRILL AND COUNTERSINK

One widely used shop practice for accurately guiding a large drill is to drill a combination pilot and countersunk hole. The pilot and countersunk hole provides a bearing surface for 60° angle machine centers. This practice is known as *center drilling.*

A *combination drill and countersink* is used for center drilling. The point of the cutting tool is positioned in a center-punched hole. The small-diameter point end produces a pilot hole. The angular surface of the countersink portion enlarges the hole and forms a 60° cone-shaped surface. The countersunk hole is drilled to the depth at which the outside diameter is smaller than the required finished hole size. 90° countersinks and countersunk holes are used in aircraft riveting.

Combination drill and countersinks are available either with a single cutting end or with double cutting ends. Center drill tips are also available. These drill tips fit into adapters from which they may be easily removed and replaced. A single-point center drill tip and an adapter are illustrated in Figure 18–2. Center drills are made of carbon tool steel, high-speed steel, or carbide. The small-diameter point ends are fragile. Care must be taken not to exert excessive force on them during the drilling process. The feed per revolution is less on the countersink portion than for normal drilling.

Combination drill and countersinks are produced with many sizes of points and outside diameters.

COUNTERBORES AND COUNTERBORING

Counterboring refers to the process of enlarging and recessing a hole with square shoulders.

Single-point center drill tip   Adapter

Figure 18–2 A Center Drill Tip and Adapter

An enlarged and recessed hole that has square shoulders is known as a *counterbored hole.* Holes are counterbored to permit pins with square shoulders, screw heads, or other mating parts to fit *below* the surface of a workpiece. The cutting tool that is used for enlarging a hole is called a *counterbore.* A counterbore consists of a shank, body, and pilot.

The shank may be ground straight or to a standard Morse taper. A taper shank standard Morse taper counterbore with pilot are illustrated in Figure 18–3.

Different sizes of pilots may be used with one counterbore. Other counterbores that are used for only one size of screw head are made solid. Counterbores are usually three or four fluted. Counterbores are available in outside diameter sizes that vary, by increment of 1/16″, from 1/4″ to 2″. Similar metric sizes range from 6mm to 50mm.

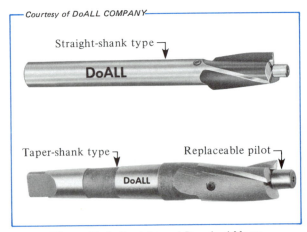

Straight-shank type

DoALL

Taper-shank type   Replaceable pilot

DoALL

Figure 18–3 Straight and Standard Morse Taper Counterbores

The cutting action of the counterbore is produced by the cutting lips. These lips are formed on the end (face) of each land. The cutting lips are radial. The lips are relieved to form the cutting angle. The flutes are spiral and provide for the flow of chips out of the workpiece.

Counterbores made of high-speed steel are widely used. Carbide-tipped counterbores are practical where tough materials are to be counterbored. Rough operations, where the outer surface of a workpiece is uneven or hard (as in the case of scale on castings), require a carbide-tipped counterbore.

General counterboring operations are performed at a slower speed than the speed used for drilling. A cutting fluid must be used on materials that require a lubricant.

DRILL/COUNTERBORE COMBINATION

Drilling and counterboring for socket-head and other standard fasteners may be performed in one operation. A short-length *drill/counterbore combination* is designed for such an operation. The drill/counterbore may be a solid type with a body of regular length or it may be a tip type.

The tip type fits into an adapter body. The body has a slightly tapered hole that holds and secures the tip. The adapter has a hole drilled through the body. A knockout rod is inserted into the hole and tapped against the tip to remove it. A typical set of combination drill/counterbore tips (Figure 18–4) accommodates socket-head cap screw sizes #5, 6, 8, and 10, and 1/4″, 5/16″, 3/8″, 7/16″, and 1/2″. Holes may be drilled and counterbored to diameter tolerances of +0.005″.

SPOTFACERS AND SPOTFACING

Spotfacing refers to the machining of a flat surface that is at right angles (90°) to a hole. Spotfacing produces an area against which a square-shouldered part may fit accurately. The operation may be performed on a drilling machine. The end-cutting tool is called a *spotfacer*. The surface is said to be *spotfaced*.

A counterbore may be used as a spotfacer when its diameter is large enough to produce a spotfaced area of the required dimension.

Courtesy of DoALL COMPANY

Figure 18–4 Set of Drill/Counterbore Tips in a Holder

BORING AND BORING TOOLS

Boring describes the process of enlarging a hole by using a *single-point cutting tool* (Figure 18–5). The cutting edge may be adjusted to accommodate different work sizes. Holes that have been formed by casting, punching, drilling, or any other process may be bored.

The cutting tool may be held in a *boring bar* or a *boring head*. The amount the tool is adjusted is indicated on a graduated scale. The

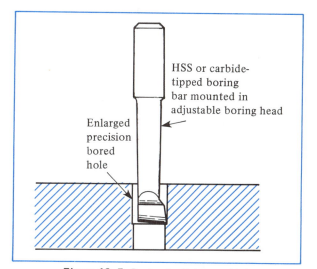

HSS or carbide-tipped boring bar mounted in adjustable boring head

Enlarged precision bored hole

Figure 18–5 Boring to Enlarge a Hole

boring process permits the taking of heavier finish cuts than is possible with reamers. The single-point cutting tool is adjustable to bore any diameter within the range of the boring head.

The taper shank of a boring head may be inserted directly in the taper bore of the drill press spindle. The cutter is secured in the offset section of the boring head. Small holes are usually bored on drilling machines. The same tables of recommended cutting speeds that are used for similar boring operations on lathes and boring mills are used. The cutting speed (sfpm) or m (min) is governed by the:

- Kind of material to be cut,
- Nature of the operation,
- Type of cutting tool that is used,
- Rigidity of the setup,
- Design of the drilling machine.

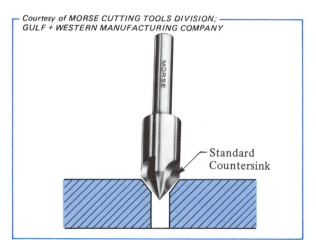

Standard Countersink

Figure 18–6 Countersinking to a Required Depth

How to Countersink a Hole

STEP 1 Mount the workpiece in a holding device. Drill a hole of the required diameter.

STEP 2 Select a countersink with the correct cutting angle and outside diameter.

STEP 3 Determine the cutting speed and RPM. Set the spindle speed (RPM).

Note: The speed is largely governed by the maximum outside diameter of the countersunk hole.

STEP 4 Replace the drill with the countersink. Bring the cutting edges carefully to the drilled hole. Note the reading on the graduated quill of the machine. Set the depth stop.

STEP 5 Start the machine. Brush or flow on the cutting fluid. Slowly cut to the required depth (Figure 18–6).

STEP 6 Stop the machine. Clean the workpiece. Check the depth.

STEP 7 Remove any burred edge.

How to Counterbore a Hole

STEP 1 Select a counterbore with a pilot that is a few thousandths of an inch smaller than the drilled or reamed hole (Figure 18–7).

STEP 2 Determine the cutting speed, spindle RPM, and feed. Set the machine speed and feed.

Note: Reamer speeds for the same diameter may be used.

STEP 3 Position the pilot in the hole. Bring the cutting edges carefully to the work surface. Set the spindle stop to the required depth.

STEP 4 Apply cutting fluid to the cutting area. Start to counterbore. Continue until the hole is counterbored to the required depth.

Note: It is good practice during the process to periodically withdraw the counterbore and check the cutting action and quality of surface finish.

STEP 5 Clean the workpiece. Burr the edge of the counterbored hole.

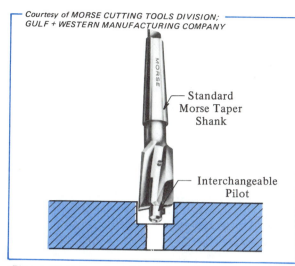

Courtesy of MORSE CUTTING TOOLS DIVISION;
GULF + WESTERN MANUFACTURING COMPANY

Standard
Morse Taper
Shank

Interchangeable
Pilot

Figure 18–7 Required Dimensions Met by the Pilot and Outside Diameter of a Carbide-Tipped Counterbore

How to Drill a Center Hole

STEP 1 Select the required size of combination drill and countersink. Chuck the cutting tool.

STEP 2 Position the center-punched hole in line with the pilot drill point.

STEP 3 Determine the cutting speed and RPM for the pilot drill. Set the spindle speed (RPM) to accommodate the diameter of the drill.

> **Note:** This spindle speed may need to be reduced when the full depth of the point is reached and the angular (countersink) portion begins to cut. Reducing speed avoids chattering and produces a fine surface finish.

> **Note:** Brush on a small quantity of cutting fluid to help to produce a better cutting action and a smooth-finished angular surface.

STEP 4 Continue to drill and countersink to the required depth. Remove any burrs.

How to Bore a Hole Using a Drill Press

STEP 1 Select a boring bar and boring head. The boring bar diameter must be small enough to permit adjustments of the cutting tool.

STEP 2 Examine the cutting edge of the boring tool. Sharpen if necessary.

> **Note:** Check the diameter that the combination of cutting tool and boring head will bore. The diameter should permit boring the hole to the specified size.

STEP 3 Determine the RPM for the recommended sfpm of a single-point cutting tool. Set the spindle at the required RPM.

STEP 4 Position the cutting tool over the pre-drilled or formed hole. Determine the additional amount of material to be removed.

STEP 5 Move the cutting tool to depth. Take a light cut for about 1/16″. Stop the machine. Measure the diameter.

STEP 6 Adjust the cutter if necessary. Take the cut. Bore through to the required depth.

STEP 7 Continue to take successive cuts.

> **Note:** If the hole is to be machined to close tolerances, more than one roughing cut may be required. The final finish cut is taken using a finer feed and a light cut of a few thousandths of an inch.

STEP 8 Burr the bored hole.

How to Spotface a Hole

STEP 1 Select a spotfacer with a pilot that is slightly smaller than the hole in the workpiece. The outside diameter of the cutter must meet the job requirements.

STEP 2 Set the machine speed and feed. Since there is a large cutting area, the speed should be estimated by the largest outside diameter to be cut. The feed should be less than the speed for drilling a hole to this same diameter.

STEP 3 Spotface to just clean the surface of the workpiece. Examine the quality of finish. Determine the depth dimension. Set the machine stop.

STEP 4 Continue to spotface to the required depth. Apply a cutting fluid, either by nozzle or brush, where required.

Note: If chatter marks appear, reduce the cutting speed and examine the angle of the cutting edges. The edges may be ground at too steep an angle.

STEP 5 Clean the workpiece and burr it.

Safe Practices in Countersinking, Counterboring, Boring, and Spotfacing

- Fit the spindle taper and cutting tool shank and tang properly. Securely seat them.
- Rotate all right-hand cutting tools in a clockwise cutting direction.
- Examine the cutting edges of the cutting tool and stone any burrs.
- Stop the machine if there is chattering. Correct the causes of chatter.
- Feed the small-diameter cutting point of a combination drill and countersink slowly and carefully into a workpiece. The point may be fractured if a heavy or jarring force is applied or if the speed is too fast.
- Use a brush to apply cutting fluid to the point where the cutting action is taking place.
- Remove wiping cloths from the work area whenever a cutting tool or spindle is to revolve.
- Secure loose clothing before operating any rotating machinery.

TERMS USED IN COUNTERSINKING, COUNTERBORING, BORING, AND SPOTFACING

Countersinking	The process of machining a conical surface at the end of a hole.
Countersink	A cutting tool with flutes and cutting edges ground at a specified included angle. An angular cutting tool used to produce a countersunk hole.
Counterboring	The process of enlarging the diameter of a hole. Recessing a hole with a shoulder to a required depth.
Counterbore	A multiple-fluted end-cutting tool. (The outside diameter accommodates the diameter of a mating part that is to be inserted in a counterbored hole.) A piloted end-cutting tool that produces a concentric, shouldered, recessed hole.
Spotfacing	The process of producing a surface of a given diameter that is square with the axis of a hole.
Spotfacer	A double-edged or multiple-fluted cutting tool of a required diameter. An end-cutting tool that produces a flat, round surface.
Center drilling	The process of drilling a pilot and countersunk hole with a combination pilot and center drill.
Boring	The process of enlarging a hole with a single-point cutting tool. Machining a hole round. Accurately positioning the location of a hole.
Boring head	An accessory for holding and accurately positioning a cutting tool. Used in machining a hole to a close-tolerance dimension.

SUMMARY

- Recessed areas for angular and shouldered screws, bolts, and pins are produced by countersinking or counterboring.
    - Spotfacing produces a flat, circular machined surface. The surface is perpendicular to the axis of a hole.
- Holes may be enlarged by boring. Boring tools may be adjusted to produce holes of various diameters. Bored holes may be machined to precise limits—for example, ±0.0005″ and ±0.02mm.
    - The 60°, 82°, and 90° (for aircraft applications) included angles of countersink cutting lips are standard. Standard countersinks accommodate screw heads, angular rivets, and parts that are commonly manufactured to these angles.
- Counterbores are made of high-speed steel for general machining. Tungsten carbide tips are used for heavy-duty counterboring and on abrasive, hard, tough materials.
    - Counterbores are designed solid or with a removable pilot. The pilot is slightly undersized to avoid scoring a previously drilled or reamed hole.
- The spotfacer is an end-cutting tool. A multiple-lip counterbore of the required outside diameter may be used for spotfacing. A spotfacing tool having a two-lip cutter that is secured to the cutting bar is also commonly used.
    - The cutting speed for spotfacing and counterboring depends on the outside diameter and the material in a workpiece.
- The cutting speed for countersinking and counterboring is slower than the speed used for drilling.
    - Cutting fluids are used in countersinking, counterboring, and spotfacing materials that require a lubricant and/or cutting compound.
- General safety precautions relating to rotating spindles, cutting tools, burrs, and the use of eye protection devices must be observed.

UNIT 18 REVIEW AND SELF-TEST

1. State two main functions served in the machine shop by holes that are countersunk to either 60° or 82°.

2. Differentiate between a combination drill and countersink and a center drill tip.

3. Indicate two functions that are served by an interchangeable pilot on a counterbore.

4. Tell what the difference is between spotfacing and boring with respect to the cutting tools that are used for each process.

5. List the steps for boring a hole when using a drill press.

6. List three safety precautions to observe when working with countersinks, counterbores, spotfacers, and boring tools.

SECTION ONE

Engine Lathes: Design Features and Operations

UNIT 19

Functions, Controls, and Maintenance

OBJECTIVES

A. FUNCTIONS, TYPES, AND FEATURES

After satisfactorily completing this unit, you will be able to:

- Understand lathe construction in terms of (1) power source and motion-producing mechanisms, (2) work-holding and work-rotating accessories, (3) components for producing tool and work movements, and coolant systems.
- Identify the principal lathe components and systems and their operation, including work-holding devices, tool-holding setups, and general lathe accessories.
- Visualize functions, sizes, and capacities of manufacturing and special form-turning lathes from specifications.
- Analyze the reasons for following recommended *Safe Practices*.
- Communicate in *Terms* which are common to lathe work.

B. LATHE CONTROLS AND MAINTENANCE

- Identify and use headstock controls on various head drives.
- Apply feed and thread controls in relation to the quick-change gear box, apron, and thread-cutting attachment.
- Understand the operation of overload safety devices and the care and maintenance of the lubricating and cutting fluid system.
- Interpret common machining problems, causes, and preventive actions.
- Perform the following processes.
  - Align the Tailstock.
  - Clean and Lubricate the Lathe.
  - Clean the Coolant System.
- Follow recommended *Safe Practices* dealing with lathe controls and maintenance.

The lathe is used for *external* and *internal cylindrical machining processes.* Basic external operations include facing, turning, knurling, cutting off, threading, and polishing. Basic internal operations relate to drilling, boring, turning, undercutting, countersinking, counterboring, reaming, tapping, and threading.

All of these internal and external operations are covered in detail. Principles and applications of cutting tools, work-holding and driving devices, and machine setups are described. Cutting speeds, feeds, cutting fluids, and other machining practices are also included.

A. FUNCTIONS, TYPES, AND FEATURES

HISTORIC MACHINE TOOL DEVELOPMENTS

THE ALL-IRON SCREW-CUTTING LATHE

As stated in Unit 1, today's machine tools: lathes, mill/turning centers, and other automated machining centers and systems, are all a part of a continuous line of successive developments.

However, it is David Wilkinson of Rhode Island and the English instrument maker Henry Maudslay who are credited with establishing the basic reference point for systematic machine tool design and construction. Wilkinson designed and built a *lathe carriage* in 1794 (which he patented in 1798). Building on this development and the experiences of earlier inventors, Henry Maudslay succeeded in 1800 to combine six important design features in a lathe that was completely made of *iron.* This *all-iron screw-cutting lathe* combined a spindle, slide rest, V-ways, lead screw, change gears, and a work support (tailstock) and provided for a power drive. These same fundamental features are incorporated in the most sophisticated precision lathes of today.

Maudslay's screw-cutting metal lathe laid the groundwork for the design and manufacture of other machine tools and the mass production of interchangeable parts. The metal-cutting lathe met the growing need to industrialize. The capability of the lathe was increased. More complex parts were machined to ever higher dimensional tolerances and qualities of surface finishes.

BASIC FEATURES OF THE MAUDSLAY LATHE

The power-driven *spindle* rotated a workpiece. The *slide rest* supported cutting tools so that they could be positioned and moved along a workpiece. The turning of the *lead screw* produced movement of the slide rest at a constant rate of feed. The *change gears* made it possible to vary the distance the slide rest moved in relation to the spindle RPM. The change gears were also important in changing the nature of a cut. Various feeds were needed for rough and/or finish machine operations and to produce screw thread pitches of different sizes. The *tailstock* provided a center on which to support workpieces. One of the accessories on the Maudslay lathe was a *follower rest*, to support long workpieces.

Made all of iron and of heavier construction than previous lathes, the Maudslay lathe increased the speed and the accuracy with which parts could be machined and the cutting capability to machine metal parts. The term *engine lathe* was derived from the fact that lathes in the early 1800s were power driven by steam engines.

THE HAND-FED, EIGHT-STATION TURRET LATHE

From Maudslay's developments came the need for other machines. The milling machine, boring mill, and production screw-cutting machines are examples. The demands for metal fasteners like screws, bolts, and other threaded parts led to an adaptation of the early metal lathe.

In 1845 Stephen Fitch of Middlefield, Connecticut, designed and built an *eight-station turret lathe.* Eight tools were mounted on and positioned by a turret mechansim. Each tool or positioning device was formed by a skilled craftsperson. The tools and devices were held in the turret. Once the machine was set up, a less-skilled worker operated the turret lathe and mass-produced interchangeable parts.

In the mid-1800s the Northeast became the center of many significant machine tool developments. Gay and Silver of Chelmsford, Massachusetts, and James Hartness of Windsor, Vermont, also developed turret lathes. The Pratt and Whitney and Brown & Sharpe machine tool companies were founded to manufacture small turret lathes during the Civil War.

THE AUTOMATIC TURRET LATHE (SCREW MACHINE)

Still greater production demands for metal machine screws led to the invention of an *automatic turret lathe* in the late 1860s. This lathe combined the features of Maudslay's screw-cutting lathe and Fitch's turret lathe. An added feature, called the *brain wheel* by its designer, Christopher Spencer, was a cam-activated device. The brain wheel (cam) positioned a cutting tool to perform a particular operation, such as straight turning. At the end of a cut, a cam moved the cutting tool away from the workpiece. The die, mounted in a turret station, was brought to the workpiece. The die cut the threads in the turned workpiece. After the die was withdrawn, other cams positioned cutting tools and stops in the different turret stations.

Since the automatic turret lathe was designed principally to complete machine screw threads, it became known as the *automatic screw machine.* Spencer later developed the first multiple-spindle lathe. He was finally able to mass-produce small screws by feeding coiled wire through a three-spindle, automatic screw machine.

ENGINE LATHES AND MACHINE TURNING CENTERS

Successive improvements and additions have been made throughout industrialized nations to the work of Maudslay, Fitch, Spencer, and the early machine tool companies in the Northeast and other parts of the United States. This work has contributed significantly to the present high-ly sophisticated mass-production capacity to make precision parts and mechanisms.

The engine lathe remains a basic machine tool. A modern, conventional lathe with major features identified is shown in Figure 19–1. The parts and mechanisms of a lathe serve one of the following three major and interrelated functions:

- To provide power to all moving parts,
- To hold and rotate a workpiece,
- To position, hold, and move a cutting tool.

THE LATHE BED

The body to which other components of a lathe are fitted is called the *lathe bed.* It consists of a ruggedly designed casting or weldment. *V-ways* and other *ways* are machined on the top surface of the bed. A headstock, gearbox, lead screw and feed rod drive, and coolant system are usually attached to the lathe bed. The carriage and tailstock mechanisms are aligned on the ways of the bed and may be moved longitudinally.

MOTION-PRODUCING MECHANISMS

Motion for all moving parts of a lathe is transmitted from a power source such as a *motor.* The motor is connected by belts or gears to a *headstock* through which a spindle is driven. The speed and feed of each moving part of the lathe is directly related to the spindle movement.

End gears are used to establish the precise RPM of the spindle in relation to the *gear train* in the *quick-change gearbox.* The gear ratios, in turn, control the RPM of the *lead screw* and *feed screw* according to the spindle RPM.

The rotary motion and speed and feed of the lead screw are converted in the *apron.* Linear movements are produced for the *carriage* (longitudinal feed) and *cross slide* (transverse feed). These same movements of the carriage and cross slide may also be performed by hand feeding. On some lathes the lead screw serves the function of controlling the movement for both threading and feeding.

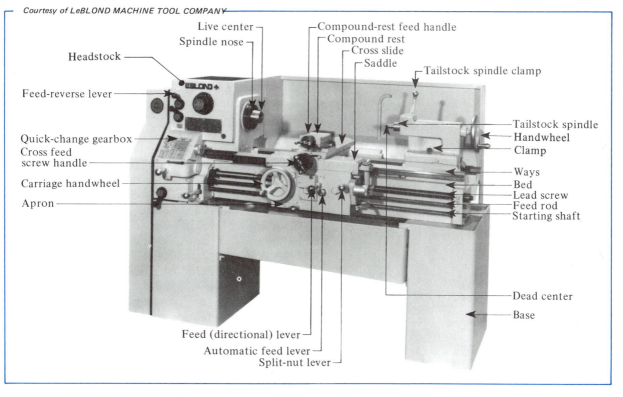

Figure 19–1 Major Features of a Conventional Lathe

COOLANT SYSTEMS

The purposes for which a cutting fluid is used in lathe work are the same as for other machining processes. The fluid serves:

- As a coolant to reduce the machining temperature of the cutting tool and workpiece,
- As a lubricant to improve the cutting action and the quality of the surface finish,
- As an agent to reduce the amount of force required during the cutting process.

Many lathes have a built-in *coolant system*. The system requires a *chip pan*, which is attached to the bed and base. Both the chips and the cutting fluid flow into the chip pan. The cutting fluid drains from the chip pan into a reservoir in the coolant system.

The coolant system consists of a circulating pump, a reservoir, piping, and a nozzle. The cutting fluid is pumped through a flexible hose to a nozzle. The nozzle is mounted on the carriage

and is positioned to deliver the fluid where it is needed. The fluid follows the cutting action. As it returns to the reservoir, it flows through cleaning meshes. The fluid is recirculated after being cleaned.

LATHE COMPONENTS FOR PRODUCING WORK AND TOOL MOVEMENTS

The headstock and the spindle serve a work-rotating function. The end gears, quick-change gears, and carriage mechanism are all related to tool movement functions. The tailstock serves a work- or tool-supporting function as well as a rotating function.

THE HEADSTOCK

The headstock is located at the head, or left-hand end, of the lathe bed. The headstock contains the spindle drive mechanism and the lathe spindle.

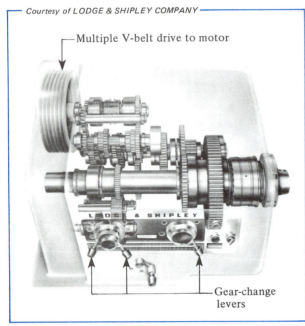

—Multiple V-belt drive to motor

Gear-change levers

Figure 19–2 A Phantom View of a
Gear-Drive Headstock

Power and motion from a motor are transmitted through the headstock drive mechanism to the lathe spindle. The spindle may be driven by *multiple V-belts,* a *variable-speed drive,* or by *gear transmission.* Figure 19–2 is a phantom view of a gear-drive headstock.

The spindle speed of small cone-pulley belt-driven lathes is controlled by the position of the belt on one of the step-cone pulleys. The drive may be direct. Speed changes on belt drives are made by shifting the belt to a different step-cone pulley combination.

Some variable-speed drives provide for a wide range of speeds by engaging a set of *back gears.* Still other variable speed drives may be controlled electrically or mechanically while the lathe is in motion.

Heavy powered lathes have completely geared headstocks. The gears are engaged in a manner similar to shifting an automobile transmission. The gear ratio on a geared-head lathe is changed by *speed-change levers.* Spindle speed changes must be made when the gears are *not* in motion to prevent stripping the gear teeth. *Gear* combinations *(trains)* provide double, triple, quadruple, and other spindle speed ranges.

On the end of the headstock there is an *end gear train.* The gears are enclosed by modern guards that are provided with a *safety electrical disconnect switch.* The switch shuts down the machine when the end guard is removed. The gears within the geared headstock are of special alloy steel. The gears are induction hardened and precision ground. All parts are lubricated with an oil bath system. The oil level is sighted on an oil level indicator.

The required spindle speed is obtained by setting the speed-selector levers. The spindle is actuated by a *forward/stop/reverse lever.* This lever is located on the apron to provide on-the-spot control by the operator from any operating position.

New machines are being designed for increases in spindle speeds. Fluid and electro-magnetic devices, variable-speed motors, and other mechanisms are being used to meet increased machine tool requirements.

THE HEADSTOCK SPINDLE

The dominant feature of the lathe is the spindle. The condition, accuracy, and rigidity of the spindle affect the precision of most machining operations and the quality of surface finish. The headstock spindle is mounted in opposed, preloaded, precision antifriction bearings. *Spindle runout,* or *eccentricity,* is reduced to a minimum. The spindle is hollow to permit small parts, bar stock, and lathe attachments to pass through it. The face end, or nose, of the spindle is accurately ground with an internal taper. Adapters are fitted to reduce the size of the spindle hole to permit accessories, like a sleeve for the live center, to be inserted. Spindles are hardened and ground.

SPINDLE NOSE TYPES

The spindle nose is accurately ground to receive nose mounting plates. These plates align chucks and faceplates in relation to the spindle. The three basic designs for the outside of a spindle nose include the

- Older threaded spindle nose with a squared shoulder,
- Long taper spindle nose, and the
- Cam-lock spindle nose.

The *threaded spindle nose* receives a threaded mounting. The shoulder helps to align the mounted device. The threads hold and secure the device in place and transfer the spindle driving force.

The *long taper spindle nose* provides a tapered surface. This surface receives the corresponding taper of a spindle nose mounting plate. A key transfers the spindle driving force to the nose mounting. The threaded ring on the spindle screws onto a matching thread. The threaded ring draws and securely holds the mounting on the taper. A spanner wrench is used to tighten or loosen the threaded ring nut.

The *cam-lock spindle nose* has a short taper. This taper accurately positions the spindle mounting. The spindle mounting has cam studs that fit into a ring of holes on the face. The spindle mount is held securely on the taper and against the face of the spindle lock nose by the cam studs. These studs are secured by turning a chuck key. The spindle mounting is driven by the cam studs.

END GEARS AND QUICK-CHANGE GEARBOX

The gears on the end of the headstock between the spindle gear and the drive gear on the gearbox are called *end gears*. End gears serve two functions:

- To transmit motion from the spindle to the lead screw and feed screw through the gearbox,
- To change the direction of the lead and feed screws.

The relationship between the spindle speed and the lead and feed screws is established through a series of gears. The gears in a *quick-change gearbox* are engaged by levers.

A gear-change mechanism for thread pitches (Unified and SI metric) is pictured on the lathe headstock in Figure 19–3. Gear shifting is done by positioning one or more of the three knobs *when the spindle is stopped.* The range of Unified threads that may be cut on this lathe is from 11 to 100 threads per inch. The metric thread combinations may be set from 0.275mm to 2.7mm pitches.

An *index plate* on a quick-change gearbox indicates the various lever positions for setting

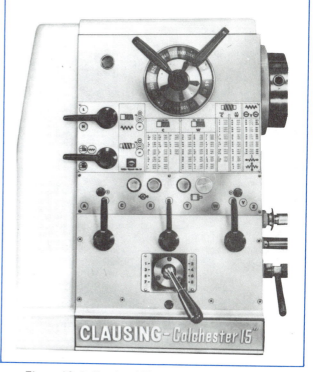

Figure 19–3 Feed and Thread Pitch Change Knobs (Unified and SI Metric Systems)

the gearbox for either feed or thread cutting. Pitches and feeds are given on the quick-change gearbox plate in terms of both inch- and metric-standard measurements.

THE CARRIAGE

The lathe cutting tool is secured, moved, and controlled by different attachments and mechanisms on the lathe carriage. The carriage consists of a *saddle*, *cross slide*, and *apron* (Figure 19–4). With this combination, a cutting tool may be fed into or away from a workpiece. The rotary motion of the lead screw and the feed screw is converted to a horizontal motion. The movement may be across the length of a workpiece and/or at a right angle to it. A *compound rest* is attached to the cross slide to produce small tapers and other angle surfaces.

The Apron. The controls and mechanisms for all movements of the carriage are housed in the apron. A selector lever positions the proper

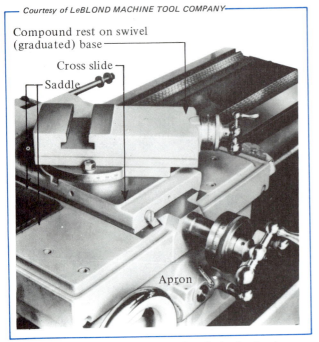

Figure 19–4 Major Design Units of a Lathe Carriage

gears for longitudinal and cross feeds. A friction clutch engages the feed. A *half-nut lever* permits engaging the lead screw for thread cutting.

The Saddle. The saddle supports the apron and the cross slide, on which a compound rest is mounted. The cross slide mechanism provides tool movement along the work axis or at right angles to it. The crosswise motion is produced whenever the cross slide feed screw is turned either by hand or power.

The Cross Slide and Compound Rest. The compound rest is mounted on the cross slide. The compound rest may be swiveled through 360° to any angular position. The base is graduated in degrees to permit angular adjustments. The compound rest may also be swung to conveniently position a cutting tool. The compound rest is fed by hand.

The basic functions of the compound rest include:

- Cutting angles and short tapers,
- Moving the cutting tool to face to close tolerances (when set parallel to the ways of the lathe),
- Feeding the tool in thread cutting.

Cross Slide (Traverse) and Compound Feed Dials. The amount the cross slide is moved may be measured on a cross slide feed dial. Similar measurements for compound rest movements are read on a compound feed dial.

The feed dials are graduated according to the precision built into the lathe and the general job requirements for which it was designed. Some feed dials have readings in thousandths of an inch (0.001″) or two-hundredths of a millimeter (0.02mm). Higher precision lathes have dials (collars) that read to one ten-thousandth of an inch (0.0001″) and one one-hundredth of a millimeter (0.01mm).

Many new lathes have dual inch-standard and SI metric-standard dials. These dials are located on the cross slide and compound rest feed screws and on other lathe micrometer measuring devices (Figure 19–5).

THE TAILSTOCK

There are two units to a tailstock: a *base* and a *head.* The base may be positioned quickly and secured *longitudinally* along the bed of the lathe. The head may be moved *transversely* at right angles to the lathe central axis. These movements position the tailstock either *on center* (the headstock and tailstock axes aligned) or *off center* (for turning tapers). A *zero index line* and other graduations are machined on the base

Figure 19–5 English/Metric Combination Dials

and head sections. These graduations are used to center or offset the tailstock.

The head includes the *tailstock spindle*. The spindle has a standard Morse internal taper. A *dead center* is secured in the tapered spindle.

LATHE ACCESSORIES FOR HOLDING WORK OR TOOLS

TOOLHOLDERS (TOOL POSTS)

Cutting tools are held and positioned in holders by three general types of cutting-tool attachments. These attachments are called *tool posts*.

Type 1 is designed for a single operation at one setting. The cutting tool is held in a holder. The holder is positioned in a *tool block* (rectangular) or *tool post* (round) holding device.

Type 2 provides for multiple operations (Figure 19–6). Each tool is adjusted in the holding device. Up to four successive operations are performed by a quick-release and positioning lever. This type of cutting-tool holder and positioning device is often called a *four-way turret*.

Type 3 is a *precision* block and toolholder that permits rapid tool changes. The cutting tools are *preset* in each holder. The cutting tool and holder are accurately positioned on a locating pin and in an accurately slotted tool block (Figure 19–6).

Each of the three basic types of toolholders has a T-shaped base. The base fits the T-slot of the cross slide. The base is tightened and held securely in the T-slot.

LATHE CENTERS

Lathe centers are available with tapered bodies. The *live center* requires a tapered adapter or sleeve. The center and sleeve are held directly in the headstock spindle. The *dead center* is mounted in the tailstock spindle. A workpiece revolves on the dead center.

Both live and dead centers may be of the solid, hardened-steel type. Carbide-tipped dead centers are available. These centers have the ability to withstand high temperatures caused by friction. However, it is often desirable to use a revolving dead center to reduce friction.

THE FACEPLATE

A *faceplate* is a common work-holding device. The elongated slots permit unusually shaped workpieces to be positioned and strapped securely against the vertical (90° to the lathe axis) plane surface of the face plate.

Care must be taken in faceplate work to use balancing weights. These weights offset any

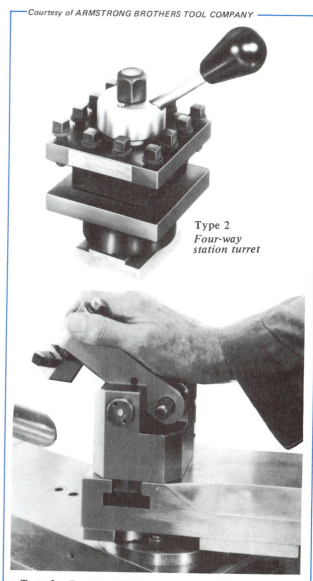

Type 2
Four-way station turret

Type 3 *Precision block and preset toolholders*

Figure 19–6 Basic Types of Tool-Post Holders

centrifugal forces that may be created by the unequal distribution of a mass.

THE DRIVER PLATE

A great deal of work is machined on a lathe between centers. A centered workpiece is driven by a *lathe dog* and *driver (dog) plate.* The flat, flanged driver plate fits on and is secured to the spindle nose. The tail of a lathe dog rides in one of the two radial slots to drive a centered workpiece.

LATHE ATTACHMENTS

There are five general lathe attachments. Their names indicate their prime functions: taper attachment, threading attachment, micrometer stop, center rest (steady rest) and follower rest, and rapid traverse mechanism.

TAPER ATTACHMENT

The *taper attachment* is one of the most accurate, fastest, and practical methods of cutting a taper on a lathe (Figure 19–7). An adjustable angle attachment is clamped against one of the back ways. The mechanism consists of an adjustable block that clamps to the cross slide.

The block moves at a fixed rate according to the angular (tapered) setting of the attachment. This produces the same movement in the cross

slide. As the longitudinal feed is engaged and the work turns, a taper is produced. There are two sets of graduations on the tapered attachment. These graduations read in *degrees of taper* and *inches of taper per foot* (or metric equivalents). On the model illustrated, graduations at the headstock end provides an accurate vernier setting.

THREADING ATTACHMENTS

The *thread-chasing dial* is one type of threading attachment. It is fitted on the side of the apron. The chasing dial rotates freely when the split nut is not engaged for threading. The dial stops rotating when the split nut is engaged.

Another common threading attachment is the *high-speed threading attachment.* The attachment shown in Figure 19–8 fits on the side of the apron. It contains its own split nut and a tripping mechanism. An adjustable stop is preset for the length of thread that is to be cut. The split nut automatically disengages each time the preset length is reached. The feed-reverse lever is used to change the direction of the lead screw to right- or left-hand threads.

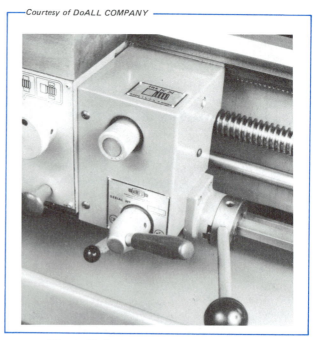

Figure 19–8 Automatic, Adjustable Stop, High-Speed Threading Attachment

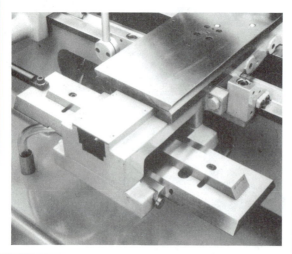

Figure 19–7 A Taper Turning Attachment

MICROMETER STOP

The *micrometer stop* is a convenient device consisting of a single or multiple stop micrometer head and an attachment for securely clamping the stop on one of the front ways. A micrometer setting provides an accurate stop. The carriage is brought to the micrometer stop when duplicating multiple pieces or when finishing parts or surfaces to exact lengths.

STEADY (CENTER) REST AND FOLLOWER REST

A long workpiece must be supported at regular intervals and continuously while a cut is being taken. The *steady rest (center rest)* performs this function. It helps to prevent springing of a workpiece under the cutting action of a cutting tool. The steady rest is secured to the ways at a point along a workpiece that provides maximum support.

The *follower rest* performs a similar function except that it follows the cutting action. The follower rest is attached to the saddle and therefore moves along the work (Figure 19–9). The work-supporting shoes of the follower rest support a revolving workpiece at two places. The shoes prevent the forces of the cutting tool against a workpiece from springing the workpiece.

RAPID POWER TRAVERSE

Considerable time is saved, particularly on heavy models of lathes, by the addition of a *rapid traverse attachment* (Figure 19–10). The attachment relates to increased speed in making forward and reverse longitudinal movements of the carriage and the cross slide.

Rapid power traverse is controlled through longitudinal and cross feed friction levers. A safety clutch is incorporated to automatically disengage one of the feeds and thus prevent overloading.

LATHE SIZES

Lathe models are usually designated by specifications such as *swing over bed* and the *center-to-center distance between the headstock and tailstock* (Figure 19–11). For example, a 13″ (330mm) by 25″ (635mm) general-purpose lathe will swing a 13″ (330mm) diameter workpiece over the bed. However, only an 8 1/4″ diameter (210mm) part will clear the cross slide. The 25″ (635mm) center distance will accommodate a maximum length of 25″ between the centers of this lathe.

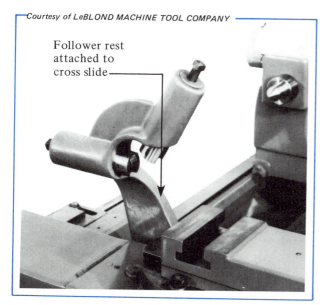

Follower rest attached to cross slide

Figure 19–9 A Follower Rest

Figure 19–10 A Rapid Traverse Attachment

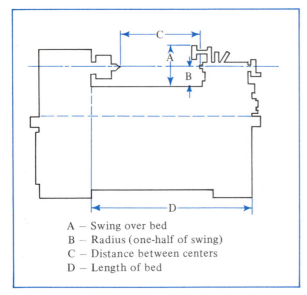

A — Swing over bed
B — Radius (one-half of swing)
C — Distance between centers
D — Length of bed

Figure 19–11 Designations of Lathe Size and Swing

CATEGORIES OF LATHES

Lathe models may be grouped in three broad categories: all-purpose lathes,   production lathes, and special production lathes.

ALL-PURPOSE LATHES

Small sizes of *all-purpose* lathes are mounted on benches and are designated as *bench* lathes. Lathes of still smaller size that are used by instrument makers are called *instrument* lathes. Larger-sized (diameter and length) lathes like 10″ to 48″ (250mm to 1200mm) models are known as *engine* lathes.

A high-precision model may be specified as a *toolroom*, or *toolmaker's*, lathe. The toolroom lathe has a wide range of speeds, especially high speeds, and is usually equipped with a complete set of accessories. The tolerances of all parts of the toolroom lathe are held to closer limits than the limits of a standard lathe. The toolroom lathe makes it possible to machine to the high degree of accuracy required in toolmaking, instrument making, and other precision work.

A lathe that has a swing of 20″ or larger and is used for roughing (hogging) and finishing cuts is often referred to as a *heavy-duty* lathe. The swing (capacity) of a lathe may be increased. A

section of the lathe bed adjacent to the headstock may be cut away to permit large diameter workpieces to be turned. In other words, there is a gap in the bed next to the headstock. This lathe is referred to as a *gap* lathe.

PRODUCTION LATHES

Large quantities of duplicated parts that require lathe operations to be performed at high-production rates are produced on adaptations of the lathe. Horizontal, vertical, and slant bed bar and chucker turning machines (turret head machines) and numerically controlled turning centers and combination machining centers are examples.

The design of these machine tools, multiple tooling setups, and simultaneous machining processes and technology are treated in detail in the *Advanced Machine Tool Technology and Manufacturing Processes* book.

SPECIAL LATHE EXAMPLES

Two examples of *special production* lathes are the *duplicating* lathe and the *floturn* lathe. These lathes perform processes (contour turning and roll flowing) that increase the range of common lathe work.

Duplicating Lathe. The duplicating, or contour, lathe is used in machining irregular contours and blending sections. The cutting tool movement is guided by air, hydraulic, mechanical, or electrical devices. The shape that is produced is the result of a tracing device. This device follows the contour of a *template* and controls the movement of the cross slide tool.

Floturn Lathe. The floturn lathe is used for *roll flowing* as a method of cold-forming metal. A heavy pressure is applied spirally with two hardened rollers against a metal block. The lathe and the floturn process may be used with flat plates and forgings and castings and for extruded and other shapes.

The metals may range in hardness from semi-hard steels to softer ferrous metals. In operation, the mandrel and workpiece turn. A roll carriage forces the metal ahead of it. The squeezing and stretching of the metal cause the workpiece to

SPECIFICATIONS OF BASIC LATHE MODELS

The design and construction of engine lathes cover a wide range of design features and mechanisms. For example: the diameter, bore, and nose shape of the spindle mounting; gear trains for spindle speeds, carriage feed rates, and threading capacity, etc.

Specifications of three basic lathe models (general purpose, precision toolroom, and heavy-duty), manufactured by different machine tool builders, are featured in Table 19–1.

Table 19–1 Selected Specifications of Basic Engine Lathe Models

	13″ (330mm) General-Purpose Lathe		*(250mm) 10″ Precision Toolroom Lathe*		*48″ (1220mm) Heavy-Duty Turning Lathe*	
	Inch Standard	*Metric Standard (millimeters)*	*Inch Standard*	*Metric Standard (millimeters)*	*Inch Standard*	*Metric Standard (millimeters)*
Capacity (General)						
Swing over bed	13″	330mm	12 1/2″	315mm	48″	1220mm
Swing over cross slide	8 1/4″	210mm	7 1/4″	180mm	36″	915mm
Center distance	25″	635mm	20″	510mm	60″	1520mm
Headstock						
Hole through spindle (diameter)	1 1/2″	38mm	1 13/32″	35mm	4 1/16″	103mm
Spindle range	40 to 2500 RPM		High 40 to 4000 RPM Low 8 to 800 RPM		6 to 750 RPM	
Number of spindle speeds (forward and reverse)	24		Infinite		36	
Taper of spindle center	#3 Morse taper		#2 Morse taper		#7 Morse taper	
Motor (two speed)	3 hp (2.25 kw)		5 hp (3.75 kw)		60 hp (45 kw)	
Cross Slide						
Cross slide travel	7 1/2″	190mm				
Compound rest travel	5 1/2″	140mm	2″	50mm	7″	45mm
Thread and Feed Ranges						
Thread changes (UN)	35		60		48	
Thread range (TPI)	2 to 56		3 to 184	*English/Metric Gear Box*	1/2 to 28	
Number of metric pitches	39			25		48
Range of metric pitches	0.2 to 14mm			0.25 to 11mm		0.75 to 42.0mm
Range of feeds (longitudinal)	0.001″ to 0.040″	0.03 to 1.0mm	0.0005″ to 0.016″	0.013 to 0.406mm	0.0035″ to 0.196″	0.09 to 5mm
Range of feeds (cross)	0.005″ to 0.020″	0.013 to 0.50mm				
Lead screw	1 1/8″ × 4 TPI	M28 × 4 TPI	1″ × 8 TPI	1″ X 8 TPI	2 1/2″ × 2 TPI	M63 × 12.7
Tailstock						
Spindle travel (traverse)	4 3/8″	110mm	3 1/2″	85mm	12″	300mm
Spindle taper	#3 Morse taper		#2 MT	#2 MT	#7 Morse taper	
Setover	± 1/2″	± 12mm	± 1/2″	± 12mm	± 1/2″	± 12mm

take the contour of the mandrel. The wall thicknesses can be held accurately to within ±0.002″ (0.05mm).

B. LATHE CONTROLS AND MAINTENANCE

HEADSTOCK CONTROLS FOR SPINDLE SPEEDS

STEP-CONE PULLEY DRIVES

Spindle speed (RPM) changes on lathes with step-cone pulley drives are made by stopping the lathe. The belt tension lever is then moved to loosen the belt. The manufacturer's spindle speed table is usually mounted on the headstock cover. The table (index plate) gives the full range of spindle speeds.

BACK-GEAR DRIVES

On some designs of lathes there is a back-gear drive for slower than normal speeds. The back gears increase the force the spindle may exert for heavier cuts. The speed is reduced to provide a slower range of RPM. Back gears should be changed only when the power is off.

The back gear mechanism is engaged or disengaged by moving the lever or control knob.

VARIABLE-SPEED DRIVE CONTROLS

Spindle speed changes are made on variable-speed drives while the motor is running. The speeds are changed hydraulically by positioning the speed-control dial at a desired speed. The spindle RPM is read directly from the dial at the

Courtesy of CLAUSING CORPORATION

Figure 19–12 A Variable-Speed Control Dial for High and Low Spindle Speed Ranges

index line. Figure 19–12 shows a variable-speed control dial. The speed range for the back-gear drive is from 52 to 280 RPM. The direct drive range (open belt) is from 360 to 2,000 RPM.

Back-gear (low spindle) speed combinations on variable-speed drives are engaged by a second lever on the headstock while the lathe spindle is stopped.

GEARED-HEAD CONTROLS

The geared-head lathe incorporates all the features of direct spindle drive and back-gear reduction. Spindle RPM changes are made when the spindle is stopped and different sets of gears are engaged.

The spindle speed changes appear on an index plate that is attached to the headstock. The positions of the gear-change levers are indicated for each speed.

MACHINE CONTROLS FOR FEEDS AND THREADS

QUICK-CHANGE GEARBOX CONTROLS

The *quick-change gearbox* serves the two-fold function of turning the feed screw and the lead screw. An index plate on the gearbox indicates the range of feeds per revolution and the pitch of inch-standard or metric threads.

The positions and combinations of the quick-change gear levers for a particular feed or screw thread are given on the index plate. Two levers control the feed and thread pitch ranges. Another lever known as the *feed-reverse lever* changes the direction of rotation of the lead and feed screws.

APRON CONTROLS

The actual automatic feeding of the cutting tool on the carriage is controlled by a *feed-change lever* or automatic feed knob. Usually, when the lever is in the up position, the cutting action is toward the headstock. The lever is moved to a neutral position to stop the feed. The downward position of the feed lever reverses the direction of feed. Similarly, a second feed lever or knob engages the automatic cross feed. The direction of the cross feed may be changed by moving the feed-reverse lever.

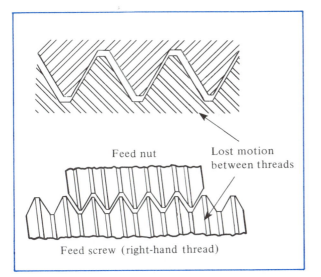

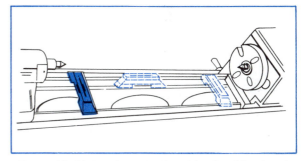

Figure 19–14 Leveling a Lathe with a Precision Level

Figure 19–13 Lost Motion between Two Threads

LOST MOTION

All machine tools and instruments have *lost motion*. Lost motion is the distance a male or female screw thread or a gear tooth turns before the mating slide, gear, or other mechanism moves. Lost motion is also known as *backlash*, *slack*, *play*, or *end play* (Figure 19–13).

Lost motion may be corrected by withdrawing or turning (backing off) the screw thread or gear away from a required point. The direction is then reversed, and the cutting tool is advanced by feeding it in one direction only.

For example, a cutting tool that is mounted in a tool post on a compound rest is advanced by turning the cross feed handwheel clockwise. If the cutting tool is moved too far, the correct diameter cannot be machined by simply reversing the direction of motion and bringing the cutting tool out of the workpiece a given distance. Instead, the tool must be backed away from the workpiece. The required distance is at least a part of a turn beyond the graduated measurement to which the tool is to be set. The cross feed screw direction is then changed (reversed).

The cutting tool is advanced to the required graduated reading on the handwheel dial. A trial cut is taken for a short distance. The part is measured. All subsequent cuts may be accurately taken by continuously advancing the cutting

tool in the one direction. Checks are made with a micrometer or other measuring instrument.

LATHE ALIGNMENT AND ADJUSTMENTS

One of the requirements in installing any precision equipment is *leveling*. Leveling is the process of positioning a machine surface so it falls in a true horizontal (0°) plane or a right-angle (90°) vertical plane. When a machine is designed and constructed, the mating and moving surfaces are machined in relation to these planes. When a machine is installed, leveling affects all of the machine parts, mechanisms, and the actual accuracy in machining a workpiece. The leveling of a lathe is shown in Figure 19–14.

A *precision level* is used to level the lathe bed. The level is placed lengthwise. The adjusting screws in the base or leg are turned until the bubble of the level is centered in the level sight.

The level is turned 90° and placed across the ways. Sometimes parallels are used between the flat ways and the level. The leveling screws are again adjusted. The lengthwise leveling process is then repeated to ensure that the longitudinal leveling has not been changed. Once the machine is leveled, shims are placed under the legs and the machine is bolted or nested in position. The levelness is again checked. Further adjustments are made as necessary.

OVERLOAD SAFETY DEVICES

There is always the possibility of overloading gear trains, lead and feed screws, and other machine parts. Manufacturers use *shear pins* and *slip clutches* to prevent damage.

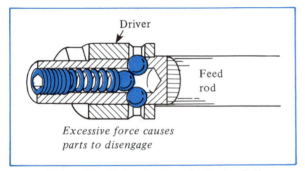

Figure 19–15 Application of a Spring-Ball Slip Clutch to a Feed Rod

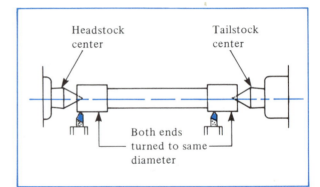

Figure 19–16 Aligning the Tailstock and Spindle Axes

A shear pin is designed to be of a particular diameter and shape. The shear pin withstands a specific force. The pin shears and the machine motion stops when the design force is exceeded. The shear pin in the end gear train is used to prevent overload and the stripping of gear teeth. A shear pin makes it almost impossible to apply an excessive force on a shaft, rod, or lever.

A slip clutch protects the feed rod and connecting mechanisms. The slip clutch is designed to release the feed rod when a specific force is exceeded. The slip clutch mechanism has an added feature that permits the feed rod to be automatically reengaged when the force is reduced. Figure 19-15 provides an interior view of a spring-ball slip clutch applied to a feed rod.

TAILSTOCK ALIGNMENT

How to Align the Tailstock

Aligning by Sight

STEP Clean the headstock and tailstock center
1 spindles. Insert the live and dead centers.

STEP Move the tailstock toward the headstock
2 until the centers almsot touch.

STEP Look down on the centers. Sight across
3 the points.

STEP Move the head section of the tailstock so
4 that the live and dead centers are aligned.

Note: Loosen the nut that binds the head and base sections of the tailstock. Then move the head section until the dead center is aligned with the live center.

STEP Tighten the binding nut to hold the tail-
5 stock head and base sections securely.

Checking Alignment by Machining

STEP Mount the workpiece between centers.
1

STEP Position the cutting tool. Take a light cut
2 for a short distance. Note the reading on the graduated dial of the cross feed.

STEP Back the cutting tool away from the
3 workpiece.

STEP Reposition the cutting tool near the head-
4 stock end of the workpiece. Move the cutting tool to the same depth, according to the graduated dial reading. Take a cut for a short distance (Figure 19–16).

STEP Measure both turned diameters.
5

Note: If the tailstock diameter is larger, it means the tailstock must be moved forward toward the operator.

STEP Position a dial indicator so that the point
6 makes contact with the tailstock spindle.

STEP Loosen the binding nut. Move the tail-
7 stock head the required distance toward the operator if the turned portion at the tailstock end is larger in diameter than at the headstock end. Move the tailstock in the opposite direction if the reverse condition exists.

STEP 8 Secure the head and base sections of the tailstock. Take a second cut on the centered workpiece. Recheck the turned diameters. Readjust the tailstock (if necessary) until the same diameter is produced on both ends of the workpiece.

Checking Alignment with a Test Bar

STEP 1 Mount the live and dead centers.

STEP 2 Place and hold a centered, cylindrical, parallel test bar between centers (Figure 19–17).

> Note: Clean the center holes of the test bar before mounting between centers.

STEP 3 Mount a dial indicator in the tool post.

STEP 4 Bring the point end of the indicator into contact with the test bar. Note the reading.

STEP 5 Move the dial indicator carefully toward the second end (Figure 19–17).

STEP 6 Note any variation in the readings.

> Note: The distance to move the tailstock is equal (approximately) to the difference in the two dial readings.

STEP 7 Return the dial indicator to the tailstock portion of the test bar. Loosen the binding nut and adjust the tailstock until the dial reading is the same as the reading at the headstock end.

STEP 8 Secure the head and base sections of the tailstock. Recheck the alignment.

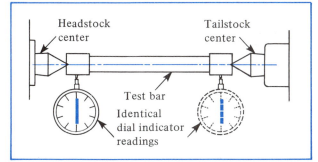

Figure 19–17 Precision Alignment of the Centers with a Test Bar and Dial Indicator

LATHE MAINTENANCE

CLEANING THE LATHE

Lathe cleaning means the safe removal of cutting chips and foreign particles from the machine, accessories, cutting tools, and workpiece. It also involves cleaning the cutting fluid reservoir and system. The purpose of cleaning, as discussed before, is to protect the worker, the machine, and the workpiece. *Chips should be removed when the machine is stopped.* A chip rake should be used when continuous chips are formed into a large mass.

When chip breakers are applied, the broken chip fragments may be brushed away from the machine areas. Comparatively large quantities of chips are shoveled out of the chip pan or tray. A chip disposal container should be readily available so that none of the cutting fluid is spilled on the floor.

After the chips are removed, the headstock, tailstock, and carriage are first brushed and then wiped with a clean cloth. The painted surfaces are wiped first; the machine surfaces, last. All accessories and cutting tools must be brushed and wiped clean.

The lead screw must also be cleaned. When the lead screw is cleaned, the machine may be started with all feeds disengaged. A hard, short-fiber brush may be moved in the threads along the entire length of the lead screw as it turns slowly.

With continuous use, gum and oil stains form on the exposed machine surfaces of the lathe. After these surfaces are wiped with a clean cloth, they should be wiped again with a cloth saturated with a small amount of kerosene or other gum or nontoxic stain-cutting solvent.

MAINTAINING THE LUBRICATION AND COOLANT SYSTEMS

CARE OF THE LUBRICANT SYSTEM

Most lathes are furnished with pressure lubricating systems. Lubricants are forced between all gears, the spindle and bearings, and other moving parts. Where necessary, sight level indicators show the amount of lubricant in an oil reservoir. These must be checked before the lathe is operated. The lubricant level must be maintained continuously. Each manufacturer provides information about the type and grade of lubricant to use. Use of the proper lubricant is important. An improper grade may cause overheating, impair the machining accuracy, and cause damage to various parts of the lathe.

In addition to machine lubrication from a pressurized automatic system, the ways and other oil holes may need lubrication with a hand oiler. They require daily cleaning and lubricating.

CARE OF THE CUTTING FLUID SYSTEM

Recommendations for completely draining the cutting fluid reservoir and system must be followed carefully. Sludge sediment and foreign particles must be removed at regular intervals.

The cutting fluid is pumped out of the reservoir into a disposable container. The sediment-collecting trays are removed and cleaned. The collection of chips and foreign particles that settle in the bottom are shoveled out. The inside container is washed clean.

The trays are replaced and the reservoir is refilled to the required level with fresh cutting fluid. The cutting fluid must be tested and appropriate for the material to be turned and the work processes.

Visual, on-the-spot tests of fluid concentration may be made with a hand-held instrument known as an *industrial fluid tester.* Such a tester is illustrated in Figure 19–18. It measures the refractive index (concentrations in the mixture) of a cutting fluid.

The instrument compensates for temperature variations in a cutting fluid. A few drops of a fluid provide the test sample. The cutting fluid concentration is quickly established by a direct reading on the fluid tester.

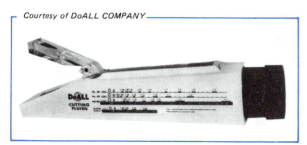

Courtesy of DoALL COMPANY

Figure 19–18 An Industrial Fluid Tester

LUBRICANTS AND LUBRICATION

Lubricants serve four main functions:

- Lubricating,
- Cooling,
- Cleansing,
- Protecting against corrosion.

Lubricating Function. A lubricant provides a *hydrostatic fluid shim* between all mating parts when the machine is in motion. In other words, the lubricant fills all clearance spaces. The metal surfaces are cushioned and slide with the least amount of friction. Less force is required to move mating parts, and friction-generated heat is decreased. Finer surface movement conditions produce smoother operation. There is less power consumption, and wear is reduced. A fluid shim also serves the important function of helping to maintain the precision of a machine tool.

Cooling Function. The heat that normally is generated in the machine tool itself is carried away by the flow of a lubricant. A cooling effect also results from the reduction of friction. Cooling also helps to prevent seizing due to expansion.

Cleansing Function. A lubricant washes away minute foreign particles. These particles are produced by the abrasion and wearing of surfaces that move against each other.

Protection against Corrosion. Exposure of unprotected ferrous metals to air, use of certain water-soluble cutting fluids, and turning of corroded surfaces may produce rust. Rust on a machined part results in a pitted and rough surface that has abrasive properties.

Rust-preventive mists are available to protect unfinished parts or machine surfaces. The rust preventive is sprayed as a mist. The rust-preventive mist has a water-displacing property. It is effective on wet or dry surfaces. Workpieces and machine parts are thus protected during idle periods or when operating without a cutting fluid.

Table 19-2 Factors Affecting Machining Conditions and Corrective Action

Problem	Probable Cause	Corrective Action
Machine vibration	—Work-holding device or setup is out of balance	—Counterbalance the unequal force —Reduce the spindle speed
	—High cutting speed	—Reduce the spindle speed
	—Floor vibration	—Secure the base firmly to a solid footing
	—Overhang of cutting tool or compound rest	—Shorten the length the toolholder extends beyond the tool post —Position the toolholder so that the compound rest has maximum support from the bottom slide surfaces
Vibration of the cross slide and compound rest	—Compound rest slide extends too far from the bottom slide support surfaces	—Reset the cutting tool so that there is greater area in contact between the top and bottom slide surfaces
	—Loose gib	—Adjust the gib to compensate for wear and lock it in the adjusted position
	—Overhang of cutting tool or compound rest	—Shorten the length of the cutting tool in the holder and the toolholder in the tool post
Play in feed screw	—Wear on the feed screw and nut	—Back the cutting tool away from the work. Take measurements by maintaining continuous contact, turning in one direction only
Chatter	—High rate of cutting speed or feed	—Reduce the spindle speed and rate of feed
	—Incorrect rake or relief angles, or form or cutting angles	—Regrind the cutting bit to the recommended rake, relief, and cutting angles and form
	—Overhang of cutting tool or toolholder	—Shorten the length the cutting tool extends from the holder and the tool post
Misalignment of headstock and tailstock axes	—Tailstock offset	—Move the tailstock head in its base until the spindle axis aligns with the headstock spindle axis

FACTORS AFFECTING MACHINING CONDITIONS

A number of common lathe problems are listed in Table 19-2. The most probable causes of each problem and the corrective actions the worker may take are also given.

How to Clean a Lathe

STEP 1 Cut off all power. Shut down the lathe.

STEP 2 Rake or brush away all chips from the machine, the accessories, and the cutting tool.

STEP 3 Wipe each accessory and cutting tool.

STEP 4 Moisten a clean wiping cloth with a thin film of oil or rust inhibitor. Wipe over all machined surfaces.

STEP 5 Replace each accessory and cutting tool in its appropriate place in the machine stand or toolroom.

STEP
6 Clean all painted and machined surfaces of the lathe with a wiping cloth.

Note: Thick gums and grease may be removed with kerosene or a manufactured solvent. Solvents cut grease without damaging adjacent surfaces.

How to Lubricate a Lathe

STEP
1 Check the level of the lubricant in the geared headstock. The level must be at or above the index line on the sight level.
STEP
2 Add lubricant by removing the filler cap. Bring the level to the index line.

Note: The lathe manufacturer's technical manual must be consulted for the correct lubricant to use.

STEP
3 Repeat the steps with the quick-change gearbox and any other gear train mechanism.
STEP
4 Fill any other oil holes with a hand oiler. Use the kind of lubricant recommended by the lathemaker.

Note: The skilled worker checks the technical manual to determine the number and types of lubrication systems. *Nonforced systems* are lubricated before operating the lathe.

How to Clean the Coolant System

STEP
1 Remove the chip and sediment trays from the coolant reservoir.
STEP
2 Clean the sludge that has accumulated. Use a trowel or metal scoop.

Note: The trays are washed in a cleaning and degreasing tank when a tank is available in the shop.

STEP
3 Siphon off the cutting fluid from the reservoir into a disposal or separating container.
STEP
4 Remove the fine particles and gum that have been deposited in the base. These particles may be shoveled out. Flush the reservoir.
STEP
5 Remove the coolant nozzle. Clean and reinstall it.
STEP
6 Remove the coolant pump if it is portable. Flush the impeller blades or immerse them in a cleansing bath.
STEP
7 Determine the cutting fluid that is to be used. Fill the reservoir to the indicated height.
STEP
8 Check the system. Turn on the coolant pump. Open the nozzle valve and check to see that the cutting fluid is circulating properly.

Safe Practices in the Care and Maintenance of the Lathe

- Secure all machine guards in place before starting the lathe.
- Make sure the lathe spindle is stopped when making speed changes on step-cone pulley, geared-head, and back-gear drives.
- Change the spindle speeds of variable-speed motor drives when the spindle is in motion.
- Check for excessive force if there is continuous slipping of a slip clutch on a feed rod. Damage may result if the condition is not corrected.
- Provide compensation for lost motion when setting a cutting tool.
- Remove chip fragments only when the lathe is shut down. Use a brush or a chip rake.
- Place chips in a chip container. Avoid splashing cutting fluid on the shop floor.
- Check all machine lubricating levels before the lathe is started.
- Use machine lubricants in forced-feed systems that meet the lathe manufacturer's specifications. Lubricants must be maintained at the level indicated on the sight indicator.

- Coat exposed machine surfaces with a rust-preventive film to protect against possible corrosion, particularly during a machine shutdown period.
- Avoid wearing loose, hanging clothing near moving parts. Also, all wiping cloths must be removed from the machine.
- Wear safety goggles when cleaning the lathe and when performing lathe operations.

A. LATHE FUNCTIONS, TYPES, AND FEATURES TERMS

Basic lathe components — The grouping of units (components) of a lathe to perform a major function. The power source and motion-producing mechanism for a workpiece; a work-holding and work-rotating device; and the motion-producing mechanism for feeds, speeds, and threads.

Geared-head lathe — A headstock design that uses gear sets to produce a wide range of spindle speeds. Gear sets are engaged by sliding gears into position. One or more speed-change levers control the speed range.

Lathe carriage — A mechanism for supporting, positioning, and securing cutting tools for longitudinal and transverse movements. A combination of apron, saddle, cross slide, and compound rest.

Threading attachment — A device that permits the exact positioning and engagement of a thread-cutting tool to track in a previously cut thread form.

Taper attachment — A control device that is set at a required taper angle and strapped to the lathe bed. A mechanism to transfer the angular movement of a slide block to the cross slide. Control of the cutting tool movement to produce a taper.

Basic lathe processes — A series of fundamental external turning processes (facing, straight and taper turning, shoulder cutting, grooving, cutting off, knurling, and threading) and a series of internal machining processes (center drilling, drilling, and countersinking; boring, counterboring, and spotfacing; reaming; recessing or undercutting, and tapping).

Production lathe — A machine tool on which all basic lathe processes may be performed. A lathe that uses a multiple-holder tool post, multiple-back carriage-position toolholder, and a multiple-station tailstock turret head for production.

Long taper and cam-lock spindle nose — Two standard types of spindle nose. A tapered spindle nose in which work-holding accessories and tools may be mounted, secured, and positioned concentrically with the spindle axis.

Longitudinal feed, or movement — The movement (feed) of the carriage between the headstock and tailstock. A movement along the ways produced either by hand or by engaging the power feed.

Lead screw — A precisely machined screw thread running the length of the lathe bed. A screw that may be engaged by an apron mechanism. A screw that changes rotary motion to longitudinal motion to reproduce a thread of a required pitch.

B. LATHE CONTROLS AND MAINTENANCE TERMS

Leveling (lathe) — A process of positioning a machined surface of the lathe so that it always lies in the same horizontal or vertical plane. A process designed to overcome lathe distortion and prevent machine inaccuracies.

Spindle speed drives	Combinations of pulley and gear drives that provide for the change of spindle speeds. Mechanisms that permit speed changes to accommodate different machining processes and materials.
Back gears	A simple gear train that, when engaged, provides a slow-speed spindle range.
Speed- or feed-change levers	A series of levers mounted on the headstock or quick-change gearbox. Positioning internal gear combinations to produce a specified spindle speed.
Starting, stopping, and reversing levers	Levers mounted on the apron to control longitudinal and cross feed direction and movement.
Slip clutch	An overload-prevention safety device. A safety device that disengages the feed rod whenever an excessive force is applied.
Lost motion, backlash, end play	Excessive clearance between two movable mating parts. A condition in which movement of one member of a mechanism fails to simultaneously move the other member because of excessive clearance between them.
Lubricating function	The filling of a void area (clearance space) between two mating parts with a hydrostatic fluid film that reduces friction between the mating moving parts.

SUMMARY

A. LATHE FUNCTIONS, TYPES AND FEATURES TERMS

- The Maudslay all-iron screw-cutting lathe refined and incorporated the spindle, slide rest, change gear, and tailstock features of earlier lathes. These features were combined with a precision lead screw, which Maudslay produced.
    - The addition of the multiple-station turret head led to higher levels of production. Individual tool setups were eliminated.
- The automatic screw machine combined design features of the engine lathe and turret lathe. The machining processes were cam actuated.
    - The three major functions of a lathe relate to:
        - Transmitting power to produce motion;
        - Holding and rotating a workpiece;
        - Positioning, holding, and moving a cutting tool at a fixed rate of feed or pitch.
- The headstock mechanism controls the work speed. The headstock spindle is used for holding and rotating the workpiece.
    - The quick-change gearbox contains gear trains. Different gear ratios are combined to obtain different feeds and leads (pitch).
- The thread-chasing dial on the apron permits the engagement of the lead screw at a proper location. With this dial, successive thread cuts are positioned accurately in the previously cut thread groove.
    - The steady rest is positioned at a fixed location to support a workpiece. A follower rest also supports a workpiece. It follows the cutting action to prevent springing of a workpiece.

- The apron of the carriage contains a half-nut mechanism, gear sets, and control levers. The levers are used to engage the longitudinal feed of the carriage, the transverse feed of the cross slide, and the rotary motion of the lead screw when thread cutting.
  - The cross slide produces transverse motion (toward or away from the center). The compound rest on a cross slide makes it possible to take angular cuts and turn a short taper.
- The tailstock provides a work- or tool-supporting surface. Cutting tools may also be held in and moved by a tailstock spindle.
  - The live center rotates with the work. The dead center may be a solid type or have a 60° cone-shaped point (live center) that rotates on anti-friction bearings.
- Inch- and metric-standard measurements may be read directly on graduated feed dials.
  - Three simple groups of lathes are the basic all-purpose engine lathes, production turret lathes (ram, saddle, bar, and chucking machines, and turning centers), and special lathes for contour turning and flowing.
- Safety practices must be followed with respect to:
  - Lifting and positioning a heavy workpiece or chuck,
  - Gathering and tying loose clothing to avoid catching in revolving machinery,
  - Placing all tools and parts on a work stand or tool cabinet,
  - Removing chips with a chip-removing rod (rake).

B. LATHE CONTROLS AND MAINTENANCE TERMS

- A precision level is used to position the lathe bed in a horizontal plane. Leveling prevents machine distortion and permits accurate machining.
  - The use of back gears extends the normal speed range to obtain slower speeds and increased torque.
- Pulley- or gear-driven speed and feed mechanisms should be stopped before changes are made in the spindle RPM.
  - Lost motion may be compensated for by backing a tool away from a workpiece. The direction is then reversed and the tool is fed to the required depth.
- Chatter results when there is excessive speed or tool overhang or when the work is not held rigidly. Incorrect rake angles, relief angles, cutting angles, or nose radius may also produce chatter.
  - Feed and pitch changes are made through the quick-change gearbox.
- Overload safety devices include shear pins and slip clutches. These devices are designed to stop the movement of mating parts whenever excessive forces are applied.
  - The tailstock may be aligned by sighting, with a test bar, or by machining a workpiece.
- Hygienic conditions must be maintained in coolant reservoirs and systems. The concentration and condition of the cutting fluid must be checked regularly and the sediment removed.
  - Operator safety requirements include the use of a safety shield or goggles. The lathe must be shut down during cleaning. All guards must be re-

placed before the machine is turned on. The operator should secure all loose clothing before operating any machine.

■ The lathe manufacturer's specifications must be followed. The proper lubricant must be used for all movable mating parts, such as gear trains and feed and speed mechanisms.

   ■ Machine lubricants must be checked daily. These must be held at or above the index line on the sight indicator.

■ The lathe must be cleaned by removing all chips, wiping all machine surfaces, and giving them a protective coating.

UNIT 19 REVIEW AND SELF-TEST

A. LATHE FUNCTIONS, TYPES, AND FEATURES

1. Indicate the relationship of end gears to the spindle RPM of a lathe.

2. Identify the design feature of a tailstock that permits offsetting for taper turning.

3. Indicate a modern safety feature that is incorporated in the design of guards that cover end gear trains.

4. State the functions served by external and internal tapers on a lathe spindle nose.

5. Give two functions that are served by the index plate on a quick-change gearbox.

6. List two advantages of the four-way turret type 2 cutting toolholder and positioning device over a type 1 rectangular tool block for producing a number of identical parts.

7. Describe briefly how a multiple-head micrometer stop works.

B. LATHE CONTROLS AND MAINTENANCE

1. State the prime function that is served by leveling.

2. Describe briefly the operation of (a) variable-speed drive controls and (b) geared-head controls on an engine lathe.

3. Explain how lost motion in mating threads or gears on a lathe is corrected.

4. Name two general protective overload devices on a lathe.

5. a. Explain the effect of corrosion on machined surfaces that slide on each other.
   b. Give a preventive step that may be taken to prevent corrosion during shut-down periods.

6. Select one method and list the steps for checking the alignment of the headstock and tailstock spindle axes.

7. Name five terms that are associated with the cleaning of the coolant system on a lathe.

8. State three safety precautions to observe in the care and maintenance of a lathe.

Lathe Cutting Tools and Toolholders

Lathe work processes begin with the selection of the most suitable cutting tool and toolholder. This section presents:

- Design characteristics and specifications of basic lathe cutting tool bits and inserts and the coding systems used to identify them,
- Design characteristics and general applications of toolholders and chip breakers
- Procedures for grinding high-speed steel tool bits.

UNIT 20

Cutting Tools, Specifications, and Holders

OBJECTIVES

A. SINGLE-POINT CUTTING TOOLS

After satisfactorily completing this unit, you will be able to:

- Apply properties and characteristics of cutting tools that are made of high-speed steel, cemented carbide, ceramic materials, and industrial diamonds.
- Identify factors that influence cutting speeds, feeds, and depth of cut, using inch and metric standards.
- Deal with continuous, discontinuous, and segemental chips, and chip control.
- Use technical charts complemented by cutting speed and cutting angle tables and formulas.
- Follow recommended *Safe Practices*.
- Apply shop *Terms* defining single-point cutting tools and holders.

B. CUTTING TOOL MATERIALS, FORMS, AND HOLDERS

- Distinguish factors that affect cutting efficiency.
- Use common terms to identify design features of single-point cutting tools.
- Interpret systems of specifying brazed-tip carbide tool bits and throwaway carbide inserts.
- Understand the function of chip breakers, cutting forces, and cutting action.
- Determine and use gages for gaging side relief, end relief, rake, and other tool bit angles.
- Select standard, carbide, and ceramic insert toolholders, a quick-change system, and heavy-duty toolholders.
- Apply data from tables, formulas, and other technical information.

C. OFF-HAND GRINDING HIGH SPEED STEEL TOOL BITS

- Prepare for and perform the following hand grinding processes.
  - Grind side-cutting and side-relief angles.
  - Grind end-cutting edge and end-relief angles.
  - Grind side- and back-rake angles; finish grind, and hone cutting edges.

217

The basic cutting tool for lathe work is a *tool bit*. Some types of tool bits have an *insert*. Both forms of tool bits are held securely by a tool-holder during machining processes. Other cutting tools are brazed onto a hardened steel shank, as in the case of cemented carbides. The tool-holder or solid shank is then positioned in relation to the work and process in a tool post, turret head, or other holding device.

A. SINGLE POINT CUTTING TOOLS: TECHNOLOGY AND PROCESSES

The efficiency of a cutting tool depends on the following factors:

— Forming the correct relief, clearance, rake, and cutting-edge angles (these angles depend on the cutting requirements and necessary dimensional accuracy and quality of surface finish);

— Proper positioning of the cutting edges with respect to the workpiece and processes;

— Rigidity of the setup used to support the cutting action;

— Properties of the cutting tool to maintain effective tool life;

— Speeds and feeds that permit the processes to be performed economically and safely.

A single-point, high-speed steel tool bit that has a square cross section is commonly used for lathe operations. Tool bits are usually ground on a bench or pedestal grinder. An aluminum oxide abrasive grinding wheel is used for grinding. Diamond-impregnated grinding wheels are used for cemented carbides. A high-speed steel tool bit is usually ground by hand. The tool is held and guided by hand to form cutting edges that have a particular shape and appropriate clearance (relief) and rake angles.

SINGLE-POINT CUTTING TOOL DESIGN FEATURES

The particular shape and the angles to which a tool bit are ground are determined by the process to be performed and the material to be cut. General terms as used in the shop in relation to single-point cutting tool bits are illustrated in Figure 20–1.

— The *cutting edge* designates the portion of a tool bit that does the cutting;

— Chips flow over the top surface, or *face*, of a tool bit;

— The surface that is adjacent to and supports the cutting edge is the *flank*;

— The end- and side-cutting edges are joined at the *nose*;

— The *point* refers to all the surfaces on the face that produce the cutting edges;

— The term *right-hand* or *left-hand*, when used, precedes the name of a cutting tool or operation and indicates the direction from which cutting action occurs—for example, the cutting action of a right-hand rough cutting, or roughing, tool is from right to left, and the movement is toward the headstock or from the center of the workpiece outward.

Cutting tools are classified as *right-* or *left-hand* and *side-cutting* or *end-cutting*. Typical side- and end-cutting operations are illustrated in Figure 20–2. The cutting edge on a right-hand tool bit is on the left side. The left-hand tool bit has the side-cutting edge ground on the right side. End-cutting edge tools cut with the end-cutting edges and are not provided with side-cutting edges. Side-cutting tools cut as the carriage moves longitudinally. End-cutting tools cut by the cross (transverse) movement of the cross slide.

In addition to the *nose radius* there are six principle tool angles. These angles include *end-* and *side-relief angles*, *back-* and *side-rake angles*, and *end-* and *side-cutting-edge angles*.

RELIEF ANGLES

A relief angle, whether for the side-cutting or end-cutting edge or the nose, permits a cutting tool to penetrate into a workpiece. The relief angle of the nose radius is a combination and blending of the end- and side-relief angles.

Tool life and tool performance depend on the relief angles. If the relief angle is too large, the cutting edge is weakened and may wear quickly, chip, or break away. Too small a relief angle may limit proper feeding. It may also limit cutting action by causing rubbing that generates considerable heat.

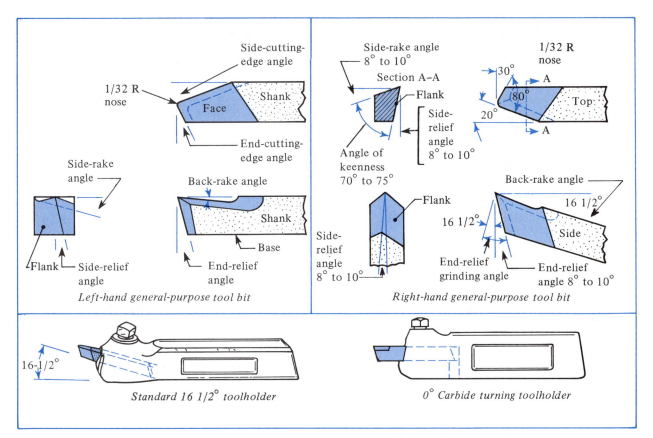

Figure 20–1 General-Purpose Single-Point Turning Tools, Toolholders, Angles, and Related Terms

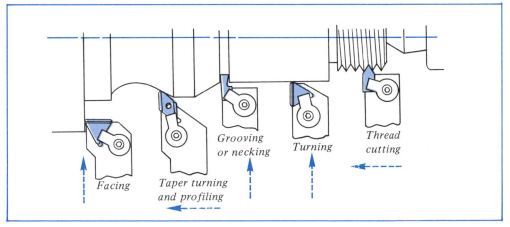

Figure 20–2 Typical Side and End-Cutting Operations Using
Carbide Inserts and Toolholders

High-speed steel cutting tools are ground with relief angles of 8° to 16°. The relief angle for general lathe processes is 10°. Smaller relief angles are used with harder materials. The harder and more brittle cemented carbide cutting tools require greater support under the cutting edge. The relief angle range for cemented carbides is from 5° to 12°.

BACK- AND SIDE-RAKE ANGLES

The side-rake angle indicates the number of degrees the face slants from the side-cutting edge. The side-rake angle is shown in Figure 20–1. It may be positive or negative. Similarly, the back-rake angle represents the number of degrees the face slants, as shown in Figure 20–1. The angle may slant downward, *positive rake*, or upward, *negative rake*, in moving along the cutting edges.

Negative side-rake and back-rake angles strengthen the cutting edge. The added strength is important when cutting hard and tough materials under severe machining conditions. Negative rake angles are used on throwaway carbide cutting tool inserts.

Manufacturers' recommendations and trade handbooks are available for determining correct rake angles for high-speed steel and carbide cutters and inserts for use with different materials of varying degrees of hardness. Sample listings in Table 20–1 are taken from the more complete Appendix Table.

SIDE- AND END-CUTTING-EDGE ANGLES

The *side-cutting edge* is ground at an angle to the side of a tool bit or tool shank. The range is from 0° to 30°, depending on the material to be cut and the depth, feed, and so on. A 20° angle is common for general lathe turning processes.

The *end-cutting angle* is formed by a perpendicular plane to the side of a tool bit and the end-cutting surface. This angle should be as small as possible and still permit the position of the side-cutting edge to be changed.

In addition to clearing the workpiece, the end-cutting angle provides a backup area for the nose and adjoining end-cutting area of the cutting tool. A 15° end-cutting-edge angle may be

Table 20–1 Recommended Rake Angles for Selected Materials (High-Speed Steel and Carbide Cutting Tools)

Material	Hardness Range (Bhn)	HSS Rake Angle (°)		Carbide Rake Angle (°)	
		Back	Side	Back	Side
Carbon Steel (plain)	100 to 200	5 to 10	10 to 20	0 to 5	7 to 15
. . .	400 to 500	−5 to 0	−5 to 0	−8 to 0	−6 to 0
Cast iron (gray)	160 to 200	5 to 10	10 to 15	0 to 5	6 to 15
Brass (free cutting) . . .		−5 to 5	0 to 10		

used for rough turning. An angle of 25° to 30° may be used for general turning processes.

LEAD ANGLE, CHIP THICKNESS, AND CUTTING FORCE

The positioning of the side-cutting edge in relation to a workpiece influences such factors as the direction the chip flows, the thickness of the chip, and the required cutting force. The *lead angle* also affects these factors. The lead angle is the angle the side-cutting edge forms with a plane that is perpendicular to the axis of the workpiece. A large lead angle has the following disadvantages:

- A greater force is required to feed the cutting tool into a workpiece,
- There is a tendency for the workpiece to spring away from the cutting tool and produce chatter,
- The vibration of the workpiece may cause the cutting edge of the tool to break down.

THE NOSE RADIUS

Cutting tools are rounded at the nose to machine a smooth, continuous, straight surface. Cutting tools with a sharp point may turn a fine helical groove in the surface. The size of the radius depends on the rate of feed and the required quality of surface finish. There is a tendency to produce chatter when too large a radius or too fast a speed is used. A nose radius of from 1/64″ to 1/8″ is common.

CUTTING SPEEDS FOR SINGLE-POINT TOOLS

Cutting speed is influenced by eight major factors:

- Material in the cutting tool;
- Material in the workpiece (its heat treatment, hardness, abrasiveness, internal hard spots, and other surface conditions);
- Rake, relief, and clearance angles and shape of the cutting areas;
- Use of a chip breaker to form and to clear away chips;
- Desired economical, efficient tool life;
- Use of a cutting fluid where required;
- Type of cut (roughing or finishing) and the depth;
- Rigidity of the work setup, the capability and construction of the lathe, and the nature of the toolholding device.

Cutting speed for lathe work refers to the rate at which a workpiece revolves past a fixed point of a cutting tool. In lathe work the cutting point is at the turned diameter. Cutting speed is measured in surface feet per minute (fpm or sfpm) or meters per minute (m/min).

Cutting speeds for metal-cutting operations are usually stated in manufacturers' tables, handbooks, and production plans. These cutting speeds are given in terms of *feet per minute (fpm)* or *meters per minute (m/min)*. In all cases, the craftsperson must judge whether there are conditions that require modifications of the recommended speeds. Associated with cutting speeds is concern for *tool life*. Tool life relates to the effective cutting time before a tool becomes dull and requires replacement.

$$\text{CS in sfpm} = \frac{(\text{Diameter in Inches} \times \pi) \times \text{RPM}}{12 \text{ Inches}}$$

When the metric system is used, the cutting speed (CS) is expressed in meters per minute (m/min). The metric equivalent of the formula is:

$$\text{CS} = \frac{(\text{Diameter in mm} \times \pi) \times \text{RPM}}{1000}$$

In actual practice the operator usually determines the RPM at which to set the spindle speed. The recommended cutting speed is found in a handbook table (Table 20–2).

Table 20–2 Cutting Speed (sfpm and m/min) Using High-Speed Steel Cutting Tools (Partial Table)

Material to Be Machined		Cutting Speed (sfpm and m/min) for Lathe Operations*		
		Turning and Boring		Screw Thread Cutting
		Roughing	Finishing	
Low-carbon (C) steel 0.05 to 0.30%C ...	sfpm	90	100	35–40
	m/min	27	30	10–12
High-carbon (C) tool steel 0.06 to 1.7%C ...	sfpm	50	70	20–25
	m/min	15	21	6–8
Brass ...	sfpm	150	300	50–60
	m/min	45	90	15–18
Aluminum ...	sfpm	200	350	50–70
	m/min	60	105	15–21

*Without using a cutting fluid

$$\text{RPM} = \frac{\text{CS} \times 12}{\text{Diameter} \times \pi} \quad \text{or} \quad \frac{\text{CS (m/sec)} \times 1000}{\text{Diameter} \times \pi}$$

(Inch Standard)   (Metric System)

B. CUTTING TOOL MATERIALS, FORMS, AND HOLDERS

Cutting tools should possess three properties: *hardenability*, *abrasion resistance*, and *crater resistance*. *Hardenability* relates to the property of a steel to harden deeply. A cutting tool must retain hardness at the high temperature developed at its cutting edge. *Crater resistance* is the ability to resist the forming of small indentations, or craters, that form on the side or top face of a cutting tool. A crater tends to undercut in back of a cutting edge. A crater weakens a cutting edge and in some cases causes it to break away under heavy cutting forces.

CUTTING TOOL MATERIALS

General cutting tools used for lathe work fall into the following groups: *high-speed steel, cast nonferrous alloy, cemented carbide, ceramic,* and *diamond*. The earlier carbon steel tool bits are seldom used for general lathe work.

HIGH-SPEED STEEL

High-speed steel is the most extensively used material for cutting tools. Its usage is based on factors such as tool and machining cost, tool wear, work finish, availability, and ease of grinding. High-speed steel single-point cutting tools are capable of withstanding machining shock and heavy cutting operations. A good cutting edge is maintained during red heat conditions.

Grades of High-Speed Steel. There are many grades of high-speed steel. Each grade depends upon the chemical composition of the steel. The grades are identified according to American Iron and Steel Institute (AISI) designations. Some of the commonly used T and M grades are indicated in Table 20–3. The addition of cobalt increases the ability of a cutting tool to retain its hardness during heavy roughing cuts and at high temperatures.

CAST ALLOYS

Cast alloy tool bits are used for machining processes at extra-high speeds. These speeds range from almost one-quarter to two times more than the speeds used for high-speed steel. Such tool bits have better wear-resistant qualities than high-speed steel. Cast alloys, like Stellite, make it possible to efficiently cut hard-to-machine cast iron and steel. However, cast alloy tool bits are brittle. They are not as tough as high-speed steel bits.

Cast alloys are manufactured as solid tool bits—for example, tips that may be brazed on tool shanks—and as cutting tool inserts. These tool bits have the advantage of maintaining a good cutting edge at high cutting speeds.

Cast alloy tool bits are popularly known by such trade names as Stellite, Rexalloy, Armaloy, and Tantung. Each trade name tool bit relates to a specific composition of chromium, tungsten, carbon, and cobalt that produces a tool bit with a high degree of hardness and high wear resistance. Cast alloy tool bits also have the ability to maintain a cutting edge while operating at red heat cutting temperatures below 1400°F.

Table 20–3 Common Grades of High-Speed Steel for Cutting Tools

Industrial Designation		
AISI	*Trade Name*	*Application*
T–1	High-speed steel	Single-point cutting tools; general purpose
T–4	Cobalt high-speed steel	Machining cast iron
T–6	Cobalt high-speed steel	Machining hard-scaled surfaces and abrasive materials
T–15*	Cobalt high-speed steel (super HSS)	Machining difficult-to-machine materials, stainless steels, high-temperature alloys, and refractory materials
M–2	General high-speed steel	Single-point and multiple-point cutting tools: milling cutters, taps, drills, and counterbores

* Extremely hard; requires slow grinding using a light grinding force

CEMENTED CARBIDES

Cemented carbides are powdered metals. They are compressed and sintered into a solid mass at temperatures around 2400°F. The composition and applications of a few selected carbides appear in Table 20–4. Cemented carbide cutting tools possess the following qualities:
- Excellent wear resistance,
- Greater hardness than high-speed steel tools,
- Ability to maintain a cutting edge under high temperature conditions,
- Tips and inserts that may be fused on steel shanks and cutter bodies for economical manufacture of the cutting tools,
- Cutting speeds that may be increased three to four times the speeds of high-speed steel cutters.

CERAMICS

The sintering of fine grains of aluminum oxide produces a ceramic material that may be formed into *cutting tool inserts*. These inserts are particularly adapted for machining highly abrasive castings and hard steels. Ceramic cutting tools are harder than cemented carbide tools and may be used at extremely high cutting speeds on special high-speed lathes. Ceramic cutting tools may be used in turning plain carbon and alloy steels, malleable iron (pearlitic), gray cast iron, and stainless steel.

Table 20-4 Applications of Selected Cemented Carbide Cutting Tools

Composition of Tool	Application
Tungsten carbides	Cuts on gray cast iron, malleable iron, stainless steel, nonferrous metals, die-cast alloys, and plastics
Tungsten carbide and cobalt (with titanium and tantalum carbide added)	Medium and light cuts on carbon and alloy steels, alloy cast iron, monel metal, and tool steels
Titanium carbide and nickel or molybdenum	Light precision cuts and precision boring cuts requiring a high-quality surface finish

INDUSTRIAL DIAMONDS

Industrial diamonds may be designed as single-point cutting tools. Diamonds are used on workpieces that must be machined to extremely close dimensional and surface finish tolerances. Diamond-tip cutting tools are used for light finish-cut turning and boring operations. The depth of cut may range between 0.003″ to 0.030″ (0.08mm to 0.8mm) using a feed of from 0.0008″ to 0.005″ (0.02mm to 0.1mm) per revolution.

Industrial diamonds may be operated at extremely high speeds for machining nonferrous and nonmetallic materials. Single-point diamond cutting tools are especially adapted for finish machining hard-to-finish metals like soft aluminum. The factors of additional cost and brittleness of industrial diamonds must be considered against the advantages.

Table 20-5 System for Designating Brazed-Tip Carbide Tool Bits

SAMPLE DESIGNATION: B R 6

Tool Characteristics

	Shape	Applications	Hand	Shank Size
A	0° lead	Shouldering, turning	R right hand	(Expressed in 1/16″)
B	15° lead	Forming to lead angle	L left hand	
C	Square nose	Grooving and facing		
D	30° lead	Undercutting and chamfering		
E	45° lead	Threading		

SYSTEM FOR DESIGNATING BRAZED-TIP CARBIDE TOOL BITS

Brazed-tip, single-point carbide tool tips are designated by a system of two letters and a numeral. As shown in Table 20-5, the first letter indicates the tool shape and the general process(es) for which the tool may be used. The second letter identifies the side-cutting edge and the direction of cut. The numeral gives the size of the square shank in sixteenths of an inch.

The designation (BR-6) of the carbide tool bit used in the example in Table 20-5 indicates that it has a lead angle of 15° (B). The tool bit may be used for any general purpose. The brazed tip is ground as a right-hand cutting tool (R). The shank (6) is 6/16″, or 3/8″, square.

SPECIFICATIONS OF THROWAWAY CARBIDE INSERTS

Carbide cutting tool inserts are preformed. A system for identifying eight design features of throwaway inserts provides complete information. The system consists of letters and numerals. Standard industrial designations for each of these sample letters and numerals are provided in handbook tables (Table 20-6).

For example, a throwaway carbide insert may be specified as SNEF-334E. This sample designation indicates the insert *shape* is square (S), has a 0° *clearance angle* (N), has a *cutting point* and *thickness class* of 0.001″ (E), is a clamp-on *type* with a chip breaker (F), is 3/8″ *square* (3), is 3/16″ *thick* (3), has a cutting point of 1/16″ *radius* (4), and is *unground* and *honed* (E).

Table 20–6 Standard Industrial Designations for Throwaway Carbide Inserts

SAMPLE DESIGNATION: (S N E F 3 3 4 E)

Shape	Clearance	Class		Type
		Cutting Point	Thickness	
R Round	N 0°	A 0.0002″	0.001″	A With hole
S Square	A 3°	B 0.0002″	0.005″	B With hole and one countersink
T Triangle	B 5°	C 0.0005″	0.001″	
P Pentagon	C 7°	D 0.0005″	0.005″	C With hole and two countersinks
D Diamond 55°	P 10°	E 0.001″	0.001″	
C Diamond 80°	D 15°	G 0.001″	0.005″	F Clamp-on type with chip breaker
O Octagon	E 20°	M**0.002″ to 0.005″	0.005″	G With hole and chip breaker
H Hexagon	F 25°			H With hole, one countersink, and chip breaker
L Rectangle	G 30°	U**0.005″ to 0.012″	0.005″	
M Diamond 86°				J With hole, two countersinks, and chip breaker
A Parallelogram 85°				
B Parallelogram 82°				
E Parallelogram 55°				
F Parallelogram 70°				
V Diamond 35°				

Size	Thickness	Cutting Point (radius of flats)	Other Conditions*
For size 1/4″ and over, use number of 1/8ths in size. For rectangle and parallelogram, use two digits: number of 1/8ths in width and number of 1/4ths in length	For size 1/4″ and over, use number of 1/16ths in thickness	0 Sharp corner 1 1/64 radius 2 1/32 radius 3 3/64 radius 4 1/16 radius 6 3/32 radius 8 1/8 radius A Square insert with 45° chamfer	A Ground all over—light honed B Ground all over—heavy honed C Ground top and bottom only—light honed D Ground top and bottom only—heavy honed E Unground-honed F Unground insert—not honed T Chamfer-cutting edge

CUTTING FORCES

There are three prime forces that act during the cutting process: *longitudinal*, *radial*, and *tangential.* As a cutting tool feeds into a workpiece, it produces the *longitudinal force* (Figure 20–3). The force of the workpiece against the front face of the cutting tool is a *radial force.* A *tangential force* results from the downward action of the workpiece against the top of the tool.

Materials are removed during machining by these three forces. The forces *wedge* or *shear away* material. On soft materials the shearing action produces a continuous ribbon or chip. The shape of the top face helps to form a spiral chip. On harder materials the wedging action

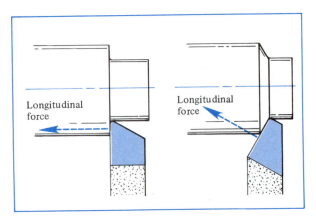

Figure 20–3 Longitudinal Cutting Force Perpendicular to the Side-Cutting Edge

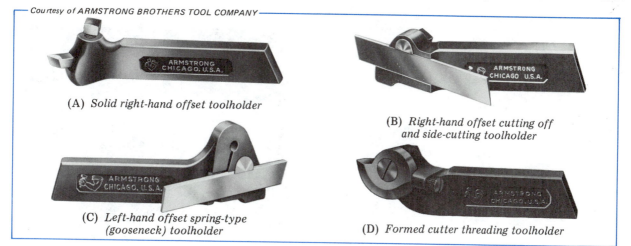

Courtesy of ARMSTRONG BROTHERS TOOL COMPANY

(A) *Solid right-hand offset toolholder*

(B) *Right-hand offset cutting off and side-cutting toolholder*

(C) *Left-hand offset spring-type (gooseneck) toolholder*

(D) *Formed cutter threading toolholder*

Figure 20–4 Basic Types of Holders for HSS Cutting Tools

causes the material to compress until a shearing point is reached. The material is then separated at the cutting edge; a chip is formed.

CUTTING-TOOL HOLDERS

There are four basic types of toolholders that are widely used for holding high-speed tool bits (Figure 20–4): *standard* toolholder (A), *cutting off* and *side-cutting* toolholder (B), spring-type *side-cutting* toolholder (C), and *threading* toolholder (D). Each of these toolholders is designed in a right-hand, left-hand (offset), or straight form. The form that is used is determined by the nature of the operation. The shape of the toolholder must permit the cutting tool and all parts of the machine setup to clear without hitting. A separate form for the threading toolholder permits the use of a circular tool blade that is especially ground for thread cutting.

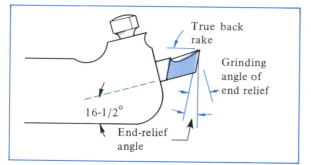

Figure 20–5 True Back-Rake Angle
(16 1/2° Angle Toolholder)

STANDARD TOOL BIT HOLDERS

A standard toolholder is designed to hold a tool bit at a fixed angle of either 0° or 16 1/2°. Figure 20–5 shows the true back-rake angle, which includes the rake angle to which the tool bit is ground plus the 16 1/2° angle of the toolholder. The end-relief angle is also increased to offset for the 16 1/2°.

A standard 0°-angle toolholder has a 0° angle formed by the bottom of the slot in the toolholder and its base. The 0° toolholder is used principally with brazed carbide-tipped tool bits. Carbide cutting tools require little or no back rake.

The 0° and 16 1/2° toolholders are furnished in five sizes to accommodate tool bits that range from 1/4″ to 5/8″ square.

ANGLE GRINDING GAGES FOR SINGLE-POINT TOOL BITS

Tables of recommended relief and rake angles for single-point cutting tools provide guidelines for the craftsperson (Appendix Table A–16). The angles may be checked for accuracy by using a flat, metal gage. Two side edges of the gage are ground to conform to specified end- and side-relief angles. Two V-grooves, one in the top and one in the bottom of the gage, are used to measure the cutting tool angles. Figure 20–6 shows how a tool bit held in a toolholder is measured for end-relief angle.

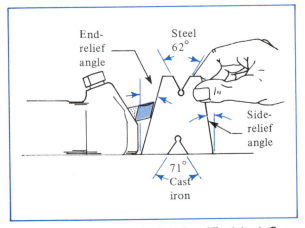

Figure 20-6 Application of a Relief- and Tool-Angle Gage

CHIP FORMATION AND CONTROL

CHIP FORMATION

The cutting action of a cutting tool produces three major types of metal chips: *continuous chip*, *discontinuous chip*, and *segmental chip*. A continuous chip represents the flow of a ductile metal (like steel) over the cutter face. The chip is in the form of a continuous spiral ribbon (Figure 20-7A).

A discontinuous chip is produced when cutting brittle metals (like cast iron). Since these metals do not flow, they fracture into many discontinuous particles (Figure 20-7B).

A segmental chip has a file tooth appearance on the outer side. The inner side that flows over the cutting tool face has a smooth, burnished surface. A segmental chip is produced when taking a heavy feed on ductile materials. The internal metal fractures spread to the outer surface of the chip. The particles are welded together by the heat of the cutting force (Figure 20-7C).

CHIP CONTROL

Chips that form continuously or segmentally at high speeds must be controlled. *Chip control* means the disposing of chips in a safe manner. The accumulation of chips near a revolving workpiece; along the bed, carriage, and cross slide; or on cutting tools is dangerous. Chips may become wrapped around a work or chucking device and thus often cause damage to the workpiece, cutting tool, or lathe. More important is the hazard of personal injury. Revolving

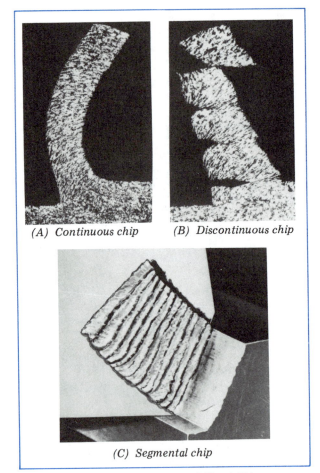

(A) Continuous chip *(B) Discontinuous chip*

(C) Segmental chip

Figure 20-7 Basic Types of Metal Chips

chips or hot metal particles may fly in a work area. Operational dangers are increased with carbide cutting tools. A great quantity of chips form and accumulate rapidly when heavy cuts are made at high speeds.

Continuous chips are broken into small, safe sizes with a *chip breaker*. The chip breaker feature may be ground as an area on the top of a tool bit, preformed on an insert, or added as a chip groove. An example of a parallel-type chip breaker for brazed carbide cutting tools is shown in Figure 20-8A. The preformed chip grooves for carbide inserts are also illustrated at (B).

Provision is made to accommodate chip breakers on toolholders. Chip breakers for throwaway insert toolholders are commercially

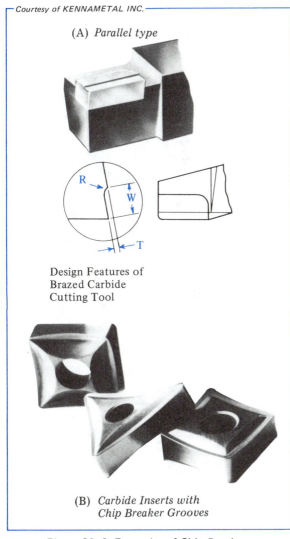

Courtesy of KENNAMETAL INC.

(A) *Parallel type*

Design Features of
Brazed Carbide
Cutting Tool

(B) *Carbide Inserts with
Chip Breaker Grooves*

Figure 20–8 Examples of Chip Breakers

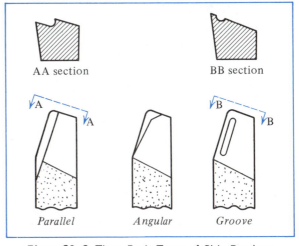

AA section   BB section

Parallel   *Angular*   *Groove*

Figure 20–9 Three Basic Types of Chip Breakers
Ground on High-Speed Steel Tool Bits

CARBIDE AND CERAMIC INSERTS

A square, diamond-point, or parallelogram-type insert has a total of eight cutting edges. When the four top cutting edges become dull, the insert is reversed. Each new cutting edge is then used. The insert is thrown away when the eight cutting edges are dull.

A toolholder for disposable carbide or ceramic inserts is designed as a negative- or positive-rake toolholder. The inserts for the positive-rake toolholder must be ground with a side-relief angle. Thus, only the top face edges may be used. The general shapes of inserts are square, rectangular, triangular, round, and diamond.

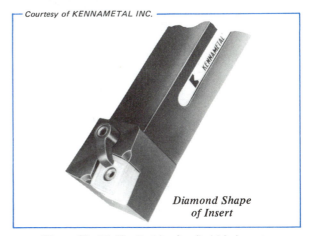

Courtesy of KENNAMETAL INC.

*Diamond Shape
of Insert*

Figure 20–10 Toolholder for Carbide Inserts

available. Chip breakers are designed for positive-, neutral-, and negative-rake toolholders and for general, light, medium, and heavy feeds.

Three general forms of chip breakers, which are ground on the top face of high-speed steel tool bits, are shown in Figure 20–9.
Cemented carbide inserts are usually brazed on a steel shank. The shank is held in a tool post or tool rest. Disposable carbide inserts are held and secured in a special type of carbide toolholder (Figure 20–10).

QUICK-CHANGE TOOL SYSTEM

Quick-Change Toolholder. The quick-change toolholder is designed with a dovetail slide. A quick-change toolholder is shown in Figure 20–11. The dovetail slide permits the holder to fit and be held in a main dovetail section of a tool post. The holder provides for cutting tool adjustments. The unit may thus be *preset*. Vertical adjustments are also possible.

The advantage of this unit is that, in quantity production, several cutting tools may be ground and preset. A preset unit may be positioned quickly to replace a worn cutting tool. The unit is held securely by a clamp.

Quick-Change Tool Post. A quick-change tool post combines the features of the quick-change tool-holder and the multiple-station turret. An individual toolholder is secured to one side of the quick-change tool post. When the tool setup is completed, each tool is positioned for a particular operation. Quick-change tool posts are held by simply turning to the next station and tightening the clamping lever.

Figure 20–11 Quick-Change Toolholder for Turning and Facing Tools

Figure 20–12 illustrates a *quick-change tool system*. The system consists of a **quick-change tool post** and a number of toolholders for different cutting tools. Each cutting tool is mounted quickly, aligned, and secured in position. The combination illustrated is used for performing general external and internal lathe operations.

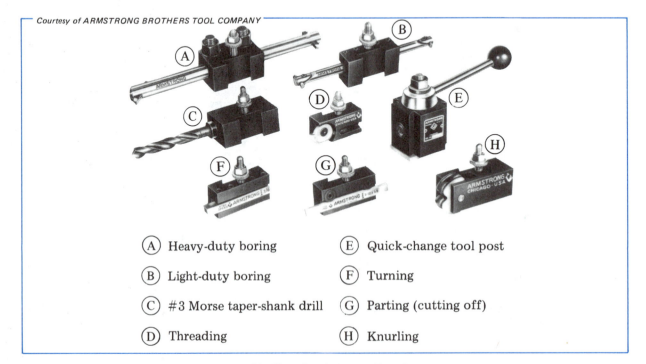

- (A) Heavy-duty boring
- (B) Light-duty boring
- (C) #3 Morse taper-shank drill
- (D) Threading
- (E) Quick-change tool post
- (F) Turning
- (G) Parting (cutting off)
- (H) Knurling

Figure 20–12 A Quick-Change Tool System

Courtesy of CINCINNATI MILACRON INC.

Figure 20–13 A Heavy-Duty, C-Shaped Block Secured on a Compound Rest

HEAVY-DUTY, OPEN-SIDE TOOL BLOCK (HOLDER)

A heavy-duty, open-side tool block or tool-holder, is used to rigidly hold a cutting tool. The holder supports the cutting tool against the tremendous forces that are exerted when heavy cuts are taken. One tool is held at one time. The C-shaped block fits in the compound rest slide (Figure 20–13). It is secured by a T-slot clamp. This type of cutting tool holder is also adapted for operations requiring a rear cross slide for a cutting tool.

Figure 20–13 shows an application of the heavy-duty tool block mounted on the compound rest.

C. OFF-HAND GRINDING HSS TOOL BITS

How to Grind High-Speed Steel Tool Bits

Note: The steps in this example apply to a right-hand turning tool (roughing cut).

Preparation

STEP 1 Determine the grade of high-speed steel that is recommended for the machining processes to be performed.

STEP 2 Consider the shape and angles of the side-, end-, and nose-cutting edges.

STEP 3 Select the appropriate toolholder.

Note: The 16 1/2° standard toolholder is used for general lathe processes.

STEP 4 Determine the correct end- and side-relief angles and side- and back-rake angles.

Note: These angles are based on the nature of the finish required, the composition of the workpiece, the nature of the process, the speed, and the feed.

STEP 5 True and dress both the coarse-grit (roughing) and fine-grit (finishing) aluminum oxide grinding wheels on the bench or floor (pedestal) grinder.

Grinding Side-Cutting Edge and Side-Relief Angles

STEP 1 Position the tool rest. Hold, position, and guide the cutting tool with the fingers of both hands (Figure 20–14). The bottom (heel) is tilted toward the grinding wheel.

STEP 2 Rough grind the side-cutting-edge and side-relief angles (Figure 20–15).

STEP 3 Check the side-relief angle. Use a bevel protractor or a side-relief angle gage.

Note: Rough grind to within approximately 1/16" from the top of the tool bit.

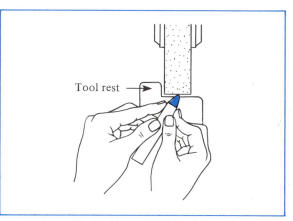

Tool rest

Figure 20–14 Holding and Guiding a Tool Bit While Grinding

Caution: Move the tool bit across the entire face of the grinding wheel to avoid uneven wearing of the wheel face. Quench the tool bit regularly. Quenching permits safe handling and protects the properties of the cutting tool.

Grinding End-Cutting Edge and End-Relief Angles

STEP 1 Raise the front end of the tool bit to increase the angle of contact with the face of the grinding wheel.

STEP 2 Rough grind the end-cutting-edge and end-relief angles to within 1/16″ of the tool bit point (Figure 20–16).

Note: The formed surface must blend the end-cutting and side-cutting angles, the flanks, and the nose radius.

Grinding Side- and Back-Rake Angles

STEP 1 Hold the tool bit against the face of the grinding wheel so that the side-cutting edge is horizontal.

STEP 2 Tilt the tool bit surface adjacent to the cutting edge inward toward the grinding wheel to grind the back-rake angle.

STEP 3 Rough grind the face to within 1/16″ of a sharp edge.

Finish Grinding

STEP 1 Repeat all of the previous grinding steps. Use the fine-grained wheel. Take light finishing cuts.

STEP 2 Grind just below the roughly ground surfaces. Finish grinding produces a fine-grained edge.

Honing (Stoning) the Cutting Edges

STEP 1 Select a medium/fine-grain abrasive stone (oilstone). Apply a small amount of kerosens or cutting fluid to the face of the stone.

STEP 2 Hold the medium-grain side of the stone against the flank of the tool bit.

STEP 3 Move the stone parallel across the side (face) of the cutting tool.

STEP 4 Repeat the honing process. Use the fine-grained side of the stone.

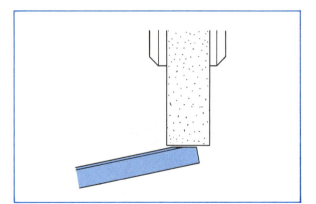

Figure 20–15 Grinding the Side-Cutting Edge and Side-Relief Angles

Note: Some workers prefer to move the tool bit when honing the cutting edges by placing the cutting tool flank on the stone and rubbing the tool bit over the stone lengthwise until a fine cutting edge is produced.

STEP 5 Hone the end-cutting edge in the same manner.

STEP 6 Repeat the honing steps for the end-relief and top-rake angles.

STEP 7 Lay the round nose on the sharpening stone. Swing the cutting tool in an arc while moving it along the stone.

STEP 8 Wipe the face of the sharpening stone to clean it and to remove the fine metal particles.

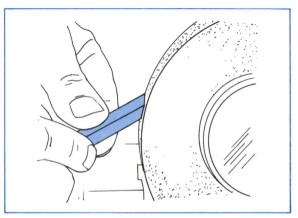

Figure 20–16 Grinding the End-Cutting-Edge and End-Relief Angles (Exaggerated)

Safe Practices in Using Cutting Tools and Toolholders

- Grind high-speed tool bits and cemented carbide and ceramic inserts to the specified clearance, rake, and cutting-edge angles. These angles affect cutting efficiency, tool life, quality of surface finish, and dimensional accuracy.

- Grind a chip breaker into a cutting tool bit or place a separate chip breaker adjacent to the cutting edges. The continuous breaking away of small chips provides safe working conditions and prevents damage to the lathe and workpiece.

- Pay attention during rough turning (taking hogging cuts) to the effect of the longitudinal, radial, and tangential cutting forces and the heat generated on a workpiece and cutting tool.

- Quench the high-speed tool bit regularly when dry grinding.

- Position the machine safety shields and use safety goggles.

- Stop the lathe to remove chips. Use a chip rake.

- Secure all loose clothing. Remove wiping cloths from the lathe before starting the machine.

TERMS USED WITH LATHE CUTTING TOOLS AND HOLDERS

Right- and left-hand turning tools	Lathe tool bits that are ground to cut from right to left (right-hand) or left to right (left-hand).
Cutting forces (lathe work)	Longitudinal, radial, and tangential forces. Forces produced by a cutting tool as it feeds into a revolving workpiece.
Quick-change toolholder	A toolholder designed to fit a dovetailed tool post. (The cutting tool is preset, and the whole unit may be replaced quickly.)
Quick-change tool system	A combination of multiple-station, quick-change cutting tool holders.
Heavy-duty toolholder	An open-side toolholder with a T-base that fits onto the compound rest.
Crater resistance	The ability to resist wear caused by abrasion and heat generated at the cutting edge and the force of chips moving over a cutting tool.
AISI designation	A standardized system of a letter and numeral for designating a cutting tool material with a specific chemical composition.
Ceramic inserts	Fine grains of aluminum oxide that are formed and sintered into regular shapes (inserts). Diamond, square, rectangular, triangular, and other round shapes of ceramic cutting tools.
Disposable insert styles	Manufacturers' designations, by letters, of cemented carbide cutting tool throwaway inserts according to design features and applications to specific processes.
Side-cutting and end-cutting edges	The edges of a single-point lathe cutting tool. Tool cutting edges that are ground to permit a tool to advance into a workpiece longitudinally (side cutting) or crosswise (end cutting).
Continuous and segmental chips	Chips that flow over a cutter as a continuous spiral (continuous chip) or as a welded file-tooth-shaped ribbon (segmental chip).
Chip breaker	An external form or a groove ground near the cutting edge to curl a chip so that it breaks continuously instead of flowing as a ribbon.

SUMMARY

■ Three important properties of cutting tools are hardenability, resistance to abrasive action, and resistance to the formation of a crater.

  ■ Single-point cutting tools are generally made of materials such as high-speed steel, cast nonferrous alloys, cemented carbides, ceramics, and diamonds.

■ The single-point cutting tool (right- or left-hand) is a side- or end-cutting tool having: a positive or negative side or back rake; side, end, and nose relief; end-cutting-edge and side-cutting-edge angles.

  ■ The four basic tool bit shapes deal with turning, facing, cutting off, and threading. Tools bits are ground with the side-cutting edge on the left side for regular turning processes. The feed is from right to left. Left-hand cutting tools are fed from left to right.

■ High-speed steel cutting bits are designated by a trade name and AISI letter and number. Varying the amounts of carbon, tungsten, vanadium, chromium, molybdenum, and cobalt affects the cutting properties of high-speed steel.

  ■ Industrial diamond cutting tools are especially suited for ferrous metals and nonferrous materials that are hard to finish to close dimensional accuracy. High-quality surface finish is possible.

■ Carbide inserts are specified according to a combination of letters and numerals. The eight symbols designate shape, clearance angle, class of cutting point and clearance, type, size, thickness, cutting point, and other conditions.

  ■ The action of a cutting tool as it is fed into a revolving workpiece produces longitudinal, radial, and tangential forces.

■ Lathe cutting tools produce three general types of cutting chips: continuous, discontinuous, and segmental.

  ■ Chip breakers are employed for safety in machining. A chip breaker controls the shape, size, and flow of chips for rapid disposal.

■ Disposable cemented carbide and ceramic inserts are available in round, diamond, square, rectangular, triangular, and other forms. Toolholders are designed to permit side or end mounting of the insert.

  ■ Cutting speeds are influenced by the material of the cutting tool and workpiece; the cutting, clearance, and rakes angles of the tool bit; the depth of cut; the speed and other conditions of the lathe; and the required tool life.

■ Cutting speeds in sfpm or m/min and spindle speeds (RPM) may be found in tables or computed by formula.

  ■ The quick-change tool post provides a series of accurately dovetailed slides. Several quick-change toolholders may be accurately positioned and held in these slides. The multiple head permits several cutting tools to be set up and moved into position.

■ The open-side, heavy-duty toolholder is used for roughing cuts that produce excessive forces.

  ■ Carbide and ceramic inserts are mounted and held to solid shanks by either pinblock, cam-action, or clamp-type tool posts.

UNIT 20 REVIEW AND SELF-TEST

A. SINGLE-POINT CUTTING TOOLS: TECHNOLOGY AND PROCESSES

1. State what effect crater resistance of a lathe cutting tool has on its cutting ability.

2. Indicate what relationship exists between the RPM of a lathe spindle and the cutting speed (sfpm).

3. Identify (a) the two cutting edges of a single-point cutting tool and (b) the purpose of relieving the flank.

4. Describe the effect of negative side- and back-rake angles on the cutting edge of a single-point cutting tool.

5. Indicate what safety precaution must be taken to control the accumulation of continuous or segmental chips at high speeds.

B. CUTTING TOOL MATERIALS, FORMS, AND HOLDERS

1. a. Give an application of a cast alloy tool bit.
   b. Tell why it is preferred in this application to a standard high-speed steel tool bit.

2. Distinguish between the manufacture of cemented carbide and ceramic inserts.

3. a. List five common shapes of cemented carbide and ceramic inserts.
   b. Give an advantage of disposable cemented carbide or ceramic inserts.

4. Cite two cutter precautions to take when selecting diamond, ceramic, or cemented carbide cutting tools.

5. List four of the eight design features that provide specifications for throw-away carbide inserts.

6. Explain the meaning of tangential force during a cutting process on a lathe.

7. a. Describe a quick-change tool system.
   b. List the advantages of such a system over a multiple-station turret.

C. OFF-HAND GRINDING HSS TOOL BITS

1. Tell how to grind the side- and end-cutting edges, the relief angles, and the face of a single-point cutting tool.

2. State two safe practices to observe in selecting, grinding, or cutting with a single-point cutting tool for lathe work.

SECTION THREE
Lathe Work between Centers

Many lathe work processes are performed with the workpiece mounted between the lathe centers. This section describes the principles of and procedures for performing the following basic between-centers processes:

- Centering and lathe setups;
- Facing and straight turning;
- Shoulder turning, chamfering, and rounding ends;
- Grooving, form turning, and cutting off;
- Filing, polishing, and knurling;
- Taper and angle turning.

UNIT 21

Center Work and Lathe Setups

OBJECTIVES

After satisfactorily completing this unit, you will be able to:

- Apply information about the functions of center holes and types and recommended sizes of combination drills and countersinks.
- True live centers and align dead centers.
- Follow general lathe setups for machining.
- Perform the following processes.
  - Lay Out and Check the Accuracy of a Center Hole.
  - Drill Center Holes (Drill Press and Lathe Techniques).
  - Mount Work Between Centers.
  - True a Live Center.
- Interpret technical data on center drill sizes.
- Follow recommended *Safe Practices* in center drilling.
- Use *Terms* defining lathe setups for center work, for drilling, and for mounting work between centers.

All basic lathe turning processes require a work-holding device and a method of turning the workpiece. Common lathe practices may be grouped as *chuck work*, *faceplate work*, and *fixture work*. Other work parts may be turned between centers.

A turned part is usually machined from a solid bar, forging, or casting. Before any turning process may be performed, the part must be prepared. The end may need to be faced to provide a flat plane reference surface. A cone-shaped (center) hole may be required. The center hole permits the workpiece to be mounted between the lathe centers or supported on one end. This unit deals with *center drilling* and *preparation of the lathe* for external turning processes.

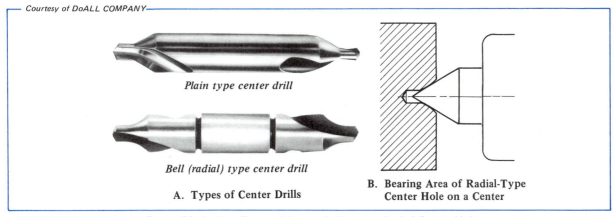

Plain type center drill

Bell (radial) type center drill

A. Types of Center Drills

B. Bearing Area of Radial-Type Center Hole on a Center

Figure 21–1 Two Types of Center Drills and a Radial Center Hole

CENTER WORK AND CENTER DRILLING

FUNCTIONS OF CENTER HOLES

A *center hole* is a cone-shaped or bell-mouthed bearing surface for the live and/or dead centers. Each center hole must be of a particular size and depth, depending on the size of the part to be machined and the nature of the operation. An improperly drilled center hole limits the bearing surface, wears away rapidly, and may result in inaccurate turning. The center drilled hole should have a fine surface finish to reduce friction on the cone- or bell-shaped bearing area as shown in Figure 21–1B.

TYPES AND SIZES OF CENTER DRILLS

Table 21–1 gives recommended sizes of *plain* center drills for different diameters of workpieces. Note in each case that the diameter to which the

hole is countersunk is always less than the body diameter of the combined drill and countersink (center drill).

A *radial* type of center drill has an 82° angle above the 60° bearing band. This type of center drill automatically produces a *safety center*. The radial center hole is lubricated easily. The radial type center drill is available in sizes ranging from #00 (0.025″ drill and 5/64″ body diameters) to #18 (1/4″ drill and 3/4″ body diameters). Similar metric sizes and styles of center drills are available. The sizes ranging from #00 to #8 are adapted for precision tool, gage, and instrument work. The heavy-duty type ranges from #11 to #18. These sizes are used for heavy production parts, especially forgings and castings. The features of a plain- and a radial-type center drill are shown in Figure 21–1A.

Table 21–1 Recommended Center Drill Sizes (Plain Type)
for Different Diameter Workpieces

Size No.	Diameter (inches)			
	Workpiece	*Countersink*	*Drill Point*	*Body*
00	3/32 to 7/64	3/64	0.025	5/64
0	1/8 to 11/64	5/64	1/32	1/8
1	3/16 to 5/16	3/32	3/64	1/8
2	3/8 to 1/2	9/64	5/64	3/16
3	5/8 to 3/4	3/16	7/64	1/4
4	1 to 1 1/2	15/64	1/8	5/16
5	2 to 3	21/64	3/16	7/16
6	3 to 4	3/8	7/32	1/2
7	4 to 5	15/32	1/4	5/8
8	6 and over	9/16	5/16	3/4

PREPARING THE LATHE FOR TURNING BETWEEN CENTERS

THE LIVE CENTER

The accuracy to which a part may be machined on centers depends on the *trueness* of the headstock (live) center and the alignment of the headstock and tailstock centers.

The trueness of the live center may be checked using a dial indicator. The indicator point is brought into contact with the surface of the live center. The spindle is revolved by hand. If the center runs out-of-true, it either may be ground in place or another live center must be inserted and tested.

TURNING A LIVE CENTER

When a chuck is already mounted, it is common practice to chuck a piece of steel and turn a center. The compound rest is swung to a 30° angle with the spindle axis. A center is roughed out and then finish turned. The side of the chuck jaw serves to drive the lathe dog and workpiece.

ALIGNING THE DEAD CENTER

The three basic methods of checking the alignment of centers were described earlier. Another fast, accurate method of aligning centers is to use a *micro-set adjustable live center* in the tailstock (Figure 21–2). The micro-set adjustable live center is used on light work.

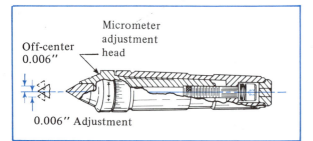

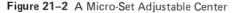

Figure 21–2 A Micro-Set Adjustable Center

How to Lay Out a Center Hole

Laying Out a Center with a Center Square

STEP 1 Cut off the workpiece. Allow sufficient stock to face the ends. File a slight chamfered edge to break the outside edge. Coat the end with layout dye.

STEP 2 Bring the V-section of the center head against the outer surface on one end of the workpiece. Scribe a line close to the blade.

STEP 3 Turn the center square head about a third of a revolution. Scribe a second line.

STEP 4 Turn the center head another one-third and scribe a third line.

STEP 5 Place the sharp point of a center punch at the intersection of the scribed lines. Strike the center punch squarely to produce a center-punched indentation.

STEP 6 Repeat steps 1 through 5 on the opposite end of the workpiece.

Note: The workpiece may also be layed out by using dividers. A height gage and V-blocks are sometimes used. The center may also be located by simply positioning and striking the punch of a bell-shaped cup. The cup centers over the workpiece end.

Checking the Layout of a Center Hole

STEP 1 Set the legs of a divider to the radius of the workpiece.

STEP 2 Insert one leg in the center-punched mark. Swing the other leg around the circumference.

Correcting an Off-Centered Punch Mark

STEP 1 Place the center punch in the center mark. Tilt the punch at an angle.

STEP 2 Tap the center punch so that the blow drives the center mark toward the accurate center.

Note: The workpiece may also be checked for accuracy of center hole layout by placing it between centers. As the workpiece is rotated by hand past a fixed point, the amount of variation (off-center) is noted. The workpiece is then removed. The center hole is repunched to move it to an exact location.

Courtesy of CLAUSING CORPORATION

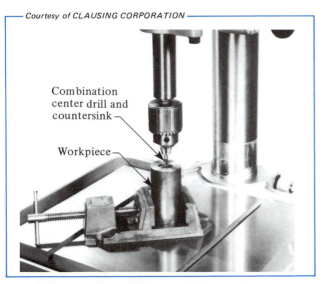

Combination center drill and countersink —

Workpiece —

Figure 21–3 Drilling a Center Hole on a Drill Press

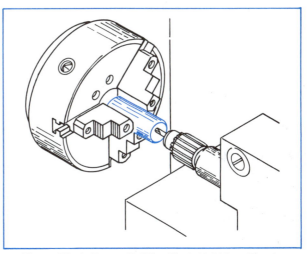

Figure 21–4 Center Drilling Work Held in a Chuck

COMMON CENTER DRILLING TECHNIQUES

How to Drill Center Holes

Drilling on a Drill Press

STEP 1 Position the workpiece using a V-block or the V-groove of a drill press vise.

STEP 2 Select the number of center drill that will produce a large enough center hole for the job requirements.

STEP 3 Chuck the center drill. Set the spindle speed. Align the drill point with the center-punched mark.

STEP 4 Brush on a thin film of cutting fluid. Bring the drill carefully into contact with the workpiece (Figure 21–3). Drill to depth.

Note: Use a fine feed and a cutting fluid to finish the center drilling.

Drilling a Center Hole on a Lathe

A. Center Drilling Work Mounted in a Chuck

STEP 1 Chuck the workpiece so that the outside surface runs concentric. Face the end.

STEP 2 Mount the drill chuck and center drill in the tailstock spindle.

STEP 3 Bring the tailstock up toward the workpiece and tighten it. Put a light pressure on the tailstock spindle by partially clamping it. Feed the center drill to start the process (Figure 21–4).

STEP 4 Continue to feed carefully to the required depth. Use a cutting fluid.

B. Center Drilling from a Lathe Spindle

STEP 1 Mount the drill chuck with a center drill in the headstock spindle.

STEP 2 Position and hold the center-punched hole on one end of the workpiece against the dead center.

STEP 3 Start the lathe. Guide the center drill in the second center-punched hole on the opposite end. Feed the workpiece toward the center drill by turning the tailstock handwheel.

STEP 4 Feed to the required depth. Use a fine feed and a cutting fluid for the final cutting action.

C. Center Drilling Long Workpieces

Long workpieces are center drilled by chucking one end. The second end is supported in a steady rest (Figure 21–5). The steady rest jaws are adjusted until the center of the workpiece is aligned with the tailstock center.

After the end is faced, the center drill is fed carefully into the revolving workpiece until the required outside diameter of the countersunk portion is reached.

Courtesy of LeBLOND MACHINE TOOL COMPANY

Figure 21-5 Using a Steady Rest to Center Drill a Long Workpiece

Caution: The center hole must be checked for adequate lubrication during each successive cutting step. The heat that is generated may cause the workpiece to expand and tighten against the centers.

STEP 7   Check the center height of the cutting tool and the position of the holder and compound rest. Be sure all parts are clear. Then proceed with the machining process.

How to Mount Work between Centers

STEP 1   Secure a driver plate on the spindle nose. Align the headstock and tailstock centers. Aligning may be done by the visual method or by one of the more precise methods that uses a dial inidcator.

STEP 2   Check the trueness of the live center.

STEP 3   Select a lathe dog that will accommodate the size and shape of the workpiece. Slide it over the workpiece to the headstock end.

STEP 4   Clean the center holes and the center. Apply a center lubricant on the tailstock center if a solid dead center is used.

STEP 5   Adjust the center so that the workpiece turns freely without end play.

STEP 6   Insert the lathe dog tail in a driver plate slot. Check to see that it moves freely and does not bind in the slot when the safety setscrew is tightened.

Safe Practices in Center Work and Lathe Setups

- Check the depth of a center-drilled hole. It must provide an angular or radial bearing surface adequate to support a workpiece and withstand cutting action.
- Use a rotating dead center when turning workpieces that revolve at high speeds.
- Use lathe dogs equipped with safety screws or safe locking devices.
- Check the bent tail of a lathe dog to see that it clears the bottom or sides of the driver plate slot. The dog position must permit a workpiece to ride on the center.
- Lubricate the angular or radial bearing surface of a workpiece that revolves on a solid dead center.
- Rotate a new workpiece and tool setup by hand to ensure that all parts clear without touching.
- Cover the lathe bed and other machined surfaces whenever a tool post grinding attachment is used.

TERMS USED IN CENTER WORK AND LATHE SETUPS

Centering   The process of drilling a small pilot hole and countersinking an angular surface in the end of a workpiece in one operation.

Center hole   A cone- or radial-shaped hole that may be used to center a drill or to provide an angular bearing surface for a lathe center.

Radial center drill	A center drill that has the added feature of an 82° angle above the 60° bearing band. A center drill that produces a safety center.
Overhang	The portion of a cutting tool, toolholder, or tool post that extends beyond a supporting surface.
Mounting work (lathe)	The process of positioning a workpiece so that it may be machined. Securing a workpiece in a chuck or between centers.
Witness marks	A light, prick-punched mark or scored line on a center, sleeve, or spindle. Position marks on mating parts. Marks made for the purpose of assembling parts in the same relative position each time.

SUMMARY

- A faced end provides a flat reference surface. Measurements may be made accurately from a faced end.
  - Center holes provide a 60°-angle bearing surface for turning work between centers. The cylindrical surfaces that are produced are concentric with the axis of the lathe centers.
- Center drills of the plain type have the countersunk portion ground to an angle of 60°. The radial type of center drill has an added countersink angle of 82°.
  - Center drills are ordered by number sizes. These sizes range from #00 (0.025″ drill and 5/64″ body diameters) to the #8 heavy-duty size (5/16″ drill and 3/4″ body diameters).
- The setting up, positioning, and securing of a tool bit, toolholder, and tool post must provide maximum support with the least possible overhang.
  - Workpieces that extend a distance from a chuck must be supported at the tailstock end with a steady rest. The steady rest jaws must be positioned to align the axes of the workpiece and tailstock spindle.
- Before the lathe is started, all loose clothing must be secured, wiping cloths must be removed, and all parts must be checked to see that they are free to rotate.

UNIT 21 REVIEW AND SELF-TEST

1. Identify (a) the functions and (b) the characteristics of a precision, radial center-drilled hole.

2. State three practices the skilled mechanic follows when using a live center.

3. Cite two advantages of the micro-set adjustable center for light turning processes as compared to using a solid dead center.

4. List five general practices for setting up a lathe and taking basic cuts.

5. State the method by which long workpieces are supported for center drilling.

6. Tell why a center is often turned from a mild-steel chucked piece rather than by using a standard, hardened live center.

7. Give four precautionary steps to take before starting to turn a workpiece between centers.

Facing and Straight Turning

The process of machining a flat plane surface on the end of a workpiece on a lathe is referred to as *facing*. The process of turning one or more diameters is known by such terms as *straight turning*, *parallel turning*, *cylindrical turning*, or *taking a straight cut*.

Facing and turning processes require differently formed cutting tools and different work-holding devices and tool setups. Some of the common shapes of high-speed steel turning tools (Figure 22–1) and principles and practices of facing and straight turning are covered in this unit.

OBJECTIVES

After satisfactorily completing this unit, you will be able to:

- Determine cutting speeds for straight turning and facing processes.
- Position the compound rest at 30°, 60°, and 84°16′ to obtain various depth of cut settings, and set the depth of cut.
- Take a trial cut to establish an initial measurement.
- Perform each of the following processes.
  - Set Up a Cutting Tool for Depth of Cut.
  - Rough and Finish Face a Workpiece (Chuck Work and Faceplate Work).
  - Do Straight (Parallel) Turning.
- Read drawings to establish unilateral, bilateral, equal, and unequal tolerances.
- Follow recommended *Safe Practices* for facing and straight turning.
- Interpret and be able to apply each new *Term*.

SETTING UP FOR FACING OR STRAIGHT TURNING

CUTTING SPEED AND FEED

The engine lathe must be set up for efficient machining. Cutting speed and the required spindle speed may need to be calculated. These speeds may also be established from handbook tables, as mentioned in an earlier unit. The lathe spindle speed is set as close as possible to the calculated RPM.

The feed for general lathe work processes usually must be determined by the operator. Consideration is given to such factors as the kind of material, the processes and cutting tools, the rigidity of the workpiece, the power (capacity) of the lathe, and the use of cutting lubricants. The selected feed is then obtained by positioning the feed-change levers in the quick-change gearbox.

POSITIONING THE COMPOUND REST

Compound Rest Set at 30°. The compound rest is graduated on the base and may be rotated through 360°. Angular settings such as 30°, 45°, and 60° are common. The 30° angle provides a convenient method of producing a fixed cutting tool movement. With the compound rest set at 30°, the amount of side (longitudinal) movement of a cutting tool is equal to one-half the distance the compound rest is moved (Figure 22–2). When the compound rest is set at 30° and the handwheel is turned 0.010″ (or 0.2mm on a metric-graduated dial), it produces a side motion of 0.005″ (0.1mm).

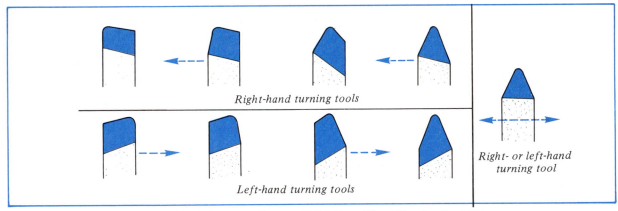

Figure 22–1 Common Shapes of High-Speed Steel Turning Tool Bits

The graduated collar on the compound rest feed screw is often used to set the tool depth. The 30° angular setting is applied to straight and to 60° included angle turning processes. This setting is especially important when cutting standard 60° thread forms. The feed per cut and the final thread depth are calculated. The cutting tool is fed to depth as shown in Figure 22–3. The graduated collar on the compound rest is used to feed the thread-cutting tool to depth.

Compound Rest Set at 90°. A 90° angular setting (parallel to the work axis) of the compound rest is practical for facing and shoulder turning. The amount the cutting tool is fed is read directly from the compound rest graduated collar setting. If the compound rest is moved 0.5mm (0.020″), the cutting tool advances the same distance (0.5mm or 0.020″).

Compound Rest Set at 84°16′. Although not a common practice, precise tool settings to approximately 0.0001″ (0.002mm) may be made with the compound rest set at an angle of 84°16′. The cutting tool feeds into the workpiece 0.0001″ (0.002mm) for each 0.001″ (0.02mm) movement of the compound rest feed screw. Because the angle graduations on compound rests are not graduated in minutes, the 16′ must be estimated.

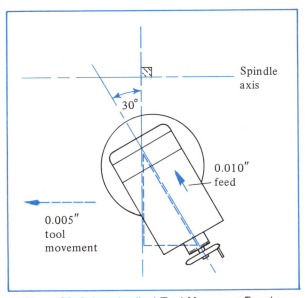

Figure 22–2 Longitudinal Tool Movement Equal to One-Half the Feed of the Compound Rest Set at 30°

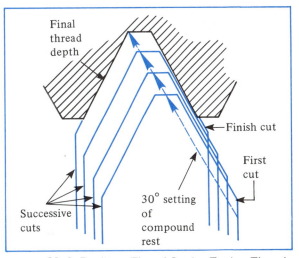

Figure 22–3 Feeding a Thread-Cutting Tool to Thread Depth by Angular Setting of the Compound Rest

PRACTICES FOR SETTING DEPTH OF CUT

The graduated micrometer collar on feed screws is a practical device for economically producing accurately machined parts. There are, however, certain procedures that must be followed to compensate for lost motion (backlash) and possible collar and cutting tool movement. Some of the steps that must be taken are:

- Turn the compound rest feed screw at least one-half turn in the direction in which the depth of cut is to be set;

- Set the locking device on the graduated feed collar before taking any readings;

- Feed the cutting tool into (not away from) the workpiece when setting the depth of cut;

- Back out the cutting tool at least one-half revolution whenever the feed handwheel has been moved too far. The cutting tool is then moved clockwise to remove the backlash and permit feeding the cutting tool to the correct setting.

How to Set a Cutting Tool for Depth of Cut

STEP 1   Select an appropriate toolholder. It must accommodate the workpiece and the nature of the cutting operation.

STEP 2   Tighten the tool bit in the toolholder. Position the tool bit for the machining operation. The cutting edge is set at center height. Solidly secure the setup to the compound rest.

STEP 3   Determine the correct spindle RPM and feed. Check the lubricating systems. Start the lathe.

STEP 4   Turn the cross feed handle until a light cut is taken and forms at least half a circle around the workpiece.

STEP 5   Move the carriage handwheel so that the tool clears the workpiece. Stop the lathe without disturbing the handwheel setting.

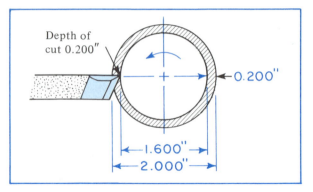

Figure 22–4 Outside Diameter Reduced by Twice the Depth of Cut

STEP 6   Measure the turned diameter section. Determine how much stock is still to be removed with one or more roughing cuts (Figure 22–4).

STEP 7   Continue to move the cutting tool to the depth of cut. The graduated collar provides an accurate measurement.

STEP 8   Start the lathe again. Take a trial cut for a short distance (usually 1/8"). Stop the lathe. Measure the diameter.

STEP 9   Take the roughing cut across the workpiece. Note the reading on the graduated collar.

STEP 10   Continue to take successive roughing cuts. Each cut is set for depth by feeding the cross feed screw in the same direction. The distance is measured by the reading on the graduated collar.

> **Note:** Some handwheels are graduated for *diameter reduction.* For example, a 0.010" movement of the graduated dial provides a 0.005" depth of cut. Other handwheels are graduated for *radius reduction.* A 0.010" movement equals a 0.010" depth of cut.

STEP 11   Increase the speed and decrease the feed for the finish cut.

> **Note:** The same procedure is followed in taking a trial final cut for a short distance. Check the diameter for accuracy and surface finish at each setting before the entire surface is turned.

FACING TOOLS, SETUP, AND PROCESSES

A workpiece is *faced* for three main reasons:

- To produce a flat plane to provide a reference surface for other measurements and operations,
- To machine a flat-finished area that is at a right-angle to the axis of the workpiece,
- To turn a surface to a required linear dimension.

THE CUTTING TOOL

Facing may be done with a side-cutting or end-cutting tool. The cutting point has a radius. The cut may be fed either toward or away from the center. The cut may be a roughing or a finishing cut. The workpiece may be held in a chuck, between centers, on a faceplate, or in a holding fixture.

Although the high-speed steel tool bit is commonly used, all factors and conditions for selecting a cutting tool must be considered. The tool bit that is selected must be the most productive for the job. Cemented-tip carbides with solid shank and carbide and ceramic inserts cut more efficiently at high speeds and on harder and tougher materials than do high-speed steel tool bits.

Two shapes of cutting tool high-speed steel bits are generally used for facing. The end-cutting tool is adapted to hogging cuts (Figure 22-5).

The facing tool bit with a 55° cutting angle and a slight radius nose is used when facing the end of a workpiece that is mounted between centers.

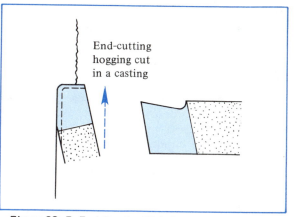

End-cutting hogging cut in a casting

Figure 22–5 End-Cutting High-Speed Steel Tool Bit Used for a Rough Facing Operation

Use a coarse feed. Feed the cutting tool across the face almost to the layout line.

Note: The end-cutting tool is fed directly into the workpiece (Figure 22–5).

Note: When a considerable amount of material is to be faced off a workpiece, a series of longitudinal cuts are first taken. The diameter is reduced for each successive cut. The cutting tool is fed to almost the required length.

STEP 4   Feed the cutting tool across the face to even the steps that are formed.

Finish Facing

STEP 1   Replace the end- or side-cutting tool with a facing tool. Position the cutting point on the center line.

STEP 2   Increase the speed. Decrease the feed. Take a light finishing cut. Start at the center and feed to the outside diameter.

Facing Work between Centers

STEP 1   Select a facing tool with a 55° cutting angle. Mount the tool bit so that the point is at center height.

STEP 2   Advance the cutting tool into the workpiece to the required depth.

STEP 3   Feed the tool from the center outward across the face of the work.

How to Face a Workpiece

Chuck and Faceplate Work

Rough Facing

STEP 1   Lay out the required length.

STEP 2   Chuck or mount the workpiece. Select and set the spindle RPM. Spindle speed is based on the largest diameter to be faced.

STEP 3   Use an end-cutting or plain side-cutting rough turning tool. Set the depth of cut.

GENERAL RULES FOR STRAIGHT TURNING

Straight, or *parallel turning* means that a workpiece is machined to produce a perfectly round cylinder of a required size. The general rules for all lathe processes apply.

- The cutting tool should be fed toward the headstock wherever possible, especially on work between centers. The cutting forces should be directed *toward* the live center.
- The number of cuts should be as few as possible. Sufficient stock should be left for a final finish cut if a fine surface finish is required.
- The size and condition of the lathe, the shape and material of the workpiece and cutting tool, and the type of cut all affect the depth of cut.
- The roughing cut must be deep enough on castings, forgings, and other materials to cut below the hard outer surface (scale).
- Deep roughing cuts should be taken using the coarsest possible feed.
- The cutting tool and holder must be set in relation to the workpiece. Any movement of the holder produced by the cutting forces should cause the cutting tool to swing *away* from the work.
- Trial cuts are taken for a short distance. Each trial cut permits the workpiece to be measured before a final cut is taken. Any undersized positioning of the cutting tool may be corrected before the part is machined.

How to Do Straight (Parallel) Turning

Taking Roughing and Finishing Cuts

STEP 1 Select the kind and shape of tool bit that is appropriate for rough turning the particular part.

STEP 2 Use a straight or left-hand toolholder that will clear the workpiece and the machine setup.

STEP 3 Position and solidly secure the tool bit and toolholder to the compound rest.

STEP 4 Determine and set the spindle speed at the required RPM.

STEP 5 Set the feed lever for a roughing cut.

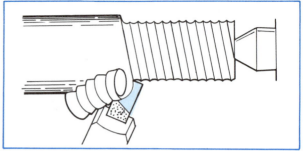

Figure 22–6 Taking a Roughing (Coarse Feed) Cut

STEP 6 Move the cutting tool to the end of the workpiece. Adjust the depth of cut until at least half of a true diameter is turned.

STEP 7 Take the trial cut for a distance of from 1/8″ to 1/4″. Leave the tool set. Return it to the starting end. Stop the lathe.

STEP 8 Measure the turned-diameter portion only.

STEP 9 Note the reading on the graduated collar. It may also be set at zero. This reading is the reference point. Subtract the measured diameter from the required diameter. An oversized allowance is made for the finish cut,

STEP 10 Move the cutting tool to depth. Start the lathe. Engage the power feed. Take the required number of roughing cuts (Figure 22–6).

STEP 11 Replace the roughing tool with a finish turning tool.

STEP 12 Increase the spindle speed and decrease the cutting feed for finish turning

STEP 13 Take the trial cut. Check the accuracy of the turned diameter. Adjust further for depth. Engage the power feed. Turn the required diameter to the specified length (Figure 22–7).

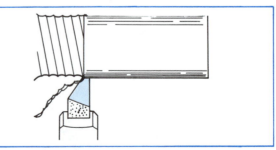

Figure 22–7 Taking a Finishing Cut

- Check the edges of the compound rest for clearance between it and the revolving workpiece or work-holding device. Special care must be taken when the compound rest is positioned at an angle to the spindle axis.

- Take up all lost motion in the cross feed screw. Then lock the graduated collar at the starting point of a measurement.

- Turn the cross feed handwheel clockwise to apply a force to set the depth of cut. The inward force causes the tool to remain at the depth of cut even when cutting forces are present.

- Position the cutting tool and holder so that any movement during the cutting process moves the cutting edge away from the workpiece.

- Wear a protective eye device and observe general machine safety practices.

COMMON TERMS USED IN FACING AND STRAIGHT TURNING

Turning (straight, parallel, cylindrical)	Common shop terminology used to describe the machining of one or more external diameters. Advancing a cutting tool longitudinally. Taking a cut across the outside diameter at a desired depth while the workpiece is revolving.
Equivalent feed of compound rest	The distance a cutting tool advances longitudinally or transversely when fed by the handwheel of a compound rest that is set at an angle.
30° compound rest setting	The position of the compound rest at 30° from either the center or cross slide axis. An angular setting that produces a longitudinal or transverse movement equal to one-half the movement of the compound rest.
Locking the graduated collar	A precautionary step of securing the cross slide or compound rest graduated collar at a desired graduation. Securing a graduated collar so that its position is not changed when other adjustments of the cutting tools or work are made.
Depth of cut (lathe turning)	A setting for a lathe cutting tool. Reducing the outside diameter by twice the actual depth of cut.
Trial cut	A common shop practice of turning a concentric area for a length of about 1/8″ to 1/4″. Establishing a reference surface (a first cut) from which other cuts may be planned in reducing a workpiece to a required size.
Reference layout points	Identifying lines or marks to show a linear dimensional limit to which a surface is to be machined.
Tolerance	A dimensional limit that guides a craftsperson in making or fitting a part in relation to a mating surface or part. The maximum variation permitted in the finished size of a machined part.
Upper and/or lower dimensional limit	An ultimate dimension for a feature of a part. The largest (upper limit) and smallest (lower limit) dimension to which a part may be machined to keep within a required degree of accuracy.
Unequal tolerances	A difference in the upper (+) and lower (−) tolerance values.

SUMMARY

- Compound rest settings of 30° and 60° are widely used in turning and facing. The compound rest feed movement may be converted into either a transverse or longitudinal movement.
  - Movements of 0.0001″ may be made by setting the compound rest at an angle of 84°16′ to the cross slide axis. As the compound rest is advanced 0.001″, the tool is fed 0.0001″.
- A trial cut permits measuring a workpiece. If the cut produces an incorrect diameter, it may be corrected. A trial cut on a tough abrasive casting or forging must be taken below the hard outer scale or surface.
  - A workpiece that is to be reduced in length by facing for a considerable distance should first be rough turned longitudinally. Each cut may then be stepped off. Finally, the cutting angle is changed and the uneven steps are faced off to produce a square, flat end.
- The dial-graduated micrometer collar is used on modern lathes for direct measurement of the cross slide and the compound rest movements. Graduations on one band are in decimal parts of an inch. On the other band the graduations are in decimal values of millimeters.
  - Once lost motion is compensated for, the graduated collar is locked to the feed screw. The graduation line becomes the reference point. New measurements are related to the reference point.
- Chuck and faceplate work may be faced with a regular side- or end-cutting tool. Centered work is usually faced with a 55° cutting point in a right-side facing tool.
  - The length of a round workpiece is sometimes scribed on a color-dyed surface for visibility during machining.
- Workpieces should be rough turned leaving 1/64″ of stock for finish turning on small diameters and 1/16″ on large diameters. The spindle speed is increased and the feed is decreased for finish turning.
  - Turn the workpiece end-for-end when the entire length is to be turned between centers. When a quantity of parts are to be machined, the roughing operations are completed first on all pieces and are followed with the final finish cut.
- All lubricating systems and moving parts of the lathe must be checked. Cloths and waste must be removed. Loose clothing is to be secured. General safety precautions for machine operation must be followed.

UNIT 22 REVIEW AND SELF-TEST

1. State why the compound rest is generally set at a 30° angle for straight turning processes.

2. List the steps to take to turn a steel part from an outside diameter of 100mm down to 80 mm.

3. Identify two general forms of facing tools.

4. State four practices the lathe operator follows that apply to all straight turning processes.

Shoulder Turning, Chamfering, and Rounding Ends

A *shoulder* is the area that connects two diameters. *Shoulder turning* is the process of machining the connecting area to conform to a particular size and shape. There are three basic shoulder shapes:

• The *square* shoulder,

• The *beveled* (angular) shoulder,
• The *filleted* (round) shoulder.

A *chamfer* is an outside edge or end that is beveled or cut away at an angle. A *rounded end* or corner differs from a chamfer in that the round shape is a radius.

OBJECTIVES

After satisfactorily completing this unit, you will be able to:

• Determine the purposes and uses of shoulders, chamfers, and rounded ends.
• Measure shoulders.
• Apply a radius gage in checking filleted shoulders and rounded ends.
• Perform each of the following processes.
  • Turning Square, Filleted, and Beveled Shoulders.
  • Chamfering.
  • Turning or Filing a Rounded Corner.
• Interpret drawings which involve shoulders and formed ends of cylindrical parts.
• Follow *Safe Practices* and use new *Terms* which apply to shoulder turning, chamfering, and rounding ends.

PURPOSES OF SHOULDERS, CHAMFERS, AND ROUNDED ENDS

CHAMFERED AND ROUNDED ENDS

Chamfered and rounded ends are turned for appearance. They also provide clearance so that the face of a mating part may seat properly against an internal shoulder (Figure 23-1A). Chamfered and rounded ends are also machined for safety and comfort in handling.

A chamfered end is especially cut on workpieces that are to be threaded. The chamfer angle provides a good face against which a single-point threading tool or die head may be started. As stated before, the chamfer reduces burrs that form easily on a sharp edge. A 30° chamfered end permits a 60° formed thread to be engaged easily.

The size of a chamfered or rounded end is usually given on a drawing. However, the craftsperson often has to judge the angle or size of radius. Many times there is a note on the drawing to BREAK ALL SHARP EDGES. The note indicates that just a light cut may be taken with a mill file against a revolving workpiece. The end may be filed to a radius or angle.

SHOULDERS

The particular shape of a shoulder depends on its use. For example, the underside of the head of a bolt and the body form a square shoulder. A square shoulder permits the head to seat flat so that there is maximum contact of the surface areas (Figure 23–1B). A square shouldered part may serve also as an end bearing or stop, as shown in Figure 23–1C. The shoulder is strengthened by a slight radius at the intersecting surfaces.

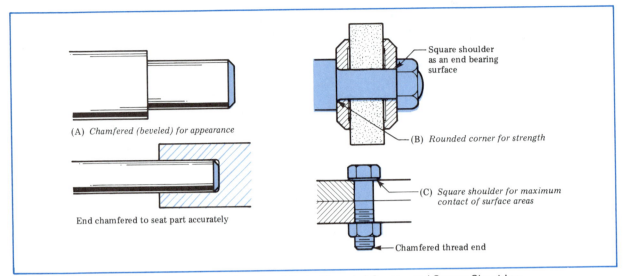

(A) *Chamfered (beveled) for appearance*

End chamfered to seat part accurately

Square shoulder as an end bearing surface

(B) *Rounded corner for strength*

(C) *Square shoulder for maximum contact of surface areas*

Chamfered thread end

Figure 23–1 Examples of Chamfered Ends, Rounded Corner, and Square Shoulders

The shoulder on a forming punch may be round or beveled to form a part with a corresponding radius (fillet) or angle. When a sharp corner is to be avoided, the filleted shoulder strengthens the part without increasing its size. Beveled shoulders also eliminate sharp corners, and they add strength, as well as improve the appearance of the part. Beveled shoulders are cut mostly at 30°, 45°, and 60° angles.

COMMON MEASUREMENT PROCEDURES FOR SHOULDER TURNING

The length of a shoulder is usually marked by cutting a light groove around the workpiece. The cutting tool is positioned by measuring the required shoulder length with a steel rule.

Shoulders that are to be machined to a precision linear dimension may be measured with a gage or depth micrometer. In precision shoulder turning, the final cut is positioned using a micrometer stop for the carriage. The compound rest feed screw may also be moved the required distance. The depth is read on the graduated collar.

The diameter and shoulder length are generally turned to within 1/64″ of the finished size to leave sufficient material at the shoulder to permit turning to the required shape and dimension. One shoulder-turning technique locates the length of a square shoulder by cutting a groove in the workpiece with a necking tool.

After grooving, the body diameter is turned to size. One or more cuts is taken until the required dimension is reached.

Another shoulder-turning technique is to *block out the shoulder* and leave enough material at the shoulder to finish turn it to the required form. The adjacent cylindrical surface is then turned to size.

APPLICATION OF THE RADIUS GAGE

The *radius gage* provides a visual check on the radius of the cutting tool or the filleted shoulder.

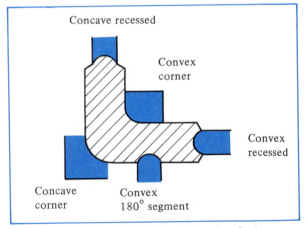

Concave recessed

Convex corner

Convex recessed

Concave corner

Convex 180° segment

Figure 23–2 Applications of the Five Gaging Areas of a Radius Gage

The accuracy of the cutting tool is checked by placing the correct radius portion of the gage over the ground radius. A radius gage and five general applications of it are pictured in Figure 23–2.

Low spots are sighted by any background light that shines between the gage and the cutting tool. The dark, high-spot areas are ground or honed if there are only a few thousandths of an inch to be removed. When no light is seen along the radius, the cutting point is formed to the correct shape and depth.

One common practice that is repeated in turning a shoulder to a *large radius* is to *step off* the shoulder. Successive facing or straight turning cuts are taken. These cuts end in a series of steps. The steps, or excess corner material, are rough turned. A stepped-off section is shown in Figure 23–3.

The lathe operator often *contour forms* a large radius by moving the longitudinal and cross feed screws simultaneously by hand. This action sweeps the cutting tool through the desired radius. Rounded corners are generally formed with a tool ground to the correct radius.

Commercial flat, metal radius gages (as previously illustrated) are available. The general sizes range in radius from 1/32″ R to 1/2″ R. There are usually five different gaging areas on each blade. These areas permit gaging the radius of a corner, a shoulder, or a recess in a circular workpiece.

REPRESENTING AND DIMENSIONING SHOULDERS, CHAMFERS, AND ROUNDED ENDS

The techniques of representing and dimensioning the three general types of shoulders and rounded and beveled ends are illustrated in Figure 23–4. The square shoulder ① is turned to an outside body diameter of 1.500″ for the 1-1/2″—12UNF thread. The length or distance to the square shoulder is 1-1/4″.

The radius of the filleted shoulder ② is indicated as 1/4 R. The beveled shoulder ③ joins the 1.250″ + 0.000″/ —0.002″ and 1.500″ ±0.001″ diameters at a 45° angle. The distance to the shoulder is 2″.

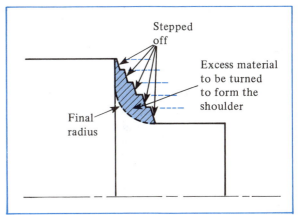

Figure 23–3 Stepping Off a Large Radius

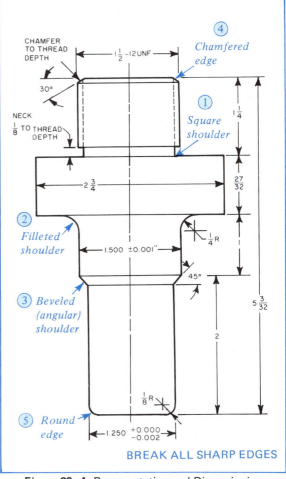

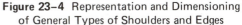

Figure 23–4 Representation and Dimensioning of General Types of Shoulders and Edges

The beveled edge ④ angle corresponds to one-half the thread angle and is 30°. While no depth is given, there is a note on the drawing to **CHAMFER TO THREAD DEPTH.** This note indicates that the chamfer is to be cut at least to the thread depth. The Unified Fine Thread (1 1/2"—12UNF) base diameter is 1.400".

The chamfered depth is one-half the difference between the outside and base diameters of the thread, or 0.050". In practice, the lathe operator cuts slightly beyond the thread depth. The drawing shows the rounded edge ⑤ is dimensioned 1/8 R. The end is turned to a 1/8" radius.

How to Turn Shoulders

Note: The procedure for turning each type of shoulder starts with a workpiece that has been layed out and rough turned to diameter, length, and shape.

Turning the Square Shoulder
Method 1: Feeding Outward

STEP 1 Reduce the excess material at the shoulder to permit taking a finishing cut.

STEP 2 Position a right-side-cutting facing tool. Set the cross feed graduated collar at zero.

STEP 3 Finish turn the outside diameter. Take the finish cut with the automatic power feed. Turn almost to the shoulder. Continue to feed the facing tool into the shoulder by hand. Then take a facing cut outward.

STEP 4 Stop the lathe. Measure the length to see whether the cut splits the layout line.

STEP 5 Remove any additional material on the shoulder face. Bring the facing tool back to the zero setting each time. The tool then is again fed toward the outside diameter.

Note: The cutting tool may also be moved to the required length either against a micrometer stop or with the compound rest. For a less precise measurement, the experienced operator often estimates how much more the facing tool should be moved into a workpiece.

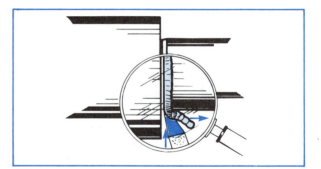

Figure 23–5 Feeding an End-Cutting Tool to Cut a Square Shoulder

Method 2: Feeding Inward

STEP 1 Move the end-cutting facing tool so that the cutting edge just touches the turned diameter of the body portion of the shoulder. Set the cross slide graduated collar at zero.

STEP 2 Back the cutting tool away from the workpiece. Start a trial cut for length on the outside diameter. Check the length with a steel rule or by micrometer measurement, depending on the accuracy needed. Lock the carriage for facing the shoulder portion.

STEP 3 Feed inward to the zero setting on the graduated collar.

STEP 4 With the cutting tool in this position, feed to the right to remove any excess material (Figure 23–5). Blend the diameter at the shoulder with the body.

Note: Some operators move the cutting point slightly beyond the zero setting for the final shoulder finishing operation. Just the corner of the shoulder is turned at this setting.

STEP 5 Repeat steps 3 and 4, if required, to machine the shoulder length to size.

Note: End-feeding and then facing is a quick way of establishing the length. However, a smoother finished surface is produced by feeding the cutting tool outward on the shoulder face.

STEP 6 Remove the tool marks. Speed up the spindle RPM to about twice the speed used for turning. Polish with an abrasive cloth.

Turning a Filleted (Round) Shoulder

STEP 1 Position a right-side round-nose turning tool.

STEP 2 Rough face the shoulder. Allow sufficient material to finish the body diameter, shoulder radius, and shoulder face.

STEP 3 Rough turn any excess material left on the rounded portion.

STEP 4 Set the cutting tool at zero on the cross feed graduated collar when the final diameter is reached.

STEP 5 Finish turn the diameter. Disengage the longitudinal automatic power feed before the shoulder is reached.

STEP 6 Feed the carriage slowly by hand. Move the tool a slight distance into the shoulder face.

STEP 7 Lock the carriage. Take a facing cut outward.

> **Note:** Decrease the spindle speed when cutting a radius to prevent chatter and digging in of the tool at the shoulder.
>
> **Note:** A cutting fluid is used to produce a fine surface finish.

STEP 8 Measure the shoulder length. Take any required additional cuts by bringing the cutting tool point to the zero setting each time. Move the tool by hand into the shoulder face to the required depth. Lock the carriage and reface.

How to Turn a Beveled (Angular) Shoulder

Method 1: Setting a Side-Cutting Tool

STEP 1 Position the side-cutting edge at the required angle. (The turning of an angular shoulder is shown in Figure 23-6.)

STEP 2 Remove excess stock from the angular shoulder.

STEP 3 Take a trial cut on the body diameter. Measure the diameter; then move the cutting tool to the required depth. Set the cross feed graduated collar at zero.

STEP 4 Use a power feed and take the cut almost to the shoulder. Disengage the power feed.

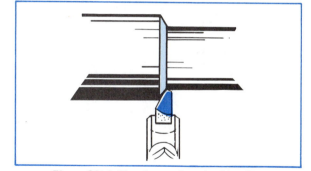

Figure 23–6 Turning an Angular Shoulder

STEP 5 Decrease the spindle speed for a large deep-cut shoulder. Reducing speed tends to reduce chatter. Continue to slowly feed the cutting tool by hand into the angular shoulder face.

STEP 6 Feed to the required length. Use a cutting fluid.

Method 2: Setting the Compound Rest

STEP 1 Position the compound rest at the required angle.

STEP 2 Set up a side-cutting tool.

STEP 3 Take cuts from the center outward. Feed the cutting tool by hand by turning the compound rest feed screw.

STEP 4 Continue to machine the bevel. The final beveled surface starts at the required length of the small diameter. It continues at the required angle to the larger diameter.

How to Turn a Round Corner

Using a Radius Cutting Tool

STEP 1 Grind an end-cutting tool to the specified inside radius. Check the accuracy with a radius or other gage.

STEP 2 Set the cutting tool point on center. Position the tool so that the radius will blend into the face.

STEP 3 Reduce the spindle speed. The larger the radius is, the slower the spindle speed.

STEP 4 Bring the inside edge of the cutting point close to the face of the workpiece. Lock the carriage. Flow on a cutting fluid. Feed the radius tool with either the cross slide or compound rest feed screw. Continue until the required radius is formed.

STEP 5 Stop the lathe. Remove all chips, and wipe the workpiece.

STEP 6 Increase the spindle speed. Start the lathe. Remove the tool marks by finish filing. Polish with an abrasive cloth.

Safe Practices in Turning Shoulders, Chamfers, and Rounded Ends

• Blend the stepped sections of a filleted or square shoulder to the desired form before a final round-nose or blunt angular-nose formed facing tool is used for the finishing cut.

• Align the cutting point of a facing tool on center.

• Reduce the spindle speed and use an appropriate cutting fluid whenever a large radius, angular shoulder, or a rounded or beveled end is to be machined. The speed reduction helps prevent chatter.

• Check the accuracy of a turned concave or convex radius with a radius gage.

• Use the left-hand method of filing a workpiece on the lathe. This method is a safety precaution to prevent hitting the revolving workpiece, lathe jaw, or chuck.

• See that the file handle fits solidly on the tang.

TERMS USED IN TURNING SHOULDERS, CHAMFERS, AND ROUNDED ENDS

Filleted shoulder (round)	The joining of two different diameters with a rounded end section. A shoulder formed by a radius. A radius turned to an inside corner.
Beveled shoulder (angular)	The joining of two different diameters with an angular surface. A shoulder formed at an angle.
Square shoulder (turned)	A plane, radial face. A face that is at right angles to the axis of a cylindrically turned workpiece.
Rounding a corner (turning)	The process of forming a radius on the end of a workpiece. Feeding a cutting tool with a round cutting face of a specified radius into a revolving workpiece.
Radius gage	A flat, L-shaped metal plate with circular internal and external shapes of a specified radius. A flat plate applied in checking the form and dimensional accuracy of inside or outside radii.
Stepping off	The process of taking a series of successive cuts. (Excess material is left at the end of each cut. A shoulder is then formed to a particular shape and size.)
Reducing the stepped-off shoulder	Removing the excess material left after rough forming a shoulder. Finish turning a shoulder to shape and size.
End-feeding	Hand or power feeding a cutting tool from an outside diameter inward toward the center.
Chamfering (turned workpiece)	Forming an angular (beveled) end. An angular surface adjoining two adjacent flat surfaces. Positioning the cutting tool at a required angle and feeding the cutting edge into the revolving workpiece.
"Chamfer to thread depth"	A drawing note that means (a) the chamfer angle is the same as the thread angle and (b) the minimum depth of the chamfer is equal to the depth of the thread.

SUMMARY

- Angular and filleted shoulders are turned to strengthen the area adjoining two diameters of a workpiece or to improve the general appearance of a part.
  - A square shoulder permits the flat seating of two mating faces. The square shoulder also serves as a bearing or positioning surface.
- Shoulder turning requires three basic processes: (1) layout (marking the length), (2) turning both diameters, and (3) machining the connecting area to a particular shape and size.
  - A chamfered thread end helps to prevent the forming of a burr. The chamfer also permits a thread form and the mating thread to be started accurately.
- A 1/64" material allowance is usually left for finish turning both diameters and the shoulder.
  - A radius gage is used to measure inside and outside turned forms. These forms are circular and of a specified radius.
- Angular shoulders are usually cut by positioning the face of the cutting tool at the required angle. The tool is fed lengthwise into the shoulder.
  - Turned shoulders and ends are dimensioned in terms of length, diameter, and the shape and size of the round or beveled area.
- Deep-filleted (round) shoulders are rough turned by stepping off each successive cut. The excess material is then rough turned to the required radius.
  - Filleted shoulders may be rough and/or finish turned with a radius cutting tool.
- A large radius is often turned by hand feeding. The cross feed and longitudinal feed handwheels are moved at the same time.
  - The final cut in shoulder turning is usually begun with the graduated collar set at zero. Subsequent shoulder forming or facing cuts are brought to or started at the zero setting.
- The spindle speed is reduced when using a radius-forming shoulder-cutting tool for the finish facing cut. A cutting lubricant helps to produce a finer cutting action and surface finish.

UNIT 23 REVIEW AND SELF-TEST

1. Indicate what the worker is expected to do when a drawing of a part has the note: BREAK ALL SHARP EDGES.
2. Explain what it means in shoulder turning to block out the shoulder.
3. List three lathe work applications of a radius gage.
4. Describe the meaning of a note on a shop sketch that reads: CHAMFER TO THREAD DEPTH.
5. State two common methods of (a) turning a square shoulder, (b) turning a beveled or angular shoulder, and (c) rounding a corner.
6. Give two safety precautions to take in turning a large radius or an angular (beveled) shoulder.

Grooving, Form Turning, and Cutting Off

Turning processes that produce a specially shaped cut-away section around a workpiece are referred to as *grooving*. A groove that is cut into an outside diameter is called a *neck*. A groove that is formed inside a hole is known as a *recess* or an *undercut*.

Form turning is used to produce a regular or irregular shape around a cylindrical surface. *Cutting off (parting)* is the separating of material by cutting through a revolving workpiece. The purposes, general cutting tools, and setups for grooving, form turning, and cutting off, are covered in this unit.

OBJECTIVES

After satisfactorily completing this unit, you will be able to:

- Interpret the functions served by grooves.
- Apply cutting tools and setups to turn square, round, and angular forms of grooves.
- Turn convex, concave, and other irregular groove forms.
- Use preformed cutting tools and form grooves by hand feeding.
- Know when to use general shapes of cutoff blades and parting processes.
- Analyze problems encountered in cutting off stock, probable causes, and recommended corrective steps.
- Perform the following processes.
  - Turn Square, Round, and V-Shaped Form Grooves.
  - Turn a Form.
  - Cut Off (Part) Stock.
- Follow *Safe Practices* for grooving, form turning, and cutting off and use *Terms* correctly.

FUNCTIONS SERVED BY GROOVES

A part is usually grooved for one of the following seven reasons:

- To permit a cutting tool to end a cut in the workpiece and be withdrawn without damage to the cutting point. One application of necking is in threading to a shoulder. The thread-cutting tool emerges from the last thread into the groove, where it may then be safely withdrawn and positioned for subsequent cuts;
- To provide a groove in which a round, spring retainer (snap-ring or o-ring) may be inserted and held for assembling and holding parts in a fixed location on a shaft;
- To permit a square-shouldered mating part to seat squarely on the shoulder face;
- To simplify other work processes, such as turning a square shoulder. A square groove is often cut to depth and to the required length; a regular side-cutting tool may then be used to turn the diameter to size;
- To form a V-shaped groove in which a belt may track and exert a driving force;
- To improve the appearance, as in the case of a knurled section;
- To provide a round recess between turned sections of a workpiece.

COMMON SHAPES OF GROOVES

There are three basic shapes of grooves: square, round, and angular (V-shaped). The cutting tools that produce these shapes are named: square-, round-, and V-nosed grooving tools.

SQUARE GROOVE

The square groove may be cut at a shoulder or at any point along the workpiece. As stated earlier, the square groove provides a channel. This channel permits a threading tool to ride into the groove or a mating part to shoulder squarely. Parts that are to be cylindrically ground are often undercut at a shoulder. The undercut makes it possible for the face of the grinding wheel to grind the diameter to size along the entire length.

The square groove may be turned with a cutoff (parting) tool. A standard high-speed steel tool bit or a carbide-tipped bit ground to width are commonly used (Figure 24–1A).

ROUND GROOVE

The round groove eliminates the sharp corners of the square groove. A round groove is particularly important on parts that are hardened or that are subjected to severe stresses. On these parts a sharp corner may crack and fracture easily.

The round groove also provides a pleasing appearance for ending such work as knurled areas. Round grooves may be produced by regular cutting tools that are formed to the required shape and size (Figure 24–2B).

V-GROOVE

Deep V-grooves are usually turned by off-setting the compound rest at the required angle. One angular side of a V-groove is cut at a time. Shallow V-grooves may be cut by using a formed V-cutting tool (Figure 24–1C).

Special care must be taken when cutting a deep groove in a long, small-diameter workpiece or in work that has been turned between centers. A follower rest should be positioned to support the workpiece behind the cutting edge.

Grooving tools are end-cutting tools. They are ground with relief for the cutting end and sides. Cutting speeds are decreased in grooving because the cutting area is greater than in straight turning. Cutting fluids and fine feeds are required to produce a fine surface finish.

FORM TURNING

METHODS OF TURNING FORMS

Convex, concave, and other irregular forms are generally turned by one of the following four methods:

- Manually controlling the movements of the carriage and cross feed,
- Turning to a required depth with a cutting tool that is preformed to the desired shape,
- Manually feeding a cutting tool until the shape that is produced conforms to a template,
- Producing the form automatically using a template and a tracer attachment.

PREFORMED CUTTING TOOLS

Form turning is a practical way of producing regular shapes as well as intricate contours.

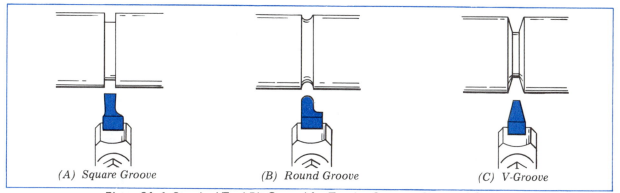

(A) Square Groove   (B) Round Groove   (C) V-Groove

Figure 24–1 Standard Tool Bit Ground for Turning Common Shapes of Grooves

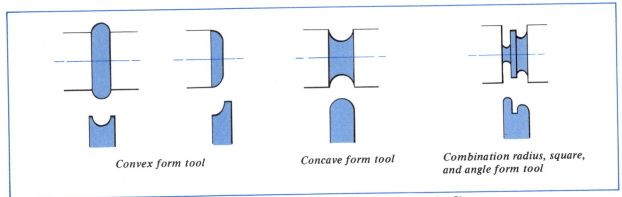

Convex form tool

Concave form tool

Combination radius, square,
and angle form tool

Figure 24–2 Examples of Form Tools for Producing Regular Shapes

Preformed tools—that is, tools that are precision ground to a required shape and size—are used in one method of form turning. Such tools are sharpened on the top cutting face to maintain the accuracy of the form. One advantage of using an accurately ground form tool is that multiple pieces may be reproduced to the same shape and size. Gaging and measuring each part thus may be eliminated.

The contours produced are usually a combination of a round, angular, or square groove or a raised section. The radius may be concave or convex. A convex radius requires the workpiece to be turned first with a raised collar section. The collar is then reduced to shape and size with a convex-radius form tool. Figure 24–2 provides examples of forms that are turned with preformed cutting tools having concave and convex radii and a contour consisting of concave radii, square, and angular sections.

The cutting action with formed turning tools takes place over a large surface area. Accordingly, the spindle speed must be reduced to about one-half the RPM for regular turning. On very large and deep contours a gooseneck toolholder is used. This holder helps to eliminate chatter and produces a fine surface finish.

PRODUCING A GROOVED FORM BY HAND

Another common practice in form turning is to form the groove by *hand*. This method is used when one part or just a few parts are to be machined and no template is available. The form is produced by the operator. A partially formed cutting tool is fed into the revolving workpiece. The longitudinal and transverse feeds are applied manually and simultaneously. The amount the carriage moves in relation to the movement of the cross slide governs the shape that is turned.

CUTTING-OFF (PARTING) PROCESSES

Cutting-off processes are usually performed with a specially formed thin blade that is rectangular in shape. The sides of the cutting tool are shaped so that there is a relief angle. The blade is held solidly against a corresponding angular surface of a straight or offset toolholder. Cutoff blades are made of high-speed steel, carbides, and cast alloy. Four common shapes of blades are shown in Figure 24–3. Blade holders are commercially available in straight and right- and left-hand offset styles.

Small workpieces are often cut off with a standard tool bit that is ground as an end-cutting parting tool. The depth to which the narrow blade or point enters the workpiece, the extent the blade must extend, and the fragile nature of this cutting-off tool requires that special precautions be taken.

PROBLEMS ENCOUNTERED IN CUTTING OFF STOCK

The lathe operator must recognize four major problems in cutting off (parting) stock. A simplified statement of each problem, the probable cause, and the recommended correction of each problem is given in Table 24–1.

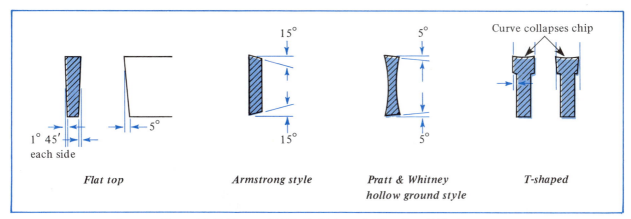

Figure 24–3 Types of HSS, Carbide, and Alloy Cutoff Blades

REPRESENTING AND DIMENSIONING GROOVES AND TURNED FORMS

The three common grooves are generally represented on drawings as shown in Figure 24–4. The dimensions give the width and depth and provide other information about the form.

The square groove in Figure 24–4A is dimensioned for width and depth. A notation is used when this groove is cut at a shoulder or to provide a channel at the end of a thread. The notation gives the width of the groove and directs that the groove be cut **TO THREAD DEPTH**. In some instances the width and depth may be given as **1/8″ WIDE × 1/8″ DEEP**.

Table 24–1 Cutting-Off Problems: Causes and Corrective Action

Cutting-Off Problem	Probable Cause	Corrective Action
Chatter	Speed too great	Decrease the cutting speed
	Tool extended too far	Clamp the cutting tool short
	Play in cross slide and compound rest	Adjust the gibs on cross slide and compound rest
	Backlash in compound rest feed screw	Back the compound rest away from the work; turn the feed screw to the left at least one-half turn
	Play in lathe spindle	Adjust the end play in the spindle bearings
Rapid dulling of cutting tool	Cutting speed too high	Decrease the cutting speed
	Too much front clearance	Decrease the front clearance angle
	Lack of cutting lubricant	Apply lard oil or required cutting compound for part being turned
	Overheating and drawing the tool temper due to insufficient clearance	Grind front, side, or back clearance
	Overheating due to excessive cutting speed	Decrease spindle speed
Cutting tool digging into work and work climbing on cutting tool	Tool set too low	Set tool at height of lathe center
	Play in compound and cross slide	Adjust gibs on cross slide and compound rest
	Backlash in compound rest feed screw	Back the compound rest away from the work; turn the feed screw to the left at least one-half turn
	Too much front clearance	Decrease the front clearance angle
	Too much top rake	Decrease the top rake
Cutting tool riding on work	Tool too high	Set tool at the height of the lathe center
	Insufficient front clearance	Increase the amount of the front clearance angle

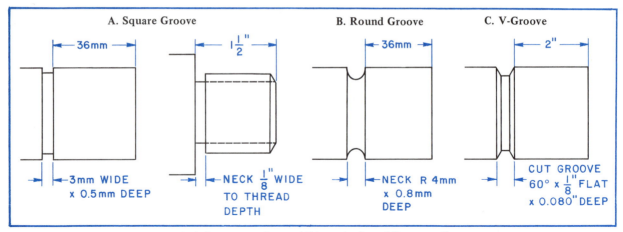

Figure 24–4 Representation and Dimensioning of Common Grooves

In Figure 24–4B the round groove (neck) is 36mm from the end of the workpiece. The 4mm radius is cut to a depth of 0.8mm. This same groove may be dimensioned by the note UNDERCUT 4mm R X 0.8mm DEEP.

The V-groove in Figure 24–4C is formed with a 60° included angle form cutter. The width of the flat end-cutting edge is ground 1/8″. The groove is cut 2″ in from the end and to a depth of 0.080″.

Sometimes, the root diameter to which a groove is to be machined is indicated by the drafting symbol ⌀ followed by the diameter as, for example ⌀ 2.50″.

How to Turn Grooves

STEP Grind the tool bit to the desired shape
1 and size. Hone the point end. Test a round-nose tool with a radius gage; test a V-shaped grooving tool with a gage or bevel protractor.

Note: The grooving processes are similar even though the shape of the cutting points differ.

STEP Mount the workpiece in a chuck or
2 between centers. Secure the tool bit in a toolholder. Position the cutting point at center height and at a right angle to the axis of the workpiece.

STEP Layout the location of the groove.
3

Note: The position of a square groove may be measured by placing a steel rule against the side of a square-nose tool. The length is read directly.

STEP Set the spindle speed at approximately
4 one-half the RPM used for turning.

STEP Start the lathe spindle. Carefully bring
5 the cutting point to the workpiece. Keep the point within the boundaries of the groove to be cut.

STEP Set the cross feed screw graduated collar
6 at zero (0).

STEP Feed the square, round, or V-shaped
7 formed grooving cutter to the required depth. The depth is read on the graduated collar. Use a fine feed and a cutting fluid.

Note: Many deep grooves are cut or widened by feeding the formed tool against the sides of the groove to relieve the force produced if the whole point is cutting and to prevent chattering.

Note: The skilled craftsperson usually checks the root diameter of the groove with a caliper, knife-edge vernier, or other precision measuring tool.

Caution: The lathe spindle and workpiece must be stopped during the check.

Note: Use a follower rest to support the workpiece when a deep groove is to be cut or the part is long and slender.

STEP 8 Widen a square groove by bringing the cutting point clear of the workpiece. Move the carriage to the right or left within the scribed lines. Again feed to the required depth.

Note: The graduated collar may be used for repositioning the cutting tool for width when the compound rest is set parallel with the center.

STEP 9 Break the sharp corners of the groove to remove any burrs. Use a smooth-cut mill file.

How to Turn a Form

Machining a Concave Form by Hand Feeding

STEP 1 Mount the workpiece. Set the spindle speed at about three-fourths the speed for turning. Mark the location and the centerline of the groove.

Note: This spindle speed is selected because only part of the cutter face will be cutting at one time.

STEP 2 Position and secure a round-nose cutting tool bit.

Note: The tool bit should be ground to the largest possible radius.

STEP 3 Grasp the carriage handwheel with one hand and the cross feed handwheel with the other hand.

STEP 4 Starting at one end of the groove, carefully feed the tool bit to cut into the workpiece. Feed to the opposite end of the groove.

STEP 5 Continue to feed in the tool. Take successive cuts from each side to step off the concave form (Figure 24–5).

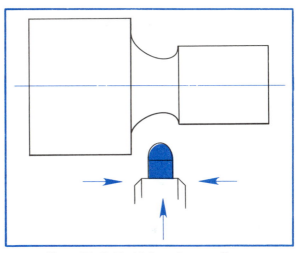

Figure 24–5 Machining a Concave Form

Note: Start the roughing cuts from each side and work toward the center. Take light roughing cuts to get the correct coordination between the carriage feed and cross feed in relation to the concave radius to be formed.

STEP 6 Check the form with a radius gage. Measure the root diameter of the groove with a caliper.

Note: A round piece of stock of the required diameter is often used as a radius gage when such a gage is not available.

STEP 7 Continue to sweep the radius tool into the concave surface. Take light finishing cuts for the final cutting of the radius.

STEP 8 Use a smooth-cut, round or half-round mill file. Finish form the radius and remove the cutting tool marks.

STEP 9 Step up the spindle RPM. Polish with an abrasive cloth.

Turning a Raised Convex Section by Hand Feeding

STEP 1 Lay out the width of the section.

STEP 2 Turn both sides of the raised radius section to the required diameter.

STEP 3 Use a round-nose tool and rough turn the radius by stepping off.

Note: The roughing cuts are usually started at the center of the convex section and fed to the outside face. One side is roughed out, then the other side.

STEP Form the roughed-out steps into a con-
4 tinuous curve. Start the finish cut either at the width of the groove or at the center. Then sweep the form by simultaneously feeding the tool while moving the carriage and the cross slide.

STEP Stop the workpiece periodically. Check
5 the convex section for size and shape. Use a radius gage or template.

STEP Take a fine finish cut to smooth out the
6 surface of the radius.

STEP Finish file the convex section to remove
7 the tool marks. The radius may be polished with an abrasive cloth.

Form Turning with a Form Cutting Tool

STEP Check the form cutting tool for shape
1 and size.

STEP Reduce the spindle speed to one-half the
2 RPM usually used for turning.

STEP Secure and position the form cutting tool
3 on center. Mark off the length of the turned section.

STEP Start the cut at the required length. Feed
4 the lead end of the form cutter until it just touches the revolving workpiece.

STEP Set the cross feed graduated collar at
5 zero (0). Start the flow of cutting fluid.

STEP Feed in to the required depth.
6

Note: The worker often carefully moves the carriage a slight amount to break the chips and give the form cutter a little more freedom.

STEP Hand file any sharp edges that result
7 from turning.

How to Cut Off (Part) Stock

STEP Select the cutoff tool blade that is recom-
1 mended for the workpiece material.

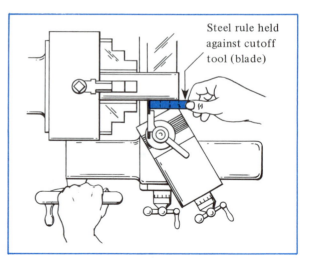

Steel rule held against cutoff tool (blade)

Figure 24–6 Locating the Cutoff Tool at a Required Length

STEP Place and secure the blade in either a
2 straight or right-hand offset toolholder. Locate the toolholder so that the cutting point is on center and the blade is at 90° to the work axis. (The settings are shown in Figure 24–6.)

Note: The blade should extend beyond the holder about 1/8" more than the radius of the part to be cut.

STEP Locate the cutoff tool at the required
3 workpiece length (Figure 24–6).

STEP Check the tool setting to be sure the
4 workpiece can be parted without interference.

STEP Set the spindle speed at one-half the
5 RPM used for straight turning.

STEP Feed the cutoff tool continuously (slow-
6 ly but constantly) into the workpiece (Figure 24–7). Use a cutting fluid for materials that require it.

STEP Continue to feed until the workpiece is
7 almost cut off.

Note: Before the work is parted, carefully remove the burrs from both sides by filing.

Caution: Use a light feed as the cutoff point approaches. The cutoff part should not be touched; it may be hot.

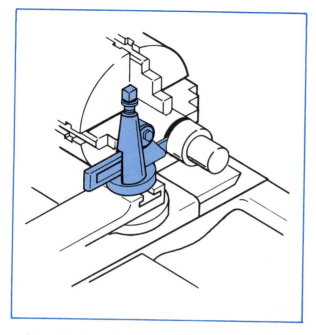

Figure 24-7 Feeding a Parting Tool Inward to Cut

Safe Practices in Grooving, Form Turning, and Cutting Off

- Set the spindle speed at about one-half the RPM that is used for straight turning with high-speed steel tool bits or cutoff blades. Apply this general rule to grooving, form turning with a preformed cutting tool, and cutting off.

- Back up a workpiece with a follower rest when turning deep grooves that may weaken the workpiece.

- Use a straight or right-hand offset toolholder. Hold the stock so that the least amount extends beyond the holding device. Cut off as close to the holding device and spindle nose as possible.

- Use a protective shield or goggles. The hot chips tend to spatter. Avoid handling hot pieces, particularly at the final stage of parting.

GROOVING, FORM TURNING, AND CUTTING-OFF TERMS

Grooving (necking)	The process of turning a cutaway surface of a particular shape and size.
Recessing (undercutting)	The process of boring an internal, circular, grooved surface to a specific size and shape.
Square, round, and angular grooves	Three basic shapes for external grooves. The cross-sectional shape of a circular indentation on the outside of a workpiece.
Form turning	The turning of an external or internal contour of a specified shape and size.
Form turning by hand	A turning process in which a formed cutting tool and manually controlled movements of the carriage and cross slide are used to produce a desired form.
Cutting off (parting, turning)	The severing of a section by cutting a groove through a revolving workpiece. The process of using a mounted parting tool to cut through a workpiece.
Cutoff (parting) blade	A comparatively thin cutting tool, commonly of rectangular cross section. The side faces taper (or are hollow ground) or are formed into a T-shape to provide clearance.
Representing and dimensioning of grooves	Graphically providing a full description of the shape, size, and other features of a groove. A line drawing with full specifications for machining a groove. Sample notations that appear on drawings. Information to a craftsperson to turn a groove of a particular shape to a specified depth.

SUMMARY

- Grooving, form turning, and cutting-off tools are made of various grades of high-speed steel, alloys, and carbides.
    - High-speed steel tool bits and cutoff (parting) blades are used for general purposes.
- Cobalt and vanadium high-speed steels, alloys, and carbide blades are practical when (a) a great amount of material must be removed as rapidly as possible, (b) the material is tough to machine, or (c) there is considerable abrasive action.
    - A groove (a) provides a recess for ending a process, (b) relieves the material around a shoulder so that a mating part will seat properly, (c) improves the appearance of a workpiece, (d) provides an angular surface on which a drive belt may ride.
- Grooves are represented and dimensioned on drawings so that the width, depth, location, and shape are specified.
    - Grooves may be cut with standard tool bits. These tool bits are especially ground to a required square, round, or angular shape or a combination of these shapes.
- Cutting fluids are used for grooving, form turning, and cutting-off processes on all materials that should be machined wet.
    - Concave and convex radii may be measured for shape and size with a radius gage. Like the square and V-shoulder, the root diameter may be measured with an outside caliper or a vernier.
- A thin rectangular-section blade with relieved sides is used for cutting off stock. The blade is set at 90° to the lathe axis, on center, and with no back rake.
    - Special precautions must be taken to see that chips do not accumulate in the groove. An accumulation of chips during a deep cutoff operation may cause the blade to bind.
- Grooving, form turning, and cutting-off operations require lower spindle speeds and a finer feed than feeds used for straight turning.

UNIT 24 REVIEW AND SELF-TEST

1. List four reasons for grooving.

2. Identify two methods of form turning on a lathe.

3. Explain why a gooseneck toolholder is used for form turning large, deep contours.

4. State four common cutting-off problems in lathe work.

5. Explain what the drawing notation CUT TO THREAD DEPTH means in relation to a groove.

6. Indicate the main work processes to turn a groove.

7. List the steps to follow in cutting off a workpiece to a specified length.

8. List four safe practices that must be observed in grooving, form turning, and cutting off.

Lathe Filing, Polishing, and Knurling

Filing and polishing are two lathe processes that produce a smooth, polished surface finish. A minimum amount of material, usually 0.002″ to 0.003″, (0.5 to 0.7mm) is removed from the outside diameter.

Knurling is another surface finishing process. By contrast with filing and polishing, knurling produces straight line, right or left diagonal, or diamond indentations and raised surfaces around the circumference of a workpiece. This unit deals with the technology related to and the processes of filing, polishing, and knurling.

OBJECTIVES

After satisfactorily completing this unit, you will be able to:

- Apply additional information about files and abrasive cloths for filing and polishing.
- Understand the functions of knurling, characteristics and specifications of knurls, and the effect on the outside diameter of knurled products.
- Use knurling tools and observe precautions in producing different knurl forms.
- Perform the following processes.
  - Finish File and Polish a Turned Surface.
  - Produce a Knurled Surface.
- Follow recommended *Safe Practices* for filing, polishing, and knurling, and use new *Technical Terms* correctly.

FILING TECHNOLOGY AND PROCESSES

TYPES OF FILES

Two common types of lathe files are used on ferrous metals. The *single-cut bastard mill file* is a general-purpose file for lathe work. A 10″ to 12″ *long-angle lathe file* is also commonly used (Figure 25–1). This file has sides (width and thickness) that are parallel. It has two uncut (safe) edges that permit finish filing to a shoulder. The long-angle lathe file has features that partially eliminate chatter, provide for rapid cleaning of chips, and reduce the probability of scoring a workpiece. Bastard, second-cut, and smooth-cut mill files produce a fine-grained surface.

Three other types of files are used on non-ferrous metals. They are the *aluminum file*, the *brass file*, and the *super-shear file*. The first two files are shown in Figure 25–2.

The teeth of the aluminum file are designed to quickly clear chips and to eliminate clogging. The combination of deep undercut and fine overcut features produces a tooth form that tends to break up the file particles. The chips are cleared during the filing process. The tooth form also helps to overcome chatter.

Brass, bronze, and other tough, ductile metals are finish filed with a brass file. The teeth on this file have a deep, short upcut angle and a fine, long-angle overcut. The design breaks up the chips so that the file clears itself.

Courtesy of NICHOLSON DIVISION; THE COOPER GROUP

Figure 25–1 A Common Type of Lathe File

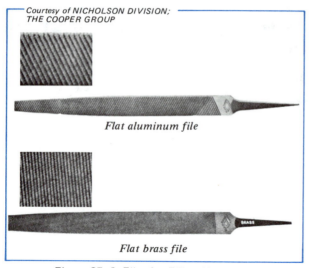

Flat aluminum file

Flat brass file

Figure 25–2 Files for Filing Aluminum
and Brass on a Lathe

Aluminum, copper, and most other non-ferrous metals may be filed on the lathe with a super-shear file. The teeth are cut in an arc (Figure 25–3). The arc is off center in relation to the file axis. The milled, double-purpose tooth permits free cutting and easy removal of chips from the teeth and produces a smooth surface finish.

FILING A TURNED SURFACE

Cutting is always done on the forward stroke when filing by hand. Strokes are overlapped to about one-half the width of the file, as is done for bench filing. The spindle speed used for filing should be about twice as fast as the speed used for turning. A revolving dead center is pre-

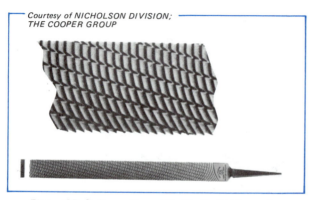

Figure 25–3 Super-Sheer File Teeth Cut in an Arc

ferred for filing and polishing work that is held between centers. A speed of 30 to 40 strokes per minute provides good control of a file in removing the limited amount of material that is left for filing. Filing is also intended to remove light tool marks.

Only a slight force should be applied to a file. Excessive force produces an out-of-round surface and tends to clog the teeth and score a workpiece. A file must be cleaned frequently with a file card and brush. The teeth are sometimes chalked to prevent clogging and to simplify cleaning. The extreme sharpness of the fine cutting edges of a new file is removed by filing the flat surface of a cast-iron block. Filing lathe work left-handed is the preferred, safer method.

After completion of the filing process, a smoother polished surface is obtained by using an abrasive cloth. Polishing is done by pressing and moving the abrasive cloth across the revolving workpiece. Aluminum oxide abrasive cloths are used for polishing steels and most ferrous metals. Silicon carbide is widely used for non-ferrous metals. Grit sizes of 80 to 100 are generally used. Finer grained and more highly polished surfaces are produced with grit sizes of 200 or finer.

High spindle speeds are required for polishing on the lathe. During the initial polishing, greater force may be applied by holding an abrasive strip against a file. Long, regular strokes are used against the revolving workpiece. A still higher gloss finish may then be produced by using a worn abrasive cloth and applying a few drops of oil.

Polishing straps are common when polishing large diameters. The abrasive cloth is held by two half sections of a form (polishing strap). Force is applied on the polishing strap to press it against the surface that is turning.

How to File a Turned Surface

STEP 1 Select a mill or lathe file of the size and cut suited for the job. Clean the file teeth. Chalk the file face. Check the handle.

STEP 2 Set the spindle speed at twice the RPM used for turning.

Caution: Move the carriage out of the way. Disengage the lead screw and feed rod before starting the lathe. Also, protect the lathe ways against metal particles produced by hand filing.

STEP 3 To file left-handed, grasp the file handle in the left hand. Hold the point of the file with the right hand (Figure 25–4).

STEP 4 Move the file with a slight pressure and at an angle across the workpiece (Figure 25–4). Take full strokes. Maintain a file rate of 30 to 40 strokes per minute.

STEP 5 Release the pressure on the return stroke and raise the file slightly off the work.

STEP 6 Move the file laterally about half of its width so that each stroke overlaps. Continue to overlap strokes across the entire face until the surface is filed.

Note: Clean the file teeth frequently to remove chips and to prevent scoring.

STEP 7 Stop the lathe at intervals and check the diameter of the work and to see that the center is properly lubricated and the center hole is not scored.

STEP 8 Continue to file. Allow from 0.0005" to 0.001" for polishing.

Note: Avoid touching the outside diameter by hand. Otherwise a slick surface will be produced.

How to Polish on the Lathe

STEP 1 Set the spindle speed at three to four times the RPM used for turning.

Note: It is assumed that the workpiece has just been filed.

STEP 2 Recheck the tailstock center and center hole and lubricate them. If possible, use a revolving tailstock center.

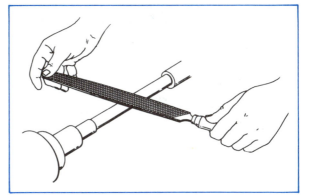

Figure 25–4 Correct Positions of the Hands and File for Filing Left-Handed on a Lathe

STEP 3 Select the appropriate abrasive grain and grit size. Tear a strip slightly wider than the width of the file and from 10" to 12" long. Place one end of the abrasive strip around the file point end. The abrasive cutting face must be toward the workpiece. Pull the abrasive cloth along the file. Hold it to the file with the left hand (Figure 25–5).

STEP 4 Start the lathe. Take long, overlapping strokes. Check the quality of surface finish and measure the diameter.

Note: Use successively finer grit sizes to obtain finer surface finishes.

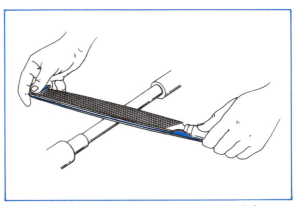

Figure 25–5 Position of an Abrasive Cloth for Polishing on the Lathe

KNURLS AND KNURLING

KNURLS AND KNURLING

A *knurl* is a raised area impressed upon the surface of a cylindrical part. *Knurling* is the process of forcing a pair of hardened rolls to form a *diamond-shaped*, *diagonal*, or *straight-line pattern*. Knurling is a *material displacement* process. It requires great force for the knurl rolls to penetrate a workpiece surface and impress a pattern.

PURPOSES FOR KNURLING

There are three general purposes for knurling:

- To provide a positive gripping surface on tools, instruments, or work parts (this surface permits ease of handling and precise adjustment);
- To raise the surface and increase the diameter to provide a press fit or an irregular surface (the knurl produces a gripping surface that prevents two mating parts from turning, as in the case of a plastic handle on the shank of a screwdriver);
- To add to the appearance of the work.

BASIC KNURL PATTERNS

- *Straight-line* (parallel-ridge) *pattern.* Formed by impressing two hardened steel knurl rolls that have grooves cut parallel to the axis of each knurl roll (Figure 25–6A).
- *Diagonal patterns. Left-hand diagonal* (helix) *pattern* formed by using a right-hand diagonal knurl. *Right-hand diagonal pattern* formed by using a left-hand diagonal knurl. The diagonal grooves and ridges are produced at a 30° angle to the axis of the knurl rolls (Figure 25–6B).
- *Diamond pattern.* Formed by combining left-hand and right-hand diagonal knurl rolls to simultaneously form right- and left-hand 30° angle grooves/ridges. The two series of knurls cross to form a diamond-shape pattern. Another forming method is to use a standard diamond-shape knurl. A *depressed* (female) *diamond pattern* is formed by using a male diamond-point knurl roll.

KNURL ROLL SIZES

Knurl rolls are available in three basic sizes that are identified by *pitch.* The three sizes and their pitches are *coarse* (14 pitch), *medium* (21 pitch), and *fine* (33 pitch). Pitch refers to the number of teeth per linear inch. A knurl roll is heat treated and rides on a hardened steel pin.

In addition to the three basic pitch sizes used for general purpose knurling, knurls ranging from 12 to 80 teeth per inch (TPI: normal teeth per inch) are available for straight-line, diagonal, and diamond knurling. The *normal teeth per inch* relate to the measurement of the knurl crests along a line that is perpendicular to the helix angle of the teeth (not the knurl axis).

EFFECT OF KNURLING ON WORK BLANK DIAMETERS

When knurling is used to increase the diameter of a work blank, the size knurl to select (in terms of the normal TPI) is established by the required increase in diameter. For example, the

(A) *Three basic pitch sizes of straight-line pattern knurl rolls*

(B) *Helical-ridge knurl rolls form diagonal or diamond patterns (three pitch sizes)*

Figure 25–6 Knurl Rolls for Producing Straight-Line, Diagonal, and Diamond Pattern Knurls

Table 25-1 Approximate Increase of Workpiece Diameter Produced by Knurling
(Circular Pitch/Threads per Inch System)
Courtesy of REED ROLLED THREAD DIE COMPANY

Normal Teeth per Inch (TPI) (Circular Pitch Knurls)	Straight and Diagonal Knurling*	Diamond Knurling* Raised Points (Male)
12	.034	.038
16	.025	.029
20	.020	.023
25	.016	.018
30	.013	.015
35	.011	.013
40	.009	.010
50	.009	.010
80	.005	.006

*30° Helix Angle

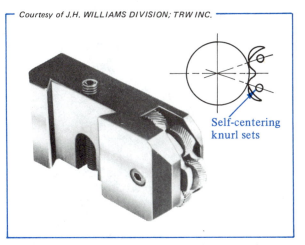

Self-centering knurl sets

Figure 25-7 Revolving Head with Six Knurl Rolls Mounted in Dovetail Slide for Quick-Change Tool Post

effects of straight-line, and 30° helix angle diagonal or diamond knurling on the outside diameter (as specified by one knurl manufacturer) is provided in Table 25-1.

Note that the range of increase in work blank diameter from .005″ to .034″ (.12 to .86mm) for straight-line and diagonal knurling and .006″ to .038″ (.15 to .97mm) for diamond knurling are *approximate*. The correct diameter of a work blank is usually determined by measuring the first knurled piece and making machining adjustments. American Standard (ANSI/ASME) B 94.6 provides general specifications and technical information tables relating to knurl pitches and work blank diameters.

TYPES OF HOLDERS

A single set of knurl rolls is usually held on a self-centering-head toolholder. Some holders are made for a single set and pitch. A multiple-head holder contains three sets of coarse, medium, and fine knurls. The knurl rolls are mounted in a revolving self-centering head that pivots on a hardened pin. The multiple heads are also designed to fit a dovetailed quick-change tool post (Figure 25-7).

In cases where the knurled surface is narrower than the width of the knurl rolls, the knurl rolls are fed in slowly to the total depth. A surface that is wider than the width of the knurl rolls requires the power longitudinal feed to be engaged immediately. The knurl rolls are moved across the required length of the workpiece. They are fed inward before the direction of feed is reversed. The process is continued until a reasonably sharp knurl pattern is produced.

KNURLING PRACTICES TO CORRECT PROBLEMS

- The knurl rolls must be kept in continuous contact with the workpiece until the process is completed.
- The automatic feed must be engaged continuously to traverse the full length of the knurled surface. Otherwise a ring is formed that produces a knurled surface with a varying pattern that is impractical to correct.
- A double-impression knurled pattern is produced when uneven pressure is applied on both knurl rolls. Double impressions may be corrected by:
  - Positioning the knurl rolls on center,
  - Tracking the teeth carefully,
  - Infeeding the knurl rolls again so that the teeth penetrate evenly.
- The knurl teeth must be cleaned before the operation and *while the machine is stopped*.
- The flow of cutting fluid must be adequate to wash any particles away from the work surface and the rolls.
- A follower rest and/or a steady rest may be required to offset the force applied during the forming of the knurl.

How to Knurl

STEP 1 Check the condition of the knurl rolls. Clean the teeth. Mount the knurl rolls on center with the faces parallel to the workpiece (Figure 25–8).

STEP 2 Set the spindle speed at about one-half the RPM used for turning. Select a carriage feed of from 0.020″ to 0.030″ (0.5mm to 0.8mm). The amount of feed depends on the pitch of the knurl. Start the lathe.

STEP 3 Move the knurling tool until about one-half of the face bears on the workpiece. Feed in the tool to a depth of 0.020″ to 0.025″ (0.5mm to 0.6mm).

STEP 4 Stop the lathe. Check the correctness of the pattern.

Note: A double impression indicates that the rolls are not centered. One roll is cutting deeper than another roll. It may be necessary to move the knurls to a new location to correct a double impression.

STEP 5 Position the cutting fluid nozzle to flow a small quantity of cutting fluid on the knurls.

Note: Use a follower rest or steady rest when support is required.

STEP 6 Engage the automatic feed. Start the lathe and feed for the full length of the knurled section.

STEP 7 Stop the lathe. Use a stiff brush to remove particles of the workpiece material from the knurls.

STEP 8 Check the knurled pattern. Continue to feed the rolls in to depth. Reverse the automatic feed. Take successive passes across the workpiece until a clean, smooth crest pattern is produced.

Note: A damaged knurl pattern results when the feed is stopped and the rolls keep impressing while the workpiece revolves.

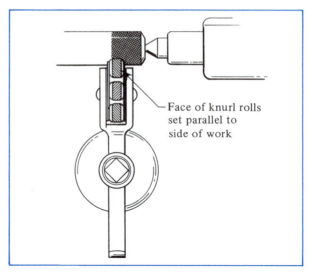

Figure 25–8 Positioning the Knurls and Holder

CALCULATING THE REQUIRED TURNED DIAMETER FOR A KNURLED SECTION (MEDIUM PITCH)

The turned diameter is calculated for *medium-pitch knurl rolls* only by using the following two simple formulas. These formulas involve the knurl pitch and the number of knurled serrations that are formed around the workpiece.

number of serrations = (medium pitch × π) × (required finished diameter of workpiece − 0.017″)

outside diameter = number of serrations × 0.015″

Example: Calculate the outside diameter for a workpiece that is to be knurled with a medium (21 pitch) diamond knurl roll. The finished diameter is to be 2″.

number of serrations = (21 × 3.1416) × (2″ − 0.017″)

= 130.82 (round off to 131)

outside diameter = 131 × 0.015″ = 1.965″

If the workpiece is turned to 1.965″, the raised outside diameter produced by a 21-pitch knurl roll will be 2″.

Safe Practices in Filing, Polishing, and Knurling

- Be sure the file handle fits correctly before beginning the lathe filing. A revolving workpiece tends to force the file toward the operator.
- File left-handed.
- Knurl rolls must be cleaned and set on center before starting the knurling process. The amount of cutting fluid must be adequate to wash away particles and to lubricate the rolls.

Lubrication permits the material to flow more easily during the process.

- Keep the fingers, wiping cloth, or brush away from a revolving workpiece and the knurl rolls.
- Check the support of a workpiece. A follower and/or steady rest may be needed to prevent bending or distortion of the workpiece during knurling.
- Disengage the lead screw and feed rod when filing and polishing to prevent them from being accidentally engaged at a high spindle speed.

FILING, POLISHING, AND KNURLING TERMS

Filing (lathe)	A finishing process for removing fine tool marks and reducing a diameter a few thousandths of an inch. Reducing a round part to a required size within a specified tolerance.
Polishing (lathe)	Producing a smoothly finished and polished surface. The cutting (abrading) action of abrasive cloth against a revolving workpiece. A final finishing process.
Knurling	The raising of a series of diamond-shaped, diagonal, or straight-line patterned surfaces. Causing a material to flow by exerting force on a revolving workpiece with a set of specially formed rolls.
Diamond knurl pattern	A series of raised areas with a square pyramid shape. Uniformly formed areas around the surface of a workpiece. A shape produced by impressing overlapped right- and left-hand helixes into a revolving cylindrical surface.
Straight-line and diagonal knurl patterns	A uniform series of grooves of triangular cross section with a slightly flat crest. A series of uniformly shaped surfaces that are impressed parallel (or at a 30° angle in the case of diagonal knurling) to the work axis of a cylindrical workpiece.
Displacement of metal (knurling)	Forcing metal to flow and form uniform series of diamond (pyramid-shaped), straight-line, or diagonal knurl raised areas.
Serrations (knurling)	Markings or formed grooves. A uniform shape (pattern) impressed into the outside diameter of a cylindrical workpiece.

SUMMARY

- A turned workpiece is often filed to produce a finely finished surface without tool marks. 0.002″ to 0.003″ (0.05mm to 0.08mm) is left for finish filing and polishing.
  - Bastard and second- or smooth-cut mill files, long-angle lathe files, or other special metal-cutting files are commonly used for lathe filing of ferrous metals. Aluminum, brass, and super-shear files are used for filing nonferrous metals and other materials. File lengths of 10″ to 12″ (or 250mm to 300mm) are common.

■ A fine surface finish may be produced by using an abrasive cloth of 80 to 120 grit. Still finer grain surfaces may be produced with 200 and finer grit sizes.

■ Knurling causes a flow of material on the outside diameter of a revolving workpiece. Diamond and straight-line knurl patterns are general. The common pitches are coarse (14), medium (21), and fine (33). Other series pitches range from 12–80 TPI.

■ Diamond, diagonal, and straight-line patterns are produced by positioning the axis of the knurl rolls parallel to the work axis. The knurl rolls are impressed into the workpiece.

■ Knurled surfaces are represented on drawings by a combination of dimensions and notes. Pitch and pattern; length, width, location, and finished diameter; and the nature of the grooves or rounded edges at the ends of the knurled section are specified.

■ The outside diameter to which a medium-pitch knurled area should be turned is calculated by first determining the number of serrations that will be formed around the finished diameter and then multiplying this number of serrations by 0.015".

■ Personal and machine safety precautions must be observed. Slow speeds are required for knurling. High speeds are used in filing and polishing. These high speeds require constant checking and lubrication of the dead center.

UNIT 25 REVIEW AND SELF-TEST

1. State the functions served by the deep undercut and the fine overcut features of an aluminum file.

2. Indicate the characteristics of a brass file.

3. Describe the shape and functions served by super-shear file teeth.

4. State what purpose is served by filing a workpiece on a lathe.

5. Indicate the kind of abrasive cloth to use (a) to polish steels and other ferrous metals and (b) to polish nonferrous metals.

6. State three guidelines to follow in polishing a workpiece on a lathe.

7. State two reasons for knurling.

8. Specify the basic knurl sizes and patterns.

9. Identify three common knurling problems and their causes.

10. List three safety precautions to observe when filing on a lathe.

11. Explain why the lead screw and feed rod on a lathe must be disengaged when filing or polishing.

Taper- and Angle-Turning Technology and Processes

A *taper* is a cone-shaped form. A taper increases or decreases uniformly in diameter along the length of a workpiece. *Taper turning* is the process of machining the uniform change in diameter. A taper may be internal or external.

A steep taper is often referred to in machine shops as either an outside or inside *angle*. Internal angles are bored. External angles are turned. A steep taper angle may be set and the cutting tool may be fed with the compound rest.

This unit deals with different types of external tapers and angles, how tapers are calculated and represented on drawings, taper- and angle-turning processes, and methods of checking tapers for dimensional accuracy. Machine tool and personal safety precautions are also covered.

OBJECTIVES

After satisfactorily completing this unit, you will be able to:

- Compute tapers and set up the lathe for taper turning.
- Gage or measure tapers.
- Use formulas to compute the angle setting of a compound rest.
- Interpret how tapers and angles are represented and dimensioned on shop and laboratory drawings.
- Perform the following processes.
    - Machine and Measure a Taper (using Taper Attachment and Tailstock Offset Methods).
    - Turn an Angle for a Short Taper with the Compound Rest.
- Follow *Safe Practices* as recommended and use new taper and angle turning *Terms* correctly.

PURPOSES AND TYPES OF TAPERS AND TAPER SYSTEMS

The taper is a commonly used design feature in machine and metal products industries. Cutting tools like twist drills, machine reamers, and end mills have taper shanks. The spindles of lathes, milling machines, boring mills, and other machine tools have internal and external tapers. Many mating parts, such as sleeves, sockets, centers, and spindles, have tapers that correspond with the tapers of the parts into which they fit.

A taper serves one of four main purposes:

- To provide a method of accurately locating two mating parts concentrically about a common axis,
- To securely hold together two mating parts (*self-holding* taper),
- To accurately align parts in relation to an axis (*self-releasing* taper) but requiring some other device to secure the parts together,
- To accurately position and hold a cutting tool, holder, spindle, or other part that revolves as a single unit around a particular axis.

Tapers permit interchangeability, accuracy in centering, and an efficient way of quickly changing a tool or machine setup. The specifications of standardized tapers are given in trade and engineering handbooks and in manufacturers' technical manuals. There are many series of standard tapers. These series meet the extensive design needs within and among many different industries such as the machine, metal manufacturing, aerospace, and maritime industries.

271

The six standard taper series that are widely used in the machine and metal industries are:

- American Standard Taper Pin,
- Jacobs Taper,
- American Standard (Morse) Taper,
- Brown & Sharpe Taper,
- Jarno Taper,
- American Standard Self-Releasing Steep Taper.

AMERICAN STANDARD TAPER PINS

American Standard Taper Pins are widely used when two mating parts must be accurately positioned and held in a fixed position. The taper pin permits easy assembling and disassembling. Taper pins are furnished in standard lengths in relation to the taper size. Taper pins have a standard *taper per foot* of either 0.2500" or 0.0208" per inch.

The general sizes are numbered from #7/0 to #2/0 and #0 to #11. Taper pin holes are usually reamed. The reamer sizes are designed so that one size overlaps the next succeeding size. The diameters at the small end of the taper range from approximately 1/16" (for the #7/0 size) to 41/64" (for the #10 size). American Standard Taper Pin tables give the large and small diameter, length of each pin, and the drill size.

JACOBS TAPER

The *Jacobs Taper* is often used with a standard Morse or Brown & Sharpe taper that fits standard machine spindles. The Jacobs is a short, self-holding taper that corresponds with a short taper bore. The bored holes are used on drill chucks or external tapers of a shaft, shank, or spindle. The Jacobs taper is widely applied on portable power-driven tools.

AMERICAN STANDARD (MORSE) TAPER

The *American Standard (Morse) Taper* has a small taper angle. This angle produces a wedging action and makes the Morse a self-holding taper. The Morse is the most extensively used taper in the machine industry. It is used on common cutting tool features such as the shanks of twist drills, reamers, counterbores, and countersinks and on machine tool features such as drill press spindles and lathe spindles.

The Morse taper series is numbered from #0 to #7. The small-end diameters of this series range from 0.252" to 2.750". The lengths vary from 2" to 10". Unlike the other taper series, the amount of taper per foot varies in the Morse series with each number except #0, #4 1/2, and #7. The approximate taper per foot is 5/8".

Table 26-1 lists the kind of information that is required for designing and machining an American Standard (Morse) taper.

BROWN & SHARPE TAPER

The *Brown & Sharpe (B & S) Taper* is another self-holding taper. The B & S taper series is numbered from #1 to #18. Within this series the diameters at the small end of the taper range from 0.200" to 3.000". The B & S taper

Table 26-1 Sample Dimensions: American Standard (Morse) Taper

Number of Taper	Diameter at Small End (d)	Standard Plug Length (L)	Diameter at Gage Point (D)	Taper per Foot	Taper per Inch
#1	0.369"	2"	0.475"	0.5986"	0.0499"
#3	0.778	3 3/16	0.938	0.6024	0.0502
#7	2.750	10	3.270	0.6240	0.0520

is 0.502″ per foot (0.0418″ per inch) for #1, #2, #3, and #13. The taper varies from 0.5024″ to 0.4997″ for numbered tapers #4 through #14. The taper for #14 through #18 is 0.5000″.

JARNO TAPER

The *Jarno Taper* series (#1 to #20) uses a uniform taper per foot of 0.600″. The number of a Jarno taper is directly related to the small diameter, the large diameter, and the length. The small-end diameters of the series range from 0.100″ to 2.000″. The following simple formulas may be used to calculate the sizes:

$$\text{Diameter at Small End} = \frac{\text{\# of Jarno Taper}}{10}$$

$$\text{Diameter at Large End} = \frac{\text{\# of Jarno Taper}}{8}$$

$$\text{Length of Jarno Taper} = \frac{\text{\# of Jarno Taper}}{2}$$

For example, a #6 Jarno taper is 6/10″ (0.600″) at the small end of the taper and 6/8″ (0.750″) at the large end and is 6/2″ (or 3″) long.

AMERICAN STANDARD SELF-RELEASING STEEP TAPER

The *American Standard Self-Releasing Steep Taper* series is similar to the earlier Milling Machine Taper series. The taper of 3.500″ per foot ensures easier release of arbors, adapters, and similar accessories from the spindles of milling and other machines. Because the steep tapers are not self-holding, they require slots or keys to drive an accessory (Figure 26–1A). The mating parts are usually drawn and held together with a draw bolt, cam-locking device, or nut. The lathe type-L spindle nose has a steep taper, a key drive, and a threaded ring nut (Figure 26–1B). The features of this nose accurately position, hold, and drive a mounted faceplate, driver plate, or chuck.

The short steep taper (3.000″ per foot) of the type D–1 lathe spindle nose centers the arbor or accessory (Figure 26–1C). The cam-lock device holds the arbor or accessory accurately and firmly on the spindle nose.

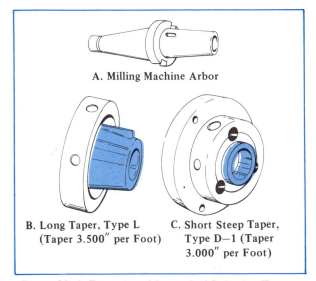

A. Milling Machine Arbor

B. Long Taper, Type L (Taper 3.500″ per Foot)

C. Short Steep Taper, Type D–1 (Taper 3.000″ per Foot)

Figure 26–1 Examples of Steep, Self-Releasing Tapers

TAPER DEFINITIONS AND CALCULATIONS

The setting up of a lathe to cut a taper requires that the taper per foot or taper per inch be known. The taper may then be turned either by setting the taper attachment, by offsetting the tailstock, or by determining the angle and then positioning the compound rest. The short taper or angle may then be turned by feeding with the compound rest handwheel.

The general terms and simple formulas for calculating the various features of a taper are examined next.

TAPER FORMULA TERMS

The following terms are used in the formulas for calculating taper features:

- T designates *taper* (the difference in size between the large diameter and the small diameter),
- T_{pi} is the *taper per inch* (the amount the workpiece diameter changes over a 1″ length),
- T_{pf} is the *taper per foot* (the change in diameter in a 1′ length),
- D refers to the *large-end diameter* of the taper,

- d denotes the *small-end diameter* of the taper,
- L_t is the *length* of the taper in inches,
- L_o is the *overall length* of the workpiece.

CALCULATING TAPER VALUES

The *taper* (T) equals the difference in the large (D) and small (d) diameters:

$$T = D - d$$

The *taper per foot* (T_{pf}) equals the difference between the large (D) and small (d) diameters in inches multiplied by 12 and divided by the required length of taper (L_t). Expressed as a formula,

$$T_{pf} = \frac{(D - d) \times 12}{L_t} \quad \text{Also, } T_{pf} = T_{pi} \times 12$$

The *taper per inch* (T_{pi}) equals the difference between the large diameter (D) and small diameters (d) divided by the length of the taper in inches (L_t):

$$T_{pi} = \frac{D - d}{L_t} \quad \text{or} \quad T_{pi} = \frac{T_{pf}}{12}$$

When the taper per inch (T_{pi}) is known, the taper per foot (T_{pf}) is found by simply multiplying by 12:

$$T_{pf} = T_{pi} \times 12$$

CALCULATING THE DIAMETERS

The *large diameter* $D = (T_{pi} \times L_t) + d$ If the taper per foot (T_{pf}) is given,

$$D = \frac{(T_{pf} \times L_t)}{12} + d$$

The *small diameter (d)* may be calculated by using the formula: $d = D - (L_t \times T_{pi})$. If the taper per foot (T_{pf}) is given,

$$d = D - \frac{(L_t \times T_{pf})}{12}$$

CALCULATING THE TAILSTOCK OFFSET

The offset tailstock method is used to cut external shallow tapers. The amount to offset a tailstock depends on the overall length of the workpiece (L_o) and the amount of taper (T_{pi} or T_{pf}). Simple formulas are used to compute the required tailstock offset (T_o).

When the taper is given as *taper per inch* (T_{pi}),

$$T_o = \frac{T_{pi} \times L_o}{2}$$

When the taper is expressed as *taper per foot* (T_{pf}),

$$T_o = \frac{T_{pf} \times L_o}{24}$$

When a taper is dimensioned with a *small diameter* (d), *large diameter* (D), *length of taper* (L_t), and the *overall length* (L_o),

$$T_o = \frac{L_o \times (D - d)}{L_t \times 2}$$

DIMENSIONING TAPERS ON CYLINDRICAL PARTS

TAPERS ON CYLINDRICAL PARTS

The taper on a round workpiece is represented by two symmetrical tapering lines. The amount of taper is usually specified in terms of taper per foot or taper per inch. For example, a standard taper pin that has a taper of one-quarter inch per foot is indicated by the note:

0.250″ (or 1/4″) TAPER PER FOOT

MEASURING AND GAGING TAPERS

Tapered surfaces must be precision machined for two mating parts to fit and align accurately and to provide maximum holding ability for small-angle tapers. There are four common methods of measuring the accuracy of the taper angle and/or the size of the large and small diameters:

- Gaging with a taper plug or ring gage,
- Measuring with a standard micrometer,
- Measuring with a taper micrometer,
- Testing with a sine bar setup.

TAPER PLUG GAGE METHOD

The *taper plug gage* is hardened and precision ground (Figure 26–2A). The face and a ground area are identified as steps (1) and (2). When the small end of a correctly turned taper

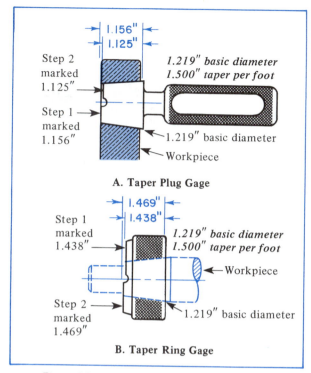

- 1.156"
- 1.125"

Step 2
marked
1.125"

Step 1
marked
1.156"

1.219" basic diameter
1.500" taper per foot

1.219" basic diameter

Workpiece

A. Taper Plug Gage

- 1.469"
- 1.438"

Step 1
marked
1.438"

1.219" basic diameter
1.500" taper per foot

Workpiece

Step 2
marked
1.469"

1.219" basic diameter

B. Taper Ring Gage

Figure 26–2 Features of Taper Plug and Ring Types of Go-Not-Go Gages (Step Style)

falls between these steps, the part is turned to size within the allowable taper limits.

TAPER RING GAGE METHOD

A hardened *taper ring gage* has a precision-ground tapered hole (Figure 26–2B). Part of the outside of the body is cut away (1). One or more index lines are marked on the flat section (2). When the taper is correct and the end of the tapered part cuts an index line, the outside diameters are accurately machined to size.

The taper must first be checked for accuracy. Usually, a fine coating of Prussian blue is applied to the machined part. The gage is then placed carefully on the taper. If the machined part is turned slightly and then removed, the high areas on the machined tapered surface are identified easily. Once the taper fits accurately, the diameters may then be checked by gage or micrometer measurement.

A few simple precautions must be taken. The mating part must be clean and free of burrs

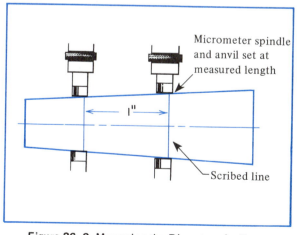

and nicks. Since a small-angle taper is self-locking when a small force is applied, the workpiece should be turned clockwise carefully. A counterclockwise withdrawing motion makes it easy to turn and remove the tapered plug or ring gage. Excessive turning causes undue wear on the gage.

STANDARD MICROMETER METHOD

When a taper ring gage is not available, the *standard outside micrometer* is often used to measure the accuracy of the taper and the diameter. The method is shown in Figure 26–3.

The turned taper is given a coating of a colored dye. Lines are scribed at reference points 1" apart. The diameters at these points are measured with a micrometer. The difference between the two diameters is then compared with the required taper per inch. It is often necessary to convert taper per foot to taper per inch and to multiply the value by the number of inches between the two scribed lines.

Any oversize variation may be corrected by adjusting the taper attachment or tailstock offset. A light cut is then taken to depth. If the variation is around 0.002", the tapered surface may be finish filed. Considerable skill is required to machine and file to precision measurements by this method.

Micrometer spindle and anvil set at measured length

1"

Scribed line

Figure 26–3 Measuring the Diameter of a Taper with a Standard Micrometer

TAPER MICROMETER

The *taper micrometer* provides a more reliable and accurate measuring instrument and method than does the standard micrometer. The taper micrometer includes an adjustable anvil and a 1″ sine bar that is attached to the frame. The sine bar is adjusted by the movement of the spindle.

The accuracy of the taper is measured by placing the taper micrometer over the workpiece. The thimble is adjusted until the jaws just touch the tapered surface. The reading on the spindle indicates the taper per inch. This reading may need to be converted to taper per foot or angle of taper.

TAPER TURNING PROCESSES

TAPER ATTACHMENT

The taper attachment provides a quick, economical, practical way of accurately turning internal and external tapers. In principle the taper attachment guides the position of the cutting tool in an angular relation to the lathe center axis. A sliding block moves along the guide bar of the taper attachment. The angular setting of the guide bar is transmitted by the sliding block to produce a similar movement of the cutting tool. There are two basic types of taper attachments. These types are the *plain taper attachment* and the *telescopic taper attachment.*

The plain taper attachment requires the removal of the binding screw that connects the cross feed screw to the cross slide. The cross slide is then moved by securing the sliding block

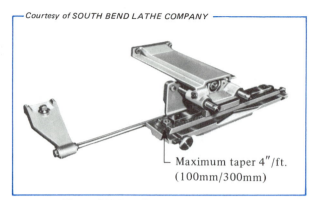

Maximum taper 4″/ft. (100mm/300mm)

Figure 26–4 A Permanently Mounted Telescopic Taper Attachment

to the extension arm on the cross slide. The compound rest feed handwheel is turned to set the depth of cut with the plain taper attachment.

The cross slide and cross feed screw and nut are not disengaged with the telescopic taper attachment (Figure 26–4). The depth of cut may be set directly with the cross feed handwheel.

ADVANTAGES OF TURNING A TAPER WITH A TAPER ATTACHMENT

One advantage of using a taper attachment is that tapers may be turned accurately, economically, and with less danger of work (center) spoilage. There are also many other advantages of turning a taper with a taper attachment:

- Alignment of the live and dead centers, required by other methods, is eliminated.
- The workpiece is supported by the full angular bearing surface at the center holes.

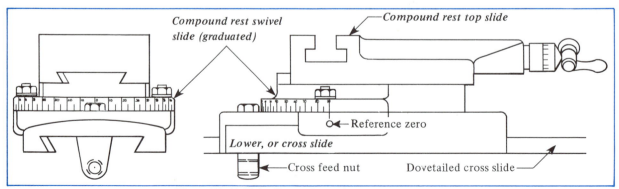

Compound rest swivel slide (graduated)
Compound rest top slide
Reference zero
Lower, or cross slide
Cross feed nut
Dovetailed cross slide

Figure 26–5 Major Parts of a Lathe Compound Rest

- Tapers may be cut on work held in a chuck, between centers, or on a faceplate, fixture, or other work-holding setup.
- A greater range of inch- and metric-standard taper sizes may be turned.
- Internal and external tapers may be cut with the same setting of the taper attachment. This one setup permits machining the matching parts accurately.
- The taper attachment may be positioned directly at the required taper or in degrees of taper. The base of the taper attachment has a scale graduated in taper per foot, millimeters of taper, and degrees of taper. Calculation of these dimensions is thus eliminated.
- The angle of taper is not affected by any variation in the length of the workpiece. The one taper setting may be used to produce multiple parts with the same taper.

ANGLE TURNING WITH THE COMPOUND REST

GRADUATIONS ON THE COMPOUND REST SWIVEL SLIDE

While all compound rest swivel slides are graduated in degrees, the arrangement is not standardized. For example, a graduated swivel slide may be marked from 90° to 0° to 90° or 0° to 90° to 0°. Two zero (0) index lines on the cross slide provide reference points.

Angles are usually stated as an *included angle* or as an *angle with a centerline* (one-half the amount of the included angle). The major parts of a compound rest with a zero reference and graduated slide are shown in Figure 26–5.

COMPOUND REST SETUP FOR ANGLE TURNING

The angle at which the compound rest is set usually corresponds to the angle with a center line. Thus the angle produced is the included angle. The degree reading on the swivel slide depends on the axis from which the compound rest is set and the arrangements of the numbered graduations.

Short, steep angles are usually turned by swiveling the compound rest. Three common setups are described. If an included angle is

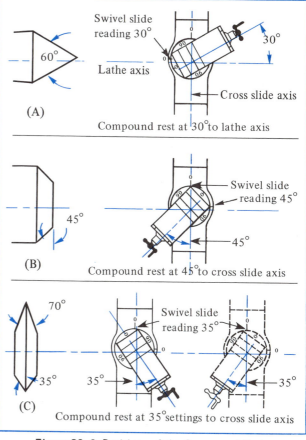

Figure 26–6 Positions of the Compound Rest for Angle Turning

shown on a drawing, the compound rest is positioned at one-half of this included (required) angle. For example, to turn the 60° center, the compound rest is swiveled to 30° (Figure 26–6A). A 45° angle may be produced by swiveling the compound rest 45° to the transverse axis (Figure 26–6B). The two sides of a 70° included angle (Figure 26–6C) are turned by swiveling the compound rest to a 35° angle and turning one side. The compound rest is then swiveled 35° in the opposite direction. The turned included angle is then 70°.

The position of the compound rest for feeding in relation to the lathe center or cross slide axis is influenced by safety and convenience considerations. Feeding with the compound rest handwheel from a position in front of the workpiece is safer and most convenient.

COMPUTING COMPOUND REST ANGLE SETTING

The compound rest is used occasionally to turn a short taper. The dimension of the taper is usually stated in terms of taper per foot (T_{pf}). This dimension must be converted to degrees and minutes to correspond with the graduations on the swivel slide. The angle may be computed by one of the following two methods.

CONVERTING T_{pf} TO ANGLE WITH CENTERLINE SETTING

The angle to which the compound rest must be set in relation to the centerline is calculated by using the following formula:

Compound Rest Angle = T_{pf} × 2.383

Example: Calculate the compound rest angle in relation to the centerline for turning a short taper of 1″ T_{pf}.

Compound Rest Angle = T_{pf} × 2.383

= 1″ × 2.383 = 2.383°

Note: The graduations on the compound rest are in divisions of one degree. In this example the worker must estimate 3/8 of one degree (0.383°). After the compound rest is set, a trial cut is taken. The T_{pf} is measured. Adjustment of the compound rest then is made if needed.

TANGENT OF THE ANGLE WITH THE CENTER LINE

The following formula is used to calculate the tangent (tan) of the angle to which the compound rest must be set in relation to the center line:

$$\tan = \frac{T_{pf}}{24}$$

Example: Using a T_{pf} of 1″, the tangent is equal to 1/24, or 0.04167. A table of natural tangent values shows that 0.04167 represents an angle of 2°23′.

Since the swivel slide graduations are in degrees, the 23′ must be estimated. Consequently, after a trial cut the taper must be checked to determine whether the compound rest setting requires further adjustment.

How to Turn and Measure a Taper

Turning with a Taper Attachment

Setting the Telescopic Taper Attachment

STEP 1 Mount the workpiece. Set the cutting tool on center.

STEP 2 Check the guide bar and sliding block on the taper attachment for free play. Adjust the gibs as needed.

STEP 3 Clean and lubricate all sliding surfaces.

STEP 4 Adjust the guide bar. Read the taper per foot (or taper angle in degrees) on the graduated base plate.

STEP 5 Tighten the lock screws. These screws hold the guide bar to the base plate.

STEP 6 Center the base plate with the cross slide. Position the taper attachment.

Note: The cutting tool must be able to traverse the workpiece the length of taper that is to be turned.

STEP 7 Secure the taper attachment in position. Tighten the clamping bracket to the lathe bed.

Note: Steps 1 through 7 are used for setting either the telescopic or the plain taper attachment.

Setting the Plain Taper Attachment

STEP 1 Set the compound rest at zero to align it with the cross slide axis.

STEP 2 Remove the binding screw to disconnect the cross slide and the cross feed screw nut.

STEP 3 Tighten the binding screw to secure the slide block and cross slide extension arm.

STEP 4 Take out the end play in the taper attachment setup. Move the carriage clear of the workpiece after each cut. Then move it back to feed longitudinally into the work at the beginning of each new cut.

STEP 5 Set the cutting tool for a light trial cut for about 1/8″.

STEP 6 Measure the small diameter. Set the depth for a roughing cut. Use the same cutting speed and feed as for regular turning. Take a tapered cut for approximately 1 1/4″.

STEP 7 Check the accuracy of the taper and the diameter. The taper usually is measured with a standard micrometer.

STEP 8 Readjust the taper attachment if necessary. Take a very light cut and recheck the taper.

STEP 9 Finish turn the taper to size. The usual 0.002″ to 0.003″ should be left for fine finish filing and polishing.

Taper Turning by the Offset Tailstock Method

STEP 1 Determine the required tailstock offset. Check the centers for alignment.

STEP 2 Position a dial indicator pointer so that it touches the side of the tailstock spindle. Set the indicator at zero.

STEP 3 Loosen the tailstock binding nut. Turn the screw to move the head on the base.

Note: The tailstock setting is sometimes positioned by the direct reading. The amount of offset is determined from the graduations on the base and the index line on the head of the tailstock.

STEP 4 Continue to move the tailstock head to the required offset. The amount is represented by the number recorded on the dial indicator.

Note: The range of movement of the dial indicator pointer must be within the limits of the instrument.

STEP 5 Tighten the binding nut. Position and secure the cutting tool on center.

STEP 6 Check the centers on the workpiece. Lubricate the dead center hole. Test the setup to see that the workpiece rotates freely.

STEP 7 Take a trial cut. Check the large and small diameters against the required taper.

Note: Checking the diameters is particularly important because the length of the workpiece affects the taper. Adjust the tailstock with a dial indicator if further adjustment is required.

STEP 8 Take a roughing cut. Test the taper for accuracy. Use a gage, micrometer, or other method for checking the taper and the accuracies of the small and large diameters.

STEP 9 Take a finish cut. Allow 0.002″ to 0.003″ on a workpiece that ranges up to 1 1/4″ diameter. The part may then be brought to size by finish filing and polishing.

How to Turn an Angle for a Short Taper Using the Compound Rest

STEP 1 Determine the required angle for setting the compound rest.

STEP 2 Loosen the lock nuts on the compound rest.

STEP 3 Swivel the compound rest to the required angle. Lock the compound rest in this position.

Note: The compound rest is set at one-half the included angle and is positioned so that the cutting tool may be fed toward the headstock.

STEP 4 Set the cutting tool on center. Start the lathe. Move the carriage longitudinally so that the cutting tool is positioned to take a trial cut.

STEP 5 Hand feed the tool bit by turning the compound rest feed screw.

STEP 6 Check the turned angle with a plug or angle gage, a protractor, or a template.

STEP 7 Take a roughing cut. Check the angle for dimensional size and accuracy.

STEP 8 Follow with other roughing cuts as required and a final finish cut.

Note: The angular surface may be filed and polished to meet the required surface finish.

STEP 9 *When two angles are to be cut, as for a bevel gear:* Reset the compound rest to turn the complementary angle. Repeat the steps for rough and finish turning and angle checking (steps 1–8).

Note: A short taper may be turned with a compound rest. The compound rest is set at the computed angle to the centerline. The angle corresponds to the taper per foot or metric equivalent. Steps 2 through 8 are then followed.

Safe Practices in Turning Tapers and Angles

- Check the size of the bearing surface of the center holes. They must be large enough to permit the workpiece to be moved out of alignment when using the offset tailstock method.
- Lubricate the dead center hole. Carefully adjust the workpiece so that it may turn freely.
- Test the lathe dog to see that the tail is free to ride in the driver plate slot.

- Move the cutting tool away from the workpiece and then position it for the next cut. This movement takes up the lost motion of the taper attachment and setup.

- Turn the tapered workpiece clockwise carefully when testing for accuracy. To release the tapered surfaces, reverse the turning direction (counterclockwise) and gently pull outward to release the tapered surfaces without damaging them.

TERMS USED IN TAPER AND ANGLE TURNING

Self-holding taper	A slightly tapered part. Two mating tapered parts that are held securely together by the tapered surfaces.
Self-releasing taper	A steeply tapered part that requires a clamping device to hold it securely on a mating taper.
Standard taper systems	Different series of tapers that are nationally or internationally accepted as a specific standard. (The tapers within a series extend over a wide range of diameters.)
Taper per foot (T_{pf})	The amount a large diameter varies from a small diameter for each 12″ length.
Tailstock offset (T_o)	The amount a tailstock is moved from a zero index position.
Taper ring gage	A gage for measuring the accuracy of an external taper. A tapered ring that is precision ground to a close tolerance.
Taper micrometer	A reliable, accurate, precision-measuring micrometer that provides for minutely measuring a taper.
Plain taper attachment	An accessory fastened to a lathe bed. An attachment that produces an angular movement of the cross slide.
Telescopic taper attachment	A mechanism that permits the transfer of motion. Transferring the angle setting of the taper attachment directly through the cross slide.
Steep taper (angle) turning	The process of setting the compound rest at one-half the included angle. Feeding the cutting tool at a preset angle by turning the compound rest handwheel.

SUMMARY

- The American Standard (Morse) Taper falls in the small-taper-angle, self-holding taper group. The taper per foot varies within the series from almost 0.600″ to 0.625″.
  - The American Standard Taper Pin series ranges from #7/0 to #11. The standard taper per foot is 0.2500″ or 0.0208″ per inch. The pins come in a variety of standard lengths and are commercially produced.

- Self-holding short tapers are used to align and hold mating tapers rigidly.
  - The Brown & Sharpe (B & S) Taper series ranges from #1 to #18, with small-end diameters from 0.200″ to 3.000″. The taper per foot varies among the different numbers. The T_{pf} is approximately 0.5000″.
- The Jacobs Taper is usually a short, self-holding taper. A principle application is the centering and holding of a drill chuck.
  - The Jarno Taper is a 0.600″-per-foot, self-holding taper. The number of the Jarno taper, divided by 10, 8, or 2, determines the small and large diameters and the length.
- The American Standard Self-Releasing Steep Taper combines features of quick, precision alignment and easy release. The long steep taper, like the type-L spindle nose, has a 3.500″ taper per foot. The steep taper of the type D–1 spindle nose is 3.000″ taper per foot.
  - Tapers may be turned by offsetting the tailstock, by using a plain or telescopic taper attachment, or by using the compound rest.
- The compound rest set at the required angle provides a practical method of turning steep angles.
  - Tapers may be gaged with a taper ring gage or standard or taper micrometer.

UNIT 26 REVIEW AND SELF-TEST

1. Identify four main purposes that are served by tapers.

2. a. Give the general range of numbered sizes of taper pins.
   b. State two major applications of American Standard Taper Pins.

3. Distinguish between a Jacobs taper and an American Standard (Morse) taper.

4. State the taper per inch for (a) Brown & Sharpe tapers and (b) Jarno tapers.

5. Differentiate between the holding devices for an American Standard self-releasing steep taper and a type D–1 short steep taper (3.000″ per foot).

6. Define each of the following taper terms or give the formula for calculating each value: (a) taper, (b) taper per foot, (c) large diameter, and (d) small diameter.

7. a. Identify two common methods of turning shallow tapers.
   b. Tell how a taper is represented and dimensioned on a drawing.

8. List three general methods of measuring the accuracy of the taper angle and/or the large and small diameters.

9. State four advantages of turning a taper with a taper attachment compared to the offset tailstock method.

10. List the steps required to turn an included angle of 60° on a workpiece between centers and using a compound rest.

11. Tell how to turn two tapered mating parts when testing the accuracy of the tapers.

12. State two safe practices the lathe worker must observe when turning tapers.

Lathe Work Held in a Chuck

Boring, reaming, drilling, internal and external thread cutting, and other lathe processes generally require that the workpiece be mounted in a chuck or a collet. This section opens with descriptions of the types of chucks typically used for these processes and the methods of mounting and truing work and centering or off-centering for eccentric turning. Safety practices are stressed to prevent personal injury or damage to the machine, accessories, or tools. Subsequent units deal with the principles of and procedures for performing the following lathe processes:

- Centering, drilling, countersinking, and reaming;
- Straight hole boring, counterboring, recessing (undercutting), taper boring, and mandrel work;
- Cutting and measuring 60° form internal and external threads;

UNIT 27

Chucks and Chucking Technology and Processes

OBJECTIVES

After satisfactorily completing this unit, you will be able to:

- Apply additional information on lathe spindle accessories for chucking workpieces, universal and independent chucks, standard and hollow core drill chucks, and different types of collets and rubber-flex chucks.
- Perform the following processes.
  - Removing Lathe Spindle Accessories (live center, faceplate, driver plate, or chuck).
  - Assembling and Mounting a Draw-In Collet Attachment or Steep-Taper Spindle Nose Collet Chuck.
  - Mounting Universal and Independent-Jaw Chucks and Changing Jaws.
  - Mounting and Truing Work in a Chuck (approximate and precision truing).
- Follow *Safe Practices* with lathe spindle accessories and chucking processes and use *Terms* correctly.

BASIC SPINDLE NOSES

TYPE-L SPINDLE NOSE

As stated earlier, there are three main types of spindle noses. The *threaded spindle nose* is the oldest type. The chuck flange and the chuck itself are centered by the accuracy of the threads, a straight-turned section, and the spindle nose shoulder.

The American Standard *type-L spindle nose* has a taper of 3.500″ per foot (Figure 27–1). The taper bore of the chuck flange centers on the corresponding taper of the spindle nose. Note also that the chuck flange has a keyway. This keyway fits the key in the tapered nose of the spindle. The key and keyway prevent the chuck from turning on the tapered surface of the spindle nose.

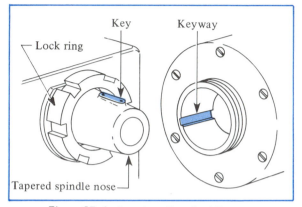

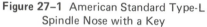

Figure 27–1 American Standard Type-L
Spindle Nose with a Key

The chuck is held securely on the taper and concentric with the spindle by a lock ring. A spanner wrench is used for both tightening and loosening the lock ring.

TYPE D–1 CAM-LOCK SPINDLE NOSE

There are five main design features of the *D–1 cam-lock spindle nose*. These features are identified in Figure 27–2. The short, steep taper of the D–1 nose is 3.000″ per foot. There are from three to six cam-lock studs ④ depending on the chuck size. The studs extend from the chuck or other accessory. The studs fit into corresponding holes in the face of the spindle nose. There are also eccentric cam locks ③ that match. As the eccentric cam lock turns against the stud, the accessory is drawn firmly onto the taper and against the face of the spindle flange.

There are register lines ① for each of the cam locks. The index line of each cam lock ② must match with the register position to mount or remove the accessory. The cam locks are given a partial turn clockwise to mount a chuck; counterclockwise, to remove it.

CHUCK TYPES AND FEATURES

UNIVERSAL THREE-JAW CHUCK

The name of the *universal three-jaw chuck* indicates that there are three accurate, self-centering jaws. These jaws are controlled by a bevel gear-driven scroll. All three jaws may be

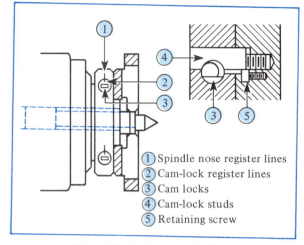

① Spindle nose register lines
② Cam-lock register lines
③ Cam locks
④ Cam-lock studs
⑤ Retaining screw

Figure 27–2 Features of a D–1
Cam-Lock Spindle Nose

actuated (moved) by turning any one of the adjusting sockets. Since the universal chuck is self-centering, the jaws do not require individual setting as do independent-jaw chucks.

Due to the shape of the screw thread on their back sides, the jaws are not reversible. Other jaw sets are available to accommodate large-diameter work and work that requires inside chucking.

FOUR-JAW INDEPENDENT CHUCK

Each jaw of the *four-jaw independent chuck* is adjusted independently (Figure 27–3).

The chuck jaws are stepped to take work of small and large diameters. They may be used to hold workpieces on either an inside or an outside diameter. The jaws are also reversible. The four-jaw independent chuck can accommodate a wider range of shapes and sizes than the universal chuck.

The four-jaw chuck may be used to grip square, round, or irregular-shaped pieces. The jaws may be positioned concentric or off center, depending on the job requirements. The work surfaces may be finished or rough as are the outer surfaces of castings and forgings or surfaces produced by flame cutting or other processes.

The face of the chuck has a series of concentric rings cut into it at regular intervals. These rings, or grooves, aid the operator in centering a

Figure 27-3 Four-Jaw Independent Chuck

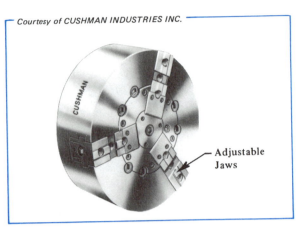

Adjustable Jaws

Figure 27-4 Power-Operated, Three-Jaw, Self-Centering Chuck

workpiece. Adjustments are made by loosening one jaw and tightening the opposite jaw. The rough setting is then checked by other approximate or precise methods of truing a workpiece.

The *combination chuck* combines the distinguishing features of both the universal and the independent chucks. The general-purpose, four-jaw combination chuck permits the jaws to be adjusted either universally or independently.

A chuck cradle of hardwood is usually used to install or to remove any type of heavy chuck. The cradle is grooved to fit and slide on the ways and under the chuck. The cradle protects the operator from injury and prevents damage to the lathe ways.

POWER-OPERATED, SELF-CENTERING CHUCKS

Power-operated, two- and three-jaw chucks with adjustable and nonadjustable jaws are widely used in production. These chucks may be *self-centering power chucks*.

The nonadjustable-jaw power chuck is recommended for general manufacturing service. It is particularly suited for repetitive operations, especially on heavy workpieces. This chuck is balanced to eliminate chatter and vibration at high spindle speeds (RPM).

The adjustable-jaw power chuck provides independent jaw action (Figure 27-4). The jaws are designed to hold irregular-shaped workpieces. Once the jaws are adjusted, the initial accuracy of positioning is maintained with successive workpieces.

STANDARD CHUCK JAW DESIGNS

Standard chuck jaws may be solid or of two-piece construction. *Solid step gripping jaws* are reversible for most chuck sizes and may be used for gripping inside or outside surfaces. Other jaws are used in sets: one set of outside gripping jaws; the other, inside gripping jaws (Figure 27-5A).

The two-piece jaws illustrated at (B) include a *master jaw* that is grooved to slide in the chuck body. The face of the master jaw is machined with a tongue and groove to accommodate a *reversible step* top jaw (which is bolted to the master jaw).

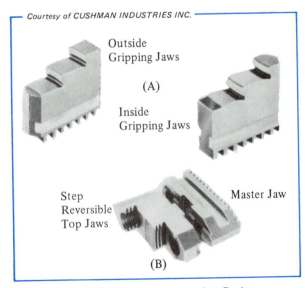

Outside Gripping Jaws

(A)

Inside Gripping Jaws

Step Reversible Top Jaws

Master Jaw

(B)

Figure 27-5 Basic Gripping Jaw Designs

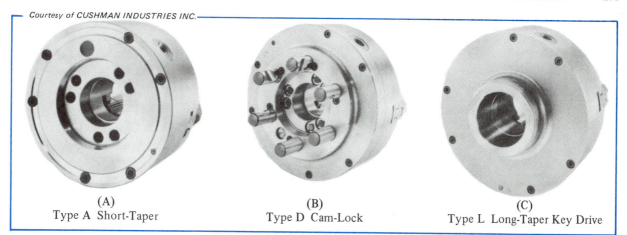

(A)
Type A Short-Taper

(B)
Type D Cam-Lock

(C)
Type L Long-Taper Key Drive

Figure 27-6 Examples of Chuck Mounted Adapter Plates for Three Basic American Standard Spindle Noses

FITTED CHUCK ADAPTER PLATE DESIGN

A chuck adapter plate is machined on one side to fit the design and assembly features of the chuck body to which the plate is to be secured. The other side accommodates the spindle nose onto which the chuck assembly is to be accurately fitted and tightened.

Three types of mounted chuck adapter plates are illustrated in Figure 27-6. The Type-A chuck adapter plate (shown at A) fits the American Standard short-taper spindle nose. The Type-D chuck adapter plate (at B) is designed for an American Standard cam-lock spindle nose. The Type-L American Standard long-taper key drive chuck adapter plate (C) fits this type of spindle nose. Chuck adapter plates are commercially available for standard threaded spindle noses, straight recessed spindle noses, and other types.

COLLET CHUCKS

Collet chucks are simple, accurate, practical work-holding devices. They are used principally for holding regular-shaped bars of stock and workpieces that are finished on the outside. The work is generally of round, square, or hexagonal shape. Two common types of collets are the *spring collet* and the *rubber-flex collet.*

SPINDLE NOSE COLLET CHUCK

The *spindle nose collet chuck* takes a wider range of work sizes than does the draw-in bar type of collet chuck. The phantom view of a spring collet in Figure 27-7 shows a bevel

gear threaded disc. This disc is mounted inside the chuck body. The bevel teeth are moved by turning the bevel socket on the chuck body with a chuck wrench. As the spring collet is drawn into the chuck body, the taper on the split jaws tightens against the workpiece. Spring collet chucks are designed to fit standard types of lathe spindle noses. Some spindle nose collet chucks use a spur gear designed to tighten or loosen the collet.

JACOBS SPINDLE NOSE COLLET CHUCK

The *Jacobs spindle nose collet chuck* takes rubber-flex collets. This chuck is also made to fit different types of spindle noses. The usual range of each collet is 1/8" (1/16" over and under a nominal size). A single rubber-flex collet and Jacobs type of spindle nose chuck serves a wide range of sizes (Figure 27-8).

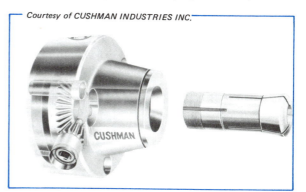

Figure 27-7 A Spring Collet Chuck and a Standard Collet

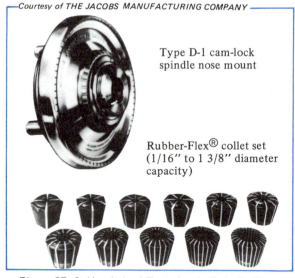

Type D-1 cam-lock
spindle nose mount

Rubber-Flex® collet set
(1/16" to 1 3/8" diameter
capacity)

Figure 27–8 Handwheel-Type Jacobs Spindle Nose
Collet Chuck with Rubber-Flex Collets

Figure 27–9 A Standard Drill Chuck
Fitted with a Taper Shank

The work is first inserted in the collet. The chuck handwheel is turned clockwise to tighten the collet jaws against the workpiece. If the workpiece is not long enough to extend into the rubber-flex collet for at least 3/4", then a plug of the same diameter is needed. This plug is placed in the back of the collet. The plug helps to ensure that the collet grips the work securely. It prevents the work from springing away if a heavy force is applied during the machining operation.

DRILL CHUCK

The standard *drill press chuck* is commonly used for drilling, reaming, tapping, and other operations on the lathe (Figure 27–9). This chuck is fitted with a taper shank. The shank fits the spindle bore of the tailstock. A sleeve or taper socket may be used to accommodate the chuck in the spindle headstock.

Another type of drill chuck is designed with a hollow core. This core permits the holding of long bars of small cross section that extend through the spindle.

WORK TRUING METHODS

APPROXIMATE METHODS

There are a number of approximate and precise methods of truing chuck work. The con-

centric rings on the chuck face provide one approximate method.

Once chucked, the work may be further tested for trueness by revolving the spindle slowly. A piece of chalk is held in one hand. The hand is steadied and the chalk is brought to the workpiece so that any high spot produces a chalk mark. The jaw (or jaws) opposite the high spot is moved out. The opposing jaw (or jaws) is moved in. The process is repeated until the work is concentric (no high spots).

Trueness may also be roughly checked by using a toolholder. Reduction of the light (space) showing between the back end of the toolholder and the slowly turning workpiece indicates the high spot. A piece of white paper may be placed on the cross slide under the toolholder to show the light (space) and thus identify which jaws must be adjusted.

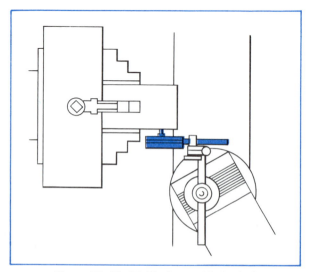

Figure 27–10 Dial Indicator Method of
Precision Truing

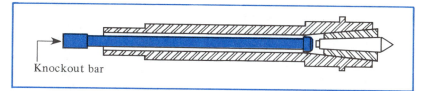

Figure 27–11 Driving a Live Center with a Knockout Bar

PRECISION TRUING BY THE DIAL INDICATOR METHOD

Out-of-trueness may be corrected by a more accurate method. A dial indicator and holder are positioned and held in the tool post (Figure 27–10). The indicator point is carefully brought into contact with the workpiece. The work is revolved slowly by hand. The amount that the work runs out-of-true is indicated by movement of the dial indicator pointer. The difference between the high and low spots is noted. The jaw opposite the high spot is moved out one-half the distance. The opposing jaw is moved in the same distance. The workpiece is again revolved and further adjustments are made. Each chuck jaw is then checked for tightness. The work is concentric when there is no movement of the dial indicator pointer.

How to Remove Lathe Spindle Accessories

Removing a Live Center

STEP 1 Insert a solid-metal knockout bar in the spindle bore (Figure 27–11).

STEP 2 Hold the end of the live center with the right hand.

STEP 3 Move the bar with a quick stroke to firmly tap the center and free the tapered surface.

Note: The live center is sometimes left in place if it does not interfere with any of the chuck work processes.

Removing a Faceplate, Driver Plate, or Chuck

Note: For mounting and removal of faceplates, driver plates, and chucks that are large and heavy, engage the back gears and slide a wooden cradle of a size and form suitable for the particular accessory to be mounted or removed.

Threaded Spindle Nose Faceplate or Driver Plate

STEP 1 Select a flat or square metal bar that fits the slots of the faceplate or driver plate. Remove burrs. Then center the bar in one of the plate slots.

STEP 2 Tap the bar firmly in a counterclockwise direction. Retap if the mating shoulders of the accessory and the spindle nose are not freed.

STEP 3 Continue to slowly turn the faceplate or driver plate by hand in a counterclockwise direction.

Note: Extra care must be taken as the last threads are reached. It is important to avoid having the driver plate or faceplate move off the spindle with the full weight riding on just one or two threads.

Threaded Spindle Nose Universal or Independent-Jaw Chuck

STEP 1 Extend a chuck jaw nearly to the outside diameter of the chuck.

STEP 2 Place a hardwood block over the back ways.

STEP 3 Reverse the direction of the spindle to revolve counterclockwise. Turn the spindle so that the extended jaw strikes the hardwood block sharply.

Note: The hardwood, plastic, or soft metal block must be strong enough to withstand the force of the jaw striking it.

STEP 4 Move the cradle under the chuck if it is large and heavy. Continue to turn the spindle counterclockwise. Exercise additional caution as the last few threads are reached.

Steep-Taper Type-L Spindle Nose Plates and Chucks

STEP 1 Select the correct size of C-spanner wrench. Place it around the lock ring. Rotate the lathe so that the lock ring slot permits the spanner wrench to be held in an upright position.

STEP 2 Engage the back gears. Place a wooden cradle under the accessory if it is heavy.

STEP 3 Place the spanner wrench around the lock ring. Hold it in place with one hand.

STEP 4 Tap the wrench handle sharply in a counterclockwise direction to free up the threads and unseat the tapered surfaces. Continue to turn the lock ring counterclockwise until it disengages.

STEP 5 Use the cradle on a heavy chuck. Carefully slide the accessory away from the spindle nose.

Cam-Lock Spindle Nose Plates and Chucks

STEP 1 Slide the wooden cradle under the accessory.

STEP 2 Turn each cam lock in a clockwise direction. Stop when the index line and the register marks coincide.

STEP 3 Tap the chuck sharply but lightly to break the taper contact if necessary.

STEP 4 Slide the accessory on the cradle until it clears the lathe spindle nose.

STEP 5 Store the accessory in its proper compartment.

Removing a Draw-In Collet Attachment

STEP 1 Remove the spring collet and the drawbar.

STEP 2 Turn the spindle nose cap counterclockwise to free the taper sleeve from the spindle bore.

STEP 3 Place each part in its proper storage compartment.

How to Mount Lathe Spindle Accessories

Assembling the Threaded Spindle Nose Draw-In Collet Attachment

STEP 1 Remove any nicks or burrs from the spindle bore or the taper sleeve and other parts. Wipe all surfaces to remove any foreign particles.

STEP 2 Place the spindle nose cap in position.

STEP 3 Insert the taper sleeve. Move the sleeve in with a quick motion to set the mating tapered surfaces. The parts are aligned and the sleeve is also held securely in the spindle bore.

STEP 4 Lock the lathe spindle by engaging the back gears. On small bench lathes engage the spindle stop pin.

STEP 5 Select a collet of the shape and size of the workpiece.

> **Note:** Accuracy and efficiency may be impaired if the collet segments are not able to bear evenly along the length of the workpiece.

STEP 6 Insert the collet in the taper sleeve. Turn the collet until the key in the taper sleeve engages the keyway of the collet.

STEP 7 Extend the drawbar through the lathe spindle.

STEP 8 Press against the face of the collet with the palm of the right hand to prevent the collet from being pushed out of the sleeve.

STEP 9 Turn the handwheel of the drawbar slowly clockwise to engage the threaded end of the collet. Continue to turn the handwheel clockwise until most of the end play is taken up.

STEP 10 Insert the burr-free workpiece. Extend it the distance necessary to perform the lathe operations.

STEP 11 Continue to turn the drawbar clockwise by hand. The force produced by the handwheel is adequate for the collet to hold the workpiece securely.

Assembling a Steep-Taper Spindle Nose Collet Chuck

STEP 1 Select a spring collet chuck with an adapter plate. The chuck and plate must fit either the type-L or D–1 spindle nose on the lathe.

STEP 2 Proceed to assemble the collet chuck on the tapered spindle nose. Secure the collet chuck to the spindle nose. Tighten either the lock ring or the cam locks.

STEP 3 Insert the collet and align the keyway and key. Engage a few threads.

STEP 4 Place the burr-free workpiece in the collet.

STEP 5 Turn the chuck handwheel. Draw the collet against the taper portion until the work is held securely.

Mounting Universal and Independent-Jaw Chucks

Cam-Lock Spindle Nose

STEP 1 Align the index line on each cam-lock stud with the register line on the spindle.

STEP 2 Slide the cleaned and burr-free chuck on a cradle. Position it on the nose spindle.

STEP 3 Turn the lathe spindle slowly by hand. Align the cam-lock studs with the clearance holes in the spindle nose.

STEP 4 Slide the chuck onto the tapered portion of the spindle and up to the shoulder.

STEP 5 Turn each cam lock in a counterclockwise direction. Apply equal force to each cam lock. The chuck should be drawn tightly against the spindle shoulder. The short tapered surfaces ensure that the chuck is correctly aligned and runs concentrically.

American Standard Taper Type-L Spindle Nose

STEP 1 Turn the lathe spindle by hand until the key on the nose and the keyway in the chuck adapter plate are aligned.

STEP 2 Slide the cradle so that the tapered bore fits onto the spindle nose.

STEP 3 Turn the lock ring to engage the threads on the chuck adapter plate.

Note: If the threads do not engage easily, turn the lock ring in the reverse direction and remove the chuck. Remove any dirt or chips that were not cleaned out the first time.

STEP 4 Remount the chuck. Tighten the lock ring by hand.

STEP 5 Place a spanner wrench around the lock ring. Set the tapered and shoulder surfaces by striking the spanner wrench sharply with the palm of one hand.

How to Change the Jaws on a Universal Chuck

STEP 1 Remove the installed set of jaws. Clean and store them in an appropriate container.

STEP 2 Select a matched set of jaws that will accommodate the work size and the nature of the operations.

STEP 3 Start assembling each jaw. Begin with jaw #1. Turn the scroll with the chuck key until the jaw is engaged.

STEP 4 Insert jaw #2 while continuing to turn the scroll. As soon as this jaw is engaged, proceed with jaw #3 and the remaining jaws in the set.

Note: The jaws must be inserted and engaged in numerical sequence. Otherwise they will not be concentric.

STEP 5 Check the concentricity of the jaws. Chuck a piece of finished stock. Test its trueness with a dial indicator.

STEP 6 Correct any inaccuracy either by grinding the jaws or by adjusting them.

Note: An adjusting screw is provided on each jaw of some universal chucks. The inaccuracy is corrected by turning the adjusting screw. The jaws may then be moved in or out as needed.

Safe Practices with Chucks and Chuck Work

- Shut off the power to the lathe when mounting or removing a chuck.
- Obtain assistance when handling a heavy chuck or other machine accessory.
- Use a wooden cradle to slide a heavy chuck onto or off a spindle nose.
- Keep the fingers out from under the chuck and the cradle.

- Stone away any burrs on the spindle nose, chuck adapter plate, or other machined part.
- Wipe all mating surfaces with a clean wiping cloth. Carefully move the palm and fingers over the parts to feel if there are any foreign particles left. Apply a drop of oil on the taper or thread of the spindle nose and the shoulder.
- Revolve the spindle by hand before starting any operation. Check to see that the chuck jaws, the workpiece, and the tool setup clear the carriage.

TERMS USED WITH CHUCKS AND CHUCK WORK

Chuck	A work-holding device for mounting and holding a workpiece in a particular position. A device that transfers motion and force from a lathe spindle, through a workpiece.
Chuck adapter plate	A circular, flanged plate designed to fit a particular type and size of spindle nose. A plate fastened to the backside of a chuck. A plate that accurately positions and holds a chuck on a lathe spindle.
Type-L spindle nose	A steep-angle taper lathe spindle nose. A spindle nose with an American Standard taper of 3.500″ T_{pf}. A tapered lathe spindle nose having a key and a lock ring. (The lock ring draws and holds a chuck securely on the taper and against the shoulder of the spindle.)
Type D-1 cam-lock spindle chuck adapter	A short, steep-angle, taper-bored adapter plate. A chuck plate bored to a 3.000″ T_{pf}. An adapter plate designed to secure a chuck on a cam-lock spindle nose.
Universal chuck (lathe)	A work-holding device with jaws that move radially and together. A device to position and hold a workpiece concentric.
Independent jaw chuck	A work-holding device for positioning a workpiece centrally or off center. A chuck with jaws that are moved independently.
Inside chucking	The process of positioning a chuck to hold a workpiece from an inside diameter or relieved area. Exerting an outward force against a workpiece.
Spindle nose collet chuck	A work-holding device mounted on a lathe spindle. Usually a hand-operated chucking mechanism that moves a metal spring collet or rubber-flex collet. A device for adjusting a collet to hold a workpiece so that it may run concentric.
Precision testing for trueness	Truing a finished machined surface to run concentric within ±0.0005″ (0.01mm) or to a finer tolerance. Using a dial indicator to true a workpiece until no movement (out-of-trueness) is recorded on the instrument.
Spindle nose accessory	A chuck, driver plate, faceplate, live center, or other device that is secured to the spindle nose. A device that positions and holds a workpiece or a cutting tool in a spindle nose.

SUMMARY

- Chuck adapter plates are designed for mounting universal, independent-jaw, and combination chucks.
    - Spindle nose accessories are accurately positioned and secured on a type-L American Standard steep taper (3.500″ T_{pf}) with a lock ring.
- The universal three-jaw chuck is a common shop chuck. It is recommended for bar stock and other workpieces that are produced with a fairly accurate round or hexagonal finished surface.
    - The combination chuck combines the features of the independent-jaw and the universal chuck. A workpiece may be set in a desired location using the independent jaws. Duplicate parts may then be chucked by turning the sockets of the universal jaws.
- Cam-lock studs fit corresponding holes in the D–1 short steep-taper (3.000″ T_{pf}) spindle nose. The taper and shoulder surfaces are held securely by each cam lock.
    - Collets provide a quick, accurate, dependable method for holding machined bars and other finished-surface products.
- The spring collet may be drawn in with a drawbar attachment. A spindle nose collet chuck may also be used with the spring collet.
    - Rubber-flex collets are designed for a spindle nose chuck. The tapered face of the chuck cap forces the metal inserts against the inside tapered walls of the chuck.
- Standard or cored drill press chucks are used for chucking and driving small-diameter cutting tools and workpieces.
    - Heavy-work accessories require extra care. Machine or tool damage and personal strain or other injury may be avoided by safe handling.

UNIT 27 REVIEW AND SELF-TEST

1. Explain how a D–1 cam lock spindle nose operates.

2. List the basic differences in the operation of a universal three-jaw chuck and the operation of an independent four-jaw chuck.

3. Tell why rubber-flex collets mounted in a Jacobs or Sjogren spindle nose collet chuck are more flexible than spring collet chucks.

4. Identify (a) two rough methods and (b) one precision method of truing work held in a chuck.

5. List the steps in removing a chuck or plate from a steep-taper (type-L) spindle nose.

6. State two precautions to take when changing the jaws on a universal chuck.

7. List four safe practices to observe when handling chucks and doing chuck work.

Centering, Drilling, Countersinking, and Reaming on the Lathe

Multiple-point cutting tools like the center drill, countersink, twist drill, and machine reamer are commonly used for producing holes. The tool design features, terms, applications, cutting speeds and feeds, and materials of which the tools are manufactured were covered in earlier units. The cutting tools and processes were related to bench work and drill press work.

This unit deals with the same cutting tools, but as applied to lathe work. One main difference from bench and drill work is that in lathe work the cutting may be done by holding the cutting tool stationary (nonrevolving). It is then fed into a revolving workpiece. The same process may also be carried on by feeding a stationary workpiece into a revolving cutting tool. Lathe hole-producing methods, additional toolholders, and machine setups are covered in this unit.

OBJECTIVES

After satisfactorily completing this unit, you will be able to:

- Select cutting tools and accessories and make setups for hole-forming processes on the lathe.
- Apply technical information from cutting speed, feed, and cutting fluid tables to practical shop problems.
- Perform the following lathe operations.
  - Center Drilling, Countersinking, and Drilling Holes.
  - Reaming Holes on a Lathe.
  - Spotting and Truing a Center Hole.
- Follow *Safe Practices* in center drilling, drilling, and countersinking and use technical *Terms* correctly.

CENTERING AND CENTER DRILLING WORK HELD IN A CHUCK

Center holes provide a bearing surface for lathe centers. A center hole may also be used as a pilot hole for subsequent operations. A common and accurate method of drilling a center hole is to use a combination center drill (drill and countersink). The small pilot drill point is especially important for work that is to be mounted between centers.

Sometimes just a central starting hole needs to be spotted in the end of a workpiece. Either a specially ground tool bit or a flat drill may be used for spotting a center.

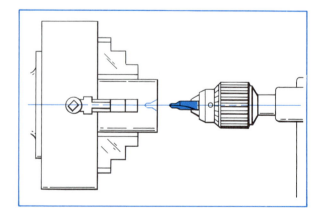

Figure 28-1 Setup for Center Drilling a Workpiece Held in a Chuck

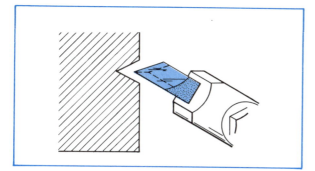

Figure 28–2 Spotting, Truing, or Centering with a Specially Ground Tool Bit

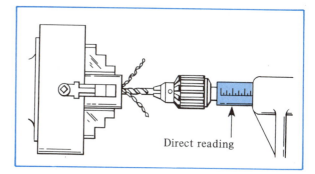

Direct reading

Figure 28–3 Measuring the Drill Depth on the Graduated Tailstock Spindle

Work to be centered is usually held in a universal chuck or collet chuck. Irregular and rough surfaces are centered using an independent-jaw chuck. The center drill is held in a standard drill chuck (Figure 28–1).

After the workpiece end has been faced, the center drill is fed into the revolving workpiece. Although the cutting speed remains the same, the diameter of the angular body increases. Care must be taken to feed the center drill slowly, particularly when feeding the small-diameter pilot drill portion at the start. Excessive force may cause the center drill to fracture.

While not common practice, a center starting hole may also be spotted by grinding a tool bit to a steep-angle point (Figure 28–2). The area in back of the cutting face is ground away sharply to permit the cutting edge to cut without interference from rubbing against the angle hole.

A special 60° countersink may also be used when a center hole needs to be trued quickly and fairly accurately. Only one lip of the countersink is ground to do the cutting. The countersink is held in a drill chuck and is mounted in the tailstock. A high spindle speed is used. The tailstock spindle is brought back as far as possible to cut down on the overhang. The spindle clamp screw is tightened lightly. The countersink is then fed slowly into the revolving workpiece. These steps help prevent play in the tailstock spindle. Play may be caused by the interrupted, uneven cut that exists until a centered hole is produced.

DRILLING PRACTICES ON LATHE WORK

A drill is held on a lathe in a number of different holding devices. Straight-shank drills may be held in a drill chuck, especially if they are smaller than one-half inch in diameter. The drill chuck with taper arbor may fit directly into the matching tapered tailstock spindle (Figure 28–3). With an adapter the chuck may also fit the spindle nose. Large-diameter, straight-shank drills are usually held from turning by a drill holder. Sometimes a drill is gripped in a regular lathe chuck. Long workpieces, which extend a considerable distance from the chuck, are supported by a steady rest (Figure 28–4).

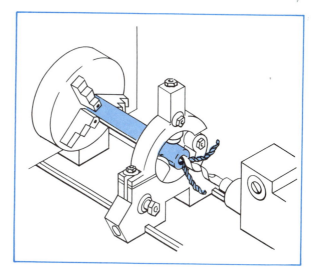

Figure 28–4 Drilling a Long Workpiece Supported by a Steady Rest, with the Drill Mounted Directly in the Tailstock Spindle

REAMING PRACTICES ON LATHE WORK

Jobber's shell, fluted-chuck, expansion, and adjustable reamers are all used in lathe work. Two or more step reamers and taper reamers are also used. All of these were described earlier under machine reamers and reaming processes.

In reaming, the cutting angles, speeds, feeds, and fluids are similar to those used for drilling, boring, and other machine tool processes. Tables of recommended operating conditions for reaming different materials are included in the Appendix.

The selection of a high-speed steel, cobalt high-speed steel, carbide-tipped, or other machine reamer depends on:

- The material of the workpiece,
- Production requirements,
- Whether the hole is interrupted,
- Other factors similar to the factors that affect hand reamers.

The workpiece requirements and machine conditions determine whether a straight-fluted or right-hand or left-hand spiral-fluted reamer is the most practical. The reamer may be of solid, expansion, or adjustable design. The machine reamer may have a straight shank that permits direct chucking. Or, it may have a taper shank that is adaptable for holding in a taper socket or sleeve or directly in a tailstock or headstock spindle (Figure 28–5). The same machine and tool safety precautions described in earlier units must be followed to prevent breakage and excessive reamer wear.

Extremely concentric holes require that the hole first be bored, leaving a minimum amount for reaming. Precise dimensional accuracy also may require both boring and hand reaming to meet the specified tolerances.

Another type of carbide-tipped reamer has flutes that are carbide tipped for the full length. These carbide tips provide a good bearing surface in the reamed hole. An inserted plug may be driven into the body to permit fine adjustments of the reamer to within 0.0001″. The reamer may also be reground to size numerous times.

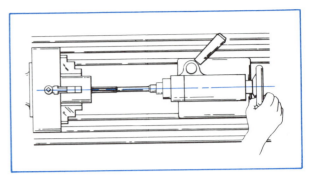

Figure 28–5 A Typical Machine Reaming Setup

How to Center Drill, Countersink, and Drill Holes on the Lathe

Cutting Tools Mounted in a Tailstock Spindle

Center Drill and Countersink Held in a Drill Chuck

STEP 1 Select the center drill or countersink that meets the job specifications. Secure a drill chuck with a taper shank that fits the tailstock spindle. Check the shank for burrs. Remove burrs and wipe the taper surface clean.

STEP 2 Mount the chuck in the tailstock spindle. Sharply snap the chuck into position to set the tapers.

STEP 3 Insert the center drill or countersink in the drill chuck and tighten it. Determine the required spindle (RPM). Set this speed.

STEP 4 Move the tailstock spindle back as far as possible. Slide the tailstock up toward the workpiece until the point of the center drill or countersink nearly touches the workpiece end. Clamp the tailstock.

STEP 5 Start the lathe. Turn the tailstock handwheel and feed slowly. Check the center drill point or the countersink to see that the cutting tool is centered.

Note: Since the workpiece was faced prior to center drilling, the facing tool may not have cut completely through the center. The facing tool may need to be brought back to the center to make a small concentric indentation for center drilling. The same step may be taken to ensure that a countersink is correctly centered.

Caution: The point of the center drill may fracture if it is fed into the workpiece off center.

STEP 6 Apply a few drops of cutting fluid. Continue to feed slowly to the required depth.

Note: In instances where long parts must be drilled or reamed on the lathe, the end of the workpiece can be supported by a steady rest. The drill is usually held in the tailstock spindle.

Drill Held in a Drill Chuck

STEP 1 Examine the shank of the twist drill. Stone any burrs. Wipe the drill clean. Insert and tighten it in the drill chuck.

STEP 2 Determine the correct cutting speed (RPM) for the material to be drilled and the cutting conditions (wet or dry) of the drill. Set the lathe spindle speed to the correct RPM.

STEP 3 Check the position of the tailstock spindle. It should be back far enough to permit drilling the hole to depth. Bring the tailstock toward the workpiece. Clamp it in position.

STEP 4 Start the drill point carefully. It must cut concentrically with the center hole or center spot. An out-of-true center spot must be corrected before proceeding further with the drilling operation.

Note: To start the drill true, turn the back end of the toolholder toward the drill. Move the toolholder in until it just touches the drill body (Figure 28–6).

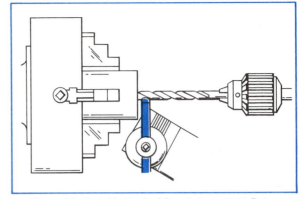

Figure 28–6 A Method of Steadying a Drill Point to Start a True Hole

Slowly feed the drill point to the depth of the cutting lips. Then remove the toolholder from the side of the drill.

STEP 5 Feed the drill into the workpiece until the point enters to the full diameter. Note the reading on the graduated tailstock spindle.

Note: The drill depth may be measured on the graduated tailstock spindle. The initial reading (before drilling) is subtracted from the final reading to establish the drill depth.

STEP 6 Direct the cutting fluid between the drill and the workpiece. Feed to the required depth.

Note: Back the drill out of the work repeatedly during deep-hole drilling. The flutes may need to be brushed to remove the chips. A continuous flow of cutting fluid must reach the drill point at all times. If great forces build up during deep-hole drilling of large-sized holes, the spindle speed should be reduced several times as the depth increases.

Caution: Remove the drill from the drill chuck immediately after completing the lathe operation to avoid personal injury caused by brushing against the drill point.

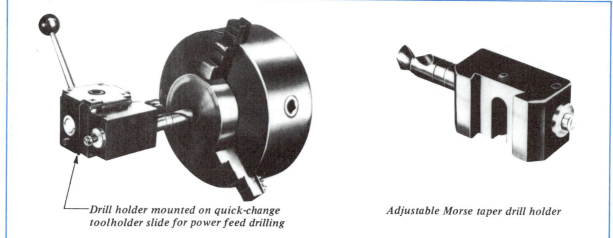

Drill holder mounted on quick-change toolholder slide for power feed drilling

Adjustable Morse taper drill holder

Figure 28–7 Quick-Change Setup for Using Lathe and Carriage Feed in Drilling

Drill Held in a Quick-Change Toolholder

STEP 1 Select the correct type and size of twist drill for the hole to be drilled. Stone any burrs from the shank.

STEP 2 Select an adjustable drill holder (Figure 28–7) that will accommodate the drill taper shank. Insert and tighten the drill in the holder.

STEP 3 Slide the drill holder on the quick-change toolholder. Lock it in position with the drill point at center height.

STEP 4 Determine the required spindle speed (RPM). Set the spindle speed. Start the lathe.

STEP 5 Set the feed rod at the recommended cutting feed. Start the flow of cutting fluid.

STEP 6 Move the drill into the center spot or center drilled hole. Feed the drill by hand until its outside diameter contacts the workpiece.

STEP 7 Stop the lathe if a blind hole is to be drilled. Locate and adjust a carriage stop at the required depth.

STEP 8 Start the lathe. Engage the power feed. Feed a through hole automatically. Discontinue the feed on a blind hole when the drill has almost reached full depth.

STEP 9 Feed to final depth by hand on a blind hole. Clear the chips. Test for depth with a depth gage.

Note: Deep-hole drilling requires the same precautions for chip removal as in drilling by other methods.

STEP 10 Remove burrs caused by the drilling operation.

How to Ream Holes on a Lathe

Note: Setups for machine reaming are similar to drilling and countersinking. The machine reamer may be held in the headstock spindle and turned. Or, the work may rotate and the reamer may be held stationary in the tailstock spindle or in a quick-change tool post. Holes that are to be reamed concentric, parallel to a fixed axis, and to within precise tolerances should first be bored.

STEP 1 Select a reamer that will produce a hole within the specified tolerance.

STEP 2 Chuck the reamer if it is a straight-shank type. Mount a taper-shank reamer in the headstock or tailstock spindle.

STEP 3 Determine the cutting speed and cutting feed, depending on the material in the workpiece, the amount of material to be removed, the reamer type, and the use of a cutting fluid.

STEP 4 Set the spindle speed (RPM). Set the feed if the setup permits machine feeding. Direct the flow of cutting fluid. Start the lathe.

STEP 5 Bring the reamer to the workpiece so that it starts to cut. Take a cut for about 1/8″. Withdraw the reamer. Stop the machine. Clean the workpiece. Measure the hole size with a gage or micrometer.

Note: If an adjustable machine reamer is used, make a + or − adjustment as needed. If a solid reamer is being used and it is cutting over size, it should be replaced.

STEP 6 Continue to ream to the required depth.

STEP 7 Withdraw the reamer. Stop the lathe. Remove the fine sharp burr. Clean the workpiece. Check for final size. Ream if necessary.

Safe Practices in Center Drilling, Drilling, Countersinking, and Reaming on the Lathe

- Remove any raised point that may be left from facing at the center before attempting to center drill.

- Reduce the speed and feed when drilling large-diameter holes to a depth of more than twice the drill size.

- Remove a tool bit or other cutting tool when it is not in use to prevent injury caused by brushing against the cutting point and edges.

- Grind the area steeply beyond the cutting point of a center-hole-spotting tool bit. The angular cutting edge must be able to cut to center hole depth without rubbing.

- Stop the lathe to clean out a drilled or reamed hole.

- Burr the edges of each drilled or reamed hole if the job permits.

TERMS USED IN CENTERING, DRILLING, COUNTERSINKING, AND REAMING ON THE LATHE

Spotting a center	Cutting an indentation at an angle in the center of a workpiece. Machining a center concentric with the axis of a revolving workpiece.
Special centering countersink	A 60°-included-angle countersink with only one lip ground to cut a center hole.
Drill holder	A device for securely holding a drill.
Straight-shank holder	A holding device that clamps on the straight shank of a drill or reamer. (The holder permits the drill to be centered, fed into, and removed from the workpiece.)
Threading into the work (breaking through)	The action of a cutting tool as it cuts through a workpiece. A drill or other cutting tool that feeds into a workpiece at the rate of the helix angle. The rapid movement through a workpiece when a drill or reamer cuts faster than the cutting feed.
Truing a center	Recutting a damaged center hole so that it is concentric with the axis of a workpiece. The process of truing or centering a hole with a center tool bit or specially ground countersink.

─────────── SUMMARY ───────────

- Center drills, drills, countersinks, and machine reamers are the same for lathe work as they are for drilling and other machine processes. The size, classification system, design features, and materials of construction of hole-producing cutting tools are similar regardless of the machine tool.
  - Center drills and small sizes of straight-shank drills, countersinks, and reamers may be held in a drill chuck. A taper arbor permits mounting in a headstock, sleeve, socket, or in the tailstock spindle.
- Center drilling, drilling, countersinking, and reaming may be performed by rotating either the cutting tool or the workpiece.
  - Holes may be produced on the lathe by feeding either the workpiece or the cutting tool to the required depth.
- Cutting tools used for producing holes on the lathe may be mounted and driven from the headstock spindle.
  - Cutting tools also may be positioned, held in, and fed with the tailstock spindle.
- Taper-shank center tools may be held in taper sleeves, sockets, the taper spindle nose, or a tailstock.
  - In lathe work, the cutting speeds and feeds for center drilling, drilling, countersinking, and reaming are similar to those used with drilling machines.
- The cutting action is easier, friction and heat are reduced, and a finer surface finish is produced when the correct cutting fluid is used.
  - Workpieces that extend a distance from the spindle nose or chuck should be supported with a steady rest. The steady rest should be located near the right-hand end of the workpiece for facing, centering, drilling, and reaming operations.
- Safe operating procedures must be observed for holding the workpiece, setting up the cutting tools, and performing the machining operations.

─────────── UNIT 28 REVIEW AND SELF-TEST ───────────

1. Tell what functions are served by (a) the pilot drill and (b) the angle cutting faces of a center drill.

2. State two differences between drilling and reaming on a lathe.

3. Identify three different (a) reamer types, (b) materials of which reamers are made, and (c) factors influencing the selection of a reamer.

4. Tell how an out-of-true center spot may be corrected before a hole is drilled.

5. Indicate the advantages of a quick-change toolholder for holding cutting tools.

6. List two cautions to observe with reamed holes.

7. List three functions that are served by a cutting fluid for deep-hole drilling.

8. State two safety precautions to follow in through drilling or reaming.

Boring Processes and Mandrel Work

This unit deals with common types and forms of boring tools, boring bars, and holders that are all specially designed for lathe work. A few adaptations of boring processes, like counterboring, recessing, and taper boring, are also described. The unit concentrates on applications of bored holes to other work processes. The arbor press and mandrels are covered in relation to force fits and mandrel work.

OBJECTIVES

After satisfactorily completing this unit, you will be able to:

- Determine allowances in boring for finish reaming.
- Understand the purposes of boring and the characteristics of forged boring bits, standard high-speed tool bits, carbide-tipped cutters, and ceramic inserts.
- Know when to use the web bar and other rigid boring bars for holding throwaway ceramic inserts and positioning angles on round boring bars with interchangeable tips.
- Select cutting tools and setups for counterboring, recessing, and cutting internal tapers.
- Use solid, expansion, gang, and threaded mandrels and the arbor press.
- Perform the following processes.
  - Boring a Straight, Counterbored, Recessed, and Taper Hole.
  - Mounting, Removing, and Machining Work On a Mandrel.
- Follow recommended *Safe Practices* and apply *Terms* appropriate to boring processes and mandrel work.

PURPOSES OF BORING PROCESSES

Boring serves four main functions:

- To enlarge the diameter of a hole, particularly very large holes;
- To true up a hole. The surface is bored concentric and straight in relation to the axis of the workpiece;
- To produce an accurate, high-quality surface finish in an odd-sized hole;
- To machine a true hole as a pilot for subsequent cutting tools.

Holes that are to be reamed concentric and to size with a hand or machine reamer should be bored whenever possible. A hole is usually bored to within a tolerance (allowance) of 0.005″ (0.1mm) to 0.007″ (0.2mm) for machine reaming. This tolerance may be increased to 0.010″ (0.3mm) for 1/2″ (12.5mm) diameter reamed holes; 0.016″ (0.4mm) for 1″ (25mm) diameter; 0.030″ (0.8mm) for 2″ (50mm) diameter; and up to 0.045″ (1.1mm) for a 3″ (76mm) diameter. Tolerances (allowances) for hand reaming on diameters up to 1″ (following a boring operation) range from 0.002″ (0.5mm) to 0.005″ (0.1mm).

BORING TOOLS, BARS, AND HOLDERS

Four groups of cutting tools are in common use for boring processes. Forged boring bits, standard-size square high-speed steel cutting tool bits, carbide-tipped bits, and ceramic inserts are widely used. Each may be ground or formed for internal boring, grooving, threading, or form turning.

FORGED BORING TOOL AND HOLDER

The *forged single-point boring tool* has an offset end (Figure 29–1). It is generally used for light boring operations. The cutting tool may be made of high-speed steel, or the cutting end may be tipped with a carbide insert. The forged tool is secured in an offset V-grooved toolholder. The toolholder is reversible to permit the holder to be used as both a right- and a left-hand toolholder. A single toolholder can accommodate a number of different cutting tool diameters. The forged cutting tool is used in small-diameter holes, particularly for sizes from 1/8″ (3mm) to 1/2″ (12mm).

The cutting end of a forged boring bit is ground similar to the end of a left-hand turning tool. The cutting end must have front clearance, side clearance, and side rake like a turning tool (Figure 29–1). The amount of front clearance increases sharply for smaller diameters. The front clearance angle must be adequate to permit the cutting edges to cut freely without rubbing at the heel. However, the amount of front clearance should not be excessive. Otherwise the cutting edge will not be properly supported and it will break away.

The shape of the cutting edges depends on the operations to be performed. For example, a forged boring bar ground for shoulder turning is similar to the shape of a left-hand facing tool. The cutting edge of a cutting tool for boring a straight hole is shaped like a regular left-hand turning tool.

BORING BARS AND HOLDERS

A *boring bar* is a round steel bar. The bar positions and holds a tool bit. Boring bars are positioned for maximum rigidity. One type has a broached square hole in one end. This hole accommodates a regular-sized, square, high-speed steel or carbide-tipped tool bit. Boring bits may be held in a boring bar at 90°, 45°, and 30° angles. The tool bit may be secured in either end by means of a setscrew.

Another type is called an *end-cap* boring bar. The cutting tool is held in position by the wedging action of a hardened plug.

WEB-BAR BORING TOOLHOLDER

The *web-bar* boring toolholder is slotted on one end. A cutting tool is slipped into the slot. The nut on the opposite end is tightened to draw a tapered drawbar that transmits a tremendous clamping force on the boring bar slots. These slots grip the cutting tool so securely that the cutting tool will break before it moves in the bar.

Correct adjustment of a boring tool is particularly important. The boring bar and/or cutting tool must be clamped as short as possible to minimize the overhang. Greater tool rigidity is thus provided for the cutting process.

Courtesy of ARMSTRONG BROTHERS TOOL COMPANY

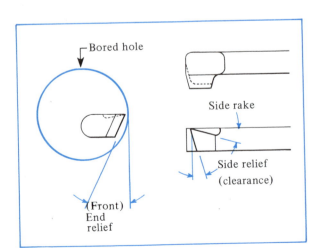

Figure 29–1 Rake and Relief Angles of a Solid, Forged Boring Tool

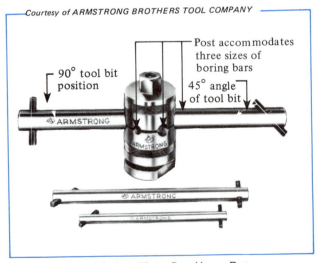

Figure 29–2 A Three-Bar, Heavy-Duty Boring Toolholder and Boring Bars

HEAVY-DUTY BORING BAR SET

The *heavy-duty boring bar set* consists of a combination toolholder/tool post and three sizes of boring bars. The availability of three different sizes permits the operator to use the largest (and strongest) size possible for a particular job. With greater strength in the boring bar, it is possible to take longer cuts at increased speeds.

Figure 29–2 shows a heavy-duty boring bar set. Note that the cutter may be positioned at 90°, 45°, and 30° in the three bars. The boring bar and cutting tool are positioned and secured by turning the nut. This nut clamps the boring bar in the holder body and secures the body in the compound rest T-slot.

BORING BAR AND
QUICK-CHANGE TOOLHOLDER

A boring bar may easily be set in a quick-change toolholder. One design of bar has three angular ends. These ends permit boring with the tool bit held at an angle or at 90° to the work axis. The quick-change toolholder may be slid

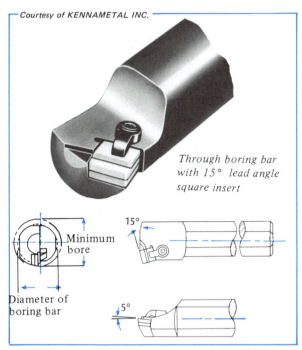

Courtesy of KENNAMETAL INC.

Through boring bar with 15° lead angle square insert

Minimum bore

Diameter of boring bar

15°

5°

Figure 29–3 Fixed-Head, Steel-Shank Boring Bar for Carbide Inserts

quickly in the dovetailed slots. The cutting tool may be adjusted vertically and locked in position.

BORING BARS FOR PRODUCTION WORK

Different types of boring bars are used with turret lathes and other semiautomatic turning machines. Such boring bars require throwaway carbide or ceramic inserts.

A fixed-head boring bar is illustrated in Figure 29–3. The boring bar uses square inserts for through boring. Another boring bar may be used for both threading and grooving by just changing the precision insert.

Other types of boring bars are also available. Holes may be bored to a square shoulder by using boring bars with triangular inserts. The shanks of these boring bars are made of steel or tungsten carbide. Tungsten carbide provides greater rigidity than standard steel boring bars.

CARBIDE INSERT SETUPS FOR BORING

The 90° slot position for the cutting tool is used for straight through boring. The cutting tool is set on center. Feeding is (right-hand) toward the spindle. A side rake of 5° to 7° provides a good cutting-edge angle. The amount of end relief depends on the inside diameter. The side-relief angle should be between 12° and 15°.

COUNTERBORING, RECESSING,
AND BORING TAPERS

Three other common boring operations, in addition to internal threading, are counterboring, recessing (undercutting), and taper boring.

For most workpieces *counterbored holes* are formed with a counterbore. On larger diameters the counterbored hole is formed by boring. The shoulder of a counterbored hole is machined either square or with a small corner radius.

An *internal recess* or *groove* requires a formed cutting tool. Usually a standard tool bit is ground to shape and used with a boring bar. Small-diameter recesses may be formed with solid, forged cutting tools.

Tapered holes are generally bored using a taper attachment. The process is similar to regular taper turning but with two major exceptions: (1) Although the tool is fed toward

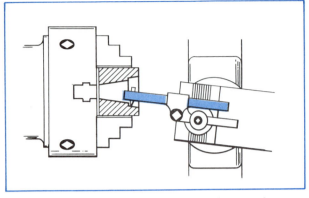

Figure 29-4 Boring Tool and Angle Setting of Compound Rest for Boring Short, Steep Tapers

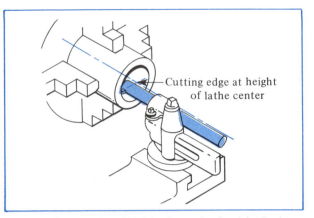

—Cutting edge at height of lathe center

Figure 29-5 Tool and Holder Setup for Straight Boring

the spindle, it is ground for turning left-hand; (2) the end-relief angle on smaller diameters must be increased to prevent the end of the cutting tool from rubbing.

The boring of *short steep tapers* combines angular turning with boring. The compound rest is positioned at the taper angle (Figure 29-4). The carriage is set in position. Angular feeding is done with the compound rest handwheel. The length of taper that can be bored is limited to the movement of the compound rest slide.

Large steep tapers are often cut using a standard right-hand offset toolholder. A standard toolholder provides greater rigidity than a boring bar. It is preferred wherever the work size and operation permit boring with it.

How to Bore a Hole

Setting Up the Lathe and Workpiece

STEP 1 Mount the workpiece. Position the workpiece out from the chuck or faceplate to bore a through hole.

STEP 2 Face the end. Center drill to within 1/32″ (0.8mm) of the finished diameter.

Note: Large-diameter holes are often cast or roughly cut out with a torch. Such surfaces are often machined directly by boring. In the case of cast iron, the first roughing cut must be below the scale.

STEP 3 Set the spindle speed. The same RPM should be used as for outside turning the same diameter.

STEP 4 Set the feed. For rough boring the feed is decreased because of the nature of the boring operation and the tendency of the boring tool to spring under heavy cuts.

Straight-Hole Boring

STEP 1 Mount the boring bar in the toolholder, with the cutting tool as close to the toolholder as possible. The boring bar length should permit clearing the depth of the hole to be bored.

STEP 2 Set the cutting edge at center height (Figure 29-5). Check to see that the heel of the cutting tool clears the diameter.

STEP 3 Start the lathe. Move the cutting tool to take a trial cut for a short distance.

STEP 4 Stop the lathe. Measure the bored diameter. Measurements may be taken with a steel rule, inside caliper, telescoping gage, inside micrometer, or a vernier caliper (Figure 29-6).

STEP 5 Position the boring tool to take the deepest possible cut with the coarsest feed.

Note: If chatter marks are produced, check the correctness of the cutting and relief angles of the boring tool. Reduce the speed, feed, and depth of cut if necessary.

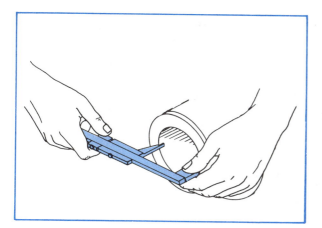

Figure 29–6 Measuring the Diameter of a Bored Hole with a Vernier Caliper

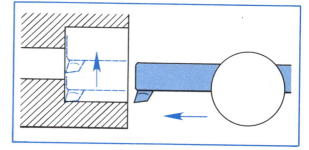

Figure 29–7 Finish Boring the Diameter and Face

STEP 6 Take a trial finish cut with a fine feed for about 1/8″. Measure the diameter. Adjust the depth of cut if required.

STEP 7 Apply cutting fluid and take the finish cut.

Note: The diameter of the bored hole should be checked at several places along its length to be sure the hole is not bell mouthed. A bell-mouthed hole indicates the tool is springing away from the work. Take an additional cut or two at the same setting to produce a parallel bored hole.

STEP 8 Break the sharp edge of the bored hole with a triangular hand scraper.

Boring a Counterbored Hole

STEP 1 Use the largest practical diameter of boring bar. Grip the boring bar short.

STEP 2 Insert a left-hand boring tool. Rough bore the counterbored area.

Note: The cutting tool is usually held in the 30° or 45° position. These positions permit the cutting tool to cut close to the shoulder.

STEP 3 Step off the area adjacent to the shoulder with a smaller-radius tool point.

Note: Stepping off reduces the amount left for final machining of the shoulder area.

STEP 4 Measure the diameter and depth of the counterbored hole with either a steel rule, depth gage, depth micrometer, or a vernier depth gage, depending on the required tolerance.

Note: The depth is sometimes reached using a carriage stop. Also the compound rest axis may be set parallel to the lathe axis. The boring bar is then fed by the compound rest handwheel to the specified depth.

STEP 5 Stone the nose and cutting point for the final cut. Increase the speed; decrease the feed. Take a light finish cut along the inside diameter (Figure 29–7). Disengage the power feed before the shoulder is reached.

STEP 6 Feed the tool to the shoulder by hand. Then feed inwardly to take a light finish cut across the face of the shoulder.

STEP 7 Remove the burr on the outside edge. Recheck the diameter and the depth.

Undercutting (Internal Recessing)

STEP 1 Grind a boring tool to the shape of the required groove or recess.

STEP 2 Swivel the compound rest to 0°. Check the setting. The compound rest should be parallel to the lathe axis.

STEP 3 Secure the tool bit in the boring bar. The boring bar should project far enough from the holder to clear the workpiece.

STEP 4   Set the spindle speed for the diameter of the workpiece and material to be bored. Use the same cutting speed as for a cutting-off process.

Note: To avoid chatter, the speed may need to be decreased if the boring bar or solid, forged tool bit overhangs the workpiece for any distance.

STEP 5   Position the left side of the cutting tool so that it almost grazes the flat face of the workpiece.

STEP 6   Set the graduated collar on the compound rest at zero. Lock the carriage in position.

STEP 7   Start the lathe. Move in the compound rest the required distance.

STEP 8   Apply a cutting fluid. Feed the cutting tool until it touches the inside diameter. Set the cross slide micrometer dial to 0°. Move the cutting tool in to depth.

Note: The cutting tool is moved sideways 0.002″–0.003″ in the groove to free up the sides the same as in cutting an external groove.

STEP 9   Use a triangular hand scraper. Remove any burrs formed on the edges of the undercut section and the outside diameter.

MANDRELS AND MANDREL WORK

Many workpieces that are bored or reamed require further machining on the outside surface. Such workpieces may be positioned accurately in relation to the work axis by using a *mandrel*. A mandrel is a hardened, cylindrical steel bar that is pressed into the finished hole of a workpiece.

Two common types of mandrels are the *solid* and the *expansion* mandrel. Other types include the *gang*, *threaded*, and *taper-shank* mandrel.

THE SOLID MANDREL

A mandrel is a hardened, cylindrical steel bar. The solid mandrel has a recessed center hole in each end. The ends are turned smaller than the body size. A flat is machined to provide a positive clamping surface. The setscrew of a lathe dog may be tightened against this surface. The large end has the mandrel size stamped on it. This marking also indicates the end on which the lathe dog should be clamped.

The body is ground with a slight taper of 0.0005″ per inch of length. The small end of mandrels under 1/2″ diameter (12mm) is usually a half-thousandth of an inch under the standard diameter. On large mandrels the small end is ground up to 0.001″ undersize. Due to the taper the large end is a few thousandths of an inch larger than the normal diameter. The accuracy of a mandrel depends on the condition and accuracy of the center holes.

THE EXPANSION MANDREL

The *expansion mandrel* accommodates a wider variation in hole sizes than the solid mandrel (Figure 29–8). The expansion mandrel consists of a taper mandrel and a slotted sleeve. The taper bore of the sleeve corresponds to the taper of the mandrel. The expansion mandrel may be expanded from 0.005″ to 0.008″ (0.1mm to 0.2mm) over the nominal size for diameters up to 1″ (25mm). The slotted sleeves come in different diameters. The same taper mandrel may be used with more than one diameter sleeve.

THE GANG MANDREL

The *gang mandrel* provides for the machining of multiple pieces. The gang mandrel has a flanged, parallel-ground body and a threaded end. Workpieces are placed side by side, and a collar and nut are tightened to hold them in place for machining.

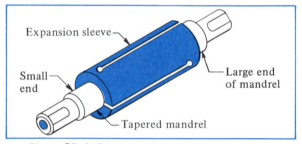

Figure 29–8 Example of An Expansion Mandrel

The gang mandrel is mounted between centers. A lathe dog, tightened against the flat surface on one end, provides the drive force.

THREADED AND TAPER-SHANK MANDRELS

The *threaded mandrel* is used for mounting threaded parts that are to be turned. The workpiece is screwed onto the threaded end of the mandrel. The recessed thread and square flange permit the workpiece to be mounted squarely.

The threaded mandrel may be designed for chuck work or for machining work between centers. The *taper-shank mandrel* may be fitted to the headstock spindle by using an adapter. An adaptation of the taper-shank mandrel has the end turned straight. This end is slotted. A special flat-head screw applies a force against the slotted segments. These segments, in turn, hold the workpiece.

MOUNTING AND REMOVING WORK ON A MANDREL

THE ARBOR PRESS

A mandrel is pressed into the finished hole of a workpiece. The force is sufficient to permit the machining of the outside surfaces of the workpiece. Sometimes the mandrel is driven into position by a soft-faced hammer. The workpiece is placed on a mandrel block and the mandrel is driven in.

When work is turned on a mandrel, the cutting tool is set so that the cutting force is directed toward the large end of the mandrel. Light cuts are taken on large-diameter work to prevent the work from turning on the mandrel.

An *arbor press* is designed for mounting work on mandrels and arbors. Figure 29–9 shows one common type of arbor press. The workpiece is mounted on the table plate. Force is applied to the lever. This force is multiplied through a pinion gear to the rack teeth on the ram. The force is applied by the ram to move the mandrel into the bored or reamed hole. The arbor press is also used widely in the assembling of shafts, pins, bushings, and other parts that require a force fit.

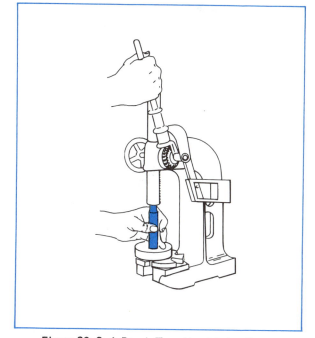

Figure 29–9 A Bench Type Hand Arbor Press

How to Remove Work from a Mandrel

STEP 1 Place the workpiece on the table plate. Use parallels or a disk between the work face and table plate if added support is needed.

Note: The large diameter (stamped size end) of the mandrel must face downward.

STEP 2 Center the small end of the mandrel under the ram. Check to be sure that the mandrel and workpiece are positioned squarely.

STEP 3 Hold the mandrel with the left hand. Apply force on the lever with the right hand.

Safe Practices in Internal Boring

- Clamp the boring bar so that the cutting tool is as close to the holder as possible. Check to see that the head end clears without rubbing into the chuck or the end of the workpiece.
- Grind the cutting point (front) relief angle steep enough so that the heel will not rub on the workpiece.
- Stop the lathe before taking a measurement.

- Position the nozzle so that the lubricant flows inside a bored hole.
- Shorten the holder length if chatter is produced. Also decrease any or all of the following: speed, feed, and depth of cut.
- Burr the edge of a bored hole or recessed area with a triangular scraper.
- Stop the lathe to remove chips and clean a bored hole.
- Use a protective shield or safety goggles and observe all machine safety practices.

TERMS USED IN BORING PROCESSES AND MANDREL WORK

Boring tool	A standard tool bit ground for an internal turning process. A solid-shank, forged cutting tool that is offset and ground to form a cutting tool.
Boring bar	A round or square-shaped steel bar. A bar for positioning and holding high-speed steel, carbide-tipped, or ceramic inserts. A steel bar that is secured in a tool post or other holder. A steel bar that is slotted or has a square-shaped hole to receive a cutting tool.
Bored tapered hole	A hole that regularly increases in diameter a given amount over a specified distance. A steep angle or a small taper angle produced by turning the inside of a hole.
Counterboring (turning)	Producing an enlarged bored hole. A bored hole that is separated by a shoulder from a smaller but concentric hole.
Undercutting	Boring a groove or recess internally on a workpiece.
Mandrel	A work-holding device on which a bored part may be mounted and secured for subsequent turning operations. Generally a hardened, ground, and centered tool having a body that tapers 0.0005″ per inch.
Arbor press	A device consisting of a ribbed frame, table, table plate, and movable ram. A device for applying force to assemble or disassemble mating parts.

SUMMARY

- Boring is the process of enlarging and producing a concentric and straight hole.
  - Boring produces a quality-finished internal surface. This surface may be machined accurately to any standard or odd-sized diameter.
- Forged and regular square-shaped high-speed and carbide-tipped tool bits are widely used for general-purpose boring processes. The forged-type solid cutting point and shank are largely used for boring small diameters.

■ The end of a standard boring bar holder is broached to receive a tool bit. The bit may be positioned at a 30°, 45°, or 90° angle to the boring bar.

■ Boring bars and holders are fitted with interchangeable ends. The bars are of different diameters.

■ Boring bars may also be positioned for rapid tool changes. The cutter and bar may be held in a quick-change toolholder.

■ Heavy-duty boring bars have a clamping device on a fixed head. Square, triangular, and rectangular inserts may be inserted and clamped in position.

■ Solid tungsten carbide boring tools provide maximum rigidity.

■ Long, tapered holes are bored with a taper attachment. Short and steep tapers are bored by swinging the compound rest to the required angle. The cutting tool is fed by the compound rest handwheel.

■ Internal recessing and undercutting require cutting tools with shapes similar to the tools used for external processes.

■ The external surfaces of a bored part may be machined by first mounting the part on a solid or expansion mandrel. Multiple pieces may be machined on a gang mandrel. Threaded mandrels are used for mounting finish threaded parts.

UNIT 29 REVIEW AND SELF-TEST

1. Determine and state the amount of material to be allowed for machine reaming the following bored holes: (a) 25mm diameter, (b) 2″, and (c) 90mm.

2. Name four groups of cutting tools that are used for boring processes.

3. Give two advantages of a heavy-duty boring bar set compared to a plain boring bar.

4. a. Identify three different boring operations performed with throwaway carbide or ceramic inserts.
b. Name the shape of the insert that is used for each operation.

5. State two differences between grinding a cutting tool for taper boring and for regular turning.

6. List the steps for boring a counterbored hole.

7. Describe undercutting (internal recessing).

8. a. List five common types of lathe mandrels.
b. Cite the advantage of a multiple-piece work-holding mandrel compared to a mandrel used for an individual part.

9. Indicate two different applications of arbor presses.

10. State two safe practices that must be followed to prevent damage to work that is mounted on a mandrel.

11. Indicate corrective steps to take if chatter is produced during boring.

Cutting and Measuring 60° Form External and Internal Screw Threads

Thread cutting was related in earlier units to the hand tapping of internal threads and the cutting of external threads with a die. Many threads are machined on a lathe with a single-point thread-cutting tool. This unit covers the gages needed to measure the different cutting-tool angles or the thread pitch. The lathe set-ups and tools; the cutting of single-pitch, right- and left-hand (internal and external) threads; and new terminology are also included. Thread measurements are applied. Formulas are used to compute required thread dimensions.

OBJECTIVES

After satisfactorily completing this unit, you will be able to:

- Relate American National, Unified, and SI Metric (60° thread angle) forms to machine setups and cutting of screw threads on a lathe.
- Set up thread cutting tools, attachments, gearing, and spindle speeds.
- Understand thread measurement and inspection with thread gages, micrometers, optical comparators, and toolmaker's microscope.
- Infeed straight and at an angle to cut a thread to depth using the compound rest.
- Apply principles of outside and inside threading to cut taper threads.
- Solve shop problems using formulas of one- and three-wire measurements, helix angles, and total cutting tool clearance.
- Perform the following processes.
    - Setting Up a Lathe for Right- and Left-Hand Thread Cutting.
    - Cutting American National, Unified, and SI Metric (60°) Form Threads.
    - Cutting Internal Threads.
- Follow recommended *Safe Practices* and apply new *Terms* on cutting and measuring internal and external threads.

GAGES, CUTTING TOOLS, AND HOLDERS

CENTER GAGE

A small, flat gage called a *center gage* is used to check the accuracy of the sides of a thread-cutting tool. This gage has a series of 60° angles. The parallel edges have a number of fractional graduations. The graduations are used to measure thread pitches. The center gage is also used to position a thread-cutting tool in relation to the axis of a workpiece.

SCREW THREAD GAGE

Another type of *screw thread gage* is a circular disc with a series of V-shaped (thread form) openings around the circumference (Figure 30–1). The thread sizes on this gage conform to the standards for American National or Unified Threads. It is used for checking Acme threading tools. A similar flat plate gage is available for checking SI metric thread-cutting tools.

Figure 30–1 Circular (60°) V-Thread Form Gage

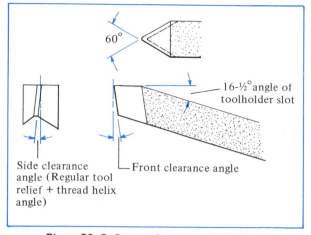

Figure 30–2 Correct Clearance Angles for a Right-Hand Thread-Cutting Tool

THREADING (THREAD-CUTTING) TOOL

A *threading tool* is a single-point cutting tool that has an included angle and a point shape that meet a specific thread form standard. The thread-cutting tool requires that the flank of the side-cutting edge be ground with additional clearance. The cutting edges are therefore ground at an angle equal to the *thread helix angle* plus the *regular relief angle*. The clearance angle prevents the flank of the cutting tool from rubbing against the sides of the threads as the tool advances along a workpiece.

Figure 30–2 shows a correctly ground right-hand thread-cutting tool. The point is rounded slightly for a Unified thread. The front clearance angle permits the cutting tool to clear the diameter of a revolving workpiece. The top face is ground at the angle of the cutting-tool holder. When secured in the toolholder at center height for threading, the top face is horizontal. The clearance angle for cutting a left-hand thread is formed on a left (side) cutting flank.

SPRING-HEAD THREAD-CUTTING TOOLHOLDER

A *spring-head* thread toolholder (Figure 30–3) is generally used with a standard high-speed steel square tool bit. The spring head may be tightened with a locking nut to permit the taking of heavy and roughing cuts. When loosened, the holder has a spring feature. This feature is especially desirable for finishing threads.

FORMED THREADING TOOL

A circular blade ground to a particular thread form is called a *formed threading tool*. The blade has appropriate relief and clearance angles. The circular-form cutter is held securely against the side face of the toolholder. A hardened stop screw is used to adjust the cutter at center height.

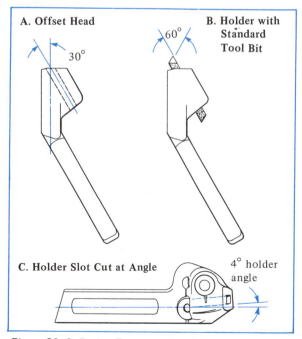

Figure 30–3 Design Features of an Offset Spring-Head Holder for a Thread-Cutting Tool

Table 30–1 Positions of the Thread-Chasing Dial to Engage the Split Nut for Cutting Threads

Nature of Threads per Inch to Be Cut	Graduation for Engaging the Thread-Chasing Dial	
Even number of threads	0 (dial graduated 1, 2, 3, 4)	Any graduation 1/2 1 1 1/2 2 2 1/2 3 3 1/2 4
Odd number of threads	0 (dial graduated 1, 2, 3, 4)	Any numbered graduation 1 2 3 4
Fractional number of threads	0 (dial graduated 1, 2, 3, 4)	Every other odd or even graduation 1/3 or 2/4
Threads that are a multiple of the lead screw pitch		Any point where the split nut meshes

THREAD-CHASING ATTACHMENT

A *thread-chasing attachment* is a threading device that is attached to the lathe carriage. Its function is to locate a position at which a lead screw may be engaged or disengaged. The exact point of engagement permits the threading tool to follow in the helix of the previously cut groove.

The device has a dial with lines and numbers. Table 30–1 gives the even and odd lines on the chasing dial. The correct engagement of the split nut on the lead screw is indicated for even, odd, and half threads.

SETTING THE LEAD SCREW

The amount a lead screw moves in relation to each revolution of the lathe spindle is controlled by gear combinations in the quick-change gearbox. The correct lead screw setting may be made by positioning the gears as indicated on the index plate. There are usually two levers to be positioned. A third lever controls the direction the lead screws turns. One position of this lever is for cutting right-hand threads. The second position disengages the lead screw. The third position reverses the direction of rotation to cut left-hand threads.

Caution: The feed rod must be disengaged when cutting threads. Otherwise the feed may be accidentally engaged, thus producing excessive forces on the feed screw. Damage may result.

SPINDLE SPEEDS FOR THREAD CUTTING

The spindle speeds required for threading are slower than the speeds used for turning. The spindle speed for cutting coarse threads on 3/4″ (18mm) and larger diameters is one-fourth the speed for turning. A faster speed, about one-third to one-half the speed for turning, is used for fine pitches and smaller diameters. The cutting speed may be increased still further when machining brass, aluminum, and other soft materials. Steels that are tougher and harder than low-carbon (soft-cutting) steels require slower speeds.

DESIGN FEATURES AND THREAD FORM CALCULATIONS

THREAD CALCULATIONS

A series of formulas is used to compute the different dimensions of screw threads. While the letters designating a design feature may be different, the same values are included in the formulas. Table 30–2 shows four standard basic thread forms and a few of the formulas for each form. There are five important dimensions not shown in Table 30–2: major, minor, and pitch diameters and normal and actual size.

The *pitch diameter* is important in design and measurement. It represents the diameter at that point of the thread where the groove and the thread widths are equal. The pitch diameter is equal to the *major diameter* minus a single thread depth. Thread tolerances and allowances are given at the pitch diameter.

The *minor diameter*, formerly called *root diameter*, is the smallest thread diameter. Minor diameter applies to both external and internal threads.

Table 30–2 Partial Set of Formulas for Common Thread Forms

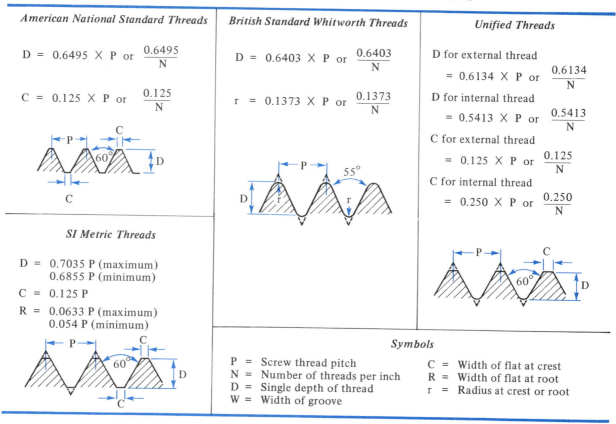

American National Standard Threads

$$D = 0.6495 \times P \text{ or } \frac{0.6495}{N}$$

$$C = 0.125 \times P \text{ or } \frac{0.125}{N}$$

British Standard Whitworth Threads

$$D = 0.6403 \times P \text{ or } \frac{0.6403}{N}$$

$$r = 0.1373 \times P \text{ or } \frac{0.1373}{N}$$

Unified Threads

D for external thread
$$= 0.6134 \times P \text{ or } \frac{0.6134}{N}$$

D for internal thread
$$= 0.5413 \times P \text{ or } \frac{0.5413}{N}$$

C for external thread
$$= 0.125 \times P \text{ or } \frac{0.125}{N}$$

C for internal thread
$$= 0.250 \times P \text{ or } \frac{0.250}{N}$$

SI Metric Threads

$$D = \begin{array}{l} 0.7035 \text{ P (maximum)} \\ 0.6855 \text{ P (minimum)} \end{array}$$

$$C = 0.125 \text{ P}$$

$$R = \begin{array}{l} 0.0633 \text{ P (maximum)} \\ 0.054 \text{ P (minimum)} \end{array}$$

Symbols

P = Screw thread pitch
N = Number of threads per inch
D = Single depth of thread
W = Width of groove

C = Width of flat at crest
R = Width of flat at root
r = Radius at crest or root

ALLOWANCE, TOLERANCE, LIMITS, AND SIZE

Screw threads are machined to various specifications. These specifications depend on the application, the material from which the part is made, the method of generating the thread form, and other factors. The size and fit depend on *allowance*, *tolerance*, and *limits*.

Allowance refers to the difference allowed between the largest external thread and the smallest internal thread. Allowance is an intentional difference between mating parts. The allowance may be positive (clearance) so that parts will fit freely. A negative allowance is specified for parts that must be assembled with force (force fit). Thus, a given allowance produces the tightest acceptable fit. Reference tables are used to establish allowances. Maximum and minimum pitch diameters are given for any classification of fit.

Example: The allowance for a 1″–8UNC class 2A (outside) and 2B (inside) fit is the difference between the minimum pitch diameter of the inside thread and the maximum pitch diameter of the outside thread.

Tolerance, as defined earlier, is the acceptable amount a dimension may vary in one direction or in both directions. The allowable tolerance for threads in the American National and the Unified thread systems is plus (+) on outside threads and minus (−) on inside threads.

Example: The tolerance for the 1″–8UNC class 2A (external) thread equals the:

maximum pitch diameter	=	0.9168″
minimum pitch diameter	=	0.9100″
tolerance (or acceptable variation)	=	0.0068″

Limits represent maximum and minimum dimensions. In the previous example the upper limit of the pitch diameter for a class 2A fit is 0.9168″. The lower limit is 0.9100″.

The *basic size* is the theoretical exact size of the designated thread. It is from the basic size that size limitations are made. The basic size of a 1 3/8″–6NC thread is 1.375″. The thread notation on a part drawing indicates the 1 3/8″ size.

The *actual size* is the measured size.

THREAD MEASUREMENT AND INSPECTION
SCREW THREAD MICROMETER

The *screw thread micrometer* measures the pitch diameter (Figure 30–4A). These micrometers are designed to measure 60°-angle threads in the inch-standard or SI metric-standard systems. Screw thread micrometers have a 1″ (25mm) range. Different micrometers are required in measuring American National and Unified threads. For instance, four micrometers of one manufacturer cover the range from 8 to 40 threads per inch (T_{pi}).

- 8 to 13 T_{pi}
- 14 to 20 T_{pi}
- 22 to 30 T_{pi}
- 32 to 40 T_{pi}

The screw thread micrometer spindle has a 60° conical point. The swivel anvil has a corresponding cone shape. The line drawing in Figure 30–4B shows the shapes and the point of contact. The micrometer measurement is taken at the pitch diameter of the thread, along the helix plane. A slightly inaccurate measurement is produced. For extremely precise threads the helix angle must be considered. For these threads the micrometer is usually set with a master *thread plug gage*. The threaded part may then be measured precisely.

THREAD COMPARATOR MICROMETER

The *thread comparator micrometer* compares a thread measurement against a thread standard. The micrometer has a 60° cone-shaped spindle and anvil. The micrometer reading is established by first gaging a standard thread plug gage.

OPTICAL COMPARATOR

The common *optical comparator* magnifies a part from 5 to 250 times and projects it upon a screen. Design features of a thread are measured in one step. The measurements include: form; major, minor, and pitch diameters; pitch; and lead error. The surface illuminator of the comparator permits examining the surface finish of the thread.

Measurements are established by using a micrometer or end measuring rods on the comparator. Angles are measured by rotating the hairlines on the viewing screen. A vernier scale permits the making of angular measurements to an accuracy of one minute of one degree. Sometimes a chart is substituted for the viewing screen.

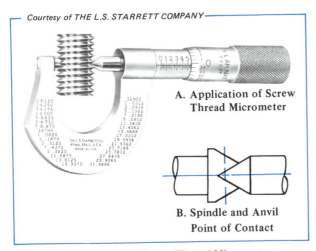

Courtesy of THE L.S. STARRETT COMPANY

A. Application of Screw Thread Micrometer

B. Spindle and Anvil Point of Contact

Figure 30–4 Using a Screw Thread Micrometer to Measure the Pitch Diameter of a Screw Thread

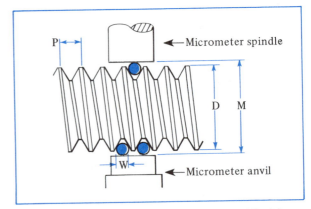

Figure 30–5 Symbols, Wire Placement, and Position of the Micrometer Spindle for Three-Wire Measurement

TOOLMAKER's MICROSCOPE

The *toolmaker's microscope* makes it possible to measure the same thread features as can be measured with the optical comparator. With this microscope the table is moved by precision lead screws. A micrometer thimble is used to turn each lead screw. Measurements, read directly from the graduations on the thimble, are within 0.0001″ (0.002mm).

THREE-WIRE METHOD

The *three-wire method* is used to check the pitch diameter of 60°-angle screw threads in the American National, Unified, or SI Metric systems.

The three-wire method is recommended by the National Bureau of Standards and the National Screw Thread Commission. The method provides an excellent way of checking the pitch diameter. Any error in the included thread angle has a very limited effect on the pitch diameter.

Three wires of the same diameter are required. Two of these wires are placed in thread grooves on one side. The third wire is placed in a thread groove on the opposite side. The wires and threads are positioned between the anvil and spindle of a standard micrometer (Figure 30–5).

Best Wire Size. The wire size to use depends on the thread pitch. There are three possible wire sizes that may be used: *largest, smallest,* and *best.* The best wire size is recommended.

Table 30–3 Formulas for Wire Size and Three-Wire Measurement

Dimension	Formula
Measurement over wires (M)	$M = D + 3W_b - \dfrac{1.5155}{N}$
Largest wire size (W_l)	$W_l = \dfrac{0.900}{N}$ or 0.900P
Best wire size (W_b)	$W_b = \dfrac{0.5155}{N}$ or 0.5155P
Smallest size wire (W_s)	$W_s = \dfrac{0.540}{N}$ or 0.540P

Symbols

D = Major thread diameter
W_l = Largest diameter of wire
W_b = Best wire diameter
W_s = Smallest wire diameter
N = Number of threads per inch
M = Three wire measurement

Calculating Wire Size and Thread Measurement. Four formulas for measuring American National and Unified threads are given in Table 30–3.

CALCULATING THE LEADING AND FOLLOWING SIDE ANGLES FOR CUTTING TOOLS

Tables are usually used to establish the clearance angles for the leading and following side-cutting edges for all forms of threads. There are times, however, when these angles must be calculated.

The helix angle of the *leading side* is represented by the

$$\text{tan of leading-side angle} = \frac{\text{lead of thread}}{\text{circumference of } \textbf{minor} \text{ diameter}}$$

The tangent of the *following side* is equal to the lead divided by the circumference of the major diameter:

$$\text{tan of following-side angle} = \frac{\text{lead of thread}}{\text{circumference of } \textbf{major} \text{ diameter}}$$

The helix angle is found by using a table of natural trigonometric functions to convert the numerical value of the tangent to the angle equivalent, which is usually given in degrees (°) and minutes (′).

Usually a relief angle of 1° is *added to* the helix angle for the leading side. One degree is *subtracted from* the helix angle for the following side.

DEPTH SETTINGS FOR AMERICAN NATIONAL FORM (60°) THREADS

The feeding of a thread-cutting tool is usually done by turning the compound rest handwheel. Some threads are cut by directly infeeding the cutting tool with the compound rest set at 0″. Where a precision form thread is to be produced, the thread-cutting tool may be fed with the compound rest set at 30°. The tool is fed at the 30° angle to thread depth. However, common shop practice is to set the compound rest

Table 30–4 Examples of Depth Settings for American National Form Threads

Threads per Inch	Compound Rest Angle Setting and Depth of Feed* (in thousandths of an inch)		
	0°	29°	30°
4	0.1625	0.1858	0.1876
6	0.1080	0.1235	0.1247
56	0.0116	0.0133	0.0134
64	0.0101	0.0115	0.0117

at 29° to provide a slight angular clearance. The right side of the cutting tool just shaves the thread to produce a fine finish.

Table 30–4 gives examples of the depth-of-feed for each thread for compound rest settings of 0°, 29°, and 30°.

Note: The thread depths are based on the correct width (0.125P) of the flat cutting-tool point. Unless this width is accurate, the depth readings will not be correct.

How to Set Up a Lathe for Thread Cutting

Preparing the Workpiece

STEP 1 Machine the workpiece to the major diameter of the thread.

STEP 2 Chamfer the end to a 30° or 45° angle.

STEP 3 Measure and mark the length of the threaded portion.

STEP 4 Undercut the length (if possible). Cut the groove to thread depth (minor diameter).

Note: Sometimes the groove is cut a few thousandths of an inch deeper. The groove simplifies the threading process. Ending the thread in a groove ensures that the mating part fits over the required length or shoulders properly.

STEP 5 Check the maximum diameter, chamfered end, and undercut for thread length and depth.

Preparation for Thread Cutting

STEP 1 Check the angle of the thread-cutting tool with a center gage (60°).

STEP 2 Check the clearance angles with a gage or protractor.

Note: The leading and following sides must compensate for the helix angles.

STEP 3 Select the appropriate type of thread-cutting tool and holder.

STEP 4 Secure the cutting tool in a holder. Clamp the toolbit as close to the holder as possible.

STEP 5 Set the cutting edge at center height.

STEP 6 Place the center gage against the workpiece.

STEP 7 Position the thread-cutting tool in the center gage to set the point at right angles to the workpiece (Figure 30–6).

Note: A light paper placed on the cross slide makes it easier to see any spacing (light) between the cutting edges and the center gage. If necessary, gently tap the tool to center it.

STEP 8 Tighten the toolholder in the tool post. Recheck the accuracy of the tool setting to see that all moving parts clear.

STEP 9 Check the clearance of all moving parts. The clearance must be adequate to prevent interference at both the start and the finish of each cut.

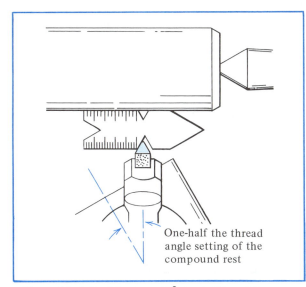

One-half the thread angle setting of the compound rest

Figure 30–6 Positioning a 60°-Angle Thread-Cutting Tool by Using a Center Gage

STEP 10 Determine the cutting speed and spindle RPM. Set the spindle at this speed.

STEP 11 Engage the lead screw gears.

STEP 12 Position the lead screw lever to turn in the correct direction for a right- or left-hand thread.

STEP 13 Disengage the feed rod by moving the feed-change lever to the neutral position.

How to Cut an Outside Right- or Left-Hand Thread

Cutting an Outside Right-Hand Thread

STEP 1 Back the thread-cutting tool away from the workpiece.

STEP 2 Start the lathe. Bring the cutting tool in until it just touches the workpiece.

Caution: A beginner should use a slow spindle speed until all steps are coordinated.

STEP 3 Set the micrometer collars on the cross feed and compound rest feed screws at the zero setting. Lock the collars at this setting.

STEP 4 Move the carriage until the cutting tool clears the workpiece.

STEP 5 Turn the compound rest handwheel to feed the tool to a depth of 0.002″ to 0.003″.

STEP 6 Engage the half nut when the correct line on the chasing dial reaches the index line.

STEP 7 Turn the cross feed handwheel counterclockwise as quickly as possible at the end of the cut to clear the tool of the thread groove.

STEP 8 Disengage the split nut at the same time as the cross feed handwheel is turned counterclockwise.

STEP 9 Stop the lathe. Return the carriage to the starting position. Check the pitch.

STEP 10 Start the lathe. Feed the threading tool for a roughing cut. Set the depth of cut with the compound rest.

Note: Engage the half nut when the chasing dial reaches the index line. Apply a cutting fluid over the work surface with a brush.

STEP 11 Continue to take roughing cuts. The depth of each successive cut should be decreased.

Note: Determine in advance the final depth reading on the compound rest micrometer collar.

STEP 12 Take the last two cuts as finish cuts. The depth may be set for 0.003″ (0.08mm), then 0.002″ (0.05mm).

Note: In some instances the finish thread-cutting tool is fed in the last 0.001″ (0.025mm) by infeeding at 30° or 90° (Figure 30-7).

STEP 13 Check or measure the thread size.

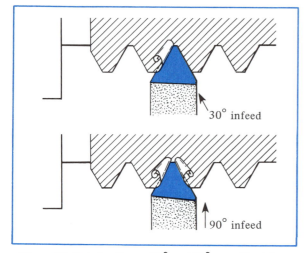

Figure 30–7 Infeeding at 30° and 90° on a Final Cut

Cutting an Outside Left-Hand Thread

STEP 1 Prepare the workpiece and set the cutting tool as for cutting a right-hand thread.

STEP 2 Set the compound rest at an angle of 30° (or 29°) for a 60°-angle thread, a 14 1/2° angle for an Acme thread. Set the compound rest axis at 0° (parallel to the lathe axis) for a square thread. The compound rest is set to the left of the cross slide.

STEP 3 Position the lead screw lever for reverse. As the split nut is engaged, the cutting tool feeds from left to right.

STEP 4 Position the cutting tool to take a light trial cut. Set the graduated collars at zero. Move the cutting tool to the left end of the threaded section.

STEP 5 Engage the half nut at the appropriate mark on the chasing dial. Take the trial cut.

STEP 6 Proceed to rough out and finish cut the thread. Check it with the same procedures used for checking a right-hand thread.

Resetting the Thread-Cutting Tool

Note: Whenever the workpiece is removed from a chuck, holding fixture, or lathe dog or the cutting tool is changed, the thread-cutting tool must be reset.

STEP 1 Repeat the steps for setting up the workpiece and tool in preparation for cutting the thread.

STEP 2 Close the split nut at the correct chasing dial position. Feed the carriage a short distance along the workpiece.

STEP 3 Stop the lathe. Move the cross feed and compound rest handwheels. Keep adjusting the cutting-tool position until the sides and point are congruent with (match) the previously cut thread.

Note: If the thread cutting tool edges touch the thread surfaces, the new cutting edges will be damaged.

STEP 4 Reset the compound rest and cross feed micrometer collars at zero. Back the cutting tool out of the thread groove. Disengage the split nut. Return the cutting tool to the beginning of the thread.

STEP 5 Continue to take roughing and/or finishing cuts.

Note: The skilled craftsperson often determines the depth of cut by revolving the workpiece and turning the compound rest handwheel until the point of the cutting tool just grazes the major diameter. The micrometer collar reading is then subtracted from the zero to which the tool was set for depth. The resultant micrometer collar reading represents the depth within an accuracy of approximately 0.002″ to 0.003″ (0.05mm to 0.08mm).

How to Clean Up the Back Side of the Thread

The cutting tool is fed to depth for each successive cut by turning the compound rest handwheel. Since the compound rest is set at a 29° or 30° angle, one side-cutting edge and the nose of the tool do the cutting. One chip is formed and flows freely over the top of the tool. If both side-cutting edges and the point of a cross slide infeed were used, the cutting tool would tend to tear the thread and produce a rough surface finish.

Often, the side opposite the thread flank that is being cut (the back side) is rough. The second side-cutting edge of the threading tool is then used to clean up the back side. The process of cleaning up the back side of a right-hand thread is shown in Figure 30-8.

LATHE SETUP FOR CUTTING SI METRIC THREADS

INCH-STANDARD AND SI METRIC LATHES

The ratios in the gear trains of late-model lathes permit thread cutting for the inch-standard and SI metric systems. The spindle and lead screw gears in the end gear train do not need to be replaced to establish the 50/127 ratio between the inch-standard and metric systems of measurement (1″ to 2.54 cm).

The levers used to obtain different spindle speeds, feeds, and lead screw movements and to reverse the direction are mounted on the headstock and quick-change gearbox. The lever for starting, stopping, and reversing the lathe is located on the apron.

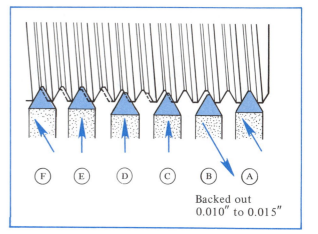

Figure 30-8 Successive Infeeding Cuts Taken to Clean Up the Back Side of a Thread

Backed out
0.010″ to 0.015″

How to Cut Right- or Left-Hand Metric Threads

STEP 1   Position the quick-change gearbox levers to produce the required SI metric pitch.

STEP 2   Prepare the workpiece and set up the cutting tool. Follow the same procedures used for cutting right- and left-hand American National and Unified threads.

STEP 3   Take the usual light trial cut. At the end of the cut, back the cutting tool out of the workpiece. Keep the half nut engaged throughout the thread-cutting process.

STEP 4   Reverse the direction of the spindle to return the cutting tool to the starting position. Stop the lathe.

STEP 5   Feed the cutting tool to take successive roughing and finishing cuts. Follow the same steps used for cutting and measuring nonmetric threads.

INTERNAL THREADING

Two common methods of cutting internal threads include *tapping* and *machine threading*. Larger and coarser thread sizes are sometimes roughed out with a single-point thread-cutting tool. A tap is then used to cut the thread to size.

THREADING USING A SINGLE-POINT CUTTING TOOL

A center gage or Acme thread gage, boring bar or other form of toolholder, and a forged or standard cutting tool are needed for internal threading. A bored hole is preferred because it is concentric.

A groove (recess) should be cut a few thousandths of an inch below the maximum thread diameter for blind holes. The groove width should be about 1 1/2 times the thread pitch. If the part permits, a shallow recess may be cut to the same depth on the end of the workpiece. The outside of the bored or recessed hole should be chamfered 14 1/2° to accommodate the Acme thread form and 30° for 60°-angle thread forms. The chamfer reduces burrs. Also, the mating threads fit easily. The inside and end recesses and the chamfered end of a bored workpiece are illustrated in Figure 30-9.

THE INTERNAL THREAD-CUTTING TOOL BIT

The cutting tool is ground to the required included angle and width. Both the leading and following cutting edges must have regular relief plus clearance to compensate for the thread helix angle. The required clearances vary with the major and minor diameters of the threads. The top face is flat, without back rake.

The largest size of boring bar possible is used to provide maximum rigidity, particularly when cutting long threads. However, the boring bar diameter must be small enough to permit backing the tool out of the recess and returning it to the starting position.

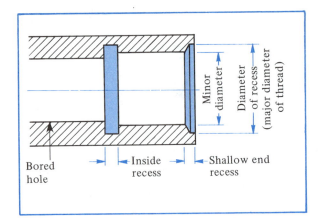

Figure 30-9 Recessing for Internal Threading

How to Cut an Internal Thread

Tapping an Internal Thread

STEP 1 Drill the hole to the required diameter.

Note: If thread concentricity is required, the thread should be chased, not tapped.

STEP 2 Mount a square (or angular) cutting tool in the boring bar. Position the tool at thread depth. Start the lathe. Cut an inside recess to about 0.003″ larger than the major diameter.

STEP 3 Turn a 30° chamfer on the end of the workpiece.

STEP 4 Position the taper tap and guide it against the tailstock center.

STEP 5 Start the taper tap by turning the tap holder and feeding the dead center.

STEP 6 Turn the chuck slowly by hand for a partial turn. Continue to thread and turn. Use a cutting lubricant throughout the threading process.

Note: Whenever possible, the tap should be turned with a tap wrench. A chuck does not provide the feel of the tapping process that a tap wrench does.

STEP 7 Repeat steps 1-6 until the threads are cut to full depth.

Note: On through holes the spindle may be run at a very slow speed. The dead center is kept continuously in the center hole of the tap.

Caution: The tailstock handwheel should be moved with the right hand while the left hand is kept on the control lever. The lever then can be moved quickly to instantly stop all motion in an emergency.

STEP 8 Use a plug tap as a second tap for through holes. Use a bottoming tap to cut threads that shoulder.

Caution: Power should not be used when tapping a blind hole.

STEP 9 Turn the tap in the reverse direction to remove it. Also, move the dead center at the same rate. The dead center should continue to serve as a guide.

STEP 10 Turn off any burrs from the ends. Clean the workpiece. Test for fit with the mating part or a gage.

Cutting an Internal Thread with a Single-Point Threading Tool

STEP 1 Check the diameter of the bored hole for size.

STEP Undercut a recess (equal to the major di-
2 ameter) at thread length. Cut a chamfer
on the end of the workpiece. If possible,
bore to the major diameter for about
1/16″.
STEP Set the lathe at the correct speed for
3 threading.
STEP Position the lead screw levers to produce
4 the specified thread pitch.

Caution: Be sure the feed rod is disengaged.

STEP Select an appropriate size of boring bar.
5 Grind the tool bit to accommodate the
thread size. Secure the tool bit in the
boring bar.

Note: Be sure the tool bit extends a short
distance beyond the full depth of thread.
The length of the tool bit must be
short enough to clear the inside bore
when the bit is moved out of the thread
groove.

STEP Set the compound rest at the correct
6 number of degrees (one-half the included
angle).

Note: The compound rest is positioned
to the left for right-hand threads and to
the right for left-hand threads.

STEP Secure the boring bar in a holder. Avoid
7 excessive overhang. Position the cutting
edges at center height.
STEP Select a thread gage appropriate for the
8 thread form. Place one side of the gage
against the workpiece or a parallel
straight edge.
STEP Bring the tool cutting edges into the gage.
9 Sight between the mating edges of the
gage and the tool bit. Gently tap the
toolholder until no spacing (light) is
seen between the mating surfaces. Secure
the setting.

Note: Recheck the center height and the
squareness of the threading tool.

STEP Place a mark on the boring bar at the
10 length of the threaded portion.

Note: This mark is important. The boring
bar mark indicates when to disengage
the split nut.

STEP Turn the compound rest handwheel to
11 take out all play. Set the micrometer
collar at zero.
STEP Start the lathe. Turn the cross feed han-
12 dle to remove end play. Reset the micrometer collar at zero when the cutting
tool just grazes the bored hole.
STEP Move the compound rest to take a rough-
13 ing cut.
STEP Hold the split-nut lever with the right
14 hand and the cross feed handwheel with
the left hand.
STEP Engage the split nut at the correct index
15 line on the chasing dial.
STEP Cut the thread. Watch the boring bar. Dis-
16 engage the split nut and move the boring
bar away from the internal thread when
the specified thread length is reached.
STEP Move the boring bar out of the workpiece.
17 Continue to chase the thread at successive
depths until the required depth is reached.

MACHINE SETUP FOR TAPER THREAD CUTTING

The recommended method of cutting an accurate
taper thread is to use a taper attachment. Taper
threads that are cut by offsetting the tailstock
center are not as accurate as taper threads cut by
the taper attachment method.

The thread-cutting tool is set perpendicular
to the axis of the workpiece. If there is no
cylindrical, solid surface on the workpiece from
which the cutting tool can be set squarely, it
may be necessary to hold the edge of a center
gage against the tailstock spindle. Once the
thread-cutting tool is positioned and secured,
the center gage is replaced with the workpiece.

Safe Practices in Cutting Internal and External Threads

- Disengage the feed rod before engaging the feed screw. Otherwise, the feed and lead forces opposing each other may damage apron parts and the accuracy of the lead screw.
- Use slower speeds for thread cutting than speeds used for other turning operations.

- Avoid excessive overhang of the cutting tool or boring bar.
- Grind the thread-cutting tool with clearance adequate to compensate for the thread helix angles.
- Grind the flat or round cutting point accurately. Thread depth measurements depend on the width of the point.
- Keep fingers and wiping cloths away from the revolving workpiece and cutting tool.

TERMS USED IN EXTERNAL AND INTERNAL THREADING

Lead	The axial movement (distance) of one mating threaded part from another part for one revolution. (Lead and pitch are equal for a standard single-lead screw thread.)
Multiple lead	The number of times a male and female thread may be engaged around the circumference. (Common leads are double (2P), triple (3P), and quadruple (4P). The depth of multiple-lead threads is only a fraction of a single-lead thread.)
Thread chasing	The process of engaging the split nut at a particular position. Precisely locating a threading tool for successive cuts to follow in the same thread groove.
Chasing dial	A rotation dial on the threading attachment. (The dial is graduated to indicate the position at which a half nut may be engaged to chase a thread.)
End gears	The train of gears between the spindle and the lead screw.
Pitch diameter	An imaginary diameter midway between the major and minor diameters of a thread. The diameter where the thread and thread groove dimensions are the same. An important thread diameter from which tolerances are measured to provide a specified fit.
Nominal size	The size used for general thread identification. (For example, the nominal size of a 1″–8NC thread is 1″. By contrast, the basic size is 1.000″–8NC.)
Limit	The upper and/or lower acceptable dimension.
Three-wire method of measuring threads	The use of three precision-ground wires of a special size to measure the pitch diameter of a screw thread.
Best wire size	The diameter of three wires that contact the thread sides (angle) at the pitch diameter.
Tangent of thread helix angle	The trigonometric value of the thread lead divided by the circumference of the thread diameter.
Side (edge) cutting angle of a threading tool	An angle equal to the regular clearance angle plus the additional helix angle of the thread.
Optical comparator	A device by which each design feature of a form may be compared against a master and with which measurements may be taken.

SUMMARY

- Many outside and inside threads are cut on a lathe with a single-point cutting tool.
  - The common thread forms include threads with 60° and 29° included angles and square threads. These thread forms may be in the American National, Unified, and SI Metric systems.
- The regular clearance angle for each cutting edge of a thread-cutting tool must be increased. The additional clearance must compensate for the thread helix angle.
  - Threads are cut on the lathe using the thread-chasing dial and threading attachment. This attachment permits engaging the split nut. The thread-cutting tool is tracked in the same groove for successive cuts.
- The spindle speeds for thread cutting are influenced by the skill of the operator, the specifications of the material, and the conditions under which the threads are machined. Speeds of one-quarter to one-half the speeds used for turning are applied to thread cutting.
  - Successive roughing cuts are taken as deeply as possible. Allow about 0.006″ (0.01mm) for finish cuts on diameters up to 1″.
- Thread measurements are taken at the pitch diameter. Internal threads are cut to the specified major diameter; outside threads are cut to the minor diameter.
  - Allowance, according to class of fit, is the specified (international) difference between mating parts or surfaces.
- The tolerance given on a drawing establishes the acceptable amount a dimension may vary from the basic size.
  - A thread may be checked for fit with a bolt, nut, or other mating part. Ring and plug thread gages may also be used. Some precise measurements may be taken with thread micrometers and optical comparators.
- Precision wires are used for V-thread (60°-angle) measurements. The best wire size in the three-wire method permits the measurement to be taken at the pitch diameter.
  - The helix angle of threads is found in handbooks or it may be calculated. A cutting tool must include additional clearance for the cutting angle.
- The compound rest is set at 30° to the right for American National, Unified, and SI Metric outside right-hand threads.
  - The angle setting is to the left of the perpendicular axis of the compound rest for cutting internal threads.
- Where possible, the threading tool is fed to depth with the compound rest handwheel. Thread depths may be read on the handwheel micrometer collar.
  - Mathematical formulas are used for calculating wire sizes and the measurement over wires.
- The cutting speed is reduced for cutting coarse threads to great depths. The depth for each successive roughing cut may be reduced.
  - The cutting tools for standard threads are generally positioned with a thread gage. These tools are set on center.
- A thread-cutting tool is reset by turning the handwheels on the cross slide and the compound rest. The movements permit changing the depth and position.

- The directions of the lead screw and the threading tool are reversed for cutting left-hand threads. End play must be taken out to permit feeding inwardly.

■ Internal threads may be tapped or machine cut on the lathe. A single-point cutting tool ground to the required thread form is used for machine cutting.

- The largest size boring bar that the diameter of the hole will accommodate is preferred.

UNIT 30 REVIEW AND SELF-TEST

1. Cite the different functions served by a center gage and a circular screw pitch gage.

2. Tell how the cutting-edge clearance angles of a thread-cutting tool are established.

3. Explain the function of a thread-cutting attachment.

4. State why the pitch diameter is important in thread calculations and measurements.

5. Tell what the differences are between the allowance, tolerance, and limits of screw threads.

6. Identify two thread (a) measuring tools and (b) inspection instruments.

7. Explain why a relief angle of $1°$ is added to the helix angle for the leading side and is subtracted from the helix angle of the following side in thread cutting.

8. List four settings or dimensions that the lathe operator must check before cutting a thread.

9. Explain the meaning of cleaning up the back side of a thread (that is rough).

10. Explain how the cross feed and compound rest handwheels are used to reset a threadcutting tool.

11. Explain the relationship between the 50-tooth spindle gear and the 127-tooth gear on the lead screw.

12. State two characteristics of internal threads cut with a single-point threading tool that make such threads superior to die-cut threads.

13. Indicate why a taper thread may be cut more accurately by using a taper attachment than by offsetting the tailstock.

14. List three safety precautions to observe when thread cutting on a lathe.

PART 7 Milling Machines: Technology and Processes

SECTION ONE

Horizontal Milling Machines, Cutters, and Accessories

This section presents introductory information about milling machines and milling processes, as follows.

- Designs and functions of principal features and accessories;
- Maintenance procedures;
- Design features and applications of standard milling cutters;
- Speeds, feeds, and cutting fluids used for milling processes; and
- Cutter-holding and work-holding devices and accessories.

UNIT 31

Milling Machines and Accessories

OBJECTIVES

A. FUNCTIONS AND MAINTENANCE OF HORIZONTAL MILLING MACHINES

After satisfactorily completing this unit, you will be able to:

- Identify five major types of milling machines.
- Describe the functions and give applications of general-purpose hand, plain, universal, omniversal, and vertical-spindle milling machines.
- Explain the function of each major part from the base to the overarm.
- State characteristics, functions, and operation of the dividing head.
- Carry on daily and periodic lubrication and maintenance checks.
- Follow recommended *Safe Practices* and correctly use *Terms* relating to design features, lubrication, and machine maintenance, (and in area B to cutter- and work-holding devices).

B. MILLING CUTTER- AND WORK-HOLDING DEVICES

- Understand and apply design features of National Standard milling machine (steep) taper spindles.
- Determine when to use vises, angle plates, and strap clamps.
- Perform each of the following processes.
  - Mounting and Removing a Milling Machine Arbor.
  - Mounting and Removing a Cam-Lock Adapter.
  - Mounting and Dismounting a Plain Vise.
  - Positioning and Checking Swivel Vise Jaws.
  - Use Work-Holding Accessories.

A. FUNCTIONS AND MAINTENANCE OF MILLING MACHINES AND ACCESSORIES

Milling is the process of cutting away material by feeding a workpiece past a rotating multiple-tooth cutter. The cutting action of the many teeth around the *milling cutter* provides a fast method of machining. The machined surface may be flat, angular, or curved. The surface may also be milled to any combination of shapes. The machine for holding the workpiece, rotating the cutter, and feeding it is known as a *milling machine*.

BRIEF HISTORY OF SIGNIFICANT MILLING MACHINE DEVELOPMENTS

The rifle manufacturing industry of the early 1800s is responsible for many developments and improvements in the early designs of milling machines. In 1818 Eli Whitney promoted the manufacture of interchangeable rifle parts. Whether Whitney built the milling machine in the United States at the time is not known.

Whitney's contributions were followed in the years prior to 1840 by the contributions of Robbins and Lawrence of Windsor, Vermont. Their milling machine used a rack-and-pinion feed mechanism. The design was later improved by George Lincoln of Hartford, Connecticut. Lincoln's worm and worm wheel were incorporated in the feed mechanism. Lincoln's design reduced chatter, added significantly to the accuracy and quality of surface finishes, and made possible further applications of the milling machine. The Lincoln machine was produced in quantity.

During the Civil War F.A. Pratt and Amos Whitney formed the P & W Machine Tool Company. The Pratt and Whitney Company still bears their names. The main products of their company were small turret lathes and milling machines. At the same time, Brown and Sharpe also produced these two basic machine tools. In addition, Brown and Sharpe contributed the design and construction of a universal milling machine and an omniversal milling machine that made it possible to machine tapered spiral grooves, helical bevel gear teeth, and other complex milling machine processes.

The Brown & Sharpe and Pratt and Whitney companies have contributed significantly to the design and development of machine tools and instruments and other precision tools and to manufacturing and production technology.

Essentially the milling machines of today incorporate features that date back to the machines of Robbins and Lawrence.

Current milling machine developments include:
- Preloaded antifriction spindle bearings that maintain great dimensional accuracy of the spindle,
- Functional coolant systems with more efficient coolants,
- Force-feed lubrication systems,
- Great accuracy of all moving and mating parts,
- Wide ranges of speeds and feeds and a high degree of control,
- Applications of carbides and ceramic cutting tools,
- Single micrometer collars graduated for direct reading in either the customary inch or the metric measurement system.

TYPES OF MILLING MACHINES

The major types of general-purpose and common manufacturing machines are:
- Knee-and-column,
- Fixed-bed,
- Rotary-table,
- Planetary,
- Tracer-controlled.

These types and the design features of numerically controlled machines are described briefly in this unit.

There are other types of milling machines (millers) for particular machining processes. For example, the *thread miller* is used for milling lead screws. The *cam miller* is designed for machining disc cams. The *skin miller* is adapted to special milling processes in the aircraft industry where, for example, the outer covering (skin) of wings is tapered to accommodate the stresses within each wing section.

The range of milling machine sizes is also extensive. Table lengths and movements extend from the small bed lengths of bench millers to bed lengths of 100 feet to 150 feet. Some millers have stationary spindles. On other millers the spindle carriers may be swiveled about a vertical or horizontal plane. Together with the cross feed (transverse), longitudinal, and vertical movements, these millers each have five axes of control. Many milling machines are programmed and are numerically controlled.

KNEE-AND-COLUMN TYPES OF MILLING MACHINES

There are five basic types of *knee-and-column* milling machines:

- Hand,
- Plain,
- Universal,
- Omniversal,
- Vertical-spindle.

Figure 31-1 A Hand Milling Machine

Hand Milling Machine. As the name implies, the hand milling machine is small. It is entirely hand operated. The hand miller pictured in Figure 31–1 may be mounted on a bench or on a floor base. This miller is particularly useful in small milling machine manufacturing operations. Slotting, single- and multiple-cutter milling, direct indexing, and other simple milling operations may be performed economically. The hand feed permits a rapid feed approach and withdrawal after a cut.

Plain Horizontal Milling Machine. The plain miller has three principal straight-line movements. The table moves in a longitudinal path. The movement is parallel to the face of the machine column. The table may be moved (cross-fed) in or out and/or up and down (vertically). The spindle is mounted in a horizontal position. The cutter is rotated by the spindle movement. The feed is in a straight line. The plain miller is a general-purpose, standard machine tool.

Universal Milling Machine. In addition to straight-line movements, the universal milling machine has the added movement of a swivel table. A standard universal miller with the main features identified is shown in Figure 31–2. The table is mounted on a saddle. The base of the table is graduated and designed for setting the table either straight (0°) or at an angle. The angle setting permits the table to travel at an angle to the column face. The universal miller is especially adapted to produce straight and spiral cuts. Common examples are the spiral flutes on reamers, drills, and other cutters and the spiral teeth on helical-form gears.

Omniversal Milling Machine. The omniversal milling machine is primarily a precision tool-room and laboratory miller. The rotation of the knee about the column face is in a perpendicular axis to the knee. This rotation makes it possible to machine tapered spirals and tapered holes.

Vertical-Spindle Milling Machine. The spindle on the vertical-spindle milling machine is mounted vertically. The two basic types of vertical-spindle millers are classified as: (1) fixed-bed and (2) knee-and-column. A knee-

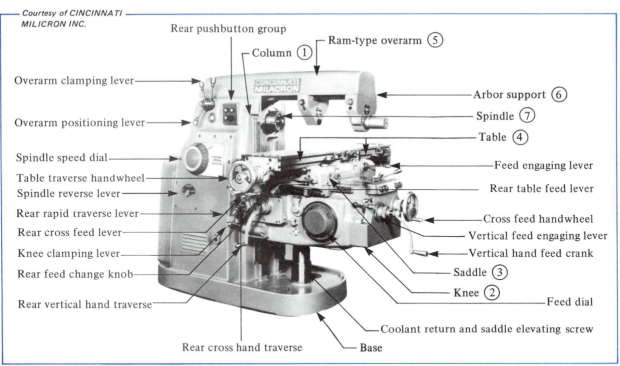

Courtesy of CINCINNATI MILICRON INC.

Rear pushbutton group

Column ①

Ram-type overarm ⑤

Overarm clamping lever

Overarm positioning lever

Spindle speed dial

Table traverse handwheel

Spindle reverse lever

Rear rapid traverse lever

Rear cross feed lever

Knee clamping lever

Rear feed change knob

Rear vertical hand traverse

Rear cross hand traverse

Arbor support ⑥

Spindle ⑦

Table ④

Feed engaging lever

Rear table feed lever

Cross feed handwheel

Vertical feed engaging lever

Vertical hand feed crank

Saddle ③

Knee ②

Feed dial

Coolant return and saddle elevating screw

Base

Figure 31-2 Main Features of a Knee-and-Column Type of Milling Machine

Courtesy of BRIDGEPORT MACHINES; DIVISION OF TEXTRON INC.

Figure 31-3 A Vertical Milling Machine

and-column type of universal toolroom vertical-spindle machine is shown in Figure 31–3. The machine accessories, attachments, cutters, set-ups, processes, and maintenance are covered in detail in successive units.

FIXED-BED MILLING MACHINE

The *fixed-bed miller* is usually a heavier, more rigidly constructed machine tool than the knee-and-column type of miller. A fixed-bed milling machine has an adjustable head that permits adjustment of the cutter in relation to the fixed position of a workpiece and table. Only the table travels longitudinally along the fixed ways of the bed.

The fixed-bed milling machine is primarily an automatically-cycled manufacturing type of machine. Milling cutters such as face mills, shell end mills, and other cutters mounted on arbors are used. Face milling, slotting, and straddle milling are common operations performed on this machine.

Some fixed-bed machines are equipped for automatic rise and fall of the spindle head to permit the milling of surfaces at different levels. These machines are also called *rise and fall millers.*

ROTARY-TABLE MILLING MACHINE

The design features of the *rotary-table milling machine* differ considerably from the features of the flatbed, plain, and universal millers. The table revolves. The worker loads and unloads workpieces in fixtures attached to the table. The table rotates under a face-type cutter. Usually there are two spindle heads.

PLANETARY MILLING MACHINE

A workpiece is held stationary on the *planetary milling machine.* One or more cutters may be held in vertical and/or horizontal spindles. Production units are also designed with magazine loading. This machine performs many operations that normally are done on the lathe. The planetary milling machine is used for workpieces that are heavy, difficult to machine, or are so unbalanced or delicate that they cannot be rotated,

TRACER-CONTROLLED MILLING MACHINE (DUPLICATOR OR DIE SINKER)

There are two basic types of *tracer controlled* milling machines. One type is called a *profile miller* (profiler). The second type has a number of names, including *duplicator*, *die sinker*, and *Keller machine.*

Profile Miller. The profile miller is similar to the vertical-spindle milling machine. Machining is usually done by small-diameter end mills. The cutting path is controlled by a *tracer (stylus)* with the same diameter and form as the end mill. The tracer follows a template of the required size and shape. The movement of the cutter is controlled by hand or automatic feed.

Die Sinking Machines (Profiler, Automatic Tracer-Controlled Miller). Many form dies and formed parts require three-dimensional machining. Die sinkers are designed for operation manually or in combination with electronic and hydraulic controls. A template is used. The template is a replica of the work to be produced.

The tracer provides three-dimensional information as it is moved over the entire surface. This information is duplicated by the movement of the corresponding profiler head and form cutter.

FUNCTIONS AND CONSTRUCTION OF MAJOR UNITS

The terms commonly used to identify the major units of all milling machines are given in Figure 31-2.

The major units of a horizontal milling machine include the knee, column, saddle (universal or standard), table, overarm, table controls, spindle, arbor support, overarm braces, drives, speed and feed controls, and the coolant system.

TABLE AND KNEE CONTROLS AND GRADUATED DIALS

Table and knee movements are made with feed screws and specially designed nuts. These screws and nuts are incorporated in the knee, saddle, and table design.

The knee is moved by an *elevating screw.* This screw extends from the inside of the knee to the base. The motion of the screw is controlled by the *vertical hand feed crank.*

A second *feed screw* is provided to move the saddle crosswise, toward or away from the column. The *cross feed handwheel* is used to turn this screw.

A third screw, known as the *table lead screw*, runs the length of the table. It is turned by a handwheel located at the end of the table.

There are micrometer collars (dials) on each of the three screw-movement controls (Figure 31-4). The collars are graduated to read in thousandths of an inch (0.001″) or in two-hundredths of a millimeter (0.02mm).

OVERARM (5), ARBOR SUPPORT (6), AND BRACES

The dovetailed ways on the top of the milling machine column accommodate the *overarm.* The overarm may be moved in a horizontal plane. The overarm axis is parallel to the spindle axis. The overarm slides toward (into) or away from the column. On heavy-duty millers a rack-and-pinion gear permits moving the overarm to position.

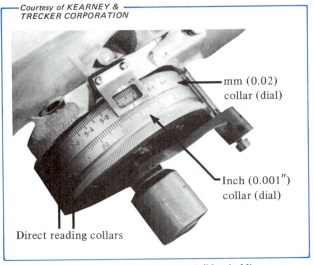

Courtesy of KEARNEY & TRECKER CORPORATION

mm (0.02) collar (dial)

Inch (0.001″) collar (dial)

Direct reading collars

Figure 31–4 Precision-Graduated Inch/Metric Micrometer Collars for Table, Knee, and Saddle Movements

Arbor supports are designed to slide on the overarm. Two common designs of arbor supports are available. The *outer arbor support* extends down below the milling machine arbor. The lower section of the support contains a bearing into which the pilot end of the arbor fits. The arbor support and overarm are used together to keep the arbor aligned and to prevent it from springing. Otherwise chatter marks, poor surface finish, and an inaccurate workpiece may be produced.

The *intermediate support* has a large center hole. A bearing sleeve rides in this center hole. The arbor fits through and is supported by the bearing sleeve. The intermediate support and bearing sleeve are used when roughing cuts and multiple cuts are to be taken.

A self-contained oiling system is incorporated in many arbor supports. The system consists of an oil reservoir, a sight gage that indicates the supply of oil, and a plunger. A small quantity of oil is forced directly into the bearing sleeve by depressing the plunger.

On heavier models the overarm is rigidly supported for heavy cuts by using *overarm braces.*

THE SPINDLE (7)

The *spindle* is housed in the upper section of the column. Four main functions are served by the spindle:

- Aligning and holding cutting tools and arbors;
- Transmitting motion and force from the power source for cutting;
- Ensuring that the machining processes are dimensionally accurate;
- Driving a slotting, vertical-spindle, or other attachment.

The spindle itself is hollow bored. The front end has a tapered hole. Standard arbors, adapters, and cutting-tool shanks fit the tapered hole. A draw-in bar is usually used to secure the tools against the taper. The front end of the spindle is fitted with two set-in lugs that extend beyond the face. Arbors, adapters, and cutting tools have identical slots. These slots match the positions of the lugs. The lugs fit into the slots and provide a positive drive. Four other holes are threaded in the spindle face. These holes are used for mounting face mills.

The direction of the spindle rotation is reversed by one of two common methods. Some models have two motor switches, one switch for clockwise rotation and a second switch for counterclockwise rotation. Other machines are equipped with a spindle-direction change lever.

MILLING MACHINE DRIVES

Spindle Speeds. Speed changes are made on a constant-speed drive by a sliding gear transmission housed in the column. The various spindle speeds (revolutions per minute) are obtained by positioning the speed-control levers. The index chart on the column gives the lever settings for a required speed.

Feeds and Feed Levers. Feed is the rate at which a workpiece is fed to a cutter. The rate is specified as so many thousandths of an inch or millimeters per revolution of the spindle. The feed rate may also be given in inches or millimeters per minute when the milling machine is equipped with a constant-speed drive.

A feed index chart on the knee or column gives the range of feeds. The position of the feed-change levers and a dial design are illustrated in Figure 31–5. The lever or dial settings regulate the lengthwise movement of the table, the cross feed movement of the saddle, and the

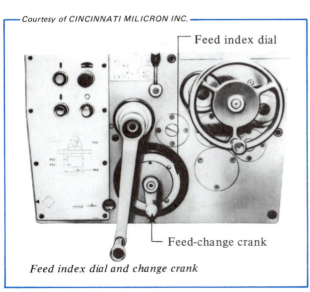

Courtesy of CINCINNATI MILICRON INC.

Feed index dial

Feed-change crank

Feed index dial and change crank

Figure 31–5 Example of Feed Controls

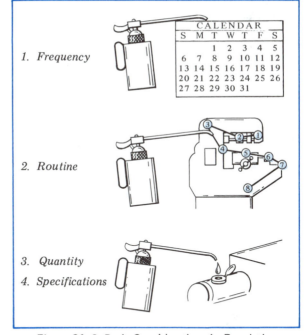

1. *Frequency*

2. *Routine*

3. *Quantity*

4. *Specifications*

Figure 31–6 Basic Considerations in Regulating Machine Lubrication

vertical travel of the knee when the automatic feed is applied.

Other levers are used to manually engage or disengage the automatic feed mechanism. These levers are located on the knee and saddle. Some machines are also designed for rapid traverse.

Trip Dogs. Trip dogs are simple adjustable guides and a safety device. They control the length of movement. The trip dog moves against a pin and causes it to move and disengage the feed. Trip dogs may be placed on the front of the table, under the saddle, and behind the knee.

MAINTENANCE OF THE LUBRICATING SYSTEMS

PRESERVING ACCURACY AND SMOOTHNESS OF OPERATION

Oiling systems are incorporated into the design of plain, universal, and other types of milling machines. The systems maintain a continuous film of oil between the mating surfaces of moving parts. The oil provides good, regular lubrication and also maintains the precision of the mating parts.

Pressure, *gravity*, and *splash systems* are widely used. The oil is forced through tubing to the surfaces, threads, and bearings that require

lubrication. Glass-faced *sight gages*, indicate whether the oil is circulating, the amount of oil in the reservoir, and the level at which it should be maintained.

REGULAR PROCEDURES FOR MACHINE LUBRICATION

Technical manuals identify the lubricating systems and points and specify the procedures to follow. The machine tool builder recommendations center around four basic considerations as illustrated in Figure 31–6.

LUBRICANT SPECIFICATIONS AND QUANTITY

The specifications (type) and quantity of lubricant depend on the application. Geared spindles are in constant rotation and produce the forces necessary for cutting. The spindle lubricant must withstand variations in temperature and must have the required viscosity to flow, cool, and lubricate under varying machining conditions. The lubricant on the column ways requires different properties than the oil used to meet the heat-removing needs of the spindle.

FORCE-FEEDING THE SPINDLE, KNEE, TABLE, AND COLUMN SYSTEMS

Oil is distributed to the spindle, spindle keyways, and gear train by pumping from the reservoir in the column.

The column, the knee, and the saddle are three units that are usually oiled by pressure and splash systems.

The gears and screws housed in the knee and the bearing surfaces in the column are usually force fed oil from the knee reservoir. A mechanically operated pump delivers the oil under pressure.

The saddle parts and the table ways are lubricated by a single-shot oiling device located at the front of the saddle. The hand-operated plunger force-feeds the oil from the reservoir to the bearing surface.

As indicated before, the pilot and bearing sleeve on the arbor are oiled directly from the arbor support reservoir. Oil cups that require filling by hand are used for other surfaces that require lubrication. Some smaller machines are lubricated almost entirely by hand. The type and the size of the oiler are indicative of the quantity of lubricant required. For instance, the cross feed screw requires a few drops of oil. Larger bearing surfaces and the gear train require a continuous flow over the gears and sliding parts.

It is important on all machine tools that the operator follow the recommended oiling procedures and routine. In practice, after the machine is set up, an inspection is made of all sight gages. There must be an adequate supply of oil in each of the oil wells. The machine is started only after all lubricating systems have been checked.

How to Lubricate and Maintain the Milling Machine

Systems Requiring Daily Lubricant Checks

STEP 1 Read the lubricant chart supplied with the machine. Note all of the lubricating systems and the types of oiling devices. Locate each lubrication point and reservoir.

STEP 2 Check the lubricant level. Use the sight gages on the column, saddle, and knee reservoirs (Figure 31-7A).

STEP 3 Locate the reservoir filler cap if the lubricant level requires that oil be added.

Note: All dirt particles should be removed and the area around the cap wiped until it is clean (Figure 31-7B). It is important to prevent dirt, chips, and grit from accumulating and entering an oil reservoir.

STEP 4 Check the recommended viscosity and other specifications if additional lubricant is needed. Add the lubricant with a spout oiler until the correct level indicated on the sight gage is reached (Figure 31-7C).

Note: Pouring in a small quantity of lubricant at a time allows the level to rise to a true indication.

Caution: Wipe up any lubricant that may have spilled on the floor or onto the machine.

STEP 5 Replace the reservoir cap. Again wipe up any lubricant on the cap and surrounding area.

STEP 6 Start the milling machine. Check the flow gage to determine that the oil pumps are operating correctly.

Note: A fine stream of lubricant appears in the flow gage when the oil pump is functioning. Periodically during the machine operation, recheck to see that the pump is circulating the lubricant.

STEP 7 Hand oil the oil cups as specified by the machine manufacturer.

Note: The number and location of all stations should be established and then oiled in sequence.

STEP 8 Brush all chips from the table and T-slots. Remove chips from between the table and column. Wipe the machined surfaces until they are clean and free of chips and cutting compound. These surfaces include the overarm and arbor supports.

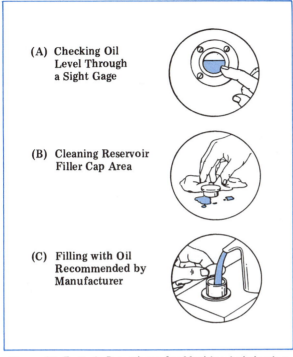

(A) Checking Oil Level Through a Sight Gage

(B) Cleaning Reservoir Filler Cap Area

(C) Filling with Oil Recommended by Manufacturer

Figure 31–7 Basic Procedures for Machine Lubrication

STEP 9   Lubricate all machined sliding surfaces that are not automatically oiled. These surfaces include the knee, column, and table ways and the overarm. Use a spray cleaner on surfaces where the lubricant has congealed.

Note: Some workers apply a small quantity of oil on the machined surfaces. This oil is then spread with a clean wiping cloth to produce a thin film of lubricant over the whole surface area.

STEP 10   Clean and lubricate the cross slide, table, and knee-elevating screws if not included in an automatic lubricating system.

Note: During a shutdown period of a few days or longer, the machined surfaces should be sprayed or coated with a rust inhibitor.

Systems Requiring Periodic Lubricant Checks

Note: The condition of oil filters, driving chain (if included in the machine design), motors, and reservoirs should be checked periodically.

STEP 1   Check the type and condition of the oil filters. Use a cleaning bath if the filter pad or screen requires cleaning.

Note: A loaded filter pad must be replaced in order to remove metal and other foreign particles.

STEP 2   Test the condition of the lubricant in each reservoir with a viscosity tester.

Note: Discoloration of the lubricant is a sign that the lubricant is dirty. It has picked up and is circulating dirt and other fine foreign particles. Under conditions of reduced viscosity and discoloration, the lubricant is not able to maintain a correct, clean film between the moving and sliding parts.

Caution: The power supply must be cut off before opening the motor compartment. On many models a power cutoff safety feature is an integral part of the machine construction.

STEP 3   Check the lubricating level of the motor bearings. If lubricant is to be added, check the specifications, then add.

STEP 4   Examine the driving chain if there is one. Use the driving chain lubricant specified by the manufacturer.

Note: If the driving chain is lubricated by a splash system, check the oil supply level. Add lubricant as required.

STEP 5   Examine the condition of each reservoir. If the lubricant requires changing, remove the drain plug. Drain the lubricant. Cleanse the reservoir to remove the sludge (thickened, gummy lubricant and foreign particles).

Note: A solvent is used to dissolve the sludge for easy removal.

STEP 6   Follow the manufacturer's recommendations and refill the reservoir to the line indicated on the sight gage.

B. MILLING CUTTER- AND WORK-HOLDING ACCESSORIES

CUTTER-HOLDING ACCESSORIES

The spindle of the milling machine provides a bearing surface that aligns the arbor, adapter, or tool shank. The power to drive the cutter is transmitted from the spindle to the cutter.

Modern milling machine spindle noses are ground to a standard steep machine taper size of 3 1/2″ per foot, or an included angle of 16°36′. The angle of taper makes the taper *self-releasing*. A locking device such as a quick-releasing collar or draw-in bar is used to secure an arbor or adapter to the spindle taper.

A milling machine cutter is generally held by one of four methods:

- Mounting the cutter on the nose of the spindle,

- Inserting a shank directly in the spindle bore,

- Using adapters and collets,

- Mounting the cutter on an arbor.

Cutter-holding devices thus include the spindle nose; arbor; keyseat and key; spacing collars, bearing sleeve, and arbor nut; draw-in bar; adapter; sleeve and collet; and cutting-tool holder.

SPINDLE NOSE

The four common steep taper sizes of spindle noses are #30, #40, #50, and #60. The range of the large diameters of the common taper sizes is given in Table 31–1.

Table 31–1 National Standard Steep Machine Tapers

Taper #	Large Diameter
30	1 1/4″
40	1 3/4″
50	2 3/4″
60	4 1/4″

ARBOR STYLES

There are three styles of arbors: A, B, and C. The *style A arbor* has a pilot (Figure 31–8A). The pilot fits into a corresponding bearing in the arbor support. Style A permits a workpiece to be brought up close to the arbor. The advantage is that smaller-diameter cutters may be used with style A arbors that can be used with style B. Where needed, a bearing sleeve and an intermediate arbor support are added to provide additional bearing support and prevent a long arbor from vibrating.

The *style B arbor* is supported by one or more bearing sleeves and the same number of intermediate supports (Figure 31–8B).

Style B arbors are used where a minimum of clearance is needed between the cutters, workpiece, and holding device.

Multiple milling operations often require the use of two arbor supports for heavy-duty milling. This setup provides maximum rigidity and

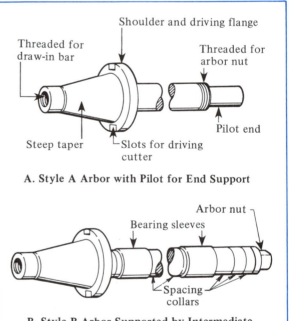

A. Style A Arbor with Pilot for End Support

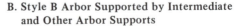

B. Style B Arbor Supported by Intermediate and Other Arbor Supports

Figure 31–8 Style A and B Milling Machine Arbors

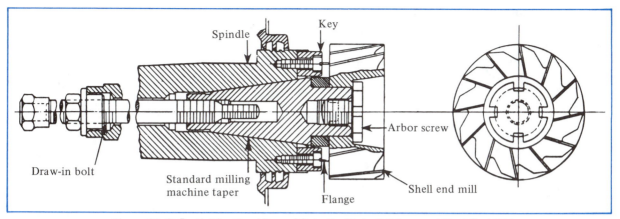

Figure 31-9 Partial Section View of Spindle, Style C Arbor, and Shell End Mill

eliminates chatter. The bushing and bearing sleeves have a free-running fit.

The *style C arbor* is a short arbor. It is sometimes called a *shell end mill arbor*. The cutter is mounted on a shouldered end. It is held against the face of the arbor by a screw. A special wrench is used to reach into the counterbored face of the shell end mill. The cutter is driven by two lugs on the arbor. These lugs mate with corresponding slots in a shell end mill (Figure 31-9).

SPACING COLLARS, BEARING SLEEVE, AND ARBOR NUT

Spacing collars are used for spacing or locating purposes. They also hold one or more cutters on an arbor. Spacing collars have parallel faces ground to accuracies as close as ±0.0005'' (±0.01mm). A *bearing sleeve* also fits on the arbor and is keyed to fit the arbor support bushing (Figure 31-8B).

An *arbor nut* is a round nut with two flat surfaces machined along part of the length. The milling machine arbor wrench fits this nut. The thread direction is opposite the direction of the milling operation.

THE DRAW-IN BAR

A *draw-in bar* has two threaded ends, an adjustment nut, and a threaded collar. The draw-in bar extends almost the length of the spindle. One threaded end screws into the tapered end of the arbor or adapter. The

draw-in nut on the other end is adjusted against the left (outside) face of the spindle. The threaded collar is tightened to draw the tapered arbor and spindle surfaces together (Figure 31-10).

The process is reversed to release the surfaces and remove the cutter-holding device. The end of the draw-in bar is tapped gently but firmly with a soft-face hammer to release the tapers.

ADAPTERS

Adapters are available to accommodate older styles of arbors and cutting tools that have self-holding tapers for Brown & Sharpe or Morse taper-shank cutting tools.

Another adapter is made for interchanging face milling cutters from one type of spindle to

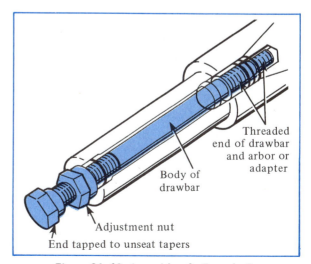

Figure 31-10 Assembly of a Draw-In Bar

a standard spindle. With this adapter shell end mills that have an internal taper designed for the old short taper spindle nose can be mated to the standard steep taper spindle nose. A special long draw-in bar is used to pull the tapered surfaces together and to secure the face milling cutter.

The *cam-lock adapter* is designed to hold cutting tools that have steep taper shanks (Figure 31–11). As the cam-lock is turned against a corresponding surface on the tool shank, there is a positive lock that prevents the cutter or holder from turning.

SLEEVES AND COLLETS

The milling machine *sleeve* serves the same function as a standard sleeve used on a drill press, lathe, or other machine tool (Figure 31–12 Ⓔ). A sleeve has an internal and external taper. The external taper surface fits into an adapter. The taper on the cutting tool corresponds with the internal taper of the sleeve. The sleeve makes it possible to use small-taper-diameter cutters in a large spindle bore.

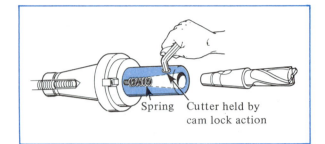

Figure 31–11 A Cam-Lock Adapter

A *collet* serves a function similar to the function of a sleeve. A collet is a precision-ground, cylindrical sleeve (Figure 31–12 Ⓕ and Ⓗ).

Examples of common cutter-holding devices like adapters, arbors, collets, and bushings are illustrated in Figure 31–12.

WORK-HOLDING DEVICES

PLAIN MILLING MACHINE VISE

One of the most widely used and practical holding devices is the *milling machine vise*. Usually this vise is heavier than the all-purpose vise

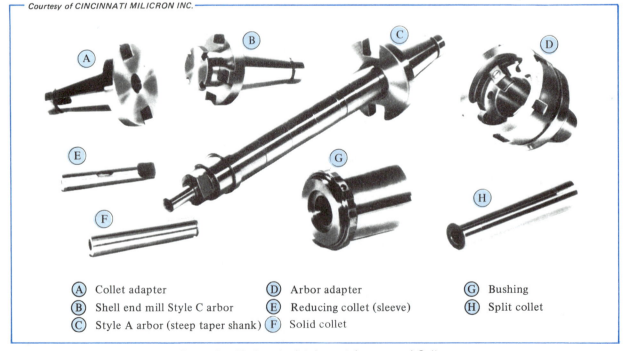

Courtesy of CINCINNATI MILICRON INC.

Ⓐ Collet adapter   Ⓓ Arbor adapter   Ⓖ Bushing
Ⓑ Shell end mill Style C arbor   Ⓔ Reducing collet (sleeve)   Ⓗ Split collet
Ⓒ Style A arbor (steep taper shank)   Ⓕ Solid collet

Figure 31–12 Standard Arbors, Adapters, and Collets

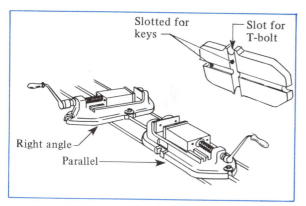

Figure 31–13 Parallel and Right-Angle Positions
of a Plain Milling Vise on the Table

used with drill presses. The milling machine vise base is slotted (grooved) in two directions. Keys fit the vise grooves and the table slot. These keys align the vise jaws either parallel or at right angles to the column face (Figure 31–13).

SWIVEL VISE

The *swivel vise* has the added feature of a movable, graduated base. The workpiece may be machined parallel to the column face or at a required angle. A *universal vise* is used when workpieces must be held at a compound angle.

Another milling machine vise is the *cam-action vise*. It is especially adapted to machining multiple pieces.

VISE JAWS AND FIXTURES

Most milling machine operations require that the workpieces be held with conventional (straight) vise jaws. However, special jaws may be added to a vise to serve as a fixture. Irregularly shaped parts may be positioned, secured, and machined accurately with a special set of jaws. For production of a large quantity of irregularly shaped parts that require precise locating, a fixture is usually used.

OTHER WORK-HOLDING DEVICES

Accessories and work-holding devices are strapped to the table with standard straps and clamps. A number of shapes and sizes are available to accommodate different workpieces and milling processes. Flat, finger, adjustable, and V-strap clamps are examples.

Standard milling machine bolts, like the bolts used on drilling machines, are of three common types. The T-heads that fit the T-slot of the table may be designed with a square-head, cutaway T-head, or a tongue block with a stud.

Flat and step blocks support the ends of clamps. Adjustable jacks are used to accommodate different work heights. Wedges and shims are used to compensate for unevenness.

Round work is generally held in *V-blocks*. The underside of the base is fitted with a tongue and may be mounted over a table slot. The V-block faces are parallel to the table. Milling machine *angle plates* provide a surface that is at 90° to the horizontal plane of the table. The base of the angle plate has two grooves. These grooves are cut at right angles to each other. A tongue is fitted to the groove. The angle plate may be positioned parallel with or at a right angle to the column face (Figure 31–14).

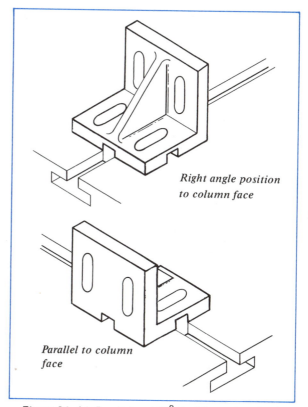

*Right angle position
to column face*

*Parallel to column
face*

Figure 31–14 Parallel and 90° Positions of an Angle
Plate on the Milling Machine Table

Parallels serve auxiliary functions in layout and work positioning. Solid and adjustable parallels are used in milling machine work.

The types, sizes, and applications of C-clamps and parallel clamps are similar to the clamps used in bench and drill press work.

How to Mount and Remove a Milling Machine Arbor

Preparing the Machine Arbor

STEP 1 Remove chips from the table top and T-slots.

Caution: The machine must be off. Use a T-handle cleaner, stiff brush, and pan to remove the chips. Wipe the table with a clean cloth.

STEP 2 Clean the nose end of the spindle with a wiping cloth.

Note: If it is necessary to use an air gun, place a cloth over the back end of the spindle.

STEP 3 Feel and visually examine the taper bore and spindle nose to be sure the surfaces are clean. They must also be free of burrs and nicks.

Note: Burrs and nicks may be removed by carefully using a round abrasive stick.

STEP 4 Clean the draw-in bar, particularly the threaded front end and the collar.

STEP 5 Insert the draw-in bar part way into the spindle.

STEP 6 Lock the spindle to prevent it from turning.

STEP 7 Clean the arbor support. Check the bearing for nicks and burrs. Remove the arbor-support if the arbor cannot be installed with it in place.

STEP 8 Select the required style and size of arbor.

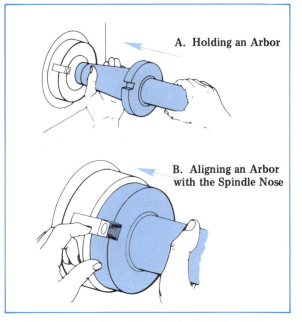

A. Holding an Arbor

B. Aligning an Arbor with the Spindle Nose

Figure 31–15 Mounting a Milling Machine Arbor

STEP 9 Recheck the arbor. Be sure the arbor, spring collars, bushing, and nut are clean. All parts, especially the tapered end and bushing, must be free of burrs and nicks.

Mounting a Milling Machine Arbor

STEP 1 Hold the tapered-shank end of the arbor in one hand and the shaft portion in the other (Figure 31–15A).

STEP 2 Insert and guide the tapered shank carefully into the spindle nose. Align the slots (Figure 31–15B). Move the arbor in as far as possible.

STEP 3 Move in the draw-in bar and turn it to engage the threads of the arbor.

STEP 4 Turn the draw-in bar nut by hand as far as possible.

STEP 5 Use an open-end wrench to turn the draw-in bar nut. Tighten the nut until the arbor is held securely in the taper and against the spindle face.

STEP 6 Unlock the spindle. Start the machine. Check the end of the arbor to see that it runs perfectly true.

Note: If the arbor does not run true, two conditions may exist. First, the arbor may not be seated correctly and is improperly mounted. Second, the arbor may be sprung out of shape. The arbor should be removed and reexamined. If it is not sprung, it can be remounted. If it is sprung, it must be replaced.

Removing a Milling Machine Arbor (Styles A, B, and C)

STEP 1 Stop the machine and clean it.

STEP 2 Remove the cutter, workpiece, and arbor support. Then lock the spindle.

STEP 3 Fit an open-end wrench to the nut on the draw-in bar.

STEP 4 Apply a force on the wrench in a counterclockwise direction. Turn the nut almost one turn.

STEP 5 Tap the end of the draw-in bar with a soft-face hammer. Deliver a sharp blow to release the steep-taper arbor from the spindle.

Note: It is often more difficult to release an arbor from a self-holding taper spindle. With such a spindle, remove the draw-in bar. Insert in its place a bar that is just slightly smaller in diameter than the spindle hole. Gently but firmly tap the end of this bar. Hold the arbor with one hand during this freeing process to prevent the arbor from falling when the tapers are freed.

STEP 6 Unscrew the draw-in bar with the left-hand. Support the arbor with the right.

STEP 7 Use both hands to remove the arbor when the draw-in bar is free.

Note: The small, style C arbors can usually be handled with one hand.

STEP 8 Wipe the arbor clean, and return it (with the collars, bearing sleeve, and nut in place) to the storage rack or cabinet.

STEP 9 Remove the draw-in bar and store it.

STEP 10 Unlock the spindle.

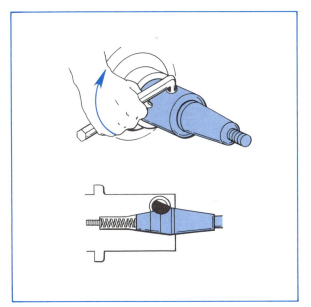

Figure 31–16 Tightening the Cam-Lock Arbor in the Adapter

How to Mount and Remove a Cam-Lock Adapter

Mounting a Cam-Lock Adapter

STEP 1 Select the adapter that fits the spindle nose taper and the shank of the cam-lock toolholder (or cutting tool).

STEP 2 Lock the spindle. Be sure all surfaces are clean and free of burrs.

STEP 3 Mount the adapter. Follow the same steps as used for mounting an arbor.

STEP 4 Recheck the taper hole of the adapter and the taper shank of the arbor to be sure that they are clean and free of burrs.

STEP 5 Insert the cam-lock arbor in the adapter. Match the cam lock and slot.

STEP 6 Turn the setscrew clockwise with a setscrew wrench.

STEP 7 Continue to turn the setscrew until the arbor is securely seated and the tapers are locked together (Figure 31–16).

STEP 8 Unlock the spindle. Start the machine and test the arbor for trueness.

How to Mount and Remove a Plain Vise

Mounting the Vise

STEP 1 Clean out the table slots with a slot cleaner. Wipe with a clean cloth.

STEP 2 Examine the slots and remove any burrs or nicks.

STEP 3 Place the vise on a bench or tool stand. Wipe it clean. Remove any burrs from the base.

STEP 4 Insert the tongues in the appropriate set of slots, depending on whether the vise is to be set with the jaws parallel or at 90° to the arbor.

STEP 5 Place the vise carefully on the table. The tongues should fit snugly into the table slots without being forced.

> Note: Position the vise as near to the middle of the table as possible. Be sure no chips fall between the vise base and table when the vise is turned.

STEP 6 Slide a T-head bolt into each T-slot (Figure 31–17). Place a washer over each bolt. Tighten the two nuts uniformly on the clamping bolts.

How to Position and Align Swivel Vise Jaws

Using Graduations on the Vise

STEP 1 Loosen the clamping bolts that hold the vise to the base.

STEP 2 Swing the vise so that the zero index line is aligned with the required angle graduation on the base.

STEP 3 Tighten the clamping nuts by hand until a slight force is applied.

STEP 4 Recheck the angle setting.

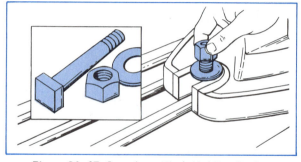

Figure 31–17 Securing a Work-Holding Device with a T-Bolt, Washer, and Nut

Using a Dial Indicator

A dial indicator is used to accurately align vise jaws either parallel or at right angles to the table movement. The accuracy of the setting is limited by the dimensional accuracy of the dial indicator. The fixed jaw can thus be set to within 0.0005" (0.01mm) or less, depending on the indicator.

The indicator setup for checking the vise jaws for parallelism with the column face is shown in Figure 31–18. When the vise jaws are to be positioned at a right angle to the column face, the dial indicator is turned 90°. The stationary jaw is then moved (transversely) past the indicator pointer by turning the cross slide handwheel. If there is a difference in the dial readings, the vise is tapped gently in a direction away from the higher reading until there is no change in the indicator readings (Positions A and B).

How to Use Work-Holding Accessories

Using Stops, Strap Clamps, and Bolts

STEP 1 Clean the table slots, table, and workpiece. Wipe with a clean cloth. Check and remove any burrs or nicks. Place the workpiece on the table.

STEP 2 Select two stops. These metal stops are the width of the table slots. Insert each stop into the table slot. Position each stop near the end of the workpiece.

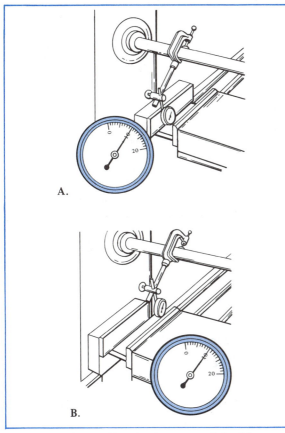

Figure 31-18 Checking the Accuracy of the Alignment of a Solid Vise Jaw (Parallel to Column)

STEP Move the workpiece against the two
3 stops.

STEP Slide the T-bolts in the table slots as
4 close as possible to the work.

Note: The cutaway T-head bolt may be dropped into position and turned.

STEP Use a clamp that accommodates the size
5 of the workpiece and the nature of the milling machine operation.

STEP Place a block under the end of the clamp
6 to provide maximum clamping leverage on the workpiece.

Note: The block size should be as near as possible to the thickness of the clamping surface.

STEP Tighten the nut securely.
7

Note: If the workpiece requires that clamps be placed on both ends or in different sides, the nuts should be tightened uniformly.

STEP Check to see that the correct force is
8 being exerted. The workpiece must be strapped flat on the table.

Using V-Blocks and Step Blocks

STEP Select the size of V-block required to
1 adequately nest the workpiece.

STEP Clean the table, workpiece, and V-blocks.
2 Remove any burrs or nicks.

STEP Position the V-blocks to provide maxi-
3 mum support for the workpiece. The blocks and clamps must also clear the milling operation.

Note: A general practice is to position the workpiece as close as practical to the center of the table and column.

STEP Select a flat or other appropriate type of
4 clamp and step block.

STEP Insert the T-bolts in the table. Slide each
5 T-bolt to the middle of the V-block.

STEP Protect the surface finish if necessary.
6 Place a metal shim between the clamp and the workpiece.

STEP Support one end of the strap clamp on
7 the step block. The height should be the same as the workpiece setup.

STEP Tighten each nut securely.
8

Using an Angle Plate, Parallel or C-Clamp, and Jack

STEP Select an angle plate that accommodates
1 the size of the workpiece. It also must permit secure clamping of the part.

STEP Clean all machined surfaces of the table
2 and angle plate. Remove any burrs or nicks.

STEP Position the angle plate on the table.
3 Locate it as near to the center as possible. The angle plate and workpiece should be

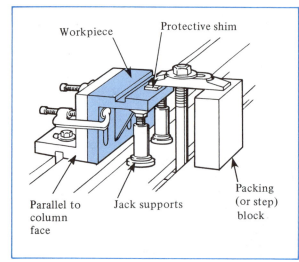

Workpiece
Protective shim
Parallel to column face
Jack supports
Packing (or step) block

Figure 31–19 Clamping Setup for Machining a Workpiece Mounted on an Angle Plate

as close as practical to the column. Secure the angle plate in position.

STEP 4 Select either a C-clamp or a parallel clamp.

Note: C-clamps are generally used to support large workpieces where heavy cuts are taken.

STEP 5 Adjust the C-clamps or the jaws of the parallel clamps. Place the workpiece against the vertical face of the angle plate. Tighten each clamp. Use only enough force to correctly position the workpiece.

Note: A protecting shim of metal, plastic, or cardboard should be placed between any rough surface and the machined face of the angle plate or under a strap clamp.

STEP 6 Select one or more support jacks. Place them under any overhanging surface that is to be milled (Figure 31–19).

Note: Sometimes the overhanging portion is also clamped. The jack should then be located under the workpiece, with one end of the clamp directly above the jack.

A. Safe Practices in Lubricating and the Maintenance of Milling Machines

- Shut down the milling machine when examining, cleaning, and lubricating moving parts or before opening the motor compartment.
- Use a T-slot metal cleaner to remove the extremely sharp chips from the T-slots. A stiff brush and small chip pan should be used to clear the chips from the table, saddle, knee, and base.
- Keep the coolant strainer on the end of the table clear to permit recirculating the cutting fluid.
- Check all sight gages. Maintain the proper level of lubricant in each reservoir. any excess oil after lubricating.
- Wipe up any lubricant that has been spilled or has overflowed onto any machine surface or the floor.
- Place oily and dirty wiping rags or cloths in a metal container.
- Observe the standard personal safety precautions against loose clothing and the use of wiping cloths around moving parts.

B. Safe Practices with Milling Cutter- and Work-Holding Devices

- Remove burrs or nicks from the spindle face and nose and the arbor adapter taper, shaft, bearing, and spacing collars; and other work-holding devices.
- Tighten the draw-in bar with enough force to pull together the tapered and face surfaces of the spindle and arbor.
- Locate milling cutters on an arbor as close to the column as possible to provide maximum support for the cutting tool during machining.
- Disengage steep tapers by striking the draw-in bar a sharp blow. The bar must not be loosened more than one thread to prevent damage to the threads.
- Mount and remove a style A arbor by holding it with two hands.
- Store arbors vertically in a rack to prevent them from hitting together or being bent.

TERMS USED WITH MILLING MACHINES

A. FUNCTIONS AND MAINTENANCE OF MILLING MACHINES

Milling	The process of removing material with a revolving multiple-tooth circular cutter. Machining flat, angular, circular, and irregular straight, tapered, and spiral surfaces with a milling cutter.
Knee-and-column (hand, plain, and universal) horizontal milling machines	Milling machines designed around three major components: a column, a knee, and a table. A classification of milling machines. Usually, a single horizontal-spindle machine on which a workpiece on the table is fed past a revolving cutter.
Fixed-bed milling machine	A milling machine adapted for manufacturing. The table moves longitudinally parallel to the face of the column along fixed bed ways.
Planetary milling machine	Performs typical lathe and milling machine operations. Especially adapted to heavy, intricately designed, or hard-to-handle workpieces that cannot be turned or machined readily from a vertical position.
Tracer-controlled milling machine	Two- and three-dimensional duplicators and profilers. Styli and duplication heads synchronized to duplicate either two- or three-dimensional irregular forms.
Automatic profiler	A complementary system of control, processing, and drive mechansims for pneumatically, electronically, and mechanically actuated units. A profiler that automatically reproduces a desired form.
Saddle (universal)	A two-section mechanism that moves on dovetailed slides of the knee. (The top slide nests the table. The two sections permit setting the table at an angle.)
Overarm, arbor support, and braces	Collectively the machine features that rigidly support a spindle, arbor, and cutters.
Feed-change levers, (milling machine)	Levers that are positioned in a fixed relationship to geared, V-belt, or variable-motor drives.
Force-feeding (lubrication)	A system of forcing a lubricant from a reservoir to rotating or sliding members.

B. MILLING CUTTER- AND WORK-HOLDING ACCESSORIES

Standard steep machine tapers	A steep-angle taper system of 16°36′, or 3 1/2″ taper per foot. A self-releasing taper. A system of designating milling machine spindle nose tapers (#30, #40, #50, and so on).
Style A, B, and C arbors	Three general designs of milling machine arbors.
Spacing collar	A round sleeve with parallel faces. These faces are accurately ground to specified lengths.
Bearing sleeve	A hollow cylinder with a keyway cut lengthwise. The outer bearing surface is ground accurately to fit the arbor support bushing.
Draw-in bar	A round bar threaded at one end to screw into an arbor or adapter. The screw end permits a shouldered nut to draw in and secure tapered surfaces.
Adapter	A cutter-holding device. A device that adapts the various sizes of cutters and shanks to the milling machine spindle size.

Sleeve	A cylindrical toolholder. The outside surface fits into an adapter. The inside bore accommodates the shank of a cutting tool.
Collet	An adjustable, slotted sleeve. The sections move to permit holding a limited range of tool shank diameters.
Cam-lock adapter	A device for quickly and securely locking and unlocking a cutting tool in an adapter. An eccentric stud and corresponding slotted cutter. A device for applying the force needed to seat and hold a cutter in an adapter.
Tongue	A rectangular-cross-sectioned metal piece. A part fitted into a machined groove in the base of a work-holding device. A rectangular block that aligns a work-holding device parallel or at right angles to the column face.

SUMMARY

A. FUNCTIONS AND MAINTENANCE OF MILLING MACHINES

- There are seven broad categories of milling machines: knee-and-column, fixed-bed, rotary-table, planetary, tracer-controlled, machining center, and numerically controlled. Milling machines are also designed for thread, cam, skin, and other specific machining processes.
    - The general, all purpose, knee-and-column type of millers include hand, plain, universal, and vertical-spindle milling machines.
- Fixed-bed millers are production machines. One or more spindles are adjusted in relation to the table. Table movement is longitudinal in a fixed position on the bed. Standard milling operations are performed. All movements may be automatically cycled for production.
    - Rotary-table milling machines are adaptable to large workpieces that are mounted on a revolving table. These workpieces are fed past one or more rotating spindles with cutters.

B. MILLING CUTTER- AND WORK-HOLDING ACCESSORIES

- The major structural members of general-purpose milling machines are the column, the knee, the standard or universal saddle, and the table.
    - Table and knee movements are made by lead screws. Graduated micrometer collars provided direct measurements.
- The arbor and cutting tools are supported by the overarm, arbor supports, and overarm braces.
    - The spindle bearings are preloaded. The spindle is designed to align, hold, and drive cutting tools and attachments. Gears and sliding parts mesh with mating members in the gearbox.
- Spindle speeds and feeds are controlled by positioning gear-change levers and power levers. Trip dogs control the table movement.
    - Standard steep machine tapers are self-releasing. They are machined to a taper of 3 1/2″ per foot, or at a 16°36′ included angle. The tapers are designated by a two-digit numbering system. Common sizes are #30, #40, #50, and #60.
- Self-releasing tapers require a set of lugs. These lugs transmit the spindle force to the cutter.

- A Style A arbor is a long arbor that is supported on the end by a pilot. The cutting action takes place close to the milling machine arbor. Small diameter cutters are used with Style A arbors.
- Style B arbors require a bearing sleeve and an arbor support. The Style B arbor is adaptable to heavy-duty milling operations.
  - Style C arbors provide for the quick release of cutting tools. These arbors are widely used for shell end mills. Sleeves, collets, and other toolholding accessories may be securely held in different designs of Style C arbors.
- A dial indicator is used to set a fixed vise jaw or the vertical surface of an angle plate perpendicular to within an accuracy of ±0.0005″ (0.01mm). The surfaces may be checked for alignment parallel or at a right angle to the column face.
  - Workpieces are commonly clamped to the table in V-blocks or on an angle plate. Flat, curved, and tongue strap clamps and C- or parallel clamps are used.
- Jacks support overhanging surfaces of workpieces that are held in a vise or on an angle plate.
  - Plain and swivel vises, angle plates, and V-blocks are aligned by tongues. These tongues fit snugly into the table slots. The graduations on the base of a swivel vise permit setting the vise jaws to a required angle.
- Sleeves permit the use of regular self-holding taper-shank tools. Morse and Brown & Sharpe tapers may be held in a steep-taper spindle adapter.
  - Collets are used for holding and changing from one size and type of round-shank cutting tool to another. The collet and cutter are held in an adapter.

UNIT 31 REVIEW AND SELF-TEST

A. FUNCTIONS AND MAINTENANCE OF MILLING MACHINES

1. Distinguish between the principal straight-line table movements of plain and universal horizontal millers and plain vertical milling machines.
2. Classify three nonstandard milling machine types.
3. Identify the major construction parts and mechanisms of a milling machine.
4. Describe the functions of (a) a dividing head and (b) a rotary table.
5. Explain the functions of the automatic lubrication systems of modern milling machines.
6. Indicate four precautionary practices to follow when lubricating and maintaining a milling machine.

B. MILLING CUTTER- AND WORK-HOLDING DEVICES

1. Describe style A, B, and C milling machine arbors.
2. Explain briefly the function served by the draw-in bar.
3. List five work-holding devices for general milling processes.
4. Give the steps to follow in removing a style A, B, or C milling machine arbor.
5. Describe three practices for positioning vise jaws on a milling machine.
6. Explain how V-blocks and other work-holding accessories are used to nest a round workpiece and hold it securely on a milling machine table.
7. State three machine or tool safety precautions the milling machine operator must observe.

Technology and Applications of Standard Milling Cutters

A *milling cutter* is a cutting tool that is used on the milling machine. Milling cutters are available in many standard and special types, forms, diameters, and widths. The teeth may be straight (parallel to the axis of rotation) or at a helix angle. The cutter may be right-hand (to turn clockwise) or left-hand (to turn counterclockwise).

Dimensional and other design standards for milling cutters have been approved by the American National Standards Institute (ANSI).

This unit covers general features of milling cutters, forms of cutter teeth, materials used in milling cutters, and standard milling cutter types. Speeds, feeds, and cutting fluids are included. The actual milling processes are covered in later units.

OBJECTIVES

A. FEATURES AND APPLICATIONS OF STANDARD MILLING CUTTERS

After satisfactorily completing this unit, you will be able to:

- Describe features of cutters in terms of ANSI terminology.
- Identify and know when to use the following cutters and characteristics.
  - Positive, zero, and negative rake angles,
  - Saw-tooth, formed-tooth, and inserted-tooth cutters,
  - Light-duty, heavy-duty, and helical plain milling cutters,
  - Plain, half-side, staggered-tooth, and interlocking cutters for side milling,
  - Plain and screw-slotting saws and saws with side and staggered teeth,
  - Single- and double-angle, T-slot, and Woodruff keyseat, and other formed cutters.
  - Shell end mills, solid-shank end mills, and inserted tooth face milling cutters, and single-point fly cutters.
- Follow *Safe Practices* and correctly use *Terms* relating to standard milling cutters.

B. SPEEDS, FEEDS, AND CUTTING FLUIDS FOR MILLING

- Solve problems involving cutting speeds and feeds using tables relating to high-speed steel, cast alloy, and carbide milling cutters.
- Interpret conditions affecting cutting speeds and feeds.
- Position spindle speed controls to regulate speed (RPM) and direction of rotation.
- Adjust longitudinal (table), transverse (saddle), and vertical (knee) feed mechanisms and controls, including rapid traverse.
- Select cutting fluid delivery systems and fluids.
- Perform the following processes.
  - Setting the Spindle Speed (RPM).
  - Setting Feeds for Table, Knee, and Saddle.
  - Selecting and Supplying Cutting Fluids.
- Follow recommended *Safe Practices* and correctly use *Terms* relating to speeds, feeds, and cutting fluids.

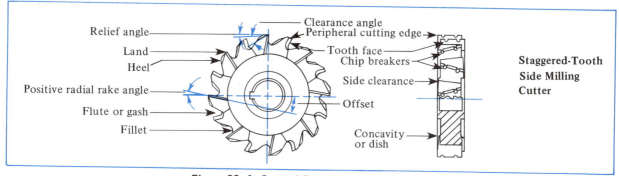

Figure 32–1 General Features of Milling Cutters

A. FEATURES AND APPLICATIONS OF MILLING CUTTERS

Milling cutters may be broadly grouped in two categories: *standard* and *special*.

Some of the terms used to identify the major features of milling cutters appear in Figure 32–1.

There are a number of other terms that are used to describe a milling cutter or to order a cutter from a toolroom. The following features are described by these terms:

- Outside diameter,
- Cutter width (measurement across the face),
- Hole size (usually +0.001″ of nominal arbor size),
- Keyway (to match arbor keyseat),
- Shank (straight or tapered: B & S or Morse self-holding taper),
- Cutter teeth.

BASIC FORMS OF CUTTER TEETH

Circular milling cutters are designed with three basic forms of teeth. These forms are the *standard tooth*, *formed tooth*, and *inserted tooth*.

STANDARD-TOOTH FORM

Some of the common design features of the standard tooth form are illustrated in Figure 32-2. Teeth are said to be *radial* when each tooth face lies along a line that cuts through the center of the cutter. Teeth are cut either with a radial tooth face or at a 0° *(zero)*, *positive*, or *negative rake* angle to the radial line.

The tool life and performance of carbide-tipped teeth, which operate at exceedingly high speeds and coarse feeds, are improved when the teeth have negative rake.

The purpose of zero or negative rake is to protect the cutting edges of carbide tools. Their edges are brittle in comparison to the edges of high-speed steel tools. With a negative rake the cutting forces fall within the cutter body. Surprisingly, carbide cutters are capable of producing a high-quality surface finish at high machining speeds.

There may be two or three clearances on the standard-tooth form. The teeth are ground at a slight *clearance angle (primary clearance)* around

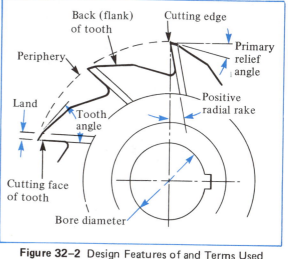

Figure 32–2 Design Features of and Terms Used with a Standard Milling Cutter

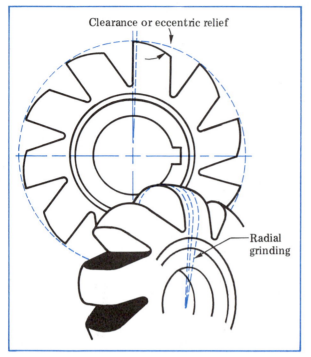

Figure 32-3 Shape of a Formed Cutter Face
Retained by Radial Grinding

the periphery. This clearance extends from the tooth face and produces a cutting edge. The narrow, flat surface directly behind the cutting edge is called the *land*. The teeth on the saw-tooth form are sharpened by grinding the land. The angle formed by the cutting face and the land is the *tooth angle*. The body of the tooth is cut at a secondary angle and at a third angle. The design provides maximum support for the cutting edge. It also provides a *chip space* for the fast removal of chips.

FORMED-TOOTH MILLING CUTTER

As the name suggests, a formed-tooth cutter has a contour, or tooth outline, of a particular shape. A concave milling cutter with a specified diameter produces a round shape on a work-piece. A gear cutter machines a gear tooth that conforms to specific design requirements. The right- or left-hand radius cutter mills a round corner. The flute cutter cuts flutes on drills, reamers, taps, and other cutting tools. A single formed cutter is used for some applications. In

applications that involve the milling of a wide contour, formed cutters may be set up in combination.

The clearance, or *eccentric relief* of a formed cutter, follows the same contour as the cutting edge. Each semicircular cutting face on a radial-tooth formed cutter retains the original shape when radially ground at a zero rake angle (Figure 32-3).

STANDARD TYPES OF MILLING CUTTERS

PLAIN MILLING CUTTERS

Two examples of *plain milling cutters* are shown in Figure 32-4. The diameter and the width of a cutter depend on whether a part is to be *slab milled* (milling a wide, flat surface) or requires a narrow-width slot. There are three broad classes of plain milling cutters: *light-duty*, *heavy-duty*, and *helical*.

Light-Duty Plain Milling Cutters. Light-duty plain milling cutters up to 3/4" (18mm) wide generally have straight teeth. These teeth are formed around the periphery only and are parallel to the cutter axis. High-speed cutters have from four to five teeth for each inch of diameter.

Cutters over 3/4" (18mm) wide usually have helical teeth. The helix angles range from 18° to 25°. These angles produce a shearing, cutting action. Less force is required, vibration and chatter are reduced, and a better quality of surface finish is produced with a helical-tooth cutter than with a straight-tooth cutter.

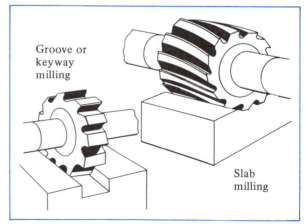

Figure 32-4 Two Examples of Plain Milling

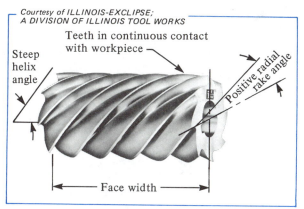

Courtesy of ILLINOIS-EXCLIPSE;
A DIVISION OF ILLINOIS TOOL WORKS

Figure 32–5 A Helical-Type Plain Milling Cutter

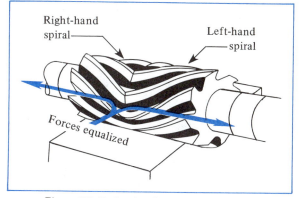

Figure 32–6 Cutting Forces Equalized by Right- and Left-Hand Helical Cutters

Heavy-Duty Plain Milling Cutters. Heavy-duty plain milling cutters are also called *coarse-tooth cutters*. These cutters average two to three teeth for each inch of diameter. For example, a 3″ heavy-duty plain milling cutter usually has eight teeth. The helix angles of the teeth of heavy-duty cutters are steeper than the angles of light-duty cutters. Different cutter manufacturers vary the helix angles from 25° to 45°.

Wide, flat surfaces are usually slab milled with a particular type of heavy-duty cutter. This cutter is called a *slab mill, slabbing cutter*, or *roughing cutter*. This type of cutter has interrupted teeth. Part of each tooth is relieved at a different place along the length to break up the chip. A slabbing cutter is able to produce a good surface finish under heavy cutting conditions.

Helical Plain Milling Cutters. Helical plain milling cutter teeth are formed at a steep helix angle (Figure 32–5). This angle ranges from 45° to 60°, or steeper. The teeth are designed so that they engage the work at a steep right- or left-hand helix angle. The cutting force is absorbed in end thrust. The teeth are continuously engaged in comparison with the intermittent cutting action of a straight-tooth cutter. The helical plain milling cutter therefore eliminates chatter.

The direction of the helix (spiral) establishes the direction of the cutting force. The angle of the helix determines the end force that is generated. The helix may be right- or left-hand. A right-hand spiral on a cutter turning clockwise produces a cutting force in the direction of the spindle. The end bearings are designed to absorb such force. Force toward the spindle is preferred to the direction of force produced by a left-hand helix. A left-hand helix tends to pull the arbor from the spindle.

Wide surfaces are often milled by interlocking a right- and a left-hand helical cutter. The forces exerted by each cutter are thereby offset (canceled) (Figure 32–6).

Helical plain milling cutters are efficient for wide, shallow cuts. They are not as practical as the heavy-duty cutters for deep cuts, slab milling, or coarse feeds.

SIDE MILLING CUTTERS

Side milling cutters are similar to plain milling cutters. However, in addition to teeth around the periphery, other teeth are formed on one or both sides. Most of the cutting is done by the teeth around the periphery. The side cutting teeth cut the side of a workpiece. These cutters are mounted on and are keyed on the arbor.

Side mills are not recommended for milling slots. There is a tendency for the side cutting teeth to mill wider than the specified cutter width.

Four types of side milling cutters are in general use. They are the *plain, half, staggered-tooth*, and *interlocking* side milling cutters.

Plain Side Milling Cutters (Figure 32–7). The plain side milling cutter has teeth on the periphery and on both sides. The teeth on the sides taper slightly toward the center of the cutter. The concavity of the teeth provides clearance, or side relief. The plain side milling cutter is adaptable for general-purpose side milling, slotting, and straddle milling.

The four milling cutters illustrated in Figure 32–7 have different design features. The most commonly used cutter is made of high-speed steel (Figure 32–7A). The three remaining plain side milling cutters have heat-treated alloy steel bodies that are stress relieved (Figure 32–7B, C, and D). The teeth consist of carbide tips that are fused to each cutter body. The extremely smooth surface finish prevents abrasive chips from adhering to the cutting edges.

The cutter in Figure 32–7B is used for nonferrous and nonmetallic materials. The four teeth on this cutter have a positive radial rake.

The cutter in Figure 32–7C has a negative radial rake. This cutter is designed for milling steels.

The cutter in Figure 32–7D has a neutral radial rake and a neutral axial rake. The cutter sides are not formed with clearance. This cutter is used for milling cast-iron and malleable iron parts.

The carbide-tipped cutters in Figure 32–7 machine at considerably higher cutting speeds and feeds than do high-speed steel cutters.

Half Side Milling Cutters. One side of the half side milling cutter has teeth. The other side is flat and resembles a plain milling cutter. The teeth around the circumference are helical. The teeth produce a shearing, cutting action that minimizes chatter. The top and side teeth are deep cut to serve two purposes: (1) to permit the free flow of chips and (2) to extend tool life. The teeth may be reground a greater number of times than can the teeth of a standard side milling cutter.

Half side milling cutters are designed for heavy-duty side milling. These cutters may also be used in pairs for slotting to a required dimension. Spacing collars are used to separate the cutters. A right-hand and a left-hand cutter are often used for straddle milling.

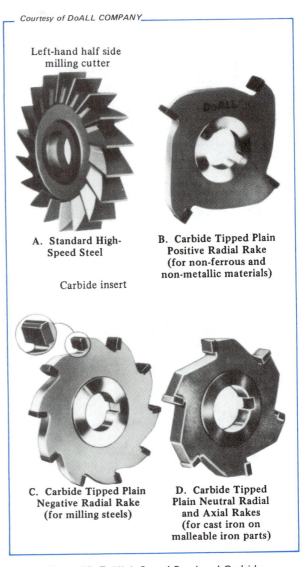

Left-hand half side milling cutter

A. Standard High-Speed Steel

Carbide insert

B. Carbide Tipped Plain Positive Radial Rake (for non-ferrous and non-metallic materials)

C. Carbide Tipped Plain Negative Radial Rake (for milling steels)

D. Carbide Tipped Plain Neutral Radial and Axial Rakes (for cast iron on malleable iron parts)

Figure 32–7 High-Speed Steel and Carbide-Tipped Plain Side Milling Cutters

Staggered-Tooth Side Milling Cutters. The staggered-tooth side milling cutter has cutting teeth with alternate right- and left-hand helixes. One side of each tooth forms a side tooth. Each land is ground from 1/64'' (0.4mm) to 1/32'' (0.8mm) wide. There is a minimum back clearance of 0.001'' per inch. This clearance prevents the cutter from binding in the cut (Figure 32–8).

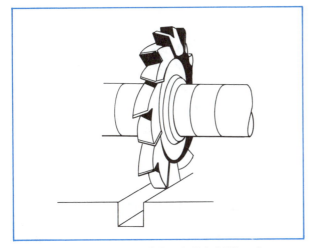

Figure 32–8 A Staggered-Tooth Side Milling Cutter

Figure 32–9 Interlocking Side Milling Cutters

The overall width of each tooth is narrower than the width of the cutter. The chip produced is thus prevented from wedging in the slot. The teeth are sharpened around the circumference. Occasionally grinding is required on the side teeth.

The staggered-tooth side milling cutter is a heavy-duty cutter. It is designed to remove large amounts of metal, to mill with a minimum of vibration and chatter, and to produce a high-quality surface finish in deep cuts.

Interlocking Side Milling Cutters. The interlocking side milling cutter is made in two halves. These halves are placed side by side and interlock (Figure 32–9). The teeth around the circumference of the interlocking cutters are alternately long and short. The alternate teeth interlock. The cutters may be separated for a specific width by using spacing washers. The amount of separation is controlled by the overlap range of the interlocking teeth.

Interlocking cutters are used for milling plain faces and other parallel surfaces to close tolerances. If the cutters are to be used as adjustable side milling cutters, it is practical to regrind the face and side teeth. The width is adjusted by inserting spacing washers on the arbor between the right- and left-hand cutters. The cutters may be reground so long as the teeth interlock.

T-SLOT AND WOODRUFF KEYSEAT CUTTERS

T-Slot Cutters. The T-slot cutter is a single-process cutter (Figure 32–10). It is used to mill a T-slot after a groove has been milled with a side or end milling cutter. It is an unusual cutter. It mills five sides of a T-slot at the same time: the bottom, the two interrupted faces, and the right and left sides.

The T-slot cutter is made in one piece. It consists of a shank that fits into an adapter. The end opposite the shank has teeth. These teeth are usually staggered and cut at a helix angle. The area between the teeth and shank is machined to a smaller diameter than the narrow width of the T-slot.

T-slot cutters are available to mill slots in tables, columns, and plates. These slots take standard T-bolts with diameters ranging from 1/4″ (6mm) to 1 1/2″ (38mm). Because of the construction and simultaneous cutting on five surfaces, light feeds are required. High-speed steel and carbide-tipped T-slot cutters are available.

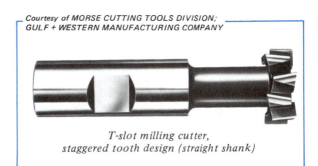

T-slot milling cutter,
staggered tooth design (straight shank)

Figure 32–10 Straight Shank T-Slot Cutter

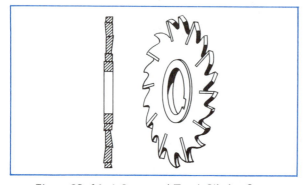

Figure 32-11 A Staggered-Tooth Slitting Saw

Woodruff Keyseat Cutters. Woodruff keys are widely used between two mating parts that are to be keyed together. The Woodruff keyseat cutter is the cutting tool used for milling a semi-circular keyseat. Cutter details and keyseating processes are covered in Unit 37.

SLITTING SAWS AND SLOTTING CUTTERS

Slitting saws are a type of plain or side milling cutter. They are designed for cutoff work on the milling machine and for cutting narrow slots. The three general-purpose types are the *plain* metal-slitting saw, metal slitting saw *with side teeth*, and *staggered-tooth* (Figure 32-11) metal-slitting saw.

Screw-Slotting Cutters. The screw-slotting cutter is designed for cutting shallow slots and screw heads. Screw-slotting cutters look like plain metal-slitting saws. However, the nature of the operations they perform requires different design features. The general cuts taken are short, the depth is shallow, and the material being cut is usually an easily machined steel. These cutters are also used on soft nonferrous metals and thin-sectioned parts. Features and applications of slitting saws and screw-slotting cutters follow in Unit 36.

ANGLE MILLING CUTTERS

There are two basic types of *angle milling cutters: single-angle* and *double-angle*. Some single-angle cutters are of the hole type (Figure 32-12). These cutters are mounted on a regular arbor. Other single-angle cutters are threaded for mounting on a threaded adapter.

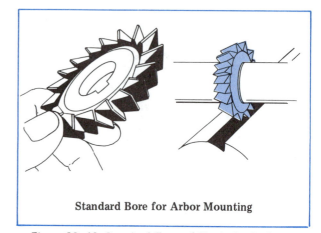

Standard Bore for Arbor Mounting

Figure 32-12 Standard Type of Single-Angle Cutter

Double-Angle Milling Cutters. The teeth of double-angle cutters are V-shaped. They are cut in the two angular faces. Double-angle milling cutters are available with standard included angles of 45°, 60°, and 90°. Double-angle milling cutters with unequal angles are also available.

END MILLING CUTTERS

End milling is the process of machining horizontal, vertical, angular, and irregular-shaped surfaces. The cutting tool is called an *end mill*. End mills are used to mill grooves, slots, keyways, and large surfaces. They are also widely used for profile milling in die making. End mills are coarse-tooth cutters. The teeth are cut on the periphery as well as on the face.

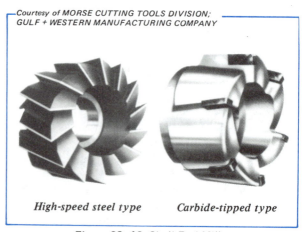

High-speed steel type   *Carbide-tipped type*

Figure 32-13 Shell End Mills

Just two examples of the larger diameter shell end mills are shown in Figure 32–13. Full descriptions, features, specifications, and processes are included later in Unit 34. Small diameter solid end mills are covered in Unit 40 relating to vertical milling machine processes.

FACE MILLING CUTTERS

The *face milling cutter* is essentially a special form of end mill. The size is usually 6″ or larger in diameter. The teeth are beveled or rounded at the periphery of the cutter. Thus only a small portion of each tooth is in contact with a workpiece. The remaining part (front face) of each tooth is ground with a clearance. This clearance reduces the width of the blade and thus eliminates scoring of the milled surface and reduces any drag against the cutter. The face of each tooth acts as a finish cutting tool and removes a small amount of material.

The body of a face milling cutter, like the cutter shown in Figure 32–14 is usually made of an alloy steel. High-speed steel, cast alloy, carbide, or carbide-tipped blades (inserted teeth) are available. The blades are adjustable. They may be either reground and resharpened or replaced.

Face milling cutters are heavy-duty cutters. Heavy cuts, coarse feeds, and high cutting speeds are essential. Hogging cuts also require cutting below the scale on castings, rough edges on parts that are cut out by burning processes, and forgings.

The blades must extend far enough to permit free flow of the large quantity of chips that are produced and must also be rigidly supported. The blades must cut without chatter and must be protected against breakage. Sometimes *gullets* (grooves) are cut into the body. These (gullets) are machined ahead of the blades. The gullets provide additional space for the chips.

The blades are serrated for easy adjustment and secure locking in position. The body may be machined so that the blades are held straight or at an angle to the radial and axial planes. Radial and axial angles give a face mill the same shearing action as the helical teeth on a plain milling cutter.

The blades are secured to the body by screws and other wedging devices. The body on some

face milling cutters is accurately positioned by the internally ground taper bore. The body is held on a stub arbor or other adapter. It may be held securely on the taper by a clamping strap and draw-in bar. Face milling cutters may be either right- or left-hand.

FLY CUTTERS

A *fly cutter* consists of an arbor or other holding device, one or more single-point cutters, and setscrews or other fasteners. The cutting tool is usually a tool bit ground to a desired shape. A fly cutter may be used for internal boring or for external operations. Plain, angular, and form milling are a few examples. Fly cutters are made of high-speed steels, cast alloys, and cemented carbides.

ROTARY FILES AND BURRS

Although they are circular in shape like a milling machine cutter, *rotary files* and *burrs* are not classified as milling cutters. The cutting edges consist of closely spaced, shallow grooves cut around the circumference of the cutter.

Rotary files and burrs are adapted to metal-removing applications using a flexible or other portable hand unit. Trimming a weld, preparing parts to be welded, and removing small amounts of metal in form dies are examples of the type of work performed with these cutting tools. Rotary files and burrs are made of high-speed steel or cemented carbide.

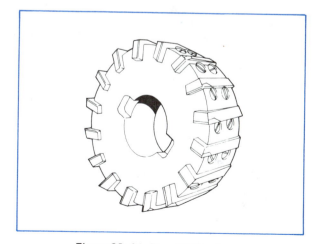

Figure 32–14 Face Milling Cutter

B. SPEEDS, FEEDS, AND CUTTING FLUIDS FOR MILLING

CUTTING SPEEDS IN MILLING MACHINE WORK

CUTTING SPEED DEFINED

Cutting speed (for a milling machine) refers to the distance the circumference of a milling cutter revolves in a fixed period of time (Figure 32–15). The *rate* of speed expresses the distance a point on the circumference of a milling cutter travels in one minute. The rate is given in surface feet per minute (sfpm) or meters per minute (m/min). Too slow a cutting speed reduces cutting efficiency. Too high a speed may cause damage to a cutter or workpiece. Operator judgment is necessary. The cutting speed and feed must be suited to the material and the work processes.

DETERMINING CUTTING SPEEDS

Parts materials vary according to hardness, structure, and machinability. The materials of which cutters are made also differ. High-speed milling cutters are practical for general-purpose operations. Cast alloy and cemented carbides are adapted to the machining of tough, abrasive materials. Cutting with carbide is done at speeds that are faster than the speeds used in the high-speed steel cutter range.

Data on cutting speeds and feeds has been gathered for many years. This data has been collected in tables that cover the ranges of cutting speeds and feeds for different milling cutters and workpiece materials. The tables represent *average* cutting speeds and/or feeds. The cutting speeds may be increased by one-third to two-thirds for finishing cuts. Cutting speed tables give the surface feet or meters per minute at which each material may be machined efficiently. Table 32–1 lists the cutting speeds for five different types of cutter materials. Soft cast iron is used as an example of a workpiece material. A more complete listing of cutting speeds appears in the Appendix.

The operator determines the cutting speed from a work order or a table. This cutting speed must then be related to the diameter of a cutter. The cutter or spindle RPM is then computed.

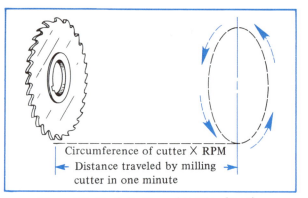

Circumference of cutter × RPM

Distance traveled by milling cutter in one minute

Figure 32–15 Definition of Cutting Speed

COMPUTING SPINDLE SPEEDS (RPM) AND CUTTING SPEEDS

The speed (RPM) at which a spindle should rotate may be calculated by one of the following formulas:

Inch-standard

$$RPM = \frac{\text{cutting speed in sfpm} \times 12}{\text{cutter circumference ('')}}$$

Metric-standard

$$RPM = \frac{\text{cutting speed in m/min}}{\text{cutter circumference (m)}}$$

The same basic formula is used whether the cutter diameter is expressed in inch-standard or SI metric dimensions. For inch-standard dimensions the cutting speed in surface feet per minute is multiplied by 12. The product is divided by the circumference of the cutter in inches. If the cutting speed is given in meters per minute and the cutter diameter is measured in millimeters, the cutter circumference in terms of its meter value is used.

Example: Table 32–1 indicates that a high-speed steel milling machine cutter may be operated wtihin a range of 50 to 80 sfpm. Assume that the highest cutting rate (80 sfpm) is to be used on a solid, soft-cutting cast-iron part. The part is to be face milled. A 3.500″-diameter, plain, high-speed steel, helical milling cutter is selected. Determine the spindle (and cutter) RPM. Use $\pi = 3$.

Step 1 cutter circumference = $\pi \times d$

$= 3 \times 3.5 = 10.5''$

Table 32–1 Example of Cutting Speed (sfpm)

Material to Be Milled	Milling Cutter Material				
	High-Speed Steel	Super High-Speed Steel	Stellite	Tantalum Carbide	Cemented Carbides
Cast Iron Soft	50 to 80	60 to 115	90 to 130		250 to 325

Step 2 cutting speed of HSS cutter = 80 sfpm

Step 3 $RPM = \dfrac{CS \times 12}{C''}$

$\qquad = \dfrac{80 \times 12}{10.5} = 91.42$

Step 4 Select the closest (next lower) spindle speed to 91.42 RPM—for instance, 90 RPM.

Sometimes it is necessary to compute the cutting speed (sfpm) when the spindle speed and cutter diameter (d) are known.

Step 1 CS in sfpm $= \dfrac{(\pi \times d) \times RPM}{12}$

Step 2 Using RPM = 90, the cutter diameter of 3.500″, and π rounded off to 3,

$\qquad CS = \dfrac{(3 \times 3.5) \times 90}{12} = 78.75$ sfpm

The variation between the initial cutter speed of 80 sfpm and the computed 78.75 sfpm is due to using 90 RPM as the closest spindle speed. In general shop practice the following simplified formula for cutting speed (sfpm) is used in the inch-standard system:

$\qquad RPM = \dfrac{sfpm \times 4}{d}$

or

$\qquad sfpm = \dfrac{RPM \times d}{4}$

FACTORS THAT INFLUENCE CUTTING SPEEDS

Tables show, in general, that the harder or more abrasive the material, the slower the cutting speed. In addition to the materials in a work-piece and cutter, a number of the following factors also affect cutting speeds:

- Depth, width, and thickness of the cut;
- Design and sharpness of the cutter teeth;
- Method of holding and supporting the cutter and workpiece;
- Nature and flow of cutting fluid;
- Required degree of surface finish and dimensional accuracy;
- Type and condition of the milling machine.

SPINDLE SPEED DRIVE AND CONTROL MECHANISMS

Spindle speed changes are usually made by arranging gears in specific combinations. Spindle speeds are indicated on the side of the column either on an index plate or a dial. Gear combinations are controlled by levers. The levers are a mechanical means of sliding (shifting) gears in the transmission to produce a desired speed. Speeds in the geared head are usually provided in two ranges: *low* and *high*. The low range uses a back-gear reduction. The back gears are not engaged in the high range.

The spindle speed on some small milling machines is controlled by shifting belts. Other small machines include a variable-speed drive. The speeds are changed hydraulically. Large, heavy machines require a gearbox. The gear combinations are housed inside the upper section of the column. The gear combinations are controlled on some large models by a single speed-selector lever or crank.

Another gear change setup for controlling spindle speeds is illustrated in Figure 32–16A. The machine has two speed-selector levers. The lower lever may be moved in an arc to any one of three positions. The top lever may be set to one of five positions: A, B, C, D, or E. With the

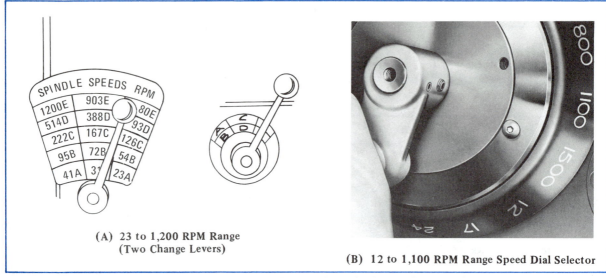

(A) 23 to 1,200 RPM Range
(Two Change Levers)

(B) 12 to 1,100 RPM Range Speed Dial Selector

Figure 32-16 Speed Index Plate and Dial for Setting Spindle RPM

top lever in position A, the bottom lever may be set at 23, 31, or 41 RPM. The two levers provide fifteen spindle speeds, ranging from 23 to 1,200 RPM. By contrast, the single dial selector in Figure 32–16B provides a series of spindle speeds from 12 to 1,100 RPM.

Spindle speed changes on geared-head milling machines are made when the spindle is stopped. Forward (clockwise) and reverse (counterclockwise) spindle rotation is controlled by a spindle starting and direction lever. A push-button combination is used on machines equipped with a reversing motor.

FACTORS THAT AFFECT CUTTING FEEDS

INCREASING AND DECREASING THE RATE OF FEED

Feeds per tooth for general milling processes are suggested in Table 32–2. The range extends from feeds for rough milling operations on hard, tough ferrous metals to light cuts on nonferrous metals. Feeds are given in 0.000″ and 0.00mm. An average starting feed for each process, in both fractional inch and millimeter measurements, appears under the "Average Starting Feed" column in the table.

Table 32-2 Suggested Feeds per Tooth for General Milling Processes

Process	Average Starting Feed		Range of Feed	
	0.000″	0.00mm	0.000″	0.00mm
Face milling	0.008	0.20	0.005–0.030	0.10–0.80
Straddle milling	0.008	0.20	0.005–0.030	0.10–0.80
Channeling or slotting	0.008	0.20	0.005–0.020	0.10–0.50
Slab milling	0.007	0.20	0.005–0.020	0.10–0.50
End milling or profiling	0.004	0.10	0.002–0.010	0.05–0.25
Sawing	0.003	0.10	0.002–0.010	0.05–0.25
Thread milling	0.002	0.05	0.001–0.005	0.02–0.10
Boring	0.007	0.20	0.005–0.020	0.10–0.50

Table 32–3 Recommended Feed per Tooth (0.000'') for Cemented Carbide Milling Cutters (Partial Table)

Material	Face Mills	Spiral Mills	Side and Slotting Mills	End Mills	Form Relieved Cutters	Circular Saws
Malleable iron and cast iron (medium hard)	0.016	0.013	0.010	0.008	0.005	0.004

Feeds may be increased for certain operations like slab milling; heavy roughing cuts; abrasive, scaled surface conditions; and easily machinable materials. On light cuts it is possible to increase both the feed and the speed.

COMPUTING CUTTING FEEDS

Feed is a combination of distance and time. The time interval is expressed as *per minute*. The distance represents how far a workpiece moves during a milling process. Feed may be given in terms of inches per minute (ipm) or millimeters per minute (mm/min). Feed may also be stated as the thickness of the chip that is cut away by each tooth.

Some feed tables list recommended feeds (0.000'' or 0.00mm) per tooth for selected materials and milling processes. Table 32–3 shows one line from the feed table in a handbook. Six milling machine processes are recorded. The recommended feed per tooth is given for using cemented carbides to mill malleable and medium-hard cast-iron parts.

Example: A six-tooth spiral mill is used on a malleable iron casting. Each tooth cuts a 0.013'' thick chip. The cutter rotates at 200 RPM. The automatic feed (inches per minute) for roughing out is calculated in the following manner:

Step 1 The feed per revolution of the spiral mill is equal to the feed per tooth (0.013'') multiplied by the number of teeth, or 0.078''.

Step 2 The feed (inches per minute) equals the feed per revolution (0.078'') multiplied by the number of revolutions of the cutter per minute (200), or 15.6 inches per minute.

FEED AND SPEED FORMULAS (MILLING PROCESSES)

Table 32–4 gives the formulas for computing different speed and feed values for milling machine processes. The earlier used Table 32-1 provided a single example from a larger table of cutting speeds. The cutting speeds in sfpm are given for milling soft cast iron using four different cutter materials.

Table 32–4 Formulas for Computing Feeds and Speeds for Milling Processes

Required Value	Formula
Feed per Tooth (F_t)	$F_t = \dfrac{F''}{n \times RPM}$
Feed per Revolution of Cutter (F_r)	$F_r = F_t \times n$ $F_r = \dfrac{F''}{RPM}$
Feed in Inches per Minute (F'')	$F'' = F \times n \times RPM$ $F'' = \dfrac{i^3 pm}{D \times W}$
Revolutions of Cutter per Minute (RPM)	$RPM = \dfrac{sfpm \times 12}{C \text{ in Inches}}$ Simplified formula: $RPM = \dfrac{sfpm \times 4}{d}$
Cutting speed (sfpm)	$sfpm\ (CS) = \dfrac{C \times RPM}{12}$ Simplified formula: $sfpm = \dfrac{d \times RPM}{4}$

Symbols

F''	=	Feed in inches per minute	sfpm	=	Surface feet per minute
F_t	=	Feed per tooth	CS	=	Cutting speed
n	=	Number of teeth in cutter	i^3 pm	=	Cubic inches of material removed per minute
C	=	Circumference of cutter			
d	=	Diameter of cutter in inches	D	=	Depth of cut
			W	=	Width of cut
RPM	=	Revolutions of cutter per minute			

FEED MECHANISMS AND CONTROLS

The designs of feed mechanisms, like the designs of speed devices, vary. Some mechanisms provide fixed feeds. Other mechanisms provide variable feeds that give an infinite number of feeds in ipm or mm/min. Some of the heavy milling machines have a range of variable feed rates from 1″ to 150″ (25mm to 3,800mm) per minute. This range permits the skilled operator to select the desired thickness of chip per tooth.

Rapid traverse speeds promote a fast operating cycle. Rapid traverse feeds up to 300 (7,600 mm/min) are used in some semiproduction machines.

Feed selection is made during the cutting cycle. A cutter is fed slowly into a workpiece. The feed is then increased or decreased. The feed is determined by the amount the cut is varied for depth and width.

Feed controls are usually independent of the spindle speed. Changes in spindle speed do not affect the rate of feed. The range of feeds may be indicated on a dial or index plate. A single feed-selector lever is usually used on a dial-type feed control.

On other controls two levers are used. One lever is moved vertically to one of a specific number of positions. Similarly, the second lever is positioned horizontally. These two levers control the gear feed combinations and therefore provide a range of different feeds; for example, 3/4″ (19mm) per minute to 30″ (760mm) per minute. Some machine feed controls have sixteen change feeds. The horizontal feeds range, in geometric progression, from 1/2″ to 60″. The vertical (knee) feeds are from 1/4″ to 30″.

Milling machines are provided with regular and rapid power feeds. The knee may be raised or lowered. The regular feed is engaged by using a directional lever (up–neutral–down).

The saddle may be moved toward the column or away from it. A directional power feed (in machines that are so equipped) provides for cross feeding.

The table may be moved longitudinally by power. The feed is engaged by a directional feed-control lever. When the lever is positioned to the right, the table feeds to the right. In the neutral (center) lever position the table remains stationary. In the left lever position the table feeds to the left.

The feed and traverse controls for the table, knee, and saddle of a heavy-duty plain milling machine are identified in Figure 32–17.

How to Set the Feed (Table, Saddle, and Knee)

STEP 1 Determine the correct feed.

STEP 2 Move the power feed lever(s) to the required feed position(s). The lever position(s) is usually shown on the feed index plate or dial.

STEP 3 Move either the longitudinal (table), transverse (saddle), or vertical (knee) feed-control lever. The lever to use depends on whether the table, saddle, or knee is to be power fed. The direction in which the lever is moved depends on the required direction of power feed.

STEP 4 Turn the appropriate clamp lever to permit the table, knee, or saddle to slide. The other clamp levers are tightened to securely hold the corresponding milling machine units.

Note: If necessary, the trip dogs should be set. The dogs will trip (stop) the power feed at set positions and thus prevent damage to the workpiece, setup, and machine.

STEP 5 Use the rapid traverse lever to quickly position the workpiece and cutter.

STEP 6 Disengage the power feed, and engage the handwheel for hand-feeding.

Note: Hand-feeding is used to start a cut. When the operator is assured that the milling process can be carried on safely, power feed is then engaged.

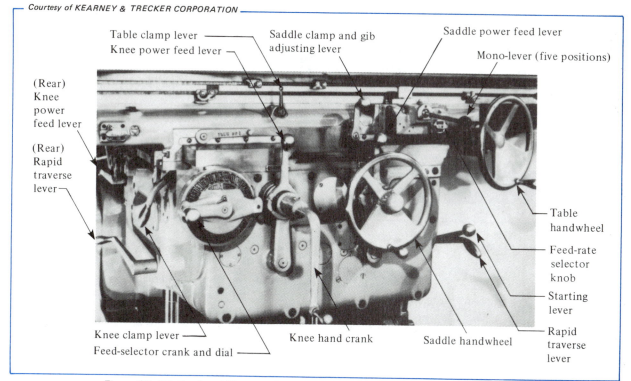

Table clamp lever
Knee power feed lever
Saddle clamp and gib adjusting lever
Saddle power feed lever
Mono-lever (five positions)

(Rear) Knee power feed lever

(Rear) Rapid traverse lever

Table handwheel
Feed-rate selector knob
Starting lever
Rapid traverse lever

Knee clamp lever
Feed-selector crank and dial
Knee hand crank
Saddle handwheel

Figure 32–17 Feed and Traverse Controls of a Heavy-Duty Plain Milling Machine

CUTTING FLUIDS AND MILLING MACHINE SYSTEMS

COOLING PROPERTIES OF CUTTING FLUIDS

A cutting fluid must have at least the following important properties:

- A high specific heat factor, which relates to the ability to absorb heat (the greater the capacity to absorb heat, the faster the cooling action);
- Sufficiently low viscosity, which regulates the property of the cutting fluid to cling and to flow (Figure 32–18).
- Maximum fluidity to penetrate rapidly to the cutting edges.
- Resist deterioration resulting from excessive heat.

LUBRICATING PROPERTIES OF CUTTING FLUIDS

Figure 32–19 illustrates the considerable force required to shear material. The friction of

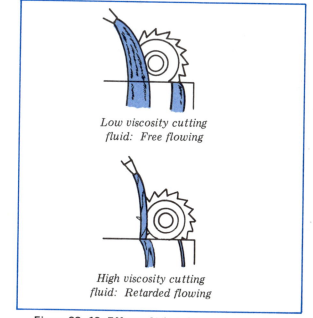

Low viscosity cutting fluid: Free flowing

High viscosity cutting fluid: Retarded flowing

Figure 32–18 Effect of Viscosity on the Flowing Properties of Cutting Fluid

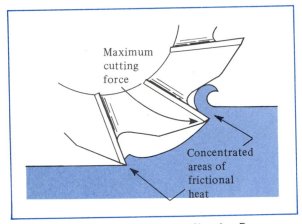

Figure 32–19 Heat-Generating Shearing Forces
of Milling Cutter Teeth

the cutting action generates heat. A good cutting fluid reduces friction in the following ways:

- By providing a film of lubricant between the contact surfaces and work.
- By lubricating the chip, the contact surfaces of the workpiece, and the milling cutter.
- By maintaining a film of lubricant between the cutting edges and faces of the cutter and the chip. The frictional heat generated by chips sliding over a cutting face may become so intense that the chips fuse to the cutter teeth.
- By quickly flowing chips away from the cutting area.

Not all metals require a lubricant. Cast iron should be cut without a lubricant because it contains *free graphite*. The free graphite lubricates the cutting tool. Any additional oily lubricant produces a gummy substance that retards chip movement. The chips tend to clog the chip spaces, and the cast iron becomes glazed.

INCREASING TOOL LIFE

A good cutting fluid prevents the development of frictional heat and resists the adhesion of chip particles to the cutting edges and faces. The cutting edge is also protected from wearing away.

A great deal of milling with carbide cutters is done dry. However, when a cutting fluid is used, there must be a large, constant flow. Interrupted flow of the cutting fluid on a carbide tool causes thermal cracking, which, in turn, produces chipping of the tool.

IMPROVING AND PRESERVING SURFACE FINISHES

A good surface finish may be produced by dry machining certain metals like cast iron and brass. The surface finish of other materials can be improved by using a cutting fluid to remove the chips.

Preserving the surface finish means selecting a cutting fluid that will not discolor or rust a finished surface. Discoloration and corrosion of nonferrous metals are produced by additives in the cutting fluid. Rusting of ferrous parts may be avoided by rapidly evaporating the water in a cutting fluid solution. The remaining oil coats and protects the surface.

CHIP ELIMINATION

To produce ideal cutting conditions, a large enough quantity of cutting fluid must flow in order to:

- Move the chips off the cutting edges and out of the cutter flutes,
- Flow chips away from the cutting area. If the chips are drawn back between the revolving cutter and the workpiece, additional cutting is required.

There are only a few teeth in continuous contact during a milling process. Therefore, the rate of flow and the selection of an appropriate cutting fluid should permit the chips to be removed during the noncutting time.

KINDS OF CUTTING LUBRICANTS AND COMPOUNDS
CUTTING OILS

Generally, mineral, animal, and vegetable cutting oils are *compounded* (blended). The blending is done for economical reasons as well as for improving certain cutting qualities. The straight cutting oils include lard oil and untreated mineral oils. Lard oil is mixed with mineral oils to prevent it from becoming rancid and clogging feed lines. Lard oil is, however, an excellent tool lubricant under severe cutting conditions.

When lard oil is mixed with cheaper mineral oils, the percentage of lard oil is determined by the nature of the cutting action and the hardness of the material to be cut. Mineral oils generally have better lubricating qualities than soluble oils. Lard oil is added when the cutting forces are too severe for straight mineral oils.

Straight cutting oils are commercially produced in three basic grades. The general *all purpose cutting oil* has low viscosity and is used on free machining steels, brass, and aluminum. *Medium-duty cutting oils* are applied to the machining of high alloy and stainless steels and other difficult-to-machine metals. *Heavy-duty straight cutting oils* are used for fast feeds, heavy cuts, and difficult machining processes on ferrous metals, high alloy steels, and stainless steels.

WATER SOLUBLE COOLANTS

Water has excellent properties to rapidly absorb and carry off heat. Unfortunately, water has two drawbacks. First, rust forms on the workpiece and machine surfaces of the miller. Second, water thins the lubricating oils needed between mating surfaces.

Some soluble oils are mixed with water to form an *emulsion*. The emulsion is milky in appearance. The soluble oils are usually a mineral oil, a vegetable oil, or an animal oil. An *emulsifying agent* is added to allow the cutting oil to form a stable mixture with water. The consistency of the emulsion depends on the ratio of water to soluble oil. The emulsion consists of fine droplets of oil that are suspended in the water, soap, and oil mixture. Other soluble oils are available in a clear, transparent mixture.

In addition to rust and corrosion inhibitors, water soluble coolants include special additives to ensure that the coolant does not break down under extreme lubricity conditions. Water soluble coolants have greater cooling ability than

straight cutting oils and machine parts and workpieces are easier to clean.

Water soluble coolants are widely used for such basic processes as: drilling, tapping, reaming, turning, milling, grinding, and sawing. A number of commercial grades are available for severe machining processes on ferrous and nonferrous metals where heavy feeds and extreme speeds are used.

SYNTHETIC AND SEMISYNTHETIC COOLANTS

Synthetic coolants reduce the surface tension of water so that the coolant may freely flow to penetrate between the cutting tool and the workpiece. Heavy-duty synthetic coolants (which include additives) are recommended for hard-to-machine applications.

Semisynthetic coolants are transparent when mixed with water and provide good visibility of machining operations. *Bacteriacides* are added to prevent the breakdown of a coolant and to add to its work life; rust and corrosion preventatives, to protect machined surfaces and workpieces. Semisynthetic coolants are applied in difficult machining operations.

SULPHURIZED OILS

The addition of sulphur to a cutting fluid permits the cutting speed to be increased significantly. The cutting fluid is also able to withstand the greater forces that accompany heavy cuts on tough materials. Sulphurized oils are dark in color. They continue to darken as more sulphur is added, until all transparency is lost.

A pale yellow, transparent cutting fluid is produced by mixing mineral oil with a *base oil* (such as lard oil) to which sulphur is added. The amount of sulphur depends on the nature of the operation.

KEROSENE

Kerosene is used primarily on nonferrous metals like aluminum, brasses, bronzes, magnesium, and zinc. The machinability ratings on these metals are above 100 percent.

AIR AS A COOLANT

Although air is not a cutting fluid, it performs two cutting-fluid functions. First, a flow of air under pressure helps to remove chips from the cutter and workpiece. Second, the air cools the cutting tool, the chips, and the part. Air is used on cast iron and other metals and materials where a cutting fluid cannot be applied. The air stream may be produced by suction or as a blast.

RUST AND CORROSION INHIBITORS

As previously stated, rust and corrosion preventative oils provide surface protection for short-term indoor as well as long-term outdoor storage. Rust and corrosion preventatives are widely used on machine tools, precision tools or instruments, punches and dies, jigs and fixtures, etc., and for the storage of other equipment and raw materials.

APPLICATION OF CUTTING FLUIDS

CIRCULATING PUMP SYSTEM

A circulating pump system provides a continuous volume of cutting fluid. The hollow base of a milling machine usually contains the cutting fluid reservoir. A pump drains the cutting fluid from the reservoir and forces it through the feed lines. A flexible tube at the cutter end permits the nozzle (distributor) to be positioned to reach any point on any style of arbor. The cutting fluid flows through the nozzle and over the cutter, workpiece, and chips. The amount of flow is controlled by a shut-off valve near the nozzle. The design and shape of the nozzle shapes the stream. The nozzle may produce a wide stream for applications such as slab milling, or separate streams for multiple cutters, or a stream confined to one narrow cutter.

The cutting fluid and chips drain through the T-slots and channels in the table. The fluid drains through strainers into troughs at the end of the table. The fluid then passes through tubing back to the reservoir to be recirculated. The rimmed base has a screened opening that permits any cutting fluid that reaches the base to be screened and returned to the reservoir.

Chips and other particles are first strained out by metal screens (sieves). Some screens are placed in the ends of the table or in the hose connection. Some machine designs include additional screening. The cutting fluid returns to the fluid tank, where it flows through a box-shaped strainer into the reservoir.

The whole circulating system and reservoir require regular cleaning. Sediment that accumulates in the bottom must be removed. The cutting fluid requires constant checking for composition, viscosity, and sediment. Once checked, the cutting fluid must be brought up to specifications. Recommended cutting fluids for various machining operations and machinability ratings of ferrous and nonferrous metals are provided by product manufacturers and technical handbook tables.

How to Supply Cutting Fluids

Using a Constant-Flow Circulating System

STEP 1   Check the composition of the cutting fluid.

Note: The comparator shown in an earlier unit may be used. A fluid sample is drawn from the reservoir and tested. The sample should be representative of the solution in the reservoir.

STEP 2   Select the distributor. The type should provide the kind and quantity of flow needed for the particular job.

Note: One of the three common types of cutting fluid distributors may be selected. The elongated type provides a uniform flow across a wide cutter. The multiple-opening type directs the fluid over a number of cutters. The single-opening type directs a limited flow over a narrow-width cutter.

STEP 3   Attach the distributor to the flexible tube or pipe.

STEP 4 Position the distributor bracket. Move it in the overarm until the distributor is centered with the cutter.

STEP 5 Adjust the distributor in relation to the face of the cutter. Move it to within one inch of the cutter.

STEP 6 Place splash guards to confine the fluid to the table area if a large volume of cutting fluid is to be used.

STEP 7 Start the circulating system pump.

STEP 8 Move the control valve slowly until the desired flow is delivered.

STEP 9 Start the milling operation.

Note: Chips may need to be removed from the table during heavy cutting operations. The operator needs to check continuously to see that there is a free circulation of the cutting fluid.

Caution: Any cutting fluid that drops on the floor must be wiped up immediately. Oily cloths and rags should be placed in a metal container.

STEP 10 Check during the cutting operation to see that there is a continuous flow of cutting fluid. Adjust the valve to control the flow.

STEP 11 Turn off the control valve and the machine at the end of the operation or to set up a new workpiece.

Note: Standard machine cleanup procedures are followed.

MIST COOLANT SYSTEMS

Mist cooling systems are designed to direct a high-velocity mist spray of lubricant between a cutting tool and the workpiece. These systems are effective for heavy cutting processes at high rates of speed and feed.

Mist coolant systems operate on the "Venturi" principle. As a mass of gas or a fluid (air in this case) that is moving at a high velocity is

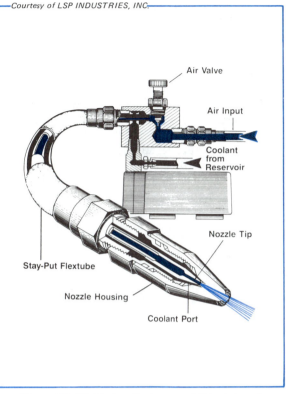

Courtesy of LSP INDUSTRIES, INC.

Figure 32-20 Design Features and Principles of Operation of a Mist Coolant System

released through the small end of a tapered nozzle, the air/coolant mixture expands and the temperature drops. In the mist coolant system illustrated in Figure 32-20, as the high velocity air stream flows to the nozzle tip it draws coolant from the reservoir to the coolant port. At this point, the air and coolant form a mist spray.

As the mist coolant spray comes in contact with heat that is produced by a machining process, the fine particles of water in the coolant evaporate rapidly to draw off heat. The powerful air stream in the spray serves three main purposes: (1) to provide fresh coolant, (2) to spread the evaporation process, and (3) to help prevent chip welding and friction buildup during machining.

Mist coolant systems are designed with *fixed* or *adjustable flexible tubing* to position the mist spray nozzle. A *control valve* is provided for regulating the spray pattern.

A. Safe Practices in the Care and Use of Milling Cutters

- Stop the milling machine before setting up or removing a workpiece, cutter, or accessory.
- Loosen the arbor support arm screw slightly. The arm should move but should still provide support while the arbor nut is loosened.
- Make sure that all chips are removed from the cutter and workpiece and that the cutter is not hot. A wiping cloth should be placed on the cutter to protect the hands. Use two hands when sliding the cutter off the arbor.
- Determine the hand of the cutter, the required direction of spindle rotation, and whether the arbor nut is right- or left-hand.
- Protect the table surface with a wiping cloth or protective tray if setup and measuring tools are to be placed on it.

B. Safe Practices in Operating Speed, Feed, and Coolant System Controls

- Reduce the spindle RPM if there is excessive wear on the lands.
- Shift the speed-selector lever to change the spindle speed in a geared head when the gears are stopped.
- Start a cut by hand-feeding. Once it is established that the work and cutter are correctly positioned and there is good cutting action, either the speed or feed, or both, may be increased.
- Set the trip dogs before engaging a power feed.
- Check clearances between the cutter, arbor support, work-holding device, and workpiece before engaging the rapid traverse or regular power feed.
- Sample the cutting fluid by testing for composition, viscosity, and sediment. Bring the cutting fluid back to strength before using it on a cut.
- Wipe up any cutting fluid that spills on the floor. Dispose of oily waste and wiping cloths in a metal container.
- Avoid skin contact with cutting fluids. Direct contact with cutting fluids may cause skin infections.
- Wear a protective shield or safety goggles. Secure all loose clothing when working with or around moving machinery.

TERMS USED WITH MILLING CUTTERS, SPEEDS, FEEDS, AND CUTTING FLUIDS

A. FEATURES AND APPLICATIONS OF STANDARD MILLING CUTTERS

Radial rake	The angle formed by the face of a cutter tooth and a radial plane. The plane extends from the outside tooth edge through the centerline.
Negative, zero, and positive rake	Radial tooth (rake) angles that provide maximum tool life and cutting efficiency. A required rake angle for a specific workpiece, machining process, and cutter application.
Axial clearance	Clearance ground on the side of a cutter. Hollow grinding side teeth to prevent binding and to reduce friction on the sides of a cutter.
Plain milling cutter (light- and heavy-duty and helical)	A class of cutters designed for light, medium, and slab (coarse, heavy) milling. Straight (parallel) teeth or teeth cut at an angle.
Side milling cutter	A combination of teeth machined on the face and side of a milling cutter. A cutter with teeth designed for side and/or face milling operations.
End milling	Milling processes using the end and/or face of a cutter. Cutting with a solid-shank or adapter-held shell end mill.

Face milling	Usually, the process of machining the face of a workpiece. Using a milling cutter that cuts on the beveled or round cutting edges on the periphery of the cutter.
Staggered teeth	Teeth on a circular cutter that alternate with a right- and left-hand helix. Additional chip clearance and tooth relief is provided on the noncutting side.
Interlocking cutters	Milling cutters with sides recessed and alternate teeth ground back on the inside faces. Two cutters that are meshed.
Single- and double-angle cutter	The included angle and angle form of a milling cutter. The number of angle faces on a cutter: one (single-angle) or two (double-angle).
Woodruff keyseat cutter	A milling cutter designed to mill a circular groove (keyseat) to a specified diameter and width.
Formed-tooth cutter	A tooth form that produces a desired profile. A cutter with teeth relieved to the same shape and size behind the cutting face.

B. SPEEDS, FEEDS, AND CUTTING FLUIDS FOR MILLING

Spindle speed (RPM)	The revolutions per minute that a spindle should rotate for a milling cutter to cut at a required cutting speed.
Cutting speed (sfpm or m/min)	The distance the circumference of a revolving cutter travels in one minute.
Constant-rate feed drive	Table, saddle, and knee feeds that are independent of the spindle speed (RPM).
Feed (per tooth)	The thickness of a chip removed by one tooth of a milling cutter. Usually the chip thickness in thousandths of an inch or metric equivalents.
Feed-control levers	Separate levers that control the direction and movements of the knee, saddle, and table.
Rate of feed (ipm or mm/min)	The distance a workpiece feeds into a milling cutter in one minute.
Rapid traverse	A mechanism for increasing feeds by moving the rapid traverse positioning lever. A rapid feed for speeding up the process of bringing a cutter and workpiece together.
Speed dial or index plate (geared head)	A circular, cone-shaped, or rectangular index plate on which speeds are indicated. A dial with speeds marked for different gear combinations.
Feed-selector dial or index plate	A circular, cone-shaped, or rectangular index plate on which the ranges of feeds in sfpm or mm/min are given.
Heat removed by radiation	Heat that moves through a body and is dissipated in the surrounding area. The natural flow of heat through a milling cutter, chips, and workpiece.
Cutting coolant	Straight cutting oils, water soluble, synthetic, and other fluids with friction/heat lubricating properties and the ability to rapidly flow chips away from a cutter and workpiece.
Mist coolant spray system	A high velocity fluid system that produces and delivers a fine-particle mist spray mixture of air, coolant, and other additives.
Circulating pump system	A mechanism for flowing the cutting fluid on a cutter and workpiece at a constant rate. A reservoir, circulating pump, tubing, flexible nozzle, valves, and applicator. A system for circulating and recirculating a cutting fluid.

─────────────────── SUMMARY ───────────────────

A. FEATURES AND APPLICATIONS OF STANDARD MILLING CUTTERS

- Common ANSI terms are used to describe the features of standard milling cutters.
    - Milling cutters range from single-point fly cutters to multiple-tooth hole and shank types.
- Milling cutter forms are available to mill flat, angular, shoulder, round, and other contours and for slitting and sawing.
    - A cutting-tool face may be designed with a positive, zero, or negative radial rake. The positive and zero rakes are generally applied to high-speed steel cutters.
- Zero and negative rakes are used with carbide inserts and carbide-tipped cutters. The negative rake gives added strength to the cutting edges. A fine surface finish is also produced.
    - Plain milling cutter teeth are ground with a primary clearance angle on the land.
- Formed-tooth milling cutters are relieved to the same tooth profile as the cutting face.
    - Standard plain milling cutters include types for light- and heavy-duty, flat (horizontal) milling. Helical cutters have a steep (45° to 60°) angle. The number of teeth per inch of diameter is also less.
- Side milling cutters are used primarily for milling steps (shoulders) and sides and for straddle milling.
    - Staggered teeth provide increased space for chip clearance. Each tooth in a set cuts to a narrower width. The teeth usually alternate in a right and left helix angle.
- Interlocking cutters are two cutters with teeth relieved on one face. The overall width of the two cutters may be adjusted with spacing washers. Alternate teeth interlock and overlap.
    - Angles may be milled with single-angle or double-angle cutters. The teeth are ground to cut on the side and angular face. Double-angle cutters may have equal or unequal angles.
- T-slot milling involves first the machining of a slot, followed by forming the remaining five surfaces of the T-slot.
    - Woodruff keyseat cutters produce standard semicircular slots (keyseats). The dimensions correspond with the dimensions of a fitted Woodruff key.
- Solid, straight-shank end mills are secured with a screw, a cam-lock fastener, or a collet. The hole-type end mill is located and held securely in an adapter. Taper-shank end mills are fitted with a self-holding taper.
    - Face milling is a practical method of taking hogging and finishing cuts on work faces that are 6″ wide and larger.
- High-speed steel, cast alloy, and carbide-tipped blades (inserts) are held in a rigidly constructed body. Usually, only the angular or round cutting tooth faces on the edge of the cutter do the cutting.

B. SPEEDS, FEEDS, AND CUTTING FLUIDS FOR MILLING

- The cutting speed of a milling cutter is expressed in surface feet per minute (sfpm) or meters per minute (m/min).
    - The cutting speed is converted to spindle speed, or the RPM that a given-diameter cutter must rotate.
- Feeds are given for general milling processes in terms of ipm or mm/min. In production milling, feeds are stated as cubic inches of material removed in one minute (i^3pm).
    - Cutting speeds and feeds are influenced by the following factors:
        - Size of cut (depth and width),
        - Style and size of cutter,
        - Sharpness of cutter teeth,
        - Cutter and work supports,
        - Type and quantity of cutting fluid,
        - Surface finish and dimensional accuracy,
        - Conditions of the milling machine.
- The higher cutting speeds in the recommended range may be used when:
    - Heat is dissipated quickly,
    - Coarse-tooth cutters with free-cutting helix angles provide fast removal of chips,
    - Side teeth reduce the friction,
    - Cutter support is at a maximum,
    - Efficient cutting fluids reduce friction and lower the cutting temperature,
    - Soft materials are to be cut or light cuts are to be taken.
        - The range of spindle speeds is indicated on a dial or index plate on the column. Large machines have a high and a low range of speeds. These speeds are controlled by a lever that sets the gear combinations on a geared-head machine. Some small machines are equipped with an infinite variable-speed drive.
- A lever usually controls the forward, neutral, and reverse direction of the spindle.
    - Feeds are indicated on a dial or feed index plate. Usually the longitudinal and transverse feeds are the same. Feed-selector levers control the direction of the longitudinal, transverse, and vertical movements and feeds.
- When the saddle, table, and/or knee are to be held stationary during a milling operation, the respective unit is clamped securely.
    - Cutting fluids remove heat that is not carried off by radiation, reduce friction and high temperatures on the cutting edges, and maintain a lubricating film between the cutter, chip, and workpiece. Excessive temperatures and forces tend to weld chip particles on the cutting face and rapidly wear away the cutting edges.
- Cutting fluids are recommended by manufacturers according to the machinability and design of the material, the milling operation, and the type and material in the cutter.
    - Cutting fluids may be applied by a continuous flow produced by a circulating pump system or a mist spray coolant system.

A. FEATURES AND APPLICATIONS OF MILLING CUTTERS

1. State the general specifications that are used to order milling cutters for style A and B arbors.

2. Define (a) rake angle and the effect on cutting using milling cutters with (b) positive rake or (c) negative rake.

3. Explain the function that is served by providing eccentric relief on formed cutters.

4. Describe how milling cutter teeth that are formed at a steep helix angle affect: (a) the direction of force and (b) the cutting forces.

5. Differentiate between a half side milling cutter and an interlocking side milling cutter.

6. a. Name three different types of slitting saws.
   b. Give an application of each slitting saw named.

7. Indicate differences between a solid end mill and a shell end mill.

8. Describe (a) the function of a face milling cutter and (b) its major design features.

9. List three safe practices to follow in the care and use of milling cutters.

B. SPEEDS, FEEDS, AND CUTTING FLUIDS FOR MILLING

1. Express how cutting speed and feed rates are specified for milling machine work in inch- and metric-standard systems.

2. Calculate the cutting speed of a 4″ diameter cutter traveling at a spindle speed of 100 RPM.

3. Explain how each of the following conditions affects cutting speeds for general milling machine processes: (a) increasing the width and depth of cut and the chip thickness, (b) changing from a standard tooth to a steep helix angle where a minimum amount of heat is generated, (c) using fine feeds on thin-sectioned workpieces, and (d) improving the efficiency of a cutting fluid.

4. Indicate how each one of the following conditions affects cutting feeds for general milling processes: (a) cutting through scaled surfaces, (b) roughing cuts, (c) deep grooving cuts, and (d) chipping of the cutter teeth at the cutting edges.

5. State three ways in which a quality cutting fluid (where applicable) reduces friction between a cutting tool, the workpiece, and the chips.

6. List three machine safety practices to follow in setting speed, feed, and coolant system controls.

SECTION TWO
Typical Milling Setups and Processes

This section describes the principles of and procedures for performing three of the most common milling processes carried out on the horizontal milling machine:

- Plain milling, face milling, and side milling.

UNIT 33

Plain Milling on the Horizontal Milling Machine

OBJECTIVES

After satisfactorily completing this unit, you will be able to:

- Apply information about chip formation and their effect on surface finish.
- Understand advantages and disadvantages of climb milling and conventional milling.
- Set up a milling machine, cutter, work, and controls for plain milling.
- Diagnose common milling problems, probable causes, and corrective steps.
- Position, layout, and check workpieces for plain milling using a square and surface gage.
- Perform each of the following processes.
  - Milling Flat Surfaces (parallel and right-angle surfaces on square and rectangular workpieces).
  - Milling Ends Square.
  - Laying Out and Checking with a Surface Gage.
- Follow recommended *Safe Practices* relating to work layout and plain milling and correctly use new *Terms*.

In plain milling the competent operator must be able to apply the technology and procedures that relate to each of the following processes:

- Preparing and laying out a workpiece;
- Determining the safest and most practical method of holding a workpiece;
- Setting up a workpiece accurately for each successive operation;
- Selecting and mounting a milling cutter appropriate to the job requirements;
- Determining the correct spindle speed, cutting feed, and cutting fluid;
- Setting up the machine;
- Taking the necessary milling cuts and judging the cutting action of a cutter;
- Measuring a workpiece for dimensional accuracy;
- Cleaning and maintaining the machine and accessories.

Plain milling relates to the machining of a flat surface. The plane of the surface is parallel to the axis of a cutter. The surface is usually produced by a plain milling cutter mounted on an arbor. The terms *peripheral* and *slab milling* are also used. Slab milling is the machining of wide, flat surfaces.

CONVENTIONAL AND CLIMB MILLING

CONVENTIONAL (UP) MILLING

The two common methods of removing metal are called *conventional* and *climb milling*. Conventional milling is also referred to as *up milling*. Figure 33-1 shows the direction a cutter turns in relation to the table feed. The cutter rotates clockwise. The workpiece is fed to the right. If the cutter rotates counterclockwise, the table feed is to the left (into the cutter).

The cutting action takes place from the bottom of the cut to the face of the workpiece. The part is fed into the revolving cutter. The chip starts at a very fine cut. It gradually increases in thickness. The maximum thickness of the chip is at the point where each cutting tooth cuts through the top surface of the workpiece.

Theoretically the cut starts at the perpendicular centerline of the cutter and work. In practice the chip is started just ahead of the center. The delayed action is caused by the teeth sliding over, or *burnishing*, the material. Burnishing occurs momentarily. When a sufficient force is produced, the cutter teeth bite into the surface to start forming the chip. There is also a minute time interval between the cutting action of two successive teeth.

The combined cutting action produces *teeth marks* in a milled surface. The use of a fine feed and a sharp milling cutter traveling at high speed (within the specified range) produces a machined plane surface that meets practical dimensional accuracy requirements.

The variations in the cross section of a chip and the teeth marks on a milled surface are exaggerated in Figure 33-2. The burnishing rather than cutting action tends to dull the cutting edges of a milling cutter.

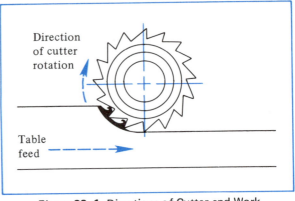

Figure 33-1 Directions of Cutter and Work for Conventional Milling

The major forces in conventional milling are upward. These forces tend to lift a workpiece. In all milling it is important to use a *work stop*. This stop prevents a part and/or work-holding device from moving on the table. A stop is bolted to the table immediately ahead of a workpiece.

CLIMB (DOWN) MILLING

Climb milling is also known as *down milling*. The work is fed in the same direction as the rotation of the cutter teeth. In Figure 33-3, the cutter rotates in a clockwise direction. The

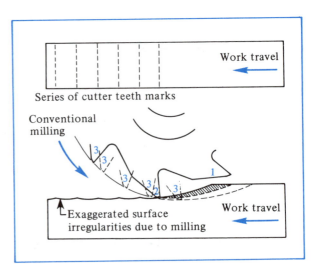

Figure 33-2 Surface Finish Generally Produced by the Cutting Action of a Milling Cutter

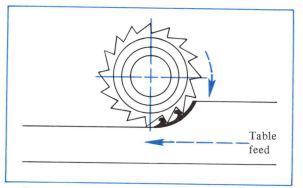

Figure 33–3 A Full Chip Milled from the Top of a Work Surface in Climb Milling

workpiece is fed into the cutter from right to left. If the cutter rotates in a counterclockwise direction, the cutting feed is from left to right.

The cutting forces in climb milling are mostly downward. A stop is usually placed ahead of a workpiece to prevent any movement during a heavy milling operation. There is also a tendency to pull the work into the revolving cutter. Therefore, climb milling is avoided unless the milling machine is rigidly constructed and/or equipped with an *automatic backlash eliminator.* Once the device is engaged, it is automatically activated during climb milling.

The chip formation in climb milling is opposite the chip formation in conventional milling. Figure 33–3 shows that the cutter tooth is almost parallel with the top surface of the workpiece. The cutter tooth begins to mill the full chip thickness. Then, the chip thickness gradually diminishes.

Climb milling has several advantages:

- Climb milling tends to eliminate surface burrs that are normally produced during conventional (up) milling.
- Because revolution and feed marks are minimized, climb milling produces a smoother cut than conventional milling.
- Climb milling is more practical than conventional milling for machining deep, narrow slots. Narrow cutters and saws have a tendency to flex and crowd sideways under the force of a cut. The cutting action in climb milling permits these cutters to cut without springing under a heavy force.

- Climb milling forces the work against the table, fixture, or surface to which it is clamped. Thus, climb milling is desirable for machining thin or hard-to-hold workpieces and for cutting off stock.
- Laboratory tests indicate that less power is required for climb milling than for conventional milling.
- Consistently parallel surfaces and dimensional accuracy may be maintained on thin-sectioned parts.
- Cutting efficiency is increased with climb milling because more efficient cutter rake angles may be used than with conventional milling.

However, climb milling has two significant disadvantages:

- Climb milling is *dangerous.* The milling machine should be equipped with a backlash eliminator. All play must be removed between the lead screw and nut. The ways and sliding surfaces must also be free of lost motion. Backlash, play, and lost motion cause the cutter to pull the work into the teeth.
- Climb milling is not recommended for castings, forgings, hot-rolled steels, or other materials that have an abrasive outer scale or surface. The continuous contact of the cutting teeth on a rough, hard surface causes the teeth to dull rapidly (Figure 33–4).

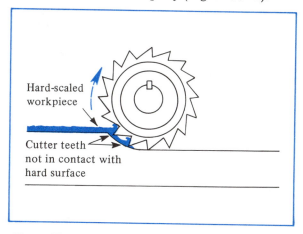

Figure 33–4 Conventional Milling Recommended for Hard-Scaled Castings, Forgings, and Rough Rolled Surfaces

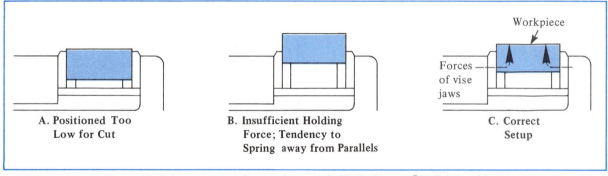

A. Positioned Too Low for Cut

B. Insufficient Holding Force; Tendency to Spring away from Parallels

C. Correct Setup

Figure 33–5 Incorrect and Correct Setups of a Workpiece on Parallels in a Vise

MILLING FLAT SURFACES

SETTING UP THE MACHINE, WORK-HOLDING DEVICE, AND WORKPIECE

Setting up the milling machine refers to the following functions:

- Clamping the work,
- Mounting the cutter,
- Setting the spindle speed,
- Adjusting the feed mechanism,
- Positioning the cutting fluid nozzle.

Small, regularly shaped workpieces are usually mounted in a vise. Large pieces may be fastened directly to the table. Irregularly shaped parts are nested and secured in fixtures.

Most jobs require a roughing cut and a finish cut. The roughing cut should be as deep as possible. The depth is determined by the rigidity of the setup, the capacity of the machine, and the surface condition. Usually 1/64″ (0.015″ to 0.020″, or 0.4mm to 0.5mm) is left for a finish cut. A finish cut is necessary for dimensional accuracy and to produce a high-quality surface finish.

Most workpieces may be conveniently held in a milling machine vise. The vise jaws are positioned accurately. The work is generally set on parallels. The height of the jaw must permit a workpiece to be held securely. The correct setup provides maximum seating and holding power. It is good practice to seat a workpiece on two narrow parallels so that the operator may determine when the part is properly seated.

The correct setup for holding a workpiece on parallels is shown at (C), Figure 33–5.

A protecting strip is placed between any rough surface, the ground jaws of the vise, and the ground faces of the parallels. Often when seating the work, the movable jaw tends to lift the part slightly off the parallels. The condition is overcome by applying a slight force with the movable jaw. The workpiece is then gently but firmly tapped. Too hard a blow tends to make the work rebound. The vise is tightened further. The work is rechecked and reseated if necessary.

THE MILLING CUTTER AND SETUP

A light-duty, heavy-duty, or helical plain milling cutter should be selected. The selection of a correct type of cutter depends on the material, the nature of the operation, and the time within which the part is to be milled.

The cutter or combination of cutters should be wide enough to mill across the width of the workpiece. If two cutters are required, it is good practice to use interlocking cutters. Interlocking cutters permit the cutting forces to be equalized. Usually cutters with the smallest possible diameter are used. The larger the diameter, the longer the feed time required to travel across the length of a workpiece.

When heavy slab cuts are to be taken, consideration must be given to selecting and mounting a cutter so that the cutting forces are directed toward the column. The arbor support must also be positioned to give the greatest possible rigidity.

How to Mill Flat Surfaces
(Plain Milling)

Milling Parallel and Right-Angle Surfaces

The surface that is milled first becomes a reference plane. If the opposite surface is to be milled parallel, the first surface is seated on parallels and secured. The usual procedure for removing burrs is followed prior to positioning. A trial cut is usually taken across the face of the second surface. The ends are measured for parallelism and thickness.

When four sides are to be milled square, the first milled side is held against the solid vise jaw. The adjacent side, if it is straight, may be placed on parallels. Otherwise the face to be milled may need to be set level by using a surface gage or other indicator. The third side is milled by seating the second side on parallels. A trial cut is taken for about 1/4". The part is measured with a steel rule, caliper, or micrometer (Figure 33–6). The knee is adjusted to produce the required dimension.

If the fourth side is rough, a round bar may be placed above center between the movable jaw and the workpiece. When the vise is tightened, the force is toward the solid jaw and parallels. This force tends to seat the workpiece. The workpiece is then measured for size. The knee is adjusted if required. The part is then milled to size.

A number of personal and machine safety precautions must be observed continuously. Once a setup is made, the knee and saddle must be locked. The cutter must rotate in the right direction. The spindle must be stopped before any measurements are taken or before the workpiece is moved. The part must be checked to see that it has not moved. There must be an adequate flow of cutting fluid.

Stopping the cutter partway through a cut causes the milling of a slight depression in the surface. Other unwanted indentations are produced if a revolving cutter is moved across a finished face.

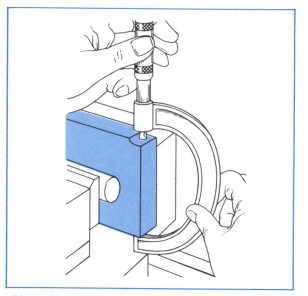

Figure 33–6 Checking Workpiece Size after a Trial Cut

How to Mill Ends Square

Vise and Workpiece Positioned Parallel to the Column Face

STEP 1 Position the vise parallel to the column face.

STEP 2 Place the workpiece on a parallel that rests on the base of the vise. Tighten the vise jaws so that the part may be tapped into position.

Note: A protective metal strip should be used to prevent the hardened vise jaws from scoring any finished surface.

STEP 3 Place the steel square on the base of the vise if possible. Otherwise use a parallel. Slide the blade up to the workpiece. Sight along the blade to see whether the workpiece side is positioned squarely.

Note: If any light shines through, move the square away. Then tap the workpiece to square it. When it is squared, securely tighten the workpiece. Recheck it with the square. A surface gage may also be used if there is a layout line on the part.

STEP Take a roughing cut. Close observation
4 of the milling operation is necessary.

Note: If the cutting force causes the workpiece to shift, immediately move the
cutter away and stop the machine. The
workpiece then needs to be reset and
held more securely.

STEP Continue with the roughing and finishing
5 cuts.
STEP Set up and mill the opposite end by fol-
6 lowing steps 1–5.

Vise Positioned Parallel with the Arbor Axis

Note: Short workpieces that have a small
cross-sectional area may be quickly and
accurately machined while held in a vise.
The vise is positioned parallel with the
arbor axis.

STEP Square the side of the workpiece relative
1 to the base of the vise. The steel square
method may be used.

Note: A layout line may be checked using
the surface gage method. A more accurate
method is to use a dial indicator. The
arm is secured to the arbor. The dial indicator point is brought into contact
with the side of the workpiece. The
knee is raised and lowered as required.
The workpiece is tapped gently until the
dial indicates that the side is vertical
(square with the vise base).

STEP Take a roughing and a finishing cut. Fol-
2 low the steps for plain milling flat surfaces.

Note: At the end of the roughing cut, the
spindle is stopped. The workpiece is
burred. The squareness of the end is
checked with the steel square.

STEP Reposition and secure the workpiece
3 to mill the opposite end. Proceed according to the steps for milling and
measuring two parallel surfaces.

Using an Angle Plate to Mill Ends Square

STEP Select an angle plate that will accommo-
1 date the length of the workpiece. Secure
the angle plate so that the cutting force
is toward the face and knee.
STEP Center the workpiece on the angle plate.
2 Clamp the workpiece so that it may be
squared.

Note: The amount the workpiece extends
beyond the top of the angle plate should
be limited. If needed for support, a packing block and shims should be used between the bottom of the workpiece and
the table.

STEP Square up the workpiece. The dial indi-
3 cator, solid steel square, or surface gage
method may be used. The selection of a
method depends on the accuracy required. Tighten the strap clamp(s). Recheck for squareness.
STEP Take a roughing cut. Clean the work-
4 piece and remove burrs. Check the end
for squareness. Adjust the workpiece if
it has shifted.
STEP Decrease the feed. Increase the speed.
5 Take a finish cut to the layout line or
required dimension.
STEP Stop the spindle. Square up the oppo-
6 site end. Follow the steps for rough and
finish milling.
STEP Measure the overall length for dimen-
7 sional accuracy.

PROBABLE CAUSES OF AND CORRECTIVE STEPS FOR COMMON MILLING PROBLEMS

Vibration of the machine, cutter, or work (or of
all three) causes unsafe machining conditions.

Table 33-1 Common Milling Problems: Probable Causes and Corrective Action

Milling Problem	Probable Cause	Corrective Action
Chatter	—Lack of rigidity in the machine, work-holding device, arbor, or workpiece.	—Increase rigidity; machine support and secure workpiece, cutter, and arbor more effectively
	—Excessive cutting load	—Use cutter with smaller number of teeth
	—Dull cutter	—Resharpen
	—Poor lubrication or wrong lubricant	—Improve lubrication and check lubricant specifications
	—Straight-tooth cutter	—Use helical-tooth cutter
	—Peripheral relief angle too great	—Decrease relief angle
Poor quality of surface finish	—Feed too high	—Decrease feed or increase speed
	—Dull cutter	—Resharpen accurately
	—Cutting speed too low	—Increase sfpm (spindle speed) or decrease feed
	—Cutter has insufficient number of teeth	—Use finer tooth cutter
Cutter digs (hogs) in	—Peripheral relief angle too great	—Decrease relief angle; use recommended angles
	—Rake angle too large	—Decrease rake angle
	—Improper speed	—Check recommended speed; adjust accordingly
	—Failure to tighten saddle or knee clamping levers	—Tighten saddle and knee clamping levers
Vibration	—Cutter rubs; insufficient clearance	—Use staggered-tooth cutter or cutter with side-relief teeth
	—Arbor size and support	—Use larger arbor and adjust support arm

Vibration also often results in damaged work. Some milling machine models have a *vibration damping unit* located inside the overarm. This unit is shown in the cutaway section of the overarm in Figure 33-7. Its function is illustrated by the vibration amplitude line drawing. The vibration (chatter) damping capability permits greater depths of cuts and higher feeds and speeds than are ordinarily possible. Cutting is more efficient and smoother. The vibration damping overarm may be used with both climb and conventional milling.

Chatter, vibration, and other common milling problems, probable causes, and corrective steps are suggested in Table 33-1.

APPLICATIONS OF THE SURFACE GAGE

Many jobs require plain milling on a surface that is irregular, warped, or varied in thickness. In milling such a surface, it is impractical to seat the part directly on parallels. Surfaces of castings, forgings, and uneven parts may be positioned (leveled) for machining by using a surface gage. The part may be held in a vise or it may be strapped to the table.

The surface gage may also be used for layouts (Figure 33-8). Guide lines may be scribed on the sides and ends of the workpiece. These lines are particularly helpful for determining whether the rough surface will clean up when the final finish cut is taken.

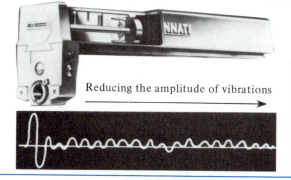

Courtesy of CINCINNATI MILACRON INC.

Reducing the amplitude of vibrations

Figure 33-7 Vibration Damping Unit Inside the Overarm

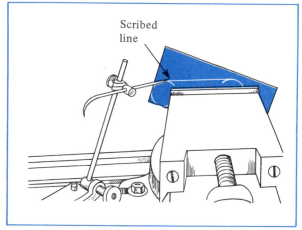

Figure 33-8 Positioning a Layed-Out
Workpiece with a Surface Gage

How to Use a Surface Gage on Milling Machine Work

Leveling a Workpiece

STEP 1 Wipe the table and surface gage. Carefully rub the palm of the hand over the table area and the base of the surface gage.

STEP 2 Adjust the scriber and surface gage spindle. The scriber should be able to reach across each corner of the workpiece.

STEP 3 Adjust the scriber so that it barely touches one corner of the work.

STEP 4 Move the surface gage to the next corner. Slide the scriber to the work. Hold the base against the table. Swivel the surface gage to move the scriber near or over the work surface.

STEP 5 Check the scriber height against the corner height. If it is higher than the scriber, tap the corner to the approximate height of the first corner.

STEP 6 Check the height of the other corners. Tap where necessary to correctly position the workpiece.

STEP 7 Use shims or jacks to support overhanging surfaces.

STEP 8 Tighten the vise jaws.

STEP 9 Recheck the height of each corner with the surface gage. Further adjust if needed.

Note: A dial indicator may be used on smooth surfaces in place of the scriber to provide for a more accurate setting.

Scribing Layout Lines

STEP 1 Chalk or apply a dye to the ends and sides of the workpiece.

STEP 2 Set the scriber point as close to the work as possible. The position must permit scribing on all required sides.

STEP 3 Set the scriber to the required height.

STEP 4 Hold the surface gage base against the table. Slide the surface gage so that the scriber is at an angle to the side to be scribed. Scribe the line (Figure 33-8).

Note: Care must be taken in handling the surface gage and scribing the line. It is good practice to check the height setting after each line is scribed.

STEP 5 Repeat the process on the remaining sides if needed. Transfer the surface gage on the machine table to a place from which the lines may be scribed.

Note: If a side is not accessible, it may be necessary to reset the spindle and scriber. The first scribed line may serve as a locating line.

Using Layout Lines to Position a Workpiece

STEP 1 Mount the layed-out workpiece in a vise, on an angle plate, or in another workholding setup. Tighten the scriber point just far enough to permit adjustment.

STEP 2 Position the scriber point so that it reaches both ends of the layout line.

STEP 3 Tap the high end of the workpiece. When the workpiece is correctly positioned, securely tighten the vise or clamps. Recheck the accuracy of the work position.

How to Complete a Milling Process

STEP 1 Stop the machine.

STEP 2 Remove, clean, and burr the workpiece.

STEP 3 Permit the splash guards to drain for a minute. Remove the chips, wipe the guards, and remove them.

STEP 4 Brush any remaining chips from the cutter, workpiece, and work-holding device onto the table.

STEP 5 Use a pan to remove any quantity of chips after the cutting fluid has drained from them for a short time. Use a T-slot cleaner and hard bristle brush. Clean the T-slots and the well area at the ends of the table.

STEP 6 Lock the spindle. Loosen the arbor nut with the arbor solid open-end wrench.

STEP 7 Loosen the arbor support clamp. Slide the support from the overarm.

STEP 8 Remove the nut, bearing sleeve, collars, key, and cutter from the arbor.

Note: The table should be protected from damage during this step.

STEP 9 Clean all parts. Remove any burrs or nicks. Replace the collars, key, bearing sleeve, and nut on the arbor.

STEP 10 Clean and store the cutter so that the teeth are protected.

Note: If the teeth are dulled or damaged, the cutter should be resharpened before storing.

STEP 11 Remove the arbor. Store type A and B arbors in vertical racks.

STEP 12 Loosen the nuts on the T-bolts of the work-holding device. Clean and examine the tongues and base. Remove any burrs. Store in the machine tender or other special storage compartment.

STEP 13 Brush away chips from all other parts of the milling machine.

STEP 14 Wipe the surrounding areas free of any cutting fluids.

Safe Practices in Plain Milling Setups and Work Processes

• Lock the spindle in the off (stopped) position so that it cannot be tripped acidentally during a machine setup or the taking of a measurement.

• Feed the work in climb milling in the same direction in which the cutter rotates.

• Allow enough material for a finish cut so that the cutter teeth can quickly penetrate the surface and remove a fine chip. Otherwise the resultant burnishing effect will tend to dull the teeth.

• Protect the ground surfaces of parallels and vise jaws from being damaged by the rough surfaces of parts.

• Avoid climb milling with a standard miller unless it is rigidly constructed or is equipped with an automatic backlash eliminator.

• Take a deep first cut below the hard outer scale on castings, forgings, and other irregular or abrasive, rough surfaces.

• Feed a milling cutter into a workpiece by hand. When it is evident that the part will not shift under the cutting force, engage the power feed.

• Mill as close as practical to the work-holding device and arbor support arm. The cutting forces should also be directed toward the spindle.

• Mill toward the solid vise jaw or the vertical leg of an angle plate.

• Lock knee and saddle before engaging the table feed.

• Stop the machine if any one of the common milling problems develops. Check the most probable cause(s). Take the suggested corrective steps.

• Store each tool and machine accessory in its correct rack, bin, or other protected location.

• Place rags and oily cloths in a metal container. The area around the machine should be dry. After wiping up oil on the floor, use a nonskid compound on the oil spots.

TERMS USED FOR LAYING OUT AND MILLING PLAIN SURFACES

Plain (peripheral milling)	Milling a flat horizontal surface parallel to the axis of a cutter.
Conventional (up) milling	Machining on a horizontal milling machine. Feeding a workpiece into the cutting edge of a rotating milling cutter. Milling a chip form that starts at the depth of the cut. (The chip increases to full thickness when the cutting edge cuts through the top surface of the work.)
Climb (down) milling	A machine cutting technique. Feeding a workpiece in the same direction that the milling cutter rotates. Milling a chip that is full thickness (equal to the feed) at the beginning of the cut. (The chip thickness diminishes to almost no thickness as the cutter tooth rotates to the depth of the cut.)
Teeth marks	Deviations from a perfect plane surface. Crests and hollows corresponding to the feed of a milling cutter in conventional milling. Fine surface indentations resulting from a burnishing rather than a cutting action of a milling cutter.
Automatic backlash eliminator	A design feature of milling machines. A device that automatically compensates for lost motion or play, usually between a feed screw and mating nut or between a rack and pinion. A control feature that makes it possible to climb mill safely.
Cleaning (up) the surface	Machining a surface to just below any low spot or rough surface. Producing a flat plane with a minimum amount of milling.
Seating the work	Firmly securing a workpiece on a parallel, packing block, or other work-positioning device. Tapping the work down so that it rests securely on a parallel surface.
Squaring a block (milling machine)	Milling four sides of a square or rectangle. Milling four adjacent sides at 90° to each other. (The opposite sides are parallel along the length.)

SUMMARY

- Milling may be performed by conventional (up) milling or climb (down) milling. Climb milling is not recommended unless a heavy-duty machine or a machine equipped with an automatic backlash eliminator is used.
  - Climb milling has the following advantages over conventional milling:
    - Surface burrs are reduced or eliminated,
    - The finished surface is almost free of cutter revolution and feed marks,
    - Cutting forces are applied against the table or work-holding device,
    - Thin-sectioned surfaces may be machined with a greater degree of dimensional accuracy,
    - Less power is required.
- Conventional milling is recommended when high-speed steel cutters are to be used on castings, forgings, and other materials with a surface scale.
  - Most workpieces that are to be plain milled are held in a standard or universal milling vise, on an angle plate, or in a fixture. Many parts are strapped on parallels directly to the table.

■ Workpieces are positioned so that the cutting force is directed toward the solid vise jaw, the work stop, the fixture, or the angle plate.

　　■ Rough and finish cuts should be taken to produce a high-quality, dimensionally accurate, finished surface.

■ Usually 0.015″ (0.4mm) to 0.020″ (0.5mm) is allowed for a finish cut. The cutter speed is increased. The feed is decreased.

　　■ A cutter is fed by hand for a short distance to permit checking the rigidity of the entire setup and measuring a workpiece. Power feed is then applied.

■ Chatter, holding to size, burnishing, and quality of surface finish are common workpiece problems. Other problems—for example, rapid dulling, burning, and breaking the cutting edges—relate to a cutter. Vibration is a machine problem. Probable causes and corrective steps are provided in table form for easy reference.

　　■ A surface gage may be used to set up a workpiece or to scribe horizontal layout lines.

■ Ground faces of parallels, vise jaws, and angle plates and the milled surfaces of a workpiece all must be protected.

　　■ Personal and machine safety rules governing the use of the milling machine, cutting fluid system, and cutters must be followed.

■ A protective shield or goggles should be worn at all times during any machine process.

UNIT 33 REVIEW AND SELF-TEST

1. Explain why climb milling is preferred over conventional milling under the following conditions: (a) machining deep, narrow slots; (b) cutting off or machining thin, hard-to-hold workpieces; and (c) eliminating surface burrs and feed marks.

2. Indicate the five major functions the operator performs in setting up a milling machine.

3. List three considerations that guide the operator in selecting the appropriate milling cutter.

4. State what advantages a vibration damping unit has over the standard over-arm in controlling chatter.

5. Identify four practical methods of milling ends square.

6. Tell how the surface gage is used to position a rough-surfaced part that is held in a milling machine vise.

7. Indicate why a deep first cut is taken below the hard outer scale on castings, forgings, and weldments.

8. State two safe practices to observe in climb milling.

Face Milling on the Horizontal Milling Machine

Three common methods of *face milling* on the horizontal milling machine are covered in this unit. These methods include applications of inserted-blade face mills, shell end mills, and solid-shank end mills. Procedures are examined for milling shoulders which are produced with shell end mills and solid end mills. Face milling as applied to these mills is referred to as *end milling*.

OBJECTIVES

After satisfactorily completing this unit, you will be able to:

- Identify design features of inserted-blade face mills, shell end mills, and solid-shank end mills.
- Know when to use cutters with positive, negative, and zero radial and axial rake angles for shop applications.
- Understand the relationship of cutter lead angle, rate of feed, and depth of cut.
- Solve problems of feed and cutter lines and relate factors that affect dimensional accuracy and surface quality in face milling.
- Apply special arbors, adapters, sleeves, and collets for mounting and using solid end mills for shoulder milling.
- Perform each of the following processes.
  - Face Mill with a Face Milling Cutter.
  - Disassemble the Work Setup and Cutter.
  - Face (End) Mill with End Mills Using Solid and Shell End Mills.
- Use shop formulas on cutting speeds, spindle RPM, feeds, and rake angles.
- Follow recommended *Safe Practices* and correctly use *Terms* related to face milling.

APPLICATIONS OF FACE MILLING CUTTERS AND END MILLS

Solid end mills are generally used to mill flat surfaces that are smaller than 1 1/2″ (38mm) wide. The range of commercial solid end mills is from 1/8″ (3mm) to 2″ (50mm) in diameter. Shell end mills are normally employed to face mill surfaces from approximately 1 1/4″ (32mm) to 5″ (130mm) wide. Face milling cutters are used in machining surface areas that are wider than 5″ (130mm).

FACE MILLING CUTTER DESIGN FEATURES

The sizes of face milling cutters require that the body be designed of a tough, durable, heat-treated steel. Grooves are cut to accurately position and hold blade inserts. The inserts are made primarily of high-speed steel and cemented carbides. The inserts may be ground, adjusted, and replaced easily and at low cost. Their design makes it possible to use the body indefinitely.

RAKE ANGLES

Figure 34–1 shows some of the design features of a face milling cutter. A face milling cutter is selected in terms of the material to be machined, the nature of the work processes, and the machine setup. Reference is made to manufacturers' tables of rake angles. From these tables the operator establishes whether to use *positive*, *zero*, or *negative radial* and *axial rake angles*. The effects of a 5° positive and a 5° negative rake angle are pictured in Figure 34–2A and B.

Positive rake angles are normally used on high-speed steel cutters. These rake angles are effective for milling tough materials and other materials that work-harden.

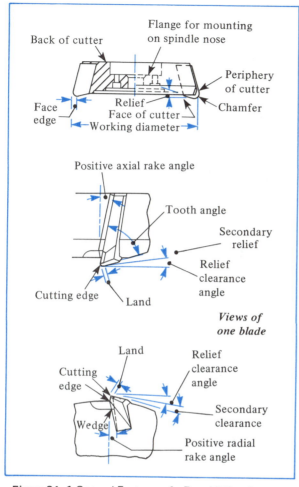

Figure 34-1 General Features of a Face Milling Cutter

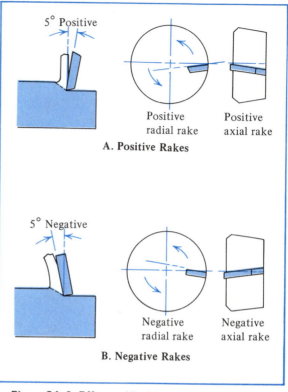

Figure 34-2 Effects of Positive and Negative Rake on Cutting (Exaggerated)

Some advantages of positive rake angles include:

- High quality of surface finish,
- Increased cutter life,
- Less power required in cutting,
- Decreased cutting forces.

Zero and negative rake angles are applied to cemented carbide inserts. These rake angles strengthen the cutting edges to withstand heavy impact and severe cutting conditions. Negative rake angles are not recommended for soft, ductile metals like copper and aluminum or work-hardening materials. Negative rake angles should be used only when a workpiece and machine set-up can withstand extreme cutting forces.

EFFECT OF LEAD ANGLE AND FEED

LEAD ANGLE

The same principle of *lead angle* that is applied in lathe work for a single-point cutter applies to face milling. Small lead angles of from 0° to 3° are often used to machine close to a square shoulder. Figure 34-3A shows that with a limited lead angle, a chip thickness of 0.015″ (0.4mm) is practically equal to the feed of 0.015″ (0.4mm).

Using the same depth of cut but increasing the lead angle to 45° changes the chip thickness (Figure 34-3B) to 0.010″ (0.25mm). Note that as the lead angle increases, the chip becomes wider. A steep lead angle limits the depth of cut. For practical purposes the maximum lead angle on face milling cutters is 30°.

There are a number of advantages to using a sizable lead angle (Figure 34-3C):

- The cutter contacts the workpiece along the blade rather than at the tip of the cutting edge,
- Cutting forces are applied where the strength of the blade is greater than at the edge,
- Only a partial chip is formed on initial impact and at the end of the chip,
- Cutting a thinner chip thickness adds to effective cutting edge life.

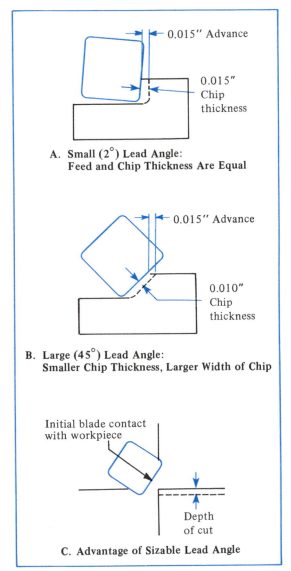

A. **Small (2°) Lead Angle:**
Feed and Chip Thickness Are Equal

B. **Large (45°) Lead Angle:**
Smaller Chip Thickness, Larger Width of Chip

C. **Advantage of Sizable Lead Angle**

Figure 34–3 Effect of Lead Angle
on Chip Thickness

FEED AND CUTTER LINES

A circular pattern of *feed* and *cutter lines* may be produced in face milling. These lines result from the:

- Size of the nose radius of the cutter blades,
- Coarseness of the feed,
- Limited rigidity in the setup,
- Nature of the machining process,
- Cutter size,
- Sharpness of the cutting edges.

The ridges produced by the face mill may be reduced by grinding the cutting edge of each blade flat. The flat should be wider than the feed lines.

FACTORS AFFECTING FACE MILLING PROCESSES

The following major factors affect dimensional accuracy and quality of surface finish:

- Free play in the table movements;
- Eccentricity, or cutter runout. The trailing edges of the cutter produce light cuts that follow the main cut;
- Lack of support of the workpiece at the point of cutting action;
- Need for more rigid stops. Additional stops may be needed to prevent any movement of the workpiece during cutting;
- Position of the table. The table must be positioned as close as possible to the spindle;
- Dull cutting edges. These edges require more power and produce excessive heat;
- Cutter revolution at too slow a speed. A slow speed causes a buildup on the cutting edge;
- Cutter revolution at too high a speed. A high speed produces excessive cutter wear;
- Too great a cutter feed. This feed causes chipping and breaking of the cutting edges;
- Positive angle of entry. Contact is made at the cutting tip, or weakest area, of the blade;
- Negative angle of entry. The initial force is applied along the cutting face, or strongest area, of the blade and away from the cutting edge;
- Failure of the cutting fluid to flood the cutting area. The intermittent cutting

*Table 34-1 General Feeds for End Mills (0.000'' feed per tooth, HSS): 1/8'' to 2''.

Diameter of End Mill	Steel				Cast Iron	Nonferrous Metals		
	Low Carbon	High Carbon	Medium Hard Alloy	Stainless		Aluminum	Brass	Bronze
1/8	.0005	.0005	.0005	.0005	.0005	.002	.001	.0005
1/4	.001	.001	.0005	.001	.001	.002	.002	.001
3/8	.002	.002	.001	.002	.002	.003	.003	.002
1/2	.003	.002	.001	.002	.0025	.005	.003	.003
3/4	.004	.003	.002	.003	.003	.006	.004	.003
1	.005	.003	.003	.004	.0035	.007	.005	.004
1 1/2	.006	.004	.003	.004	.004	.008	.005	.005
2	.007	.004	.003	.005	.005	.009	.006	.005

*These same speeds may be used for end milling operations performed on vertical milling machines.

process and the speed of carbide face mills produce a *fanning action*. A noncontinuous cooling action produces thermal, or heat, cracks in the blades. The problem is overcome by forcing the cutting fluid through a fine spray, or mist, to continuously reach the cutting edges.

FACE AND SHOULDER MILLING WITH A SOLID END MILL

Two basic processes may be performed with *solid end mills*. A plane surface may be milled, or two right-angled surfaces may be produced at the same time. These surfaces are referred to as a *step*, or *shoulder*. Solid end mills with two or more flutes may be used for face milling. These same end mills produce a shoulder when the face and peripheral teeth are set to cut the two surfaces at one time.

RATE OF FEED

Considerable judgment must be exercised by the operator in terms of cutting speeds, cutting feeds, and depth of cut. Table 34-1 provides general recommendations of feed per tooth for different end mill diameters and materials.

The small sizes of end mills are comparatively fragile. Excessive feeds cause tool breakage and chipping at the edges of the teeth. The rate of feed for solid end mills is calculated by substituting the number of flutes for the number of teeth in the formula. Slower feeds are used for deeper cuts.

CUTTING SPEEDS

Tables of cutting speeds, mentioned previously for other machining processes, apply equally to end mills. The cutting speeds given in such

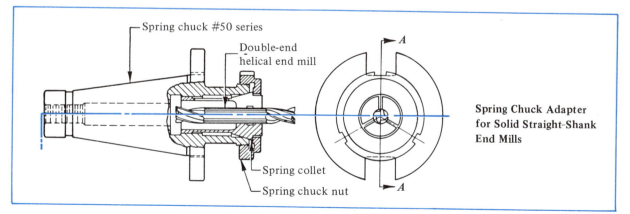

Figure 34-4 Mounting and Driving Solid End Mills

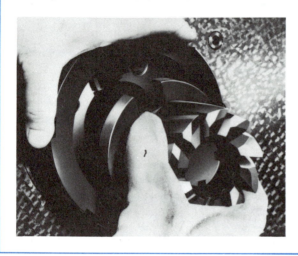

Figure 34–5 Mounting and Locking a Shell End Mill and Adapter into a Quick-Change Spindle Nose Adapter

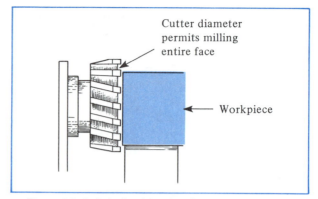

Figure 34–6 Relationship of an Inserted-Tooth Face Milling Cutter to the Width to Be Machined

tables are used to establish the spindle and cutter RPM.

HOLDING SOLID AND SHELL END MILLS

Taper-shank solid end mills that have a tang are held in a *tang-drive collet* (often called a *sleeve*). A threaded-end shank is held in a plain collet adapter. Straight-shank single and double end mills are usually mounted in a spring chuck adapter. The adapters are then secured in a type C arbor (Figure 34–4).

Figure 34–5 shows how an end mill adapter, reducing collet, and a shell end mill are held for mounting. The setup is locked into the quick-change adapter on the spindle nose by a partial turn of the clamp ring.

How to Face Mill with a Face Milling Cutter

STEP 1 Select a face milling cutter (Figure 34–6).

STEP 2 Hold the cutter with both hands and mount it on the spindle nose (Figure 34–7A). Start the fastening screws and securely tighten them (Figure 34–7B).

STEP 3 Calculate the spindle RPM. Set the speed dial or gear-change levers at the closest lower RPM. Check the direction of spindle rotation.

STEP 4 Calculate the rate of feed. Set the feed dial or feed-change lever to the required feed.

STEP 5 Set the trip dogs.

STEP 6 Check the cutting fluid. Position the distributor to flow (or spray on a cemented carbide cutter) the cutting fluid.

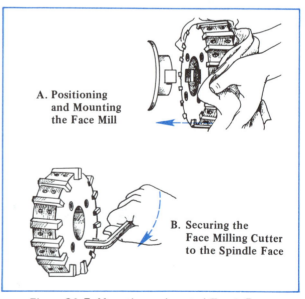

A. Positioning and Mounting the Face Mill

B. Securing the Face Milling Cutter to the Spindle Face

Figure 34–7 Mounting an Inserted-Tooth Face Milling Cutter

STEP 7 Center the cutter at the height of the center line of the workpiece. Lock the knee.

Note: If the workpiece is wider than the cutter diameter, the distance of the first cut should be not more than three-fourths of the cutter diameter. About one-fourth of the cutter should extend above the top edge of the workpiece. The second cut to clean up the face must overlap the first cut.

STEP 8 Start the spindle. Feed the saddle inward by hand until the cutter just grazes the workpiece. Set the cross feed micrometer collar at zero. Lock the collar.

STEP 9 Move the table to clear the cutter. Turn the saddle handwheel to the required depth of cut. Lock the saddle.

STEP 10 Start the flow of cutting fluid. Feed the face mill by hand. Take a trial cut for a short distance.

STEP 11 Stop the spindle and the coolant flow. Check the workpiece for size. Make whatever adjustment is needed.

STEP 12 Start the machine and the coolant flow. Engage the power feed. Take the cut across the face of the workpiece.

STEP 13 Stop the feed, cutting fluid, and spindle. Return the cutter to the starting position.

Note: Unlock the knee and move the workpiece up to overlap any second cut that may be required.

STEP 14 Increase the speed and decrease the feed for a finish cut.

How to Face (End) Mill with Solid or Shell End Mills

STEP 1 Set the workpiece in the center area of the table. Provide minimum overhang of the face to be milled. Align the face and secure the workpiece in position.

STEP 2 Select the type and largest practical size of end mill that meet the job requirements. Use the appropriate arbor or adapter and split collet, reducing or spring collet, and spring chuck.

STEP 3 Secure the type C arbor, adapter, or spring chuck in the spindle nose (Figure 34–8A).

STEP 4 Assemble the end mill.

Note: A shell end mill is mounted and held directly with a locking screw (Figure 34–8B). A straight-shank solid end mill may be tightened on an adapter or held in a spring collet. Taper-shank solid end mills are held in a tang drive or tapped-end collet.

STEP 5 Compute the cutter RPM from the known information about the cutting speed and the cutter diameter. Set the spindle speed.

STEP 6 Determine the initial feed rate. Set the feed-control dial or levers. Position and secure the trip dogs.

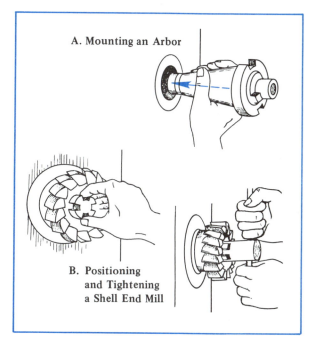

A. Mounting an Arbor

B. Positioning and Tightening a Shell End Mill

Figure 34–8 Mounting a Shell End Arbor and Securing a Shell End Mill

STEP 7 Position the distributor nozzle of the coolant system so that the required flow will be directed on the workpiece and cutter.

STEP 8 Move the face of the end mill so that it is located near the starting place of the cut. Center the cutter at the center line of the workpiece. Lock the knee.

Note: If two or more cuts are required because the face width is larger than the cutter diameter, follow the procedures in face milling.

STEP 9 Determine the depth of the cut or cuts. Start the machine. Move the cutting face of the end mill toward the workpiece until it just touches the face to be milled.

STEP 10 Set the cross feed micrometer collar at zero. Move the saddle in to the depth of the cut. Lock the saddle.

STEP 11 Start the flow of cutting fluid if it is required. Feed the cutter by hand for a short distance. Then move the cutter away from the workpiece.

STEP 12 Stop the spindle. Measure the workpiece. Make any necessary adjustments.

STEP 13 Start the spindle. Engage the longitudinal power feed. Take the cut.

Note: Some faces are milled to a required length that ends at a shoulder. In such cases, disengage the power feed near the end of the cut. Feed the cutter by hand to the layout line. Lock the table. Unlock the knee. Carefully feed the cutter down, then up. This step removes the small wedge-shaped area. The wedges are produced by the circular form milled by the cutter and the bottom and top edges of the workpiece.

STEP 14 Stop the spindle. Return the end mill to the beginning of the cut.

STEP 15 Repeat the steps for feeding the cutter to depth, measuring, and taking successive cuts.

How to Mill a Shoulder with an End Mill

STEP 1 Select the largest diameter of solid or shell end mill that is suitable for the job.

Note: The solid end mill is preferred for small-dimensioned shoulders. A shell end mill is more practical for large shoulders where a heavy cutter is required.

STEP 2 Set up the workpiece. Mount the cutter as close to the spindle as practical. Set the trip dogs, speed, feed, and coolant system nozzle.

STEP 3 Position the cutter near the workpiece. Start the spindle. Bring the cutter to the workpiece until it just touches the face.

STEP 4 Set the cross slide micrometer collar at zero. Move the cross slide in to the depth of the first cut. Lock the saddle.

STEP 5 Move the end mill over the workpiece for a short distance. Turn the knee handwheel to bring the workpiece up. Continue until the end mill just skims the surface. Set the knee micrometer collar at zero.

STEP 6 Position the workpiece at the start of the cut. Bring the knee up to the depth of the shoulder.

STEP 7 Take a trial cut for a distance equal to at least half the diameter of the end mill (Figure 34–9A). Stop the spindle and all motion. Measure the depth and width (Figure 34–9B).

Note: The width and depth may be checked against layout lines. Otherwise a steel rule may be used. More accurate measurements are taken with a micrometer depth gage.

STEP 8 Stop the spindle at the end of the cut. On long workpieces, use rapid traverse to return the end mill to the starting position.

STEP 9 Reset the end mill for other roughing or finish cutting procedures. Lock the saddle

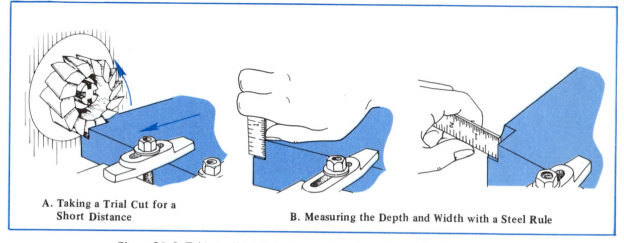

A. Taking a Trial Cut for a
   Short Distance

B. Measuring the Depth and Width with a Steel Rule

Figure 34–9 Taking a Trial Cut when Milling a Shoulder with a Shell End Mill

and knee in each procedure. Take a trial cut. Measure the shoulder. Make adjustments. Then take the finish cut.

STEP 10 Stop the machine. Follow standard personal, tool, and machine safety practices for disassembling the setup, removing burrs, cleaning up, and properly storing all items (Figure 34–10).

Safe Practices in Face, End, and Shoulder Milling

- Lock the knee and saddle before face milling.
- Handle the cutter with a cloth and only when the cutter is cool enough.
- Use work stops and heavy-duty strap clamps to rigidly hold a workpiece for deep cuts and coarse feeds.
- Remove all wiping cloths from the machine before it is placed in operation.
- Place a tray or wooden cradle under large face mills when mounting or dismounting them to prevent damage to the cutter or table.
- Engage the power feed only after it is established that the workpiece is supported rigidly enough to withstand the cutting forces.
- Stop the spindle and lock it during setup, measurement, and disassembling steps.
- Follow standard personal, tool, and machine safety precautions.

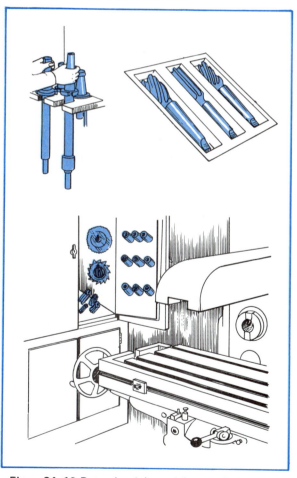

Figure 34–10 Returning Arbors, Adapters, Cutters, and Work-Holding Accessories to Their Proper Storage Places

TERMS USED WITH FACE AND END MILLING CUTTERS AND PROCESSES

Face milling (end milling)	Machining a plane surface at a right angle to the cutter axis. (End milling refers to machining a small-width surface with an end mill.)
Positive, zero, and negative radial rake angle	An angle formed by a tooth face in relation to the centerline of the cutter. The included angle formed by the cutter tooth and a centerline that extends from the cutting edge to the center of the cutter.
Positive, zero, and negative axial rake angle	An angle formed by the cutting tooth and the axis of a cutter.
Blades (inserts)	Separate cutter teeth of a face milling cutter. Blades that may be adjusted, reground, or replaced.
Feed and cutter lines	Arc-shaped, fine indentations in a face-milled surface. Surface imperfections usually machined when the nose radius and cutting edge are smaller than the rate of feed.
Lead angle effect	Differences in chip thickness and width produced by varying the angle of the cutting edge. The effect on cutter efficiency resulting from changing the lead angle of the cutting edge.
Cutter runout	An eccentric cutter motion. A condition that causes score lines (imperfections) in machining.
Positive and negative angles of entry	A positive or negative angle formed by the face of the cutting tooth and the surface of the workpiece at initial entry.
Mist spray	Flooding a cemented carbide face milling cutter with a fine spray of cutting fluid. A technique of cooling and lubricating a high-speed carbide cutter that otherwise would fan out the cutting fluid.
Overhang	Positioning a workpiece so that the face extends a minimum amount beyond the side of the table. Extending a workpiece a small distance more than the depth of the cuts to be taken.

SUMMARY

- Surfaces are face milled using three common types of milling cutters. Taper- and straight-shank end mills up to 2" (50mm) in diameter are used to face (end) mill small surfaces. Shell end mills up to 6" (150mm) in diameter are used to machine surfaces up to about 5" (125mm) wide. Inserted-tooth face mills are used for heavy-duty and wide-surface face milling.
  - The two basic cutter materials are high-speed steel and cemented carbides.
- Face milling cutters are available with blade inserts. These inserts may be formed with positive, zero, or negative radial and/or axial rakes. Rake is determined by the type of material to be cut, the processes, the work setup, the operating speed, and the feed. Generally, zero and negative rakes are used on cemented carbide inserts for machining under severe conditions.
  - The lead angle of the cutting edge affects the chip thickness and cutting efficiency. Lead angles up to $30°$ are practical.

- Feed and cutter lines may be reduced or practically eliminated. Corrective steps include: strengthening the setup so that it is more rigid, decreasing the feed or increasing the speed, and using a sharp cutter. (Reduce the nose radius and grind the face cutting edge with a flat that is wider than the feed.)
  - Use the largest possible diameter of solid end mill. The diameter of a face mill or solid or shell end mill must be larger than the width of the surface to be milled.
- End mills may also be used to cut a shoulder. Care must be taken in setting the cutter for depth and rate of feed. Cutter depth and rate of feed should not exceed values suitable for the strength of the cutter teeth and body.
  - Face milling cutters are mounted directly on the spindle nose. Shell end mills are held on a type C arbor. Straight-shank solid end mills are generally chucked in a collet. Taper-shank end mills are first inserted in a tanged or split tapered sleeve (collet).
- Stops should be used to prevent any movement of the workpiece. A trial cut is taken, using hand-feed, to test the rigidity of the setup and the cutting action. Then the power feed is engaged.
  - The milling machine operator must make judgments about increasing or decreasing the speed, feed, or depth of cut.
- Special attention must be given to the use of a mist spray. It is needed if the fanning action of a cemented carbide cutter prevents the cutting fluid from reaching the cutting teeth and machining area.
  - Face milling cutters are heavy and difficult to handle. A wooden cradle is used under the cutter to help in mounting or disassembling.
- Personal, tool, machine, and work safety precautions must be followed. All items should be returned to their proper storage places. The machine must be left clean and ready for the next job.

UNIT 34 REVIEW AND SELF-TEST

1. Give three advantages to using a steep lead angle (to a maximum of 30°) on face milling cutters.
2. State five factors that affect dimensional accuracy and quality of surface finish when face milling.
3. Identify three cutter-holding adapters for straight- and/or taper-shank solid end mills.
4. Give the steps to follow in safely mounting a face milling cutter.
5. Tell how to position a face milling cutter to mill an area that is wider than the cutter diameter.
6. Set up the series of steps to follow to end mill a shoulder with a solid or shell end mill.
7. List three safety precautions to take when disassembling a face milling setup.
8. State two safe practices to follow in end milling with small-diameter end mills.

Side Milling Cutter Applications

Side milling cutters (side mills) are widely used on horizontal milling machines. Ends and parallel steps (shoulders) and grooves are produced with various types of side mills. When two parallel surfaces are milled by two cutters mounted on the same arbor, the process is called *straddle milling*. Two or more side milling cutters mounted with other types of cutters on the same arbor are referred to as a *gang*. This gang of cutters mills at the same time. The process is known as *gang milling*. This unit deals with applications of single and multiple side milling cutters, straddle milling, and gang milling.

OBJECTIVES

After satisfactorily completing this unit, you will be able to:

- Determine uses for side milling cutters in machining plane surfaces or a step.
- Select full side milling cutters to machine one, two, or three adjacent surfaces (grooves) at one time.
- Identify functions of helical-fluted and staggered-tooth side milling cutters.
- Understand setups and principles of straddle milling surfaces and steps and the use of interlocking right-hand and left-hand helical-fluted side mills for wide groove machining.
- Plan setups for gang milling with plain, side milling, and formed cutters.
- Perform the following processes.
  - Milling Ends Square with a Side Milling Cutter.
  - Straddle Milling with Two Side Milling Cutters.
  - Milling Grooves.
  - Machining with Interlocking Cutters.
- Use formulas for cutting speeds and feeds and select cutting fluids.
- Follow *Safe Practices* and correctly use *Terms* relating to side milling processes.

SETUPS AND APPLICATIONS OF SIDE MILLING CUTTERS

Side milling cutters are of the following three basic designs:

- *Half* side milling cutters with teeth on one side and on the periphery (one, two, or three surfaces may be cut at one time);
- *Full* side milling cutters with staggered teeth (narrow, deep grooves may be milled);
- *Interlocking* cutters ground with cutting teeth on one side and on the periphery (the combination of two interlocking cutters permits the simultaneous milling of three surfaces, such as the surfaces of wide grooves).

HALF SIDE MILLING CUTTER APPLICATIONS

Side Milling an End Surface or Step. One of the common applications for a half side milling cutter is milling an end square. The workpiece is usually held in a vise. The vise is accurately aligned parallel to the arbor axis. The half side milling cutter is used with a type A, short arbor. The diameter of the cutter must permit milling the face in one cut—that is, the cutter must clear the work step (Figure 35–1).

When the cutter is mounted close to the spindle or if the workpiece is shorter than the width of the vise, the part must be turned 180° to face the second end.

388

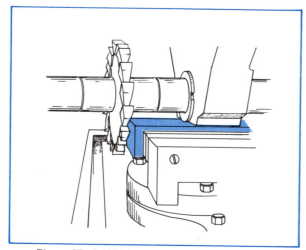

Figure 35-1 Width of a Workpiece within the Cutting Range of a Side Milling Cutter

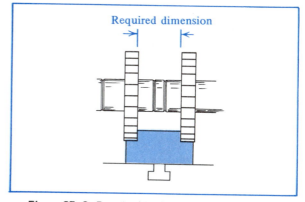

Figure 35-2 Required Width between Side Milling Cutters Produced by Spacing Collars

Milling Square Steps, or Shoulders. The half side milling cutter is also used to mill square steps, or shoulders. The side teeth mill the vertical face. The peripheral teeth mill the bottom surface. A roughing cut is usually taken, followed by a finish cut. Cutter RPM, cutting speeds, and feeds are determined in the same manner as for plain and face milling.

FULL SIDE MILLING CUTTER APPLICATIONS

Milling Two End Surfaces. The full side milling cutter is more versatile than the half side mill. With the full side mill, the setup for milling one end of a workpiece can also be used to mill the second end at the one setting. The cutter in this setup is located farther along the arbor support to permit machining with each cutting face. After the first cut the cutter is moved to the other end of the workpiece.

The cutter may be set with a steel rule. The part may be machined to a more precise linear measurement by positioning the cutter with the aid of the cross slide micrometer collar. The distance the work is moved is equal to the required dimension plus the width of the cutter.

Backlash is taken out of the cross feed screw. The saddle is then moved back to within 0.004" to 0.006" (0.1mm to 0.2mm) of the required linear dimension. The saddle is locked. A trial cut is taken. The machine is stopped. The length is measured by micrometer and the cut is

adjusted if required. This procedure helps to correct any machining error that may be produced by the cutter.

Milling a Step. A half side or full side milling cutter may be used to mill a step. The cutter is positioned to cut the vertical face to a required size. The saddle is locked in position. The knee is then raised until the cutter is set at depth. The knee is also locked. The settings may be made by using the micrometer collars or according to layout lines on the workpiece.

Milling a Groove. The sides and depth of a groove are usually layed out. The workpiece is then secured in a vise or strapped directly on the table. A full side milling cutter is used because groove milling requires cutting on the two sides and face of the cutter. A staggered-tooth side milling cutter is used for machining a deep groove.

If the groove is wider than the cutter, it is roughed out. From 0.010" to 0.015" (0.2mm to 0.4mm) is left on all sides for a finish cut. When the cutter groove and cutter size are the same, climb milling is recommended because it is easier to hold to a dimensionally accurate groove.

STRADDLE MILLING WITH SIDE MILLING CUTTERS

A straddle milling setup requires two side milling cutters and spacing collars. Spacing collars that correspond in length to the required dimension of a workpiece feature are placed between the inside hub faces of the cutters

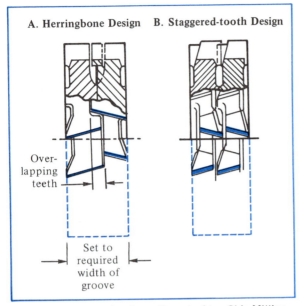

A. Herringbone Design   B. Staggered-tooth Design

Over-
lapping
teeth

Set to
required
width of
groove

Figure 35–3 Application of Interlocking Side Milling
Cutters to Machine a Groove to an Accurate Width

Figure 35–4 A Rigid Cutter, Workpiece, and
Table Setup for Gang Milling

(Figure 35–2). If necessary, a spacing collar may be ground to a needed size.

The accuracy to which the width is milled depends on how true the cutters run, how well the teeth cut, the removal of chips between the side face of the cutter and the workpiece, the nature of the cut, the cutting fluid (if required), and the work setup.

Many square, hexagonal, octagonal, and rectangular parts are machined by straddle milling. Straddle milling is particularly useful for round parts. Round parts are usually held in a chuck on a dividing head or on a rotary table. After each cut the workpiece is accurately positioned by indexing.

INTERLOCKING SIDE MILLING CUTTERS

Interlocking half side milling cutters are used in pairs for milling a wide groove or channel in one operation. The interlocking cutters may be spaced within the limits that the teeth overlap. Interlocking cutters are used to mill a groove or channel to an accurate width.

Figure 35–3A and 3B shows two sets of interlocking side milling cutters. The width is adjusted by adding thin, metal collars (shims) between the individual cutters. The left-hand and right-hand helical flutes help counteract the cutting forces.

GANG MILLING

Figure 35–4 shows a gang milling setup and process. Note how close to the spindle the operation is being performed. The table is near the column face. A type B overarm and large end bearing provide solid support for the cutters. This is a rigid setup adapted for gang milling.

A gang milling operation may include any combination of different types of cutters. The illustration shows a number of different diameters and widths of plain milling cutters and a pair of side milling cutters. Oftentimes other form milling cutters are included.

MACHINING PARALLEL STEPS

Parallel steps are milled with cutter, workpiece, and machine setups similar to the setups used to straddle mill parallel sides. However, if the parallel sides are to be milled to two different depths, the diameters of the cutters will differ. The smaller cutter diameter is equal to the larger diameter minus one-half the difference of the depth dimensions. For example, if cutters with 6″ and 4″ diameters are used, there will be a variation of 1″ between the steps that are milled.

The way in which a workpiece is held may also differ. When deep steps are to be milled, the part is either held in a fixture or clamped on the ends directly on the table to provide rigid support. Clamping straps also make it possible for the cutter to mill to the required depth without interference (Figure 35-5).

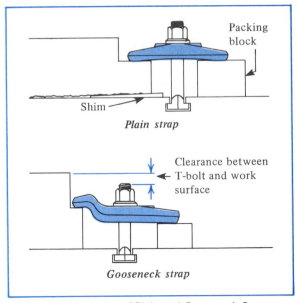

Figure 35-5 Use of Plain and Gooseneck Straps for Strapping Workpieces to the Table

How to Mill Ends Square

Using One Side Milling Cutter

STEP 1 Mount the vise centrally on the milling machine table. Align the solid jaw.

STEP 2 Support the layed-out workpiece on parallels. Position it so that the end to be milled extends just slightly beyond the depth of cut. Tighten the vise to seat the workpiece on the parallels.

STEP 3 Select a side milling cutter. The diameter must be large enough to provide clearance. The arbor bushings, overarm, and the top of the work must clear when the cutter is set to the full depth of cut.

STEP 4 Select the sturdiest possible arbor. Mount the arbor.

STEP 5 Position the cutter on the arbor close to the spindle. Assemble the collars, key, cutter, bushing, and nut by hand.

STEP 6 Move the arbor support into position and clamp it. Use the arbor wrench to securely tighten the arbor nut.

Note: The arbor nut should be tightened only after the arbor is supported in the overarm. The spindle should be locked before the nut is tightened.

STEP 7 Determine the spindle speed. Set the speed control to the required RPM.

STEP 8 Determine the feed rate. Set the table feed accordingly.

STEP 9 Move the cutter clear of the workpiece to the point where the cut is to be started. Turn the cross feed handwheel until the cutter is at the layout line. Lock the saddle.

Note: If a roughing cut and a finishing cut are to be taken, leave from 0.010″ to 0.015″ (0.2mm to 0.4mm) for a finish cut.

STEP 10 Feed by hand for a short distance. Then engage the power feed.

STEP 11 Take the finish cut. Stop the spindle. Return the workpiece to the start of the cut.

STEP 12 Clean the workpiece and area around the vise. Remove the milling burrs.

STEP 13 Turn the workpiece 180°. Seat it on parallels for the next cut.

STEP 14 Position the workpiece in relation to the cutter. If there is no layout line, measure the length with a steel rule. Lock the saddle.

STEP 15 Start the spindle and the flow of cutting fluid. Take a cut for a short distance.

STEP 16 Stop the spindle. Measure the length with a steel rule or micrometer. Loosen the saddle and make any dimensional adjustment required.

STEP
17
Continue with the cut. Repeat the cutter-setting steps if a finish cut is to be taken.

Note: The spindle speed should be increased and the feed decreased if a high-quality surface finish and high degree of accuracy are required.

STEP
18
Stop the machine. Check the length of the workpiece. Remove and burr the part. Disassemble the setup. Clean the machine. Return all tools and accessories to their proper storage areas.

How to Use Two Side Milling Cutters (Straddle Milling)

Machining Parallel Sides

STEP
1
Mount the workpiece in a vise or fixture.

STEP
2
Select two half side milling cutters (or a pair of full-side cutters); one right-hand and one left-hand cutter.

STEP
3
Mount one cutter on the arbor. Place it as close as possible to the spindle. Use a key. Check to see that the cutter direction is correct.

STEP
4
Make up the required combination of spacing collars, shims, and the arbor support bearing (Figure 35–6A). The overall length must be the same as the required linear dimension to be milled.

STEP
5
Slide half of the collars and shims over the arbor and key (Figure 35–6B). Next add the bearing.

STEP
6
Slide the arbor support arm over the arbor. Center it on the bearing. Lock the arm in place (Figure 35–6C).

STEP
7
Slide the remaining collars and shims and the cutter over the arbor and key. The direction of this cutter must be the same as the direction of the first cutter.

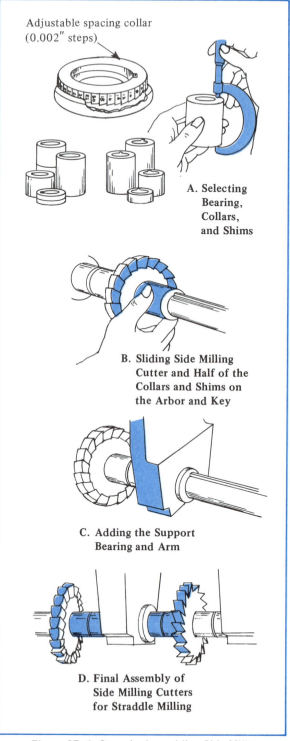

Adjustable spacing collar (0.002″ steps)

A. Selecting Bearing, Collars, and Shims

B. Sliding Side Milling Cutter and Half of the Collars and Shims on the Arbor and Key

C. Adding the Support Bearing and Arm

D. Final Assembly of Side Milling Cutters for Straddle Milling

Figure 35–6 Steps in Assembling Side Mills

STEP 8 Complete the assembly of the arbor and outer support arm. Securely tighten the arm and the arbor nut (Figure 35–6D).

STEP 9 Determine the correct spindle RPM, rate of feed, and cutting lubricant. Set the machine accordingly.

STEP 10 Position the cutters to machine to the layout lines or required dimension. Lock the saddle. Move the cutters so that they clear the bottom of the workpiece. Lock the knee.

STEP 11 Start the spindle. Feed by hand. Take a trial cut for a short distance. Stop the spindle and lock it. Measure the workpiece.

STEP 12 Restart the spindle. Continue the cut by using the power feed.

STEP 13 Return the cutters to the starting position. Stop the spindle and lock it. Recheck the finish milled surfaces for width.

STEP 14 Stop the machine. Remove the workpiece. File off any milling burrs. Clean the machine. Disassemble the setup. Replace all tools and accessories in their proper storage areas.

How to Mill Grooves

Milling with Full Side Milling Cutters

STEP 1 Lay out the width and depth of the groove.

STEP 2 Select a full side milling cutter that is equal in width to the groove size. The cutter diameter must permit clearing the setup to cut to depth. Use a staggered-tooth cutter for milling a deep groove (Figure 35–7).

STEP 3 Set the spindle speed and feed. Position the cutter according to the layout lines or measure the required distance (Figure 35–8). Lock the knee and saddle. Take a trial cut for a short distance.

STEP 4 Measure the width of the groove. If the cutter is cutting oversize, replace it.

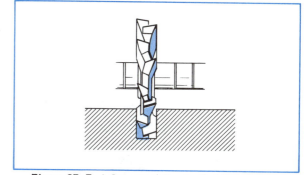

Figure 35–7 A Staggered-Tooth, Full Side Milling Cutter Used for Machining a Deep Groove

STEP 5 Take the cut when the workpiece position has been checked and the groove measurement is correct.

Note: Flood the cutter flutes and cutting area to clear the chips.

STEP 6 Stop the spindle at the end of the cut. Clean the groove and cutter. Return the cutter to the starting position if a wider groove than the cutter width is to be cut. Follow the preceding steps when taking additional cuts.

STEP 7 Disassemble the setup at the completion of the process. Follow standard practices for removing burrs, cleaning up, and replacing all tools and accessories.

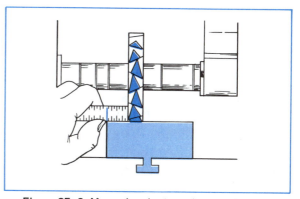

Figure 35–8 Measuring the Location of a Slot or Groove to Position a Side Milling Cutter

How to Machine with Interlocking Side Milling Cutters

STEP 1 Select a set of interlocking side milling cutters that when assembled will cut the groove to the required size.

STEP 2 Mount the cutters on an arbor as close to the spindle as conditions permit.

STEP 3 Position the cutters according to the layout lines on the workpiece.

Note: Sometimes the cutters are positioned by bringing one of the cutter teeth to within 0.001″ to 0.002″ of a side face of the work. The cross feed micrometer dial is set. The knee is moved down until the cutter clears the work. The workpiece is then moved to the crosswise position by using the micrometer collar measurement.

STEP 4 Move the workpiece up until the teeth just graze the top surface. Set the micrometer collar on the knee. Clear the cutter. Then adjust the cutter for depth of cut. Lock the knee.

STEP 5 Set the spindle RPM and feed.

Note: Use the lower RPM and feed ranges. When right- and left-hand helical-fluted cutters are used, wide and deep cuts are taken at still slower speeds and feeds.

STEP 6 Mill the groove. Follow the same procedures for trial cut, measurement, power feeding, and cutting fluid application as are used with other milling processes. Deep grooves may require one or more roughing cuts and a finishing cut.

STEP 7 Disassemble the setup at the end of the process. Remove any milling burrs. Clean the machine. Return the arbor and parts and cutters and other tools to their proper storage places.

Safe Practices for Side Milling Setups and Processes

• Check for clearance between the arbor collars or overarm and any projecting surfaces on a workpiece or work-holding device.

• Use care in positioning a side milling cutter so that the teeth are not brought against the vise or other holding device.

• Position the cutters and cutting action as close to the spindle as work conditions permit.

• Tighten the arbor nut only after the overarm is supporting the arbor.

• Use a staggered-tooth, full side mill for cutting deep grooves.

• Flood the flutes and area of cutting action to flow the chips away from a cutter and workpiece.

• Stop the spindle and lock it when taking measurements, adjusting the workpiece or cutter, or disassembling the setup.

• Provide added cutter and work support for wide, deep cutting with interlocking side milling cutters and for multiple cutter setups for gang milling.

• Observe standard safety practices related to protective goggles, setup, assembly and disassembly procedures, proper clothing, and chip removal and cleanup.

TERMS USED WITH SIDE MILLING PROCESSES AND TECHNOLOGY

Straddle milling	The process of milling two parallel surfaces or steps at the same time. The application of two side milling cutters to machine two vertical surfaces simultaneously. Milling two parallel steps in one operation.
Clear the setup	Selecting side mills with diameters that are large enough to perform the required operation. Adequate clearance between the arbor, workpiece, and work-holding setups.
Shims (arbor)	Dimensionally accurate, thin metal collars. (Shims permit adjusting the space between cutters so that they mill to a precise dimension.) Thin metal collars used alone or with standard-width arbor spacing collars.
Milling a step	The process of milling a vertical face and an adjacent horizontal face. Milling two surfaces at a right angle. (One surface is in a vertical plane and the other surface is in a horizontal plane.)
Milling a groove	Milling three surfaces at the same time. (The two vertical sides are parallel. They are at right angles to the bottom of the groove.)
Parallel steps	Right and left 90°-angle steps (shoulders) that are parallel.
Right-hand and left-hand helical flutes (interlocking cutter)	A pair of side milling cutters with teeth that interlock when assembled. A side milling cutter with flutes cut to a right-hand or left-hand helix.
Interlocking faces	The inside face of an interlocking cutter. The face against which a shim may be used to increase the width of a pair of interlocking cutters.
Gang milling	Combining a series of regular and/or special form cutters to mill a surface. The use of multiple cutters on an arbor to simultaneously mill several surfaces.

SUMMARY

■ Half, full, and interlocking side milling cutters are used to mill one or more surfaces, steps, or grooves.

    ■ The half side milling cutter is applied primarily to the milling of a plane surface or a right-angle step.

■ The full side milling cutter is more versatile than the half side milling cutter. The design features permit machining a single face, a right-angle step, or a groove.

    ■ The cutter position for side milling should be as close to the spindle as practical.

■ Layout lines are used to position a cutter and for machining to correct dimension. More accurate machining is done by feeding according to the micrometer collar settings on the cross slide and the knee. Dimensional accuracy is checked with inside, outside, or depth micrometers, as required.

- Trial cuts are taken for a short distance. Dimensions are checked with the spindle stopped. The power feed is engaged only when all dimensions are correct.

- Two side milling cutters are required to straddle mill two parallel vertical surfaces or two right-angle steps. Spacing collars of the required width are used between cutters.

  - An intermediate arbor support provides rigidity for heavy-duty cutting or when the straddle-milled cuts are widely separated.

- A dividing head or rotary table produces the indexing necessary to position a workpiece. Machining round, square, hexagonal, and octagonal forms are common applications of straddle milling.

  - Limited-depth grooves may be milled with full side milling cutters. Climb milling usually produces a more accurately milled narrow groove than does conventional milling.

- Deep-groove milling requires a staggered-tooth, full side milling cutter. The chips must be flowed out of the flutes and away from the cutting area.

  - Interlocking side milling cutters may be adjusted for width. Shims are placed between the inner faces to obtain the required dimension. The right- and left-hand helical flutes produce ideal cutting conditions.

- Plain mills, side mills, and form cutters may be combined on an arbor. A gang milling setup produces multiple forms at one time. The spindle RPM for gang milling is set for the largest diameter of cutter in the setup.

  - The rigidity of the workpiece and cutter setups, the cutting action, and the appropriate cutting fluid are safe-operating-condition checkpoints.

- Personal hygiene in handling cutting fluids and safe practices in working around moving machinery must be observed.

  - Care must be taken to safely and properly remove chips and clean work surfaces. All tools should be cleaned and stored properly.

UNIT 35 REVIEW AND SELF-TEST

1. List the major applications of (a) half side milling cutters and (b) full side milling cutters.

2. Differentiate between milling with staggered-tooth interlocking side milling cutters and gang milling.

3. Indicate the advantage of straddle milling parallel sides compared to machining each side with a single side milling cutter.

4. Tell how to machine parallel steps at different heights with a straddle milling setup.

5. State why the full side milling cutter flutes and cutting areas are flooded with cutting fluid during deep-groove milling.

6. Name two precautions to take in straddle milling.

7. List three safe practices to follow in mounting side milling cutters and cutter setups for gang milling.

Sawing and Slotting: Cutters, Setups, and Processes

Sawing and slotting operations on the horizontal milling machine require different setups and additional safety considerations than general milling processes. Since special job requirements control the form of the cutter, the teeth and cutting areas are not as strong as conventional milling cutters.

The cutters used for sawing and slotting are easily damaged and require great care to prevent fracturing. This unit describes the use of these cutters in relation to the work-holding setups, procedures, and safe practices for sawing and slotting.

OBJECTIVES

After satisfactorily completing this unit, you will be able to:

- Describe and interpret characteristics of metal-slitting saws and slotting cutters.
- Determine work-holding and cutter-holding setups.
- Perform each of the following processes.
  - Sawing on a Horizontal Milling Machine.
  - Milling Narrow, Shallow Slots Using a Screw-Slotting Cutter.
- Follow recommended personal, machine, and tool *Safe Practices* and correctly use new *Terms*.

METAL-SLITTING SAWS AND SLOTTING CUTTERS

Sawing refers to ordinary slitting, slotting, and cutting-off processes. *Slitting* relates to the use of a saw to cut through a tube or ring or to separate it into a number of parts. *Slotting* involves the machining of a narrow-width groove.

Manufacturers' tables usually provide guidelines for the number of teeth to use for slitting and sawing steel. Saws having two-thirds the number of teeth as saws used for steel are used on brass and deep slots. Copper is one of the more difficult metals to cut on the milling machine. Cutters for copper are specially shaped and have a small number of teeth compared to cutters for steel.

The cutters are thin and are called *slitting saws*. The *plain metal-slitting saw* is ground concave on both sides for clearance. The saw is made of high-speed steel with standard keyways. The cutter hub has parallel sides. The hub is the same width as the cutter teeth.

Plain slitting saws are available to fit standard arbor diameters. The general widths range from 1/32″ (0.8mm) to 3/16″ (5mm). The outside diameters vary from 2 1/2″ (64mm) in narrow widths to 8″ (200mm) for wide cutters.

The *metal-slitting saw with side teeth* (Figure 36–1) provides side chip clearance. The additional side clearance spaces between teeth and the recessed hub permit chips to flow freely away from the cutter teeth. Thus, binding and scoring of the work are prevented and the amount of heat generated is reduced. This saw is suitable for regular and deep slotting and sinking-in cuts.

Metal slitting saws with side teeth are commercially available in face widths that range from 1/16″ (1.58mm) to 3/16″ (4.76mm). The number of teeth on a cutter in this range is from 28 to 48. The outside diameters are from 2 1/2″ (64mm) to 8″ (200mm).

The *metal-slitting saw with staggered peripheral teeth* has an axial rake angle. The side teeth on this high-speed steel saw have a positive radial rake angle. The alternate helical (side) teeth help to eliminate chatter. The staggered teeth and side clearance provide the necessary chip space for deep cuts where heavy feeds are required. The side teeth are not designed for cutting. Staggered-tooth slitting saws have coarser teeth than do regular slitting saws. The staggered-tooth slitting saw is best adapted for cuts 3/16″ and wider. Standard and heavy feeds are recommended.

SCREW-SLOTTING CUTTERS

The *screw-slotting cutter* is used principally to slot screw heads and to slit tubing, thin-gage materials, piston rings, and similar work. The sides of the cutter are ground concave. Screw-slotting cutters are available in three groups of

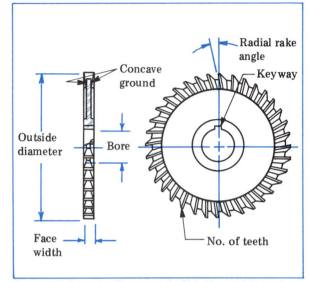

Figure 36–1 Design Features of a Side-Tooth Slitting Saw

standard sizes. These sizes and other specifications are listed in Table 36–1. Additional special screw-slotting cutters are produced to meet unusual specifications. Technical information relating to cutter diameter; bore size; number of teeth in cutter; and ranges of face width, wire gage, and screw head diameter is then furnished by the manufacturer.

WORK-HOLDING SETUPS

Rigidity is an important consideration in cutting-off, slitting, and slotting operations. Usually when screw-slotting cutters are used, the parts to be slotted are nested in a fixture.

Table 36–1 Specifications of Standard Screw-Slotting Cutters

Cutter Diameter	Bore Size	Teeth in Cutter	Ranges		
			Face Width	Wire Gage	Screw Head Diameter
2 3/4″ (69.85mm)	1″ (25.4mm)	72	0.020″ to 0.144″ (0.5mm to 3.7mm)	7–24	1/8″ and smaller through 1″ (3mm and smaller through 25mm)
2 1/4″ (57.15mm)	5/8″ (15.9mm)	60	0.020″ to 0.064″ (0.5mm to 1.5mm)	14–24	1/8″ and smaller through 3/8″
1 3/4″ (44.45mm)	5/8″ (15.9mm)	90	0.020″ to 0.064″ (0.5mm to 1.5mm)	14–24	(3mm and smaller through 9mm)

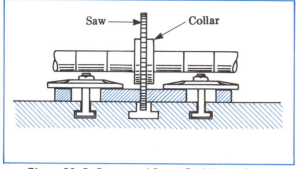

Figure 36-2 Cutter and Setup Positioned Clear of Arbor and Table T-Slot

Sections of a workpiece are often cut off on the milling machine. A solid part, where no deflection would be produced by cutting action, is often mounted directly in a vise. The area to be cut off extends beyond the vise jaw. A plain cutter may be used, or a cutter with straight peripheral and side teeth or staggered peripheral teeth and alternate side teeth may be selected. The depth of the cutting-off operation, the nature and rigidity of the setup, the cutting fluid, and the machinability of the material are factors that guide the craftsperson in cutter selection.

Another common setup is to strap the workpiece directly to the milling machine table. The position of the cut and cutter must be over one of the table T-slots. This setup is illustrated in Figure 36-2. Climb milling is then used. Added rigidity is provided in climb milling as the cutter forces the work against the table or other supporting surface.

Cast iron and other materials that have a hard outer scale may require conventional milling. Conventional milling minimizes the cutting effect that otherwise would require the teeth to be continuously cutting into the scale.

A common practice followed in the shop to overcome the tendency of the work to spring during conventional sawing is to place a strap across the face of the workpiece.

Slotting saws are sometimes used to mill narrow and other slots, where a side mill is impractical. If the slot is wider than the cutter face width, two or more cuts may be required. The saw is moved to machine the wider slot so that each cut overlaps the preceding one.

Angle pieces requiring a slot to be machined are often mounted on a right-angle plate (for surfaces at 90° to each other). The sine plate is used for cutting precise slots at any angle.

How to Saw on a Horizontal Milling Machine

Selecting and Mounting the Saw

STEP 1 Select an appropriate, sharp, high-speed steel slitting saw.

Note: The type, diameter, and width must permit the arbor to clear the work-holding straps or device and the cutter and to cut through the workpiece. The direction of the teeth depends on whether the part is to be cut off by climb or conventional milling.

STEP 2 Select the shortest length of type A arbor that will accommodate the cutter and process. Mount the arbor in the spindle.

STEP 3 Position the slitting saw as close as possible to the column. Secure it on the arbor.

Note: A proper fitting key must be used. This key also drives the cutter and the collars next to the cutter. Additional support for thin cutters must be provided. Thin collars with an outside diameter that is larger than the regular collars are used and are placed on both sides of the cutter.

STEP 4 Move the overarm and arbor support over the end of the arbor. Secure the arbor support for maximum rigidity during operation.

STEP 5 Use a table of cutting speeds. Determine the recommended starting sfpm (mm/min). Set the spindle RPM according to the sfpm and the cutter diameter.

STEP 6 Set the machine table feed for the cutting-off process.

Mounting the Work on the Table

STEP 1 Clean and remove burrs from the table and workpiece.

STEP 2 Center the workpiece on the table. Position it to permit cutting through the part at the required dimension.

STEP 3 Apply a slight force to clamp the workpiece to the table.

> Note: Straight or gooseneck clamps and the shortest possible T-bolt are used.

> Note: If sawing by the conventional method is done, a supporting piece is clamped across the top of the workpiece to prevent the workpiece from springing.

STEP 4 Set the workpiece at a right angle to the side of the table.

> Note: The workpiece is squared (aligned) by checking the position of one side with the blade of the steel square.

STEP 5 Position the workpiece squarely. Tap it with a soft-face hammer. Secure the clamps when the side of the workpiece is parallel with the blade of the square.

> Note: A flat, straight piece of material is brought against the aligned edge. This piece is strapped in position as a stop. Other pieces may be positioned against the stop without further checking.

Positioning the Cutter and Workpiece

STEP 1 Move the cutter height so that it clears the workpiece.

STEP 2 Position the saddle so that the cutter is at the line of measurement over a table T-slot. Lock the saddle.

> Note: With the spindle stopped, a steel rule is often used to measure the width of the part (Figure 36–3).

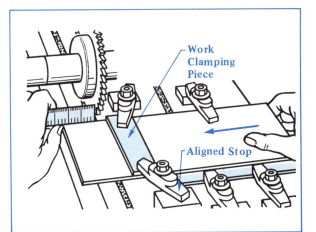

Figure 36–3 Using a Steel Rule to Measure the Width of a Part to Be Cut Off

STEP 3 Position the table and cutter lengthwise at the start of the cut for machining either in the conventional or climb milling mode.

STEP 4 Recheck to see that the cutter clears the workpiece. The cut must be taken without obstruction.

STEP 5 Start the machine. Hand-feed until a small area is milled.

STEP 6 Make sure the spindle is completely stopped. Take a measurement.

STEP 7 Continue the sawing process. Use an appropriate cutting fluid, where required. Engage the power feed when the cutting action is correct.

STEP 8 Stop the spindle. Deburr the cutoff part.

> Note: Thick parts may be held directly on parallels in a chuck.

STEP 9 Disassemble the setup when the required number of pieces are cut. Clean the machine, straps, and bolts. Return each item to a proper storage place. Wipe the milling machine clean.

How to Mill Narrow, Shallow Slots

Using a Screw-Slotting Cutter

STEP 1 Select a screw-slotting cutter to accommodate the width of slot to be cut.

STEP 2 Select a type A or B milling machine arbor. Mount the arbor securely in the spindle.

STEP 3 Place the cutter in the arbor. Use a key to drive the cutter. Secure the cutter, key, and collars on the arbor.

STEP 4 Determine the correct cutting speed for the process. Set the spindle speed to the correct RPM for the sfpm.

STEP 5 Set the machine feed and the location of the table trip dogs.

Note: The machine feed may be set at a slower rate at first. It may then be increased as conditions warrant.

STEP 6 Secure the workpieces to be slotted.

STEP 7 Locate the cutter centrally in relation to the head and body of each part.

STEP 8 Start the spindle. Turn the knee handwheel until the cutter grazes the workpiece. Set the knee graduated collar at zero.

STEP 9 Move the table lengthwise to back the cutter away from the first part.

STEP 10 Raise the knee to the required depth of cut. Lock the knee.

STEP 11 Start the coolant flow. Take a trial cut. Stop the machine. Check to see that the cut is centered and the slot is machined to the correct depth.

STEP 12 Make adjustments, if needed. Continue the process by hand-feeding. Engage the power feed when all machining conditions are satisfactory.

STEP 13 Stop the machine. Use a hard-bristle brush to remove chips.

STEP 14 Unload the slotted parts. Remove burrs.

Safe Practices for Sawing and Slotting

• Stop the spindle and lock it. The cutter must be at rest before the width and depth to which a slot is being milled are measured.

• Stand to one side and out of line of travel of a revolving saw or slotting cutter.

• Use climb milling, where practical, for sawing and slotting operations.

• Make sure there is sufficient operating space between the cutter, arbor and support arm, and the workpiece and work-holding setup.

• Drive the cutter by using a correct size key between the keyway of the arbor, the cutter, and the spacing collars. A sliding-fit key ensures a positive drive.

• Check the hand of the cutter against the direction of rotation of the spindle. Fine saw teeth and screw-slotting cutters are easily damaged when fed even for part of a revolution in the wrong direction.

• Set the position of a saw to run into a table slot. This safe practice applies to workpieces that are to be cut off using a direct table mounting setup.

• Tap the vertical feed crank by hand to control raising the work to starting position. The micrometer collar is set at zero when the revolving cutter just grazes the surface of the workpiece.

• Begin each machining process with the starting speeds and feeds recommended by the tool manufacturers. These speeds and feeds may be increased when the operator determines that the machining process may safely be carried on at higher speeds and/or feeds.

• Supply a sufficient quantity of cutting fluid, when required. The chips must flow rapidly away from the cutter teeth and workpiece.

• Make sure there is no backlash in the feed screws when a cutter is reset for additional cuts.

TERMS USED FOR SAWING AND SLOTTING

Slotting	Process of machining a narrow-width groove to divide a piece of stock (or a workpiece) into one or more parts or segments.
Sawing (milling machine)	A cutting-off process. Cutting through a workpiece to remove excess material or to machine-saw parts to specified lengths.
Metal-slitting saw	A milling cutter with straight or staggered peripheral teeth with or without side teeth. A sawing and slotting cutter relieved on the sides for clearance.
Screw-slotting cutter	A milling cutter with concave ground sides and peripheral teeth ground to a sharp edge. A milling cutter used primarily for screw slotting and slitting.

SUMMARY

■ Sawing and slitting processes are performed on the milling machine with plain metal-slitting saws, plain saws with side teeth, and slitting saws with staggered peripheral and alternate side teeth.

■ Standard metal-slitting saws with side teeth range in width from 1/16" (1.58mm) to 3/16" (4.76mm).

■ High-speed steel saws have an axial rake angle. The side teeth have a positive radial rake angle.

■ Slow hand starting feeds are important. These feeds permit maximum control, which ensures that the setup is rigid and the cutting conditions are correct.

■ Adequate flow of cutting fluid must be checked. Chips must be flowed rapidly away from the cutting teeth and the workpiece.

■ The machine must be completely stopped and the workpiece burred before any measurement is taken.

■ Each tool, work-holding device, machine accessory, and instrument is to be returned to its proper storage compartment.

■ The smallest possible diameter of slotting and slitting saws is used. The size, however, must be large enough to provide safe clearance for the cutting process.

■ Screw-slotting cutters are used principally to cut screw-head slots and for limited depth sawing and slitting processes.

   ■ Narrow slots may be milled with a screw-slotting cutter. A type A or B arbor may be used, depending on the machining requirements and cutter design.

■ Climb milling is safer and more functional than conventional milling for through slotting and slitting on table-mounted workpieces.

   ■ An extra strap is placed across a thin workpiece to provide additional rigidity.

■ Like all other machining processes, the cutting speed is determined by the process itself, size and features of the workpiece, and the rigidity of the setup. Suggested starting speeds may be increased when the operator establishes that the operation may be performed safely.

   ■ Sawing, slitting, flat keyseat, and dovetail cuts are hand fed at the start of the cut. Power feeds are engaged when cutting conditions warrant.

■ The size and shape of the cutter teeth and other design features require that great care is taken to prevent fracturing.

UNIT 36 REVIEW AND SELF-TEST

1. State the difference (a) between regular slotting and slitting on a milling machine and (b) between the cutters used for each process.

2. Name one common work-holding setup for (a) sawing off pieces from a rectangular plate on a miller and (b) cutting narrow, shallow grooves that are equally spaced around the periphery of a cylindrical part.

3. a. Secure a milling machine cutter manufacturer's catalog.
   b. Give the specifications of an appropriate slitting saw to use for each of the following jobs: (1) cutting off pieces from 1/8″ X 3″ (3mm X 76mm) flat, cold-drawn steel plates; (2) sawing sections from cast iron castings where the cuts are 30mm (1 3/16″) deep; (3) heavy-duty sawing cuts wider than 3/16″ (4.7mm) and 1″ to 1 1/2″ (25.4mm to 38.1mm) deep.

4. Identify the design features of screw-slotting cutters that make them practical for milling narrow, shallow slots.

5. List three cutter and machine safety precautions to observe when sawing or slitting on a milling machine.

Keyseat Cutting and Dovetail Milling

Keyseat cutting refers to the machining of a groove in a shaft for purposes of inserting a key that fits into a mating part and prevents it from turning. *Dovetail milling* involves accurate machining of the angular sides of a wide groove internal and external design feature to permit two precise fitting parts to slide within one another (for example, the cross slide of a lathe).

Details for cutter design, formulas for computing dimensions, machine setups, keyseat and dovetail milling processes, and important safety considerations are covered in this unit.

OBJECTIVES

After satisfactorily completing this unit you will be able to:

- Describe design features, functions, and applications of keyseats and dovetails.
- Interpret specifications and Handbook Tables relating to Woodruff keyseat cutters and dovetail milling cutters.
- Compute dimensions for machining Woodruff keyseats and internal and external dovetails.
- Perform each of the following processes.
    - Milling Keyseats with Standard and Woodruff Milling Cutters.
    - Milling Internal and External Dovetails.
- Follow recommended personal, machine, and tool *Safe Practices* and correctly use new technical *Terms.*

KEYSEAT MILLING

Three basic forms of keyseats are shown in Figure 37–1. The *plain (open) keyseat* may be milled with a single side milling cutter. When the cutter is the correct width, the keyseat is produced by taking a single cut. Other widths may be milled by taking second cuts. The *Woodruff keyseat* is produced by a fixed-sized Woodruff keyseat cutter. The *rounded-end (closed or sunk) keyseat* may be cut at any position along an arbor or drive shaft. This keyseat is produced with an end mill.

In keyseat milling, the manner in which a workpiece is held depends on the method used to produce the keyseat. Four common work-holding setups for rounded parts are illustrated in Figure 37–2. One of the simplest setups is to clamp the workpiece directly in a vise (Figure 37–2A). The workpiece is gripped at two positions that are opposite to each other. When a V-block is used in a vise, the workpiece is gripped at three places (Figure 37–2B).

A workpiece that is too large to be held in a vise is usually mounted directly on the table (Figure 37–2C). The workpiece is placed in the table groove. Straps are used to clamp the workpiece securely. A large workpiece may also be centered on V-blocks and strapped to the table (Figure 37–2D).

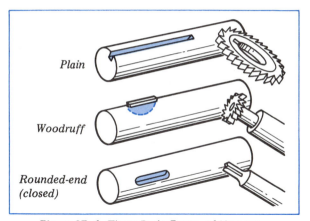

Figure 37-1 Three Basic Forms of Keyseats

COMPUTING THE DEPTH OF CUT FOR A KEYSEAT

Shop drawings for keyseats are dimensioned from the depth of cut to a theoretically sharp edge. This edge is formed where the vertical side of the keyseat and the outside diameter of the workpiece meet. In keyseat milling, the workpiece must be raised to the keyseat depth dimension plus the vertical height of the arc. The arc represents the additional material to be removed. The height of the arc varies according to the radius of the workpiece and the width of the keyseat. The following formula (illustrated in Figure 37-3) is used to compute the height of the arc.

$$A = r - \sqrt{r^2 - (1/2w)^2}$$

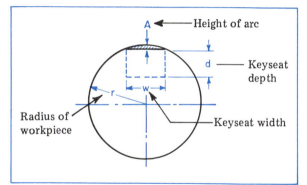

Figure 37-3 Depth of Keyseat and Height of Arc

where   A = height of arc,
        r = radius of workpiece,
        w = keyseat width.

Example: Determine the total depth of cut to mill a 1/2″ (0.500″) square keyseat on a 3″ diameter shaft. By using the arc height formula and substituting values,

$$A = r - \sqrt{r^2 - (1/2w)^2}$$
$$= 1.500 - \sqrt{1.500^2 - (1/2 \times 0.500)^2}$$
$$= 0.021'' \text{ (arc height)}$$

The total depth of cut from the top of the shaft to the bottom of the keyseat is equal to one-half the keyseat width plus the arc height. In the example,

$$\text{total depth} = 1/2 (0.500) + 0.021$$
$$= 0.271''$$

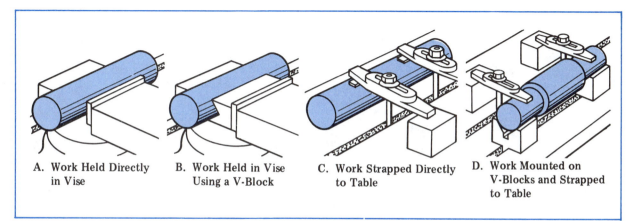

A. Work Held Directly in Vise

B. Work Held in Vise Using a V-Block

C. Work Strapped Directly to Table

D. Work Mounted on V-Blocks and Strapped to Table

Figure 37-2 Four Common Work-Holding Setups for Round Parts

How to Mill A Keyseat

Using a Standard Milling Cutter

STEP 1 Mount and secure the workpiece near the center of the table.

Note: The location must provide the least overhang and maximum rigidity for the work-holding and cutter setups.

STEP 2 Select a standard high-speed steel milling cutter.

Note: The cutter width must permit machining the keyseat width within the specified tolerance.

STEP 3 Position the cutter on the arbor so that the operation is performed under as rigid a setup as possible. Mount and secure the cutter, collars, overarm, and arbor support.

STEP 4 Align the workpiece so that it is centered with the cutter.

STEP 5 Determine the cutting speed. Set the spindle speed at the RPM closest to the recommended cutting speed.

STEP 6 Set the table feed (ipm or mm/min). Use the rate given in a feed table as a starting point.

Note: The feed rate is increased when the cutting action and setup will support a higher feed.

STEP 7 Make a final check for adequate clearance before starting the cut.

STEP 8 Start the spindle. Raise the knee carefully until the cutter grazes the top of the workpiece. Set the knee micrometer collar at zero.

STEP 9 Move the cutter off the workpiece to the starting position. Raise the knee and bring the workpiece up to the required keyseat depth.

STEP 10 Take a trial cut for a short distance to permit measuring the keyseat. Flow on an adequate supply of cutting fluid.

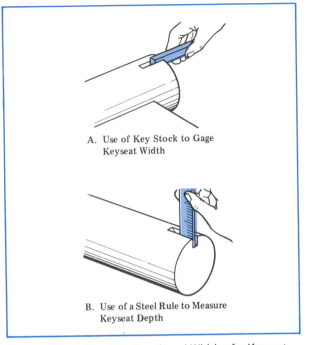

A. Use of Key Stock to Gage Keyseat Width

B. Use of a Steel Rule to Measure Keyseat Depth

Figure 37–4 Checking Depth and Width of a Keyseat

Caution: Remove the cutter from the keyseat and stop the spindle before measuring the width and depth.

Note: The width may be gaged by a GO–NO–GO gage or a piece of key stock (Figure 37–4A).

Note: The depth may be measured with a steel rule from the side of the keyseat (Figure 37–4B). A micrometer is used for a precise measurement of keyseat depth.

STEP 11 Make whatever cutter adjustments are necessary for changes in depth or width. Lock the knee.

STEP 12 Cut the keyseat to length by engaging the power feed. Near the end of the cut, disengaged the power feed. Feed by hand to the layout line. This line indicates the end of the keyseat.

STEP 13 Brush away all chips. Remove machining burrs from the workpiece. Place each item in its proper storage area.

WOODRUFF KEYSEAT CUTTER SPECIFICATIONS

NOMINAL SIZES OF WIDTH AND DIAMETER

Commercially available Woodruff cutters conform to American National Standard (ANSI) specifications. The number of the cutter and the American Standard key numbers are identified. A notation such as #506 AMERICAN STANDARD WOODRUFF CUTTER indicates both the cutter size and the nominal key dimensions of width and diameter.

The last two digits give the cutter diameter in eighths of an inch. The one or two digits preceding the last two digits give the cutter width in thirty-seconds of an inch. The #506 cutter measures 5/32″ X 6/8″. The cutter is 5/32″ wide X 3/4″ diameter.

Stock cutter sizes range from 1/16″ (1.5mm) to 3/8″ (9.5mm) wide and 1/4″ (6.35mm) to 1 1/2″ (38.1mm) diameter. These sizes are nominal dimensions. The 1/2″ (12.7mm) cutter shanks are of uniform diameter. Shank-type cutters are usually mounted in a collet.

NOMINAL HEIGHT ABOVE SHAFT

Handbook tables provide dimensional data for standard and special Woodruff keys. Key numbers are recommended for different shaft diameters. Woodruff keys project above the shaft diameter a distance equal to one-half the cutter width ±0.005″ (±0.12mm).

The depth of the keyseat in a shaft must be held to a tolerance of +0.005″/−0.000″ (+0.12mm/−0.00mm). Table 37–1 is a partial table of ANSI dimensions of Woodruff keys and keyseats. An important dimension for the machine operator is the depth (B in Table 37–1) of the keyseat from the top of the shaft. This depth represents the distance the workpiece/knee is raised from the initial zero cutter setting.

Table 37–1 ANSI Dimensions (in Inches) for Sample Woodruff Keys and Keyseats (ANSI B 17.2)

Woodruff		Keyseat Shaft					Key above Shaft	Keyseat Hub	
Key Number	Nominal Size Key	Width A*†		Depth B†	Diameter F		Height C†	Width D†	Depth E
		Min.	Max.	+0.005 −0.000	Min.	Max.	+0.005 −0.005	+0.002 −0.000	+0.005 −0.000
202	1/16 X 1/4	0.0615	0.0630	0.0728	0.250	0.268	0.0312	0.0635	0.0372
606	3/16 X 3/4	0.1863	0.1880	0.2143	0.750	0.768	0.0937	0.1885	0.0997
806	1/4 X 3/4	0.2487	0.2505	0.1830	0.750	0.768	0.1250	0.2510	0.1310

*Values for width A represent the *maximum* keyseat (shaft) width that will assure the key will stick in the keyseat (shaft). The *minimum* keyseat width permits the largest shaft distortion acceptable when assembling a maximum key in a minimum keyseat.
†Dimensions A, B, C, and D are taken at the side intersection.

How to Mill a Keyseat with a Woodruff Cutter

STEP 1 Select a Woodruff keyseat cutter with a width and a diameter that meet the dimensional requirements of the job.

STEP 2 Mount the cutter in the machine spindle. Use a spring collet or adapter for a standard Woodruff keyseat cutter. Use an arbor for hole-type cutters.

STEP 3 Align and secure the workpiece on V-blocks in a vise or other suitable work-holding device.

STEP 4 Center the cutter at the required distance from the end of the workpiece and on its vertical axis. Lock the table and saddle.

STEP 5 Start the cutter and coolant flow. Set the micrometer collar at zero at the starting position for the cut.

STEP 6 Feed the workpiece upward by hand feed to the required depth. Note this depth reading.

Caution: The force on a Woodruff keyseat cutter increases from zero to a maximum at the end of the cut. The part is hand-fed to permit the depth of cut and force on the cutter to be decreased as the cutting action increases.

STEP 7 Stop the machine. Lower the knee. Clean the cutter, workpiece, and workpiece, and working area. Remove burrs from the milled edges of the keyseat.

STEP 8 Insert a Woodruff key into the keyseat. Use a soft-face hammer to tap the key slightly to seat it.

Note: The key is removed if, after measurement, further machining is required.

CHARACTERISTICS OF DOVETAILS

Two common methods are used to machine *dovetail slides* in jobbing shops. The corresponding angular sides and the common flat base of

Figure 37–5 Milling a Dovetail Slide with a Single-Angle Cutter on a Vertical Milling Machine

mating parts may be formed by feeding a single-point cutting tool at the required angle. This method is the shaper method. The second method is to machine the dovetail on a vertical or horizontal miller. A formed single-angle *dovetail milling cutter* is used. Figure 37–5 shows a vertical milling machine setup for machining a dovetail slide.

On cast parts, dovetail slides are usually preformed. Sufficient material is provided (left on) to permit machining below (under) the scale surface. Machined dovetail slides are produced within precision limits and with low microfinish surface textures.

Other dovetail slides require milling from a solid piece. In such milling, a rectangular area is rough machined. The angle is produced by roughing out with a formed angle cutter. The sides and base are then finish machined with a dovetail cutter. The same cutter is used to machine internal and external dovetails.

COMPUTING AND MEASURING INTERNAL AND EXTERNAL DOVETAILS

The dimensional accuracy of dovetail slides may be gaged or measured. An optical compara-

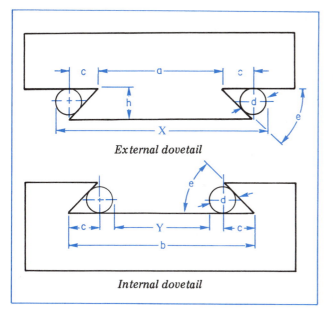

External dovetail

Internal dovetail

Figure 37-6 Dimensional Measurements of a Dovetail Slide

tor may be used to establish the accuracy of the angles and other linear dimensions. Dovetail slides may also be measured with a gage or a micrometer.

Figure 37–6 illustrates the dimensional measurements of a dovetail slide. Dimensions X on the external dovetail and Y on the internal dovetail may be measured by micrometer or vernier caliper. Dimension X for the measurement of an external dovetail may be calculated by the following formula:

$$X = d \times (1 + \cot 1/2 \text{ angle e}) + a$$

Dimension Y for the measurement of an internal dovetail is also found by formula:

$$Y = b - d \times (1 + \cot 1/2 \text{ angle e}) \text{ and}$$
$$c = h \times \cot e$$

Example: A 45° internal and external dovetail are to be milled. Calculate the overall dimensions X and Y. Use 0.500″ (12.7mm) diameter rolls and measurements (a) = 4.125″ and (h) = 0.625″. Check appropriate tables of trigonometric functions to find the cotangents of angles.

To calculate external measurement X, use the formula

$$X = d \times (1 + \cot 1/2 \text{ angle e}) + a$$

By substituting values,

$$\begin{aligned} X &= 0.500 \times (1 + \cot 22 \, 1/2°) + 4.125 \\ &= 0.500 \times (1 + 2.4142) + 4.125 \\ &= 5.8321″ \text{ (overall outside micrometer measurement of external dovetail)} \end{aligned}$$

To calculate internal measurement Y, use the formula

$$Y = b - d \times (1 + \cot 1/2 \text{ angle e})$$

By substituting values,

$$Y = b - 0.500 \times (1 + \cot 22 \, 1/2°)$$

Determining Y involves the overall width (b) of the dovetail. The overall width is equal to the smallest linear measurement (4.125 in this example) of the dovetail plus the width (c) of the right triangle multiplied by 2. That is, here

$$b = 4.125 + 2c$$

To complete the calculation for b, c must be found. The width of the right triangle (c) is equal to the dovetail height (h) multiplied by the cotangent of angle e. That is,

$$c = h \times \cot e = 0.625 \times \cot e$$

From a table of trigonometric functions,

$$\cot e = \cot 45° = 1.000$$

Therefore,

$$c = 0.625 \times 1.000 = 0.625″$$

Now that c has been found, the substitution of values in the formula for b may be completed:

$$\begin{aligned} b &= 4.125 + 2c \\ &= 4.125 + 2(0.625) = 5.375″ \end{aligned}$$

Finally, Y may be calculated by substituting 5.375″ for the length of b in the formula for Y:

$$\begin{aligned} Y &= b - d \times (1 + \cot 1/2 \text{ angle e}) \\ &= 5.375 - 0.500 \times (1 + \cot 22\text{-}1/2e) \\ &= 5.375 - 0.500 \times (1 + 2.4142) \\ &= 3.668″ \text{ (overall inside micrometer measurement of internal dovetail)} \end{aligned}$$

How to Mill Internal and External Dovetails

Using a Dovetail Milling Cutter

STEP 1 Compute the overall internal and external measurement of the angular surfaces.

STEP 2 Mount a dovetail milling cutter of the required cutting angle in the spindle.

STEP 3 Set the spindle RPM and cutting speed (table feed), depending on the machinability rating of the workpiece and other machining factors.

STEP 4 Take a roughing cut on the angular sides and across the face of the dovetail.

Note: The micrometer collar readings are recorded for the cutter setting at each angular side.

STEP 5 Stop the spindle and lock it. Clean the workpiece, working area, and measuring tools.

STEP 6 Measure the dovetail width. The two reference points for a micrometer or vernier reading are the two rolls.

STEP 7 Take finish cuts to machine the dovetail.

STEP 8 Clean the workpiece, work area, tools, and accessories. Remove burrs.

STEP 9 Remeasure the dovetail slides. Take subsequent cuts if needed.

STEP 10 Check the quality of surface finish. Select a comparator specimen according to the material, process, and required micro-finish value for surface texture.

STEP 11 Brush away all chips. Wipe the surfaces clean. Remove burrs.

STEP 12 Take down the setup. Clean and replace all tools and materials and place in proper storage areas.

Safe Practices for Milling Keyseats and Dovetails

• Examine the hand of the cutter to be sure it turns in the correct direction in relation to the direction of the spindle.

• Check the clearance between the cutter and workholding device to be sure it is adequate when the keyseat or dovetail is machined to depth.

• Check to see that the workpiece is securely mounted and strapped to the table, or in a vise or fixture, or other holding device.

• Feed the workpiece slowly by hand so that there is no sharp impact or excessive force exerted upon the cutter.

• Provide an adequate flow of coolant to quickly remove chips from the cutter and workpiece.

• Stop and lock the spindle prior to the taking of a measurement and before cleaning the cutter, work holder, and table.

• Use safety goggles or other approved eye protection during all machining processes.

• Check the quality of the cutting fluid against bacteria formation, chip particles, sludge, and other contaminants. Clean the system when required.

• Apply a liquid absorbent compound around the machine area if there is any cutting fluid on the floor.

• Discard and replace fractured or otherwise damaged cutters.

• Follow specific safety precautions provided by the machine tool manufacturer.

TERMS USED FOR KEYSEAT CUTTING AND DOVETAIL MILLING

Keyseat milling	Machining a groove in a shaft to receive one-half of a square or rectangular key.
Woodruff keyseat	Usually refers to a circular groove milled with a Woodruff keyseat cutter.
Woodruff keyseat numbers	American National Standard (ANSI) specifications for cutter widths and diameters.
Centering the cutter (Woodruff keyseat)	Aligning the vertical axis of a Woodruff cutter with the centerline of the workpiece.
Dovetail slide	An internal or external wide groove having sides machined at the same angle. Mating parts that are fitted with corresponding angular sides and a common flat base surface.
Measuring rolls	Precision-ground rolls of a specified diameter. Rolls placed against the angular surfaces of a dovetail. Rolls that permit the accurate linear measurement of a dovetail.
Dovetail milling cutter	A single-angle milling cutter that simultaneously mills the angular face and the adjoining flat surface of a dovetail.

SUMMARY

- The three basic forms of keyseats are: plain (open), Woodruff, and rounded-end (closed or sunk).

    - The depth of a keyseat is measured from the outside diameter of the part.

- Regular keyseats may be milled with a plain side milling cutter or a keyseat cutter. Keyseats that are wider than the face width of the cutter may be machined by resetting the cutter after the first cut.

    - Woodruff keyseats are cut with shank-type cutters for American National Standard (ANSI) sizes #202 through #1212.

- Arbor-type keyseat cutters are used for face widths from 3/16″ through 3/4″ (4.7mm through 19mm). The nominal diameters range from 2 1/8″ through 3 1/2″ (54mm through 88mm).

    - Woodruff keyseat cutters are hand fed to compensate for increasing cutting forces.

- Dovetails may be machined from solid or cast pieces on which sufficient stock is left to clean up the angular surfaces. Single-angle dovetail cutters permit roughing and finish machining each side and the bottom flat surface.

■ Since mating dovetails may be sliding members, the angular and adjoining areas of the flat base generally require a low microfinish surface texture. Surface finish specimens are used to compare the machined surface against the part specifications.

■ Dimensions for dovetails are calculated from drawing specifications that relate to angles, depth, and width.

■ Hardened and ground, special-diameter rolls are used in the measurement of external and internal dovetails.

UNIT 37 REVIEW AND SELF-TEST

1. Identify three common methods for accurately holding round work parts that require keyseat cutting.

2. Refer to an ANSI Table of Woodruff Key and Keyseat Dimensions for a #806 Woodruff key.
    a. Determine the *nominal sizes* for the key.
    b. State the minimum and maximum width sizes of the keyseat shaft.
    c. Give the maximum and minimum tolerances for depth of cut for the keyseat shaft.
    d. Determine the distance the knee is raised to mill the Woodruff keyseat to depth.

3. a. State what two factors must be considered when using a standard milling cutter to mill a keyseat to depth.
    b. Calculate the total depth of cut required to mill a 4″ (101.6mm) diameter shaft to accommodate a 0.500″ (12.7mm) key.
    c. Determine the overall *nominal height* correct to three decimal places for the customary inch measurement; two places for the SI metric equivalent.

4. Cite three methods that are used to check the width and depth of a keyseat that accommodates a square key.

5. State two cutter safety considerations when using shank-type Woodruff and dovetail milling cutters.

6. Identify two characteristics of milled dovetail parts.

7. Describe measuring rolls as applied to dovetail milling.

8. List the measurement steps required to measure external and internal dovetails, using steel rolls.

SECTION THREE

Indexing Devices:
Principles and Applications

UNIT 38

Direct and Simple Indexing

The *rotary* (or *circular milling*) *table* and the *dividing head* are two basic *indexing devices*. The function of each device is to accurately locate (index) workpiece features in relation to one another around a common axis. Machining operations are performed after the workpiece is rotated a required angular distance (degrees).

The functions, operation, and setting up of indexing equipment and accessories for direct and simple (plain) indexing are described in this unit. Advanced applications of the dividing head to compound (differential) indexing and to the milling of helical forms, gears, and cams are covered in the advanced book.

OBJECTIVES

After satisfactorily completing this unit, you will be able to:

- Use simple indexing devices such as the swivel vise and collet indexing fixture.
- Understand regular, ultraprecise, and vernier rotary milling tables.
- Apply standard and spiral universal dividing heads and the wide-range divider attachment.
- Interpret the functions of subassemblies of the dividing head, footstock, and center rest, and work-holding/driving accessories.
- Relate the technology of direct and simple indexing functions to actual indexing.
- Perform each of the following processes.
  - Setting Up the Dividing Head and Footstock.
  - Preparing the Dividing Head for Simple Indexing.
  - Simple indexing.
- Work through problems using shop formulas for simple indexing.
- Follow *Safe Practices* with indexing devices and for indexing and correctly use related *Terms*.

TYPES OF INDEXING DEVICES

DIRECT INDEXING DEVICES

A very simple form of indexing is often carried on with standard and compound vises. Many workpieces are indexed for a number of different cuts that are at an angle to one another by simply swiveling the vise jaw section. The angle setting is read directly from the graduations on the vise.

Workpieces held in collets are indexed directly. A *collet index fixture*, like the fixture shown in Figure 38–1, is practical for direct indexing. The two base faces are machined at a right angle. They serve as bases for mounting the fixture horizontally or vertically.

Special milling fixtures are designed with ratchet positioning devices. Once a workpiece is machined on one face, the position is unlocked. The nest in which the workpiece is held is then moved to the next angular setting.

GENERAL PRECISION INDEXING DEVICES

Rotary Tables and Attachments. The *rotary (circular milling) table* is a precision indexing device. Different models range from the line-graduated rotary table for direct indexing to extremely precise inspection devices. Power feed, footstock, and other attachments are available.

Rotary tables permit a workpiece to be turned in a horizontal plane to precise positions in degrees, minutes, and seconds. With the addition of a right-angle bracket, the rotary table is adapted for work in a vertical position as shown in Figure 38–2.

Some rotary tables are graduated in degrees. An accuracy within 30 seconds of arc is maintained through a complete rotation of the table. There is an adjustable graduated collar on the handwheel. Readings may be taken directly to each minute. Increased accuracy is provided by a vernier plate. Direct readings may be taken with a vernier plate to within 5 seconds of arc.

Ultraprecise and *optical rotary tables* are applied primarily in high-precision jig boring and inspection processes and for the *calibration* (standards setting) of master tools. Precision indexing attachments are available to measure within *one-tenth (1/10) second of arc*, or 1/36,000 part of one degree.

Figure 38--1 A Direct-Indexing Collet Index Fixture

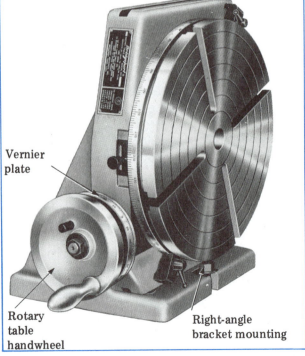

Vernier plate

Rotary table handwheel

Right-angle bracket mounting

Figure 38–2 Rotary Table with Vernier Plate Mounted on Right-Angle Bracket for Working in Vertical Position

The Standard Dividing Head. The most practical, everyday dividing device that is used in shops and laboratories is the *standard dividing head.* It is used to divide a circle, to mill one surface in relation to another, or to machine a number of surfaces around a workpiece.

Some workpieces require the machining of equal spaces—for example, four, six, or eight spaces—around the periphery. Such combinations may be positioned easily by using a *simple index head.* A workpiece is mounted in a chucking device. The chuck is connected directly to a spindle. The spindle has a direct-mounted dial plate. The work is turned (positioned) to a required location according to the divisions on the dial plate.

THE UNIVERSAL DIVIDING HEAD

The regular *universal dividing head* is a versatile, practical, and widely used dividing device. By means of a gear train, it is possible to divide a circle into thousands of combinations of equal and unequal parts. Precision linear and circular dimensions are involved.

ADVANCED MOVEMENTS AND PRECISION DIVIDERS

When cuts require a helix angle, it is necessary to swivel the milling machine table to an angle. A cutter is then advanced (fed) into a workpiece according to a fixed lead. The *universal spiral dividing head* is generally used for helical milling and is covered in detail in Unit 44.

The *wide-range divider adapter* in Figure 38–3 permits indexing from 2 to 400,000 divisions. This divider has an index plate, an index crank, and sector arms. These parts are mounted in front of the large index plate of the universal dividing head and add a ratio of 100:1 to the original 40:1 index ratio. The combination of the 40:1 ratio (40 turns of the index crank to turn a workpiece 1 complete revolution) and the 100:1 ratio increases the ratio to 4,000:1 (4,000 turns of the index crank to rotate the divider spindle, and therefore a workpiece, 1 revolution). The accuracy, as stated earlier, is within 1/10 second of arc.

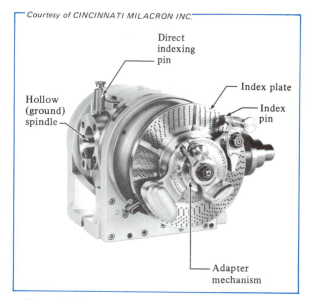

Figure 38–3 A Wide-Range (2 to 400,000 divisions) Divider Adapter on a Universal Dividing Head

DESIGN AND FUNCTIONS OF THE UNIVERSAL DIVIDING HEAD

A universal dividing head unit consists of three major components: *dividing head, footstock,* and *center rest.* The subassemblies of the dividing head include the *housing, swivel block,* and *spindle mechanism.*

MAJOR PARTS OF THE DIVIDING HEAD

The *housing* serves two main functions: (1) As a base the entire head may be bolted to the table and (2) as a bearing, the clamping straps permit the swivel block to be moved to an angle. The *swivel block* may be positioned at any angle from 5° below a horizontal position to 50° beyond a vertical position. The spindle axis of some heads may be moved through a 145° arc.

The angular setting of the swivel block (and spindle axis) is often referred to as the *position of the* (dividing) *head.* Graduations are usually cut in the swivel block and are matched with other graduations on an index plate on the housing. The swivel block is set to a required angular position and locked.

The swivel block houses the spindle, the gearing, and the indexing mechanism. The sectional drawing in Figure 38–4 shows many construction

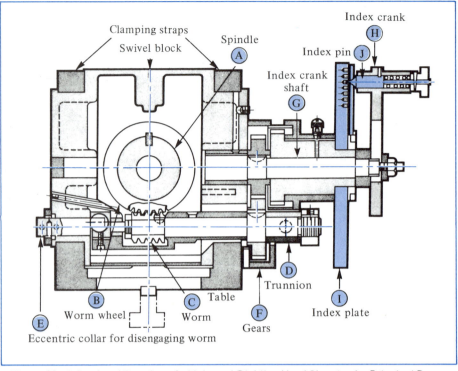

Figure 38-4 Sectional Drawing of a Universal Dividing Head Showing Its Principal Parts

features of the dividing head. Some of the principal parts are identified. These parts are referred to in the following brief description of how the dividing head works.

The spindle Ⓐ has an attached worm wheel Ⓑ. The worm wheel meshes with a worm Ⓒ on a trunnion Ⓓ. The worm may be engaged or disengaged by turning an eccentric collar Ⓔ. In a disengaged position, the spindle may be turned by hand. A workpiece may be set up to a required angle by direct indexing.

When the worm is engaged, its movement is controlled by the gears Ⓕ between the trunnion and the index crank shaft Ⓖ. As the index crank Ⓗ is moved a complete and/or partial turn, the spindle moves in a fixed relationship— that is, the spindle movement is proportional to the index ratio between the index crank and the spindle. The most commonly used ratio is 40:1.

The *dividing head spindle* is hollow ground. Generally the nose has an internal Brown & Sharpe standard, self-locking taper. Centers and other mating tapered tools and parts may be

mounted in the spindle. The nose is machined on the outside to position and hold chucks and other accessories.

The precise positioning of a workpiece is done by turning the index crank. The partial movement is indicated by the index plate and the index pin.

DIVIDING HEAD ACCESSORIES

Many parts are milled between centers and require mounting the dividing head, (spindle) center, and the footstock center. A driver is secured to the spindle center. Other workpieces are conveniently positioned and held in a *dividing head chuck*. Some dividing head chucks are of the single-step type. Other chucks include conventional or reversible three-step jaws.

The chucks are usually self-centering. The chuck itself is fitted with an adapter plate on the back side. The plate permits the chuck to

Figure 38–5 Application of a Three-Jaw Universal Vise on a Manual Indexing Spacer

be centered accurately and to be secured on the dividing head spindle.

Common work setups are made with the chuck and dividing head positioned horizontally, vertically, or at an angle. Some workpieces are held by the chuck alone. Other workpieces are also supported by the footstock.

Manual indexing spacers (an adaptation of the dividing head) are used for direct angle setting operations. The unit displayed in Figure 38–5 includes a reversible three-step jaw universal chuck mounted on a manual indexer (spacer). The workpiece is positioned for milling cuts that are at an angle to each other by directly reading the angle settings on the 360° graduated ring. Manual indexing spacing units may be operated from the horizontal or vertical position.

THE FOOTSTOCK (TAILSTOCK)

The terms *footstock* and *tailstock* are used interchangeably. The footstock (Figure 38–6) serves two basic functions: (1) to support one end of a workpiece and (2) to position the workpiece in a horizontal plane or at an angle.

THE CENTER REST

The milling machine center rest serves the same functions as a lathe steady rest. It is used principally for milling operations on long workpieces or machine setups where the cutting force would otherwise cause the workpiece to bend away from the cut. The center rest consists of a base and a V-center that may be adjusted vertically.

DIRECT INDEXING

Direct indexing is also called *rapid indexing* or *quick indexing*. When direct indexing is to be used on a universal dividing head, the worm is disengaged from the worm and wheel by an eccentric device (collar). Disengagement of the worm makes it possible for the spindle and direct indexing plate to be turned freely. This plate is mounted on the spindle. It usually has three circles of holes on the back side with 24, 30, and 36 holes in each circle. The range of direct divisions is limited to any number that is divisible into 24, 30, or 36. There are additional holes around the rim of the plate. These holes act as guides to rapidly locate some of the small divisions.

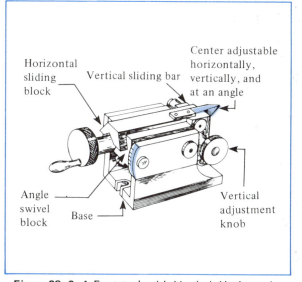

Horizontal sliding block

Vertical sliding bar

Center adjustable horizontally, vertically, and at an angle

Angle swivel block

Base

Vertical adjustment knob

Figure 38–6 A Footstock with Vertical, Horizontal, and Angular Positioning Parts

Table 38-1 Holes in Circles on Standard Index Plates

Brown & Sharpe Index Plates		
#1	*#2*	*#3*
15–16–17–18–19–20	21–23–27–29–31–33	37–39–41–43–47–49

Standard Cincinnati Index Plate	
First Side	*Reverse Side*
24–25–28–30–34–37 38–39–41–42–43	46–47–49–51–53–54 57–58–59–62–66

SIMPLE (PLAIN) INDEXING

In *simple indexing* a workpiece is positioned by using the crank, index plate, and sector arms. The dividing head spindle is connected by engaging the worm and worm wheel.

INDEX PLATES

The index plates for simple indexing, like the plates used for direct indexing, have a number of circles that are each divided into a different number of equally spaced holes (Figure 38–7). Two common series of index plate hole circles are listed in Table 38–1. There are three plates in the Brown & Sharpe (B & S) series. The standard plate for Cincinnati heads has a combination of eleven different hole circles on each side.

FUNCTIONS OF THE SECTOR ARMS, INDEX CRANK, AND INDEX PIN

Sector arms are used with index plates. The sector arms eliminate the need for counting the number of holes each time a new setting is required. Sector arms are adjustable radial arms. The arms are set and locked in position with the required number of holes between them. In operation one sector arm is brought against the plunger (index) pin. The spindle is unlocked. The pin is withdrawn and the index crank is turned carefully. When the pin is aligned with the required hole that is positioned at the second sector arm, it drops into the hole. The spindle is then locked and the milling operation is performed.

Design features are included to permit the index plate position to be changed when a cutter or a workpiece is being reset. Sometimes the ad-

justment is made by moving the index plate. At other times the index pin position may be adjusted so that the pin enters the hole in the required circle.

SIMPLE INDEXING CALCULATIONS

The standard gear ratio on the universal dividing head is 40:1. When a fractional part of a revolution is needed, the number of turns of the crank is equal to 40 divided by the number of required divisions. For example, each tooth in a 20-tooth gear is indexed by turning the crank 40 divided by 20, or 2 complete turns.

Sometimes angular dimensions are given on a drawing in place of a number of required divisions. With angular divisions the number of degrees in a circle is divided by the required angular measurement.

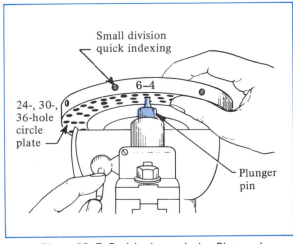

Figure 38-7 Positioning an Index Plate and Pin in Direct Indexing

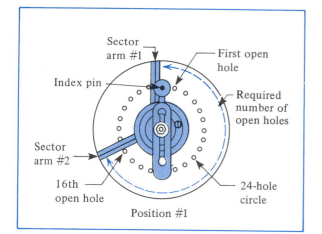

Figure 38-8 Setting the Sector Arms

The setting of the sector arms to obtain a fractional part of a revolution (or a required angular dimension) is illustrated in Figure 38-8.

Example: The surfaces of a workpiece are to be milled to an included angle of 60°. The number of required divisions must first be determined:

$$\text{required divisions} = \frac{360°}{\text{degrees in required angle}}$$
$$= \frac{360}{60} = 6$$

The 6 is then substituted in the formula:

$$\text{turns of crank} = \frac{40}{\text{required divisions}}$$
$$= \frac{40}{6} = 6\ 2/3$$

Thus, the index crank must be turned 6 complete revolutions and 2/3 of another revolution.

The 2/3 revolution may be indexed by selecting a circle of holes that is divisible by 3. The 15- and 18-hole circles on B & S plate #1 may be used. The sector arms are set for 10 holes on the 15-hole circle or 12 holes on the 18-hole circle. (It should be noted that in extremely precise angular indexing, the highest degree of accuracy is achieved by using the plate with the greatest number of holes that will accommodate the required partial turn. In this example the Cincinnati plate with 66 holes may be used.)

Each side (60° angle) may be positioned for milling by first unlocking the spindle clamp. The index crank is then turned 6 complete revolutions plus 10 holes on the 15-hole circle. The crank movement is continued slowly until the plunger pin drops into the hole at the sector arm.

If the index crank is turned too far, it first must be turned at least 1/2 revolution in the opposite direction. The index crank is then turned again in the original direction until the index hole is reached. Moving the index crank in the correct direction is important because it removes backlash that produces inaccuracies.

There is a one-step method of calculating the indexing to obtain angular dimensions. Since it takes 40 turns of the index crank to turn the spindle 1 revolution,

$$\text{one turn of the crank} = \frac{360''}{40} = 9°$$

Thus, there is a 9° movement of the spindle for each turn of the index crank. Therefore, to calculate the required indexing in the previous example, divide the 60° by 9°. Again, 6 2/3 turns of the index crank are needed to index for each 60° angle.

Not all index workpieces require that the index crank be turned a complete revolution or more.

Example: Assume that a part is to be machined with 47 divisions.

$$\text{turns of crank} = \frac{40}{\text{required divisions}}$$
$$= \frac{40}{47} \text{ of a turn}$$

In this example, the 47-hole circle plate is used. The sector arms are set to 40 holes on the 47-hole circle. Each division is indexed by turning the index crank the distance between the sector arms, or 40/47 of a revolution.

Tables and charts are provided with dividing heads. Because simple indexing is easy to compute, the technical information is most useful in compound indexing. The tables give the number of required divisions, followed by the number of whole and partial turns of the index crank, the circle to use, and the number of holes to which the sector arms are set.

How to Set Up the Dividing Head and Footstock

Setting Up the Dividing Head

STEP 1 Clean the machine and the base of the dividing head. Examine and remove any burrs.

STEP 2 Place the dividing head so that its weight and the milling operation can be carried on as near to the center of the table as practical. Use T-bolts and clamp the dividing head.

Note: The index crank should be on the operating side of the table.

STEP 3 Position the dividing head at the required milling angle. Loosen the clamping straps and position the swivel block. The angle setting is read on the graduations of the swivel block and index plate on the dividing head body. Figure 38–9 shows the reading for a horizontal setting.

Note: If there is a center in the dividing head, use a knockout bar while the spindle is in the horizontal position to free the tapers.

STEP 4 Select the appropriate work-holding device. A center and driver plate are secured to the spindle for work between centers.

Note: If a chuck is to be used, remove the center. Clean the spindle nose and the back plate on the chuck. A wooden cradle should be placed under a heavy chuck. The chuck and cradle are moved up to the spindle nose. The chuck is positioned on the spindle and locked securely.

Setting Up the Footstock

STEP 1 Clean the base of the footstock and the table. Remove any burrs.

STEP 2 Place one T-head bolt in the center slot. Position it at the place where the footstock is to be located.

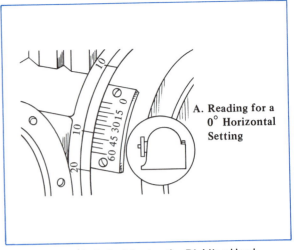

A. Reading for a 0° Horizontal Setting

Figure 38–9 Positioning the Dividing Head

STEP 3 Locate the footstock at a distance from the dividing head center. The distance must permit the center to be adjusted to the work length. Tighten the two T-bolts to secure the footstock.

STEP 4 Check the swivel block index line and the angle graduations on the frame. The zero lines should coincide for a horizontal setting.

Note: If the centers must be aligned to a greater accuracy, a test bar should be used. The horizontal alignment may then be checked with an indicator mounted on a surface gage.

STEP 5 Check the height of the center. The zero mark on the vertical sliding bar and the zero index line on the front face of the frame should be aligned.

How to Prepare the Dividing Head for Simple Indexing

Changing the Index Plate

STEP 1 Loosen the nut on the index crank shaft. Remove the washer and index crank (Figure 38–10A).

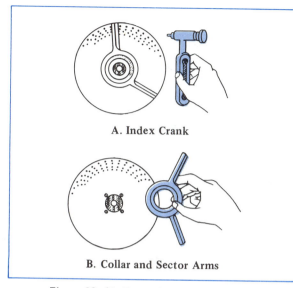

A. Index Crank

B. Collar and Sector Arms

Figure 38–10 Removing Parts to Change
an Index Plate

STEP Slip off the collar and sector arms (Fig-
2 ure 38–10B).
STEP Remove the index plate screws. Take
3 off the index plate.
STEP Screw the new index plate back in posi-
4 tion.
STEP Replace the collar, sector arms, and the
5 index crank. Select the circle of holes
that will produce the required fractional
part of a turn.

Setting the Sector Arms

STEP Position the index pin in one of the holes.
1 Tighten the index crank.

Note: The index pin should fit easily into
any one of the index holes in the circle
when the index crank is tightened.

STEP Position one sector arm with the beveled
2 edge against the index pin.
STEP Count off the required number of holes
3 on the correct circle. Adjust the second
arm to accommodate this number of
holes. Tighten the lock screw.

Note: The circle and the number of holes
are rechecked before any indexing is done.
The number of open holes must equal
the number of holes required for the partial turn of the index crank.

How to Do Simple Indexing

Full Turns

STEP Use any circle of holes on the index plate
1 for the complete turns of the index crank.
Position the sector arms to zero in one
hole on any index plate.
STEP Loosen the spindle lock. Withdraw the
2 index pin. Turn the index crank the required number of turns.
STEP Lock the dividing head spindle. Take
3 the milling cut. Return the milling cutter
to the start of the next cut.
STEP Position the workpiece for the second
4 cut. Unlock the dividing head spindle
before indexing. Lock the spindle after
indexing.

Note: The index crank must always be
turned in the same direction.

STEP Take the cut.
5

Note: These steps are repeated until the
workpiece has been turned a complete
revolution.

Full and/or Partial Turns

Note: The following steps are taken to index a part for milling graduations around
the circumference. The graduations are
60° apart. The dividing head is set up
to index each graduation.

STEP Compute the number of whole and par-
1 tial turns of the index crank.

$$\text{index crank turns} = \frac{\text{required degrees}}{9°}$$

$$= \frac{60°}{9°} = 6\,2/3$$

STEP 2 Select an index plate with a circle of holes divisible by 3.

Note: Assume a 24-hole circle plate is already mounted. Use this plate.

STEP 3 Set the index pin so that it is aligned with one of the holes. Adjust and set the index crank.

STEP 4 Move one sector arm until the beveled edge touches the index pin.

Note: The beveled edge should face in the direction in which the index crank is to be turned.

STEP 5 Loosen the lock screw on the sector arms. Count off 16 holes in the 24-hole circle. Start counting with the first hole next to the index pin.

STEP 6 Set the second sector arm close to the 16th hole. Tighten the setscrew so that the arms will not move.

Note: Recheck the number of open holes. There must be 16.

STEP 7 Lock the dividing head spindle. Proceed with the milling operation. Return the cutter to the starting position. Index for the next cut.

STEP 8 Revolve the sector arm (#1 in Figure 38–11) until it touches the index pin. Unlock the dividing head spindle.

STEP 9 Move the index crank 16 holes on the 24-hole circle. It is good practice to bring the index pin almost to the hole and tap it gently by hand until it drops into position. Lock the dividing head spindle.

Note: If the index crank is turned too far, turn the crank in the reverse direction for about half a turn. Then come forward again to take up the backlash. Turn until the index pin drops into the 16th hole.

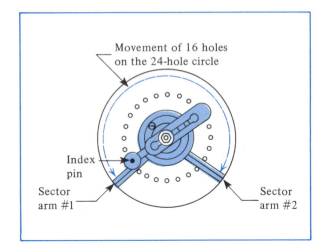

Figure 38–11 Indexing a Partial Turn

Safe Practices in Setting Up and Using Dividing Heads and Accessories

- Clean the table and base of the dividing head and footstock. Remove any burrs. Check to see that the tongues fit accurately into the table slots.
- Position the dividing head as near the center of the table as is practical.
- Tighten the milling machine dog securely to the driver plate.
- Check the alignment of the dividing head and footstock centers for milling parallel and taper surfaces and for proper seating in the center holes.
- Lock the index head spindle after each indexing and before the milling process is started.
- Remove any backlash if the index crank is moved beyond the correct hole.
- Standard safety practices must be followed for the care of the machine and accessories during setting up, operation, and disassembling. Personal safety precautions must be observed, including precautions about lifting heavy objects and wearing safety goggles.
- Help must be obtained to position any heavy dividing head on the table. A heavy chuck is mounted with a cradle.

TERMS USED WITH THE DIVIDING HEAD AND IN INDEXING

Direct indexing	The process of positioning a part for an angular dimension by direct movement of a spindle.
Simple (plain) indexing	Dividing a circle into any required number of parts in preparation for inspection, layout, or a machining process, using a dividing head.
Dividing machines	Devices used in inspection, layout, and machining to accurately position a workpiece for angular divisions. The collet index fixture, simple index head, rotary table, indexing spacers, and the universal dividing head are common examples of dividing machines.
40:1 ratio of the dividing head	The ratio between the index crank shaft and the spindle. The ratio of the gears in a dividing head. The number of turns (40) an index crank must be moved to turn a dividing head spindle 1 complete revolution. A commonly used ratio (40:1) on most indexing heads.
Universal dividing head	An indexing device, usually with a 40:1 ratio, that permits a part to be indexed directly or by controlling the full and/or partial turns of the index crank.
Driver	A forked plate that may be attached to a dividing head center to receive and hold a milling machine dog for positive (no-slip) turning of the workpiece.
Dividing head chuck	Usually a three-jaw universal chuck mounted on the spindle nose of the dividing head.
Footstock	A centering device for workpieces that require support on a centered end.
Index plate	A circular disc with a series of holes that is used on a dividing head to accurately position the spindle.
Sector arms	Two narrow, beveled arms that are set a required number of holes apart. Arms that simplify the movement and location of the index pin for each division setting.
Index pin and plunger	A sliding pin mounted on the index crank. A pin that may be positioned by a plunger to permit turning the index crank.

SUMMARY

- The rotary (circular milling) table and the universal dividing head are two basic precision devices for dividing a circle. Other direct indexing devices are also used.

  - Indexing devices extend the applications of fundamental milling processes. Plain, face, and side milling and keyway cutting, slotting, sawing, and form milling cutters and processes are all used in direct, simple, and compound (differential) indexing.

- The universal dividing head generally has a 40:1 ratio between the crank shaft and the spindle.

  - Intermediate angular dimensions are obtained by using index plates. Different numbers of holes, equally spaced on a series of circles, control the partial distance a workpiece is to be turned.

■ The spindle movement of a dividing head is controlled by the whole and partial turns of an index crank. The number of turns of the crank is computed by dividing 40 by the required number of divisions. The fractional part is computed by dividing a hole circle number by the denominator. The result is then multiplied by the numerator.

   ■ When the dimension is given in degrees, the index crank movement is calculated by dividing the required number of degrees by 9°.

■ The dividing head normally may be positioned horizontally, vertically, and at any angle within a 145° range.

   ■ The worm gears of a universal dividing head are disengaged for direct indexing. The spindle is always locked after indexing to ensure a rigid support.

■ The dividing head, footstock, and center rest are used in combination to support long workpieces.

   ■ A milling machine dog and driver are used to rotate a workpiece between centers. Universal three-jaw chucks, attached by an adapter plate to the spindle, provide a fast method of holding a workpiece.

■ Sector arms are adjustable to represent any angular division. Sector arms are set a required number of holes apart on a specific circle.

   ■ The three index plates in the Brown & Sharpe system and the single, reversible Cincinnati plate make it possible to accurately index the full range of divisions required in general machining.

■ Once a part is held securely and positioned, the milling machine processes are carried on following the procedures covered in previous units.

   ■ Care must be taken to remove any backlash in indexing. If the index crank is turned too far, it must be turned back one-half revolution, then turned in the original direction until the pin drops into the correct hole.

UNIT 38 REVIEW AND SELF-TEST

1. Distinguish between a general and a universal spiral dividing head.

2. Tell how an indexing range up to 400,000 divisions and an accuracy of 1/10 second of arc are possible with a wide-range divider.

3. Indicate what movements are possible with the horizontal sliding block and the vertical sliding bar of the footstock.

4. Differentiate between direct indexing and plain indexing.

5. Explain how to eliminate backlash caused by turning the index crank of a dividing head too far.

6. List the steps to follow in (a) changing the index plate and (b) setting the sector arms on a dividing head for simple indexing.

7. Indicate three safety precautions to take in setting up and using a dividing head and accessories.

SECTION FIVE
Vertical Milling Machines

This section deals with major design features of modern vertical milling machines. These features are related to combination vertical/horizontal machines, the two-axis tracer, the bed-type, and numerically controlled milling machines.

The versatility of the vertical milling machine is extended by attachments such as all-angle, right-angle, multiple-angle, and high-speed milling heads; horizontal milling machine attachment systems; and the cross slide milling head. These attachments as well as several tooling accessories, cutter-holding devices, and a digital readout system, are also described.

UNIT 39
Machine and Accessory Design Features

OBJECTIVES

After satisfactorily completing this unit, you will be able to:

- Discuss major machine and tool design developments from the earliest vertical milling machines up to current-day sophisticated equipment and accessories.
- Describe vertical milling machine design specifications for spindle speeds and feeds, rigid turret assembly, and handwheels and feed controls.
- Interpret the use of combination vertical/horizontal millers, profiling machines, bed-type millers, and numerically controlled machines.
- Identify and operate components of the turret head and the spindle (tool) head.
- Describe high-speed, right-angle, and rotary cross slide attachments, and an optical measurement system.
- Understand applications of such accessories as: quick-change tooling systems, shaping tool sets, boring heads and sets, holders, collets, adapters, arbors, etc.
- Follow recommended personal and machine tool *Safe Practices* and correctly use each related *Term*.

The *vertical milling machine* is used to accurately produce flat, angular, rounded, and multishaped surfaces. These surfaces may be machined in one, two, or three planes (X, Y, and Z axes). The milling processes may be vertical, horizontal, angular (single or compound angle), or spiral.

The longitudinal and cross (transverse) feeds may be by either hand or power or both. A single part may be produced or the machine may be set up for the machining of multiple, duplicate parts. A machine with numerical control may be tooled for mass production.

HISTORICAL OVERVIEW

One of the first milling machines was developed during the 1860s. It combined the vertical spindle of the drill press with the longitudinal and transverse (cross feed) movements of the milling machine. The spindle was designed to permit greater rigidity for the cutting tool. The design features of the early vertical milling machine (Figure 39–1) included a wide face over which the belt moved from the back driver pulley to the driven spindle pulley.

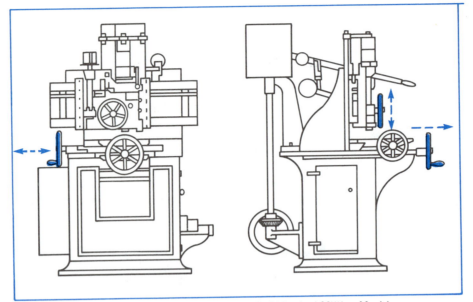

Figure 39-1 Design Features of an 1862 Vertical Milling Machine

The handwheel that actuated the spindle permitted vertical adjustment for positioning cutting tools by hand-feeding. The spindle assembly provided for increased strength to take the side forces resulting from vertical milling. Two other handwheels provided longitudinal and transverse feed movements for the fixed bed and table.

An improvement came in 1883 with the patent of the *knee-and column* design. This design provided additional space between the table and spindle. The table could be raised or lowered on the column. The spindle head was redesigned to permit setting it at an angle to the horizontal plane of the table.

Power feeds were introduced shortly after 1900 for longitudinal, transverse (lateral, crosswise), and vertical movements. The spindle design, which included a heavy-duty quill, permitted vertical power feed. By 1910, several types of vertical milling machines were available and included design features such as the fixed-position, swivel-head spindle; flat, fixed-bed, or knee-and-column tables; and hand and power feeds.

Electrical, electronic, pneumatic, and optical devices and controls were added later. They extended the applications for milling in three planes by tracing the form and simultaneously reproducing it on a workpiece.

In response to demands for more precise machining and measuring, a *jig borer* was designed for the precision layout and machining of holes. The jig borer is an adaptation of the vertical milling machine. Micrometers, standard length bar and indicator devices, and vernier scales were used to obtain exacting linear dimensions.

DESIGN FEATURES OF MODERN VERTICAL MILLING MACHINES

Modern vertical milling machines are versatile in design. Spindle speeds are usually controlled by variable-speed drives to permit a wide range of cutting speeds and feeds. Rapid traverse power feeds are available to reduce setup time in bringing the work to the cutter. The turret assembly is designed for making precision angular cuts. Handwheels and feed control levers provide for safe and convenient operator control of feeds.

Figure 39-2 shows the design features of a general-purpose vertical milling machine. The specifications for this machine give an idea of

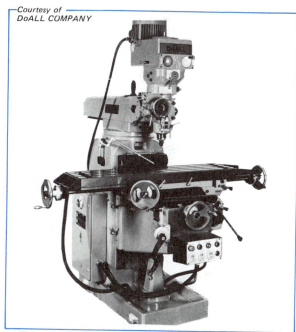

Figure 39-2 Design Features of a General-Purpose Vertical Milling Machine

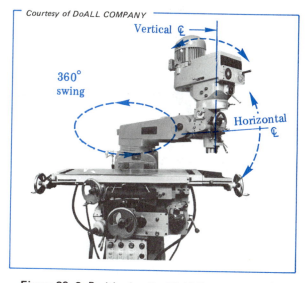

Figure 39-3 Positioning the Rigid Turret Assembly and the Swivel Head along the Major Axes

the longitudinal feed (37 1/2" or 950mm), cross feed (15 3/4" or 400mm), vertical feed (17 5/8" or 448mm). This particular quill may be fed 5 1/8" or 130mm.

SPINDLE SPEEDS AND FEEDS

The specifications also indicate that the variable-speed drive provides a low spindle speed range of 60-490 RPM and a high range of 545-4,350 RPM. There are three power feeds for the quill. These feeds range from 0.0013" to 0.004" (0.03mm to 0.10mm) per revolution. The twelve longitudinal and cross feeds (table and saddle power feeds) range from 1/2" to 20 7/8" per minute (ipm) or 12mm to 530mm per minute (mm/min). The spindle nose has a #40 National Standard taper (NS). The table size is 12" × 49" (305mm × 1,240mm). Rapid longitudinal and cross feed traverses are available at the rate of 84 5/8 ipm or 2,150 mm/min. The rapid vertical advance is half the rate of the other feeds (42 5/16 ipm or 1,075 mm/min).

There are variations among machine tool manufacturers of power feed rates for the quill. Generally, the feeds per revolution are 0.0015"

0.003", or 0.006" (0.04mm, 0.08mm, and 0.015mm). Spindle speeds may be *infinitely variable* or back gear and direct drive. Low and high speed ranges extend from 60 RPM to 5,400 RPM with a standard head. Spindles are also designed for a #7 Brown & Sharpe taper, quick-change tapers, and tapers other than National Standard (NS).

RIGID TURRET AND SWIVEL HEAD ASSEMBLY

The rigid turret assembly may be swung on its base through 360° (Figure 39-3). The head that is secured to the ram is moved by rack and pinion. The head is counterbalanced so that the operator can easily position it for angular cuts. Compound angles or single angles in the front-to-back plane may also be set (Figure 39-3). The head is easily swiveled through a worm gear arrangement. The head may be positioned to right or left of the 0° vertical axis (Figure 39-3).

HANDWHEELS AND FEED CONTROL LEVERS

Three handwheels control the longitudinal feed. Two of these handwheels are located on the sides of the table. The third longitudinal handwheel is located on the saddle. A fourth handwheel on the saddle controls the cross feed movement. The vertical movement of the knee is controlled by the crank. For safety, the handwheels are spring loaded—that is, they are *free wheeling* and remain stationary when the rapid traverse is engaged. The direction of table feed is established by the position of the feed lever. This lever operates both the rapid traverse and feed.

COMBINATION VERTICAL/HORIZONTAL MILLING MACHINE

A *combination vertical/horizontal milling machine* may be quickly and easily converted from a vertical to a horizontal mode. The term *mode* indicates an adaptation of a machine tool from one series of major processes to another. For example, by retracting (bringing back) the forearm, loosening the head bolts, and rotating the head, a vertical milling machine may be adapted to a horizontal mode for horizontal milling processes. Stops at 0° and 90° provide for vertical and horizontal alignment, respectively.

The combination vertical/horizontal milling machine shown in Figure 39–4 has a single swivel head. This head is used for milling from 0° to 180° in any plane. In the horizontal mode, an overarm support may be attached to the knee. This support serves a function similar to the overarm support of a regular horizontal milling machine. The support ensures a high degree of accuracy during heavy-duty horizontal or multiple-cutter cuts by rigidly supporting the arbor.

The combination vertical/horizontal milling machine, as the name suggests, is used for standard vertical and horizontal milling. In addition, drilling, boring, slotting, die sinking, and straddle and other heavy hogging to fine surface finish cuts are performed on this machine. The cuts may be taken by conventional or climb milling. The head has an overload protection to prevent distortion in cutting or damage to the head. Standard accessories used with this machine include the universal dividing head, rotary table with dividing plates, and swivel vise. Cutter-holding devices include arbors and arbor supports, collet holders, and collets.

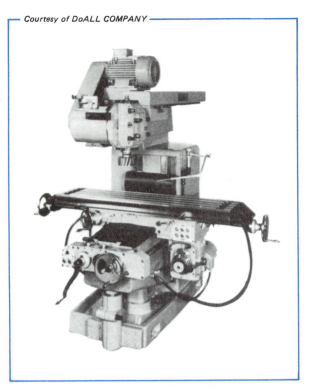

Figure 39–4 Combination Vertical/Horizontal Milling Machine

TRACER MILLING MACHINE

Tracer milling relates to the use of a template or sample part. The template serves as a guide for simultaneously controlling the movement of a cutter to reproduce a desired contour and/or form. A *tracer milling machine* has the versatility of straight-line milling (by locking the cutter at any of the axes) or profiling (by operating the cutter in two or three planes). Feeding may be manually or automatically controlled.

The manufacturer of the tracer milling machine illustrated in Figure 39–5 claims a part-to-template accuracy of ±0.002″ at 5 ipm (0.05mm at 127 mm/min). The accuracy at a higher feed rate of 30 ipm is ±0.0025″ (0.052mm at 762 mm/min).

The part-to-part accuracy is ±0.002″ at both the 5 and 30 ipm (0.05mm at 127 and 762 mm/min) feed rate. The part-to-pattern vertical axis tracing accuracy is within ±0.004″ (0.1mm).

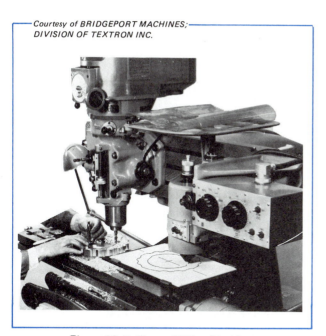

Figure 39–5 Tracer Milling Machines

Figure 39–6 Bed-Type Vertical Milling Machine

BED-TYPE VERTICAL MILLING MACHINE

A *bed-type vertical milling machine* provides for longitudinal and cross slide (transverse) movement of a workpiece under a heavy-duty head and quill. The machine pictured in Figure 39–6 has a vertical head movement of 22″ (560mm) and a quill diameter of 6″ (152.4mm). The combination of longitudinal, cross slide, and vertical travel of this machine provides it with a work capacity of 51″ × 20 5/8″ × 26 3/4″ (1300mm × 525mm × 680mm).

It is evident that the bed-type vertical milling machine with its #50 NS spindle taper is used on large workpieces where great power and heavy chip removal capacity are required.

The bed-type machine is adapted to either the machining of single parts or the manufacturing of multiple pieces at one setting. The head may be swiveled 45° to the right or left of the vertical axis for angle cuts. The machine is designed with an automatic backlash compensation system. This system extends cutter wear life and assures quality of surface finish and dimensional accuracy.

NUMERICALLY CONTROLLED VERTICAL MILLING MACHINE

Vertical milling machine control movements may be actuated from information that is stored or punched on tape (numerically controlled, NC) or that is recorded in a computer program (computerized numerically controlled, CNC). Concepts and programming of numerical control for two- or three-axes processes and four- or five-axes centers are examined in detail in the advanced book.

A *numerically controlled (NC) or computerized NC (CNC) vertical milling machine* may be used for point-to-point or contour milling with two-, three-, or more axes continuous path controls.

Figure 39–7 shows a computerized numerically controlled (CNC) vertical milling machine. This particular machine has a positioning accuracy of ±0.001″ (0.02mm) along the full table movement for all axes. The repeatability of positioning is ±0.0005″ (0.01mm) for all axes.

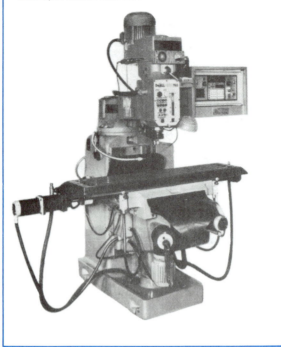

Figure 39–7 Computerized Numerically Controlled (CNC) Vertical Milling Machine

MAJOR COMPONENTS OF STANDARD VERTICAL MILLING MACHINES

The six major design components of a standard vertical milling machine are as follows:

- Column and base;
- Knee, saddle, and table;
- Turret head and ram;
- Spindle (tool) head;
- Manual, power, and rapid traverse feed drives and controls;
- Cutting fluid and lubricating systems.

The functions of each component and of the control levers for speeds and feeds are similar on standard vertical and horizontal milling machines. The exceptions are the turret head, ram, and spindle (tool) head.

FEATURES OF THE TURRET HEAD AND RAM

The turret swings on a machined base on top of the column. The turret head assembly is grad-uated through 360°. The spindle (tool) head is secured to the outer end of the ram. The ram is dovetailed to slide in the turret base. The movement of the ram, toward or away from the head, is controlled by a rack-and-pinion device.

FEATURES OF THE SPINDLE HEAD

The spindle head serves as a cutter holding unit, the power source for cutting, and houses the cutter feed control mechanism. To serve these functions, the spindle head consists of the power source, speed and feed controls, a quill, and the spindle. The main design elements of the spindle head are shown in Figure 39–8.

The spindle is usually splined to deliver maximum power without slippage between the cutter or holder shank and the spindle. A spindle lock prevents the spindle from turning when holders or tools are being inserted or removed.

Many standard taper systems and taper sizes are used by different manufacturers. The number of the taper varies with the size of the machine and spindle. The taper systems include Brown & Sharpe, American (Morse), and American Standard (NS) milling machine taper.

In addition, a Bridgeport® quick-change spindle nose is available for production purposes. By simply turning a collar, cutting tools may be changed quickly. Adapters are fitted to some spindle-nose tapers. These adapters accommodate Brown & Sharpe or Morse taper end mills. Similarly, some spindle tapers accommodate collets for holding straight-shank cutting tools.

VERTICAL MILLING MACHINE ATTACHMENTS

SPINDLE HEADS

Separate spindle heads are available as attachments. They may also be added for multiple-head machining on duplicating machines. The heads are sometimes mounted on the back of the ram. Extra heads are equipped for two-way power down-feed and up-feed. The spindle and quill are counterbalanced.

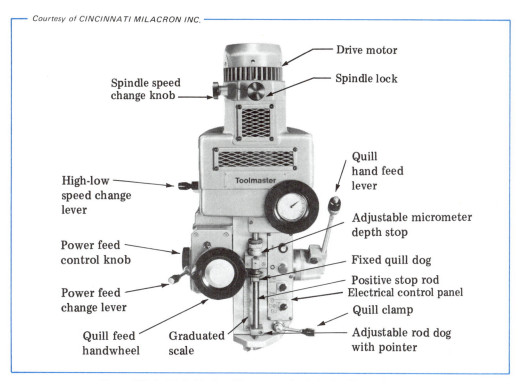

Figure 39–8 Main Design Elements of a Spindle (Swivel) Head

Spindle heads are available with a separate lever for rapid manual movement of the quill, a handwheel to obtain small movements, an integral worm and gear to permit angular positioning, a positive quill lock, a switch to reverse the spindle rotation, and a back-gear drive for maximum power. The micrometer depth stop is graduated in thousandths of an inch on standard machines and 0.02mm on metric machines.

Other spindle heads are designed for variable-speed milling, drilling, and boring. These heads have infinitely variable speeds from 60 to 500 RPM in the back-gear range and 500 to 4,200 RPM in the direct-drive range. Heavy models operate at slower speed ranges from 50 to 450 RPM and 450 to 3,500 RPM. Spindle distortion is minimized by a flow-through air cooling system on some heads.

POWER DOWN-FEED

The *power down-feed attachment* (Figure 39–9) controls the quill feed rate. The up-down feeds are infinitely variable from 0.2 to 2.5 ipm or 5 to 64 mm/min. The feed direction is controlled by the up-off-down switch. A pilot light indicates when the control is in operation.

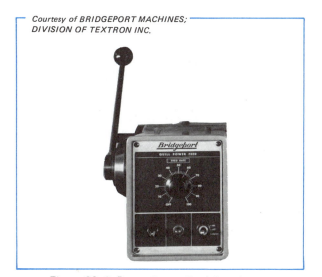

Figure 39–9 Power Down-Feed Attachment

VERTICAL SHAPER HEAD

Some toolrooms and jobbing shops use a *vertical shaper head* as a milling machine attachment. The vertical tool motion permits shaping processes that normally require special machines or broaches.

The vertical shaper head shown in Figure 39–10 may be used to form shapes in blind holes where broaches cannot be used, cut gear teeth and racks, and produce sharp internal corners. Shaping may be done at a right angle or at any simple or compound angle to the table.

This head has a stroke range of 0″ to 4″ (0mm to 102mm). The strokes per minute of a slow speed head are 35, 50, 70, 100, and 200. A standard-speed model produces twice as many strokes per minute.

ATTACHMENTS FOR VERTICAL MILLING MACHINE SPINDLE HEADS

HIGH-SPEED ATTACHMENT

The *high-speed attachment* to the standard spindle head provides for maximum rigidity and rotational speeds up to 50% higher than the usual speed range. The Quill Master® high-speed head attachment (Figure 39–11) is designed for operations requiring the efficient use of small end mills and drills. The Quill Master is available with a 1/8″ (3mm) collet, a 3/16″ (5mm) spring collet, and a 3/16″ (5mm) solid end mill holder.

One manufacturer's design of the high-speed attachment is capable of operating at 9,000 RPM. Mounted on the quill, the attachment requires one-sixth of the spindle speed. Operating at this lower spindle speed eliminates vibration. At the same time, the high speed permits small size drills and end mills to be run at proper cutting speeds. Thus, feed rates are increased over the rates used for slower speeds.

RIGHT-ANGLE ATTACHMENT (CONFINED AREAS)

Special tooling and fixtures, particularly for milling out pockets and cavities, may be eliminated by using a *right-angle attachment*. Figure 39–12 shows a vertical and two horizontal setups for end milling inside a workpiece. The right-angle attachment shown in the figure works equally well on inside or outside cuts on regular parts or irregularly shaped castings. The unit has permanently lubricated bearings and a gear housing. The minimum working space is 2″ (51mm) diameter. Milling cuts may be taken to within 1/2″ (13mm) of a walled surface.

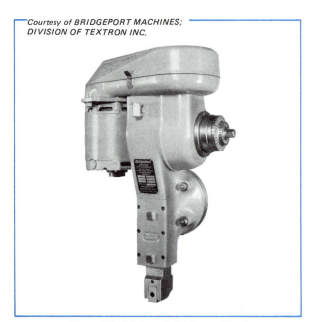

Courtesy of BRIDGEPORT MACHINES; DIVISION OF TEXTRON INC.

Figure 39–10 Vertical Shaper Head

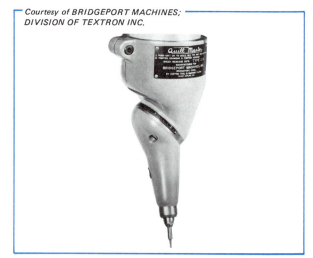

Courtesy of BRIDGEPORT MACHINES; DIVISION OF TEXTRON INC.

Figure 39–11 Vertical Miller High-Speed Head Attachment

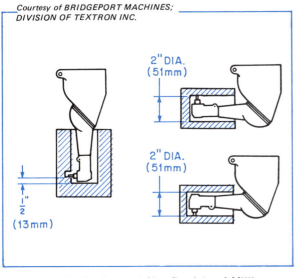

2" DIA.
(51mm)

2" DIA.
(51mm)

$\frac{1}{2}$"
(13mm)

Figure 39–12 Internal (Confined Area) Milling,
Using High-Speed and Right-Angle Attachments

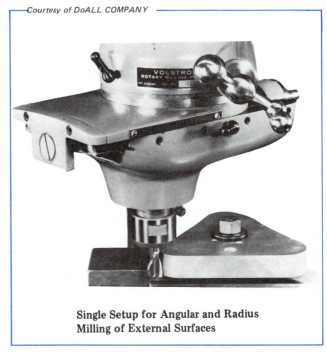

**Single Setup for Angular and Radius
Milling of External Surfaces**

Figure 39–13 Application of Rotary Cross Slide
Milling Head

A right-angle attachment may be assembled on a high-speed head for internal end milling mounted directly on the quill of a vertical spindle head. Power is transmitted by a drive shaft that is held solidly by means of a collet and drawbar. The clamping screws on the body of the right-angle attachment hold the unit firmly on the quill.

These right-angle attachments are designed with preloaded ball bearings and precision-lapped, spiral bevel gears that are grease lubricated. An auxiliary arbor support provides added rigidity for horizontal machining operations. The spindle of a right-angle attachment is designed to accommodate a collet. A supplementary right-angle attachment is used for precision milling operations in confined areas.

ROTARY CROSS SLIDE MILLING HEAD

The *rotary cross slide milling head* mounts on the vertical head quill. The attachment provides mechanical control of the cutting tool through straight, angular, and radial movements. For example, combined operations such as angles and radii, angles tangent to radii,

and one radius blending into another are possible (Figure 39–13).

The rotary cross slide milling head may be rotated in a 360° planetary motion. The degrees are subdivided into five-minute graduations. Operation of the head may be by manual or power feed.

OPTICAL MEASURING SYSTEM

Positive, accurate, and fast determination of the table position is possible by a *direct-reading optical measuring system*. Reading accuracies are within 0.0001" with the inch-standard attachment. The operator reads a single line on a scale that is calibrated every 0.010". Settings in 0.001" are obtained through a drum dial that is calibrated in 0.0001" increments. The metric-unit scales are calibrated every 0.10mm; the drum dial, in 0.002mm increments. The optical measuring system is used for longitudinal and cross travel measurements.

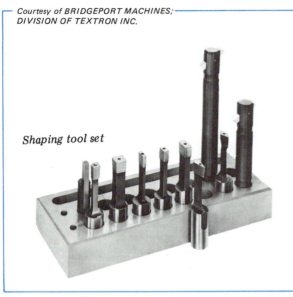

Shaping tool set

Figure 39-14 Common Shapes (Inch and Metric) of Shaping Tools

Figure 39-15 Precision Boring Chuck with Direct 0.0001'' Vernier Reading on Bore Diameter

VERTICAL MILLING MACHINE TOOLING ACCESSORIES

QUICK-CHANGE TOOLING SYSTEM

The *quick-change tooling system* is designed for fast tooling changes during production. The system is available with #30 or #40 NS taper holders. One manufacturer has the following cutter holders or adapters available for quick-change tooling (the collets are available in both inch and metric sizes):

- Quick-change spindle,
- Chucks for drills and end mills,
- Drill extension chucks,
- Non-pullout end mill collets,
- Floating reamer holders,
- Tenthset boring heads and boring bars,
- Morse taper adapters,
- Jacobs taper adapters,
- Shell end mill adapters,
- End mill adapters,
- Tap holders,
- Preset locking fixture for chucks,
- Spade blade holders,
- Spade blades,
- Fly-tool cutter holders.

SHAPING TOOL SET

Vertical shaping processes often require the use of a *shaping tool set*. The seven common shapes of cutting tools in such a set are illustrated in Figure 39-14.

BORING HEAD SETS

Precision boring is done on a vertical milling machine by using a *boring head set*. This set includes a boring head, a number of boring bars, other solid boring tools, and a container.

The shanks of boring heads are ground to #30, #40, or #50 NS tapers or to the straight style (R-8). The boring bit capacity ranges from 1/2'' for the R-8 and #30 NS shank sizes to 5/8'' for #40 NS and 3/4'' for #50 NS.

Boring heads have micrometer dials that permit settings to accuracies of 0.001'' and finer (0.0005''). Metric divisions read to 0.01mm. The tenthset boring head (Figure 39-15) has a direct 0.0001'' (0.002mm) vernier-reading adjustment of the bore diameter.

END MILL HOLDERS, COLLETS, AND ADAPTERS

While there are slight variations in construction details among manufacturers of end mill

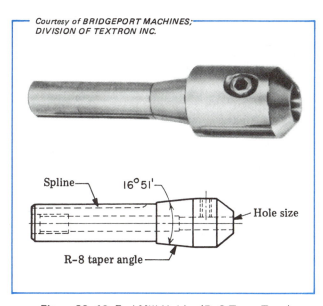

Spline

16°51'

Hole size

R-8 taper angle

Figure 39–16 End Mill Holder (R–8 Taper Type)

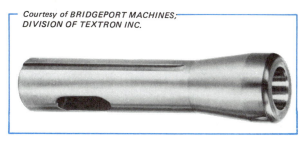

Figure 39–17 Collet Design for Vertical
Milling Machines

holders, collets, and adapters, three features are standardized: (1) the shank with its taper, (2) the threaded end to receive the drawbar, and (3) the hole to accommodate either straight or taper-shank end mills.

The shank tapers may be #30, #40, or #50; Brown & Sharpe tapers #7, #9, #10, #11, and #12; or the R–8 combination of a straight body and 16°51' included angle taper nose.

An *R–8 taper end mill holder* is seen in Figure 39–16. The diameter, taper angle, and spline are shown in the line drawing. This particular end mill holder is available with hole sizes ranging from 3/16" (5mm) to 1" (25mm).

A *Brown & Sharpe taper-shank holder* is heat treated, hardened, and ground. The hole diameters range between 3/16" to 1 1/4" (5mm to 31.75mm). The setscrew in the holder clamps securely against the flat area on the shank of the cutter. There are two setscrews on holders that accommodate 7/8" (22mm), 1" (25mm), and 1 1/4" (31.75mm) diameter mills. Most single-end mills may be held in this holder.

Collets for vertical milling machines are available for straight-shank tools in R–8 and N–2 types, Brown & Sharpe tapers of 0.500" (13mm) taper per foot, and Morse tapers of 0.599"

(15.2mm) taper per foot. One type of collet is shown in Figure 39–17.

The N–2 collet has an included face angle of 20°. The R–8 collet range is from 1/8" through 3/4" in increments of 1/64" (3mm through 19mm in 0.4mm increments). The #7 Brown & Sharpe cutter holds tools from 1/16" through 1/2" diameter in 1/64" (0.4mm) increments.

Adapters are used for taper-shank end mills and drills. These adapters are finished with an R–8 taper. Hole sizes are available for Morse or Brown & Sharpe shanks.

An *M–1 type adapter* and a cutaway section are shown in Figure 39–18. Note that the one end is threaded for a drawbar. Positive drive is provided by the keyseat and key.

Shell end mill holders extend the tooling capability to the use of shell end mills for face and side milling in one operation. These holders are manufactured for a number of different pilot diameters. The shell end mill is secured on the pilot diameter and shouldered by means of an arbor screw and end mill wrench.

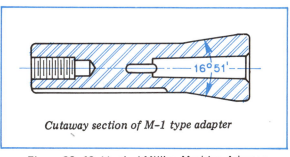

Cutaway section of M-1 type adapter

Figure 39–18 Vertical Milling Machine Adapter
for Taper-Shank Tools (M-1 Type)

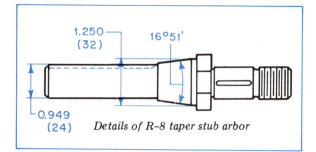

Figure 39-19 Features of a Stub Arbor
(R-8 Taper Type)

Stub arbors, as the name implies, are short-length spindle arbors. Figure 39-19 shows the features of an R-8 taper stub arbor. The arbor is provided with narrow-width spacers. Details are illustrated by the line drawing.

DIGITAL READOUT

Table movements along the horizontal (X) axis and the cross slide (Y) axis may be viewed rapidly on a *digital readout* accessory. The dual readout panel has a visual readout (separate display) for the X and Y axes (Figure 39-20). The readouts may be in the inch or metric system. Inch/metric conversion is optional.

Digital readout units are accurate within ±0.0005'' (0.01mm). Readings are displayed in color for ease of reading and viewing from any working position at the machine. The axis display shows a + or − directional sign, a decimal point, and positions for six-digit readings.

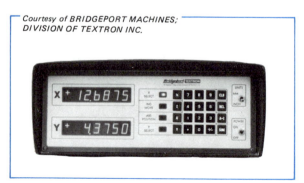

Figure 39-20 Digital Readout Panel with Display for X and Y axes

Safe Practices Using Vertical Milling Machines, Attachments, and Accessories

- Check to see that all machine guards are in place before starting the machine.
- Secure assistance for mounting heavy machine attachments and/or difficult-to-handle workpieces. Keep the back straight and vertical. Lift with the force exerted by the legs.
- Exercise care when swiveling the spindle head for angular cuts. Unless the head is counterbalanced or adjusted through gearing, the clamping bolts should be retightened lightly so that the spindle head may be moved when a force is applied to it.
- Secure cutting tools in the spindle by using an appropriate type and size of collet and/or adapter, when required.
- Check each adapter, collet and arbor, as well as collars and all other accessories, after each set-up.
- Shut down the spindle to check the bore for nicks or burrs. Stone, if necessary.
- Check the mating taper of the spindle and the shank taper of an adapter, collet, or other tool-holding accessory. Remove any burrs.
- Select the cutting speed that is appropriate for the operation, size of cutter, shape of the workpiece, and results in the required dimensional accuracy. Set the spindle RPM to the closest cutting speed.
- Determine and set the power down-feed or up-feed that is correct for the setup and operation. The feed must permit machining to the required dimensional and surface finish accuracies.
- Shut down the spindle before any measurements are taken.
- Select and use comfortably fitting safety goggles and/or a protective face shield.
- Use protective footwear.
- Observe general and specific safety precautions relating to the safe disposal of chips, condition of the coolant and coolant system, machine cleaning, and machine shutdown.

TERMS USED WITH VERTICAL MILLING MACHINES, ATTACHMENTS, AND ACCESSORIES

Vertical milling machine applications	General milling, drilling, and boring processes for machining flat, angular, circular, and rounded surfaces and contours to full or partial depth. Machine cutting in one, two, or three planes with a cutter that is mounted vertically or at a single or compound angle in a vertical spindle machine.
Vertical milling machine attachments	Mechanical, pneumatic, optical, or electronic devices, mechanisms, and movements that extend the versatility of a machine. Units that generally are actuated by the turning of the spindle.
Vertical milling machine accessories	A group of work-positioning and -holding devices such as plain and combination sine tables, rotary tables, and dividing heads with gear trains. Cutter- and tool-holding devices.
Mode	The positioning of the components of a machine tool to permit another combination of processes to be performed—for example, changing from the vertical mode to the horizontal mode.
Tracer milling	Simultaneous control of cutter movements in one, two, or three planes to reproduce a desired contour or three-dimensional form.
Bed-type vertical milling machine	A heavy-duty, primarily manufacturing type, vertical milling machine. A vertical miller that permits longitudinal and transverse movement of the workpiece.
Infinitely variable spindle speeds	Versatility to set the spindle speed at any RPM within the minimum and maximum range of speeds.
Spindle head	A major vertical milling machine component consisting of a power source, speed and feed controls, a quill for positioning cutters and holding attachments, and a spindle.
Quill feed controls	Handwheel or a hand lever for manually feeding a quill. A feed control lever for engaging the quill power feed and the power feed change lever.
Vertical shaper head	An attachment that changes rotary motion to straight-line (reciprocating) motion and permits vertical and single or compound angle shaping and slotting.
Power down-feed attachment	A mechanism for controlling the up- and down-feed rates of the quill.
High-speed head	A spindle head attachment that provides high rotational speeds for effectively and efficiently machining with small-diameter drills and end mills.

Right-angle attachment	A spindle head attachment used in confined spaces for angular, horizontal, slotting, and other milling or drilling processes.
Quick-change tooling system	A wide selection of cutter-holding devices and adapters that permits maximum tool changing flexibility.
Tenthset boring head	A boring head that permits boring diameter adjustments of 0.0001″ (0.002mm).
Digital readout	Visual display of table movements along X and Y axes in + or − directions and in inch or metric units of measure.

───────────────── SUMMARY ─────────────────

- Vertical milling machines of the 1860s combined the vertical-spindle feature of the drill press with the longitudinal and transverse (cross slide) movements of the horizontal milling machine table.
    - The redesigning in 1883 to include the knee-and-column feature permitted raising and lowering the table in relation to the column.
- The angle-setting head added machine tool capability to machine angular surfaces. A front-to-back swivel head design later made compound angle settings and milling processes possible.
    - The turret assembly of the standard general-purpose vertical milling machine may be swung through 360°. The vertical head may also be swiveled from 0° to 180° in relation to the 360° horizontal plane. An angle bracket permits swinging the head at an angle toward or away from the column.
- Quills are rigidly designed to offset heavy cutting forces. Attachments are also secured to the outside diameter. Depths may be accurately set by micrometer stops.
    - Speeds are regulated by a variable-speed control handwheel and a change lever for the high- and low-speed ranges.
- Handwheels for positioning and feeding (vertically, longitudinally, and transversely) are conveniently located on the front of the machine.
    - The combination vertical/horizontal milling machine with support arm for the horizontal mode is used for general milling machine processes. In addition, drilling, boring, slotting, and straddle milling cuts may be taken by conventional or climb milling.
- The tracer milling machine is primarily a two- and three-plane duplicating machine that follows a preestablished pattern.
    - The bed-type vertical milling machine is best adapted for heavy machining operations and production manufacturing. The bed is fixed.
- The numerically controlled vertical milling machine is activated from electronically programmed machine and tooling information. Machines may be numerically controlled (NC) or computer numerically controlled (CNC).

- The major design components of a vertical milling machine are the column and base; knee, saddle, and table; turret head and ram; spindle (tool) head; feed drives and controls for manual, power, and rapid traverse; and cutting fluid and lubricating systems.
- The spindle head is designed to hold cutters, provide the rotary cutting force, house the up- or down-feed control mechanism, and turn the spindle in a quill.
  - Attachments include the vertical shaping head; power up- and down-feeds; high-speed, right-angle, and rotary cross slide milling heads; direct-reading optical measuring system; and the inch- or metric-standard digital readout system.
- Quick-change tool systems permit fast changing of cutting tools. Each system includes a variety of designs and sizes of chucks, collets, holders, adapters, boring heads and bars, and a chuck locking fixture.
  - Boring heads and tenthset heads permit diametral settings for boring within 0.0001″ (0.002mm).
- Adapters for shank tapers accommodate NS tapers, B & S tapers, the R–8 straight body and 16°51′ taper nose, N–2, and other standard tapers.
  - Vertical-spindle arbors, stub arbors, and collars are used for standard-hole types of milling cutters.
- Reporting X and Y axes (+ or –) table movements may be viewed on a digital readout panel attachment.

UNIT 39 REVIEW AND SELF-TEST

1. List four significant design changes in vertical milling machines that have increased its versatility.

2. Give three functions of the quill on a vertical milling machine.

3. State three functions of the spindle on a vertical milling machine.

4. a. Identify two different types of attachments for vertical milling machine spindle heads.
   b. Describe briefly the function of each attachment.

5. Name six different cutter holders or adapters.

6. Tell what the difference is between an NS shank taper holder and an R-8 shank holder.

7. a. Explain briefly the purpose served by a digital readout system.
   b. Cite two advantages of a digital readout system.

8. State two vertical milling machine safety precautions to observe before turning on the power for the spindle.

Cutting Tools, Speeds and Feeds, and Basic Processes

This unit deals with cutting tools, setups, cutting speeds, feed rates, and safe practices for milling flat, angular, beveled, and round surfaces; stepped surfaces; and slots, dovetails, and keyways on the vertical milling machine.

OBJECTIVES

After satisfactorily completing this unit, you will be able to:

- Identify factors affecting the selection of flutes and end mill cutters.
- Understand positive and negative radial rake angles and applications of two-flute and four- or more flute end mills.
- Determine when to use standard length end mills, ball-end and square-end end mills, regular and high-helix end mills, and keyway and tapered end mills.
- Tell about cobalt heavy-duty roughing mills and advantages, and solid and collet chuck end mill adapters.
- Determine starting cutting speeds and feeds and select an appropriate cutting fluid.
- Perform each of the following processes.
  - End Milling Flat and Stepped External Surfaces.
  - End Milling an Angular Surface.
  - End Milling a Slot or Keyway.
  - Cutting a T-slot on a Vertical Miller.
- Apply shop formulas and manufacturers' technical data about cutting speeds, feeds, and cutting fluids.
- Follow recommended machine, tool, and personal *Safe Practices* and correctly use related *Terms.*

The cutters for the basic vertical milling processes are the standard cutters used for horizontal milling and other general machining processes. While high-speed steel cutters are common, cobalt high-speed steel, solid carbide, and carbide-tipped tools are widely used. The carbide tools provide for maximum production, efficient chip removal, and high abrasion resistance.

DESIGN FEATURES OF END MILLS

Like all other cutting tools, the main design features of end mills are designated by technical terms, as shown in Figure 40-1. Here, the tooth face is ground with a *positive radial rake angle.* End mills are also designed with a *negative radial*

rake angle. End mills are formed with *straight* or *spiral teeth* and with *right-* or *left-hand helix angles.* The angles may be *standard* for general-purpose milling or *high* for heavy-duty production milling.

CONSIDERATIONS IN SELECTING END MILLS

Single-end end mills are general-purpose mills. *Double-end* end mills are more economical production milling cutters. Fractional sizes of end mills up to 1″ (25mm) are standard stock items. Large sizes are also commercially available, but in limited sizes. Figure 40–2 illustrates *stub, regular,* and *long-length* double-end end mills.

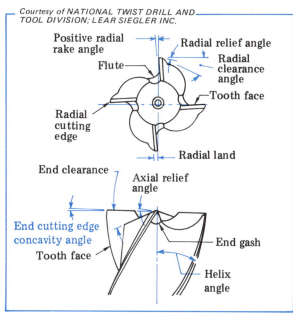

Figure 40–1 Four-Flute (Gashed End Type) End Mill Showing Design Features and End Mill Terminology

SUB-LENGTH, SMALL-DIAMETER END MILLS

Stub-length, small-diameter end mills combine positive rake angles on the flute face, primary relief angles, and short-length teeth. These end mills are designed to machine efficiently at the higher speeds that are required for their small diameters.

A feed of approximately 0.0005″ (0.01mm) per tooth is recommended for general-purpose end milling with small-diameter, high-speed steel end mills.

The flutes on stub end mills should be long enough to mill the required surface, but no longer than necessary. Most small-diameter end mills are used until they become broken or dulled. Regrinding of very small diameters is not feasible. However, the larger sizes of small-diameter end mills may be reground because these sizes have a raised margin and primary side relief.

SQUARE- AND BALL-END MULTIPLE-FLUTE END MILLS

Square-end, four-flute, single- and double-end end mills are used for general-purpose end milling, slotting, stepping, slabbing, shallow pocketing, tracer milling, and die sinking processes. When

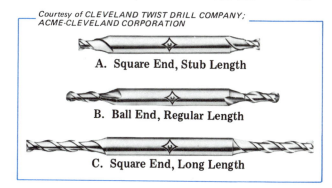

Figure 40–2 Double-End (HSS) End Mills

ground to a *ball end*, these end mills are adapted to center cutting for processes such as die sinking, fillet milling, tracer milling, and other processes requiring a radius (ball) to be formed.

Long and *extra-long*, square- and ball-end, multiple-flute end mills permit deep cavity milling due to the long flute (cutting tooth) length. An extra-long, square-end, four-tooth end mill is pictured in Figure 40–3A. A long, ball-end end mill is shown in Figure 40–3B.

HIGH-HELIX END MILLS

High-helix end mills provide the shear cutting action necessary for producing fine finish surfaces. These general-purpose end mills are used primarily for milling nonferrous and harder alloy workpieces.

When these end mills are used as *slabbing mills* for heavy stock removal, additional support for slabbing cuts is sometimes obtained by using a center in the center hole of the end mill.

Figure 40–3 Basic Types of Four-Flute, Single-End End Mills

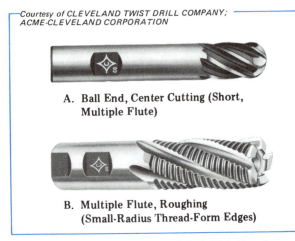

A. Ball End, Center Cutting (Short, Multiple Flute)

B. Multiple Flute, Roughing (Small-Radius Thread-Form Edges)

Figure 40–4 Heavy-Duty, Cobalt High-Speed Steel, Single-End End Mills

KEYWAY-CUTTING END MILLS

Keyway-cutting end mills have the same construction details as general-purpose end mills. One difference is that the outside diameter is ground to a tolerance of +0.0000"/-0.0015" (+0.00mm/-0.04mm). The undersize diameter compensates for the tendency of the end mill to cut oversize. The –0.0015" (–0.04mm) permits the end mill to cut a keyway to a precise, nominal size.

TAPERED END MILLS

Tapered end mills are constructed with cutting edges formed at standard included angles of 10°, 14°, and 15°. The ends are usually ball shaped. The taper end mill is adapted to machining angle surfaces at 5°, 7°, and 7 1/2° and for die sinking processes that require the sides of the relieved surfaces to be cut at an angle.

COBALT HIGH-SPEED STEEL END MILLS

Cobalt high-speed steel end mills are used for heavy-duty center cutting of high-tensile steels and high-temperature alloys. The extra-heavy-duty end mill in Figure 40–4A has a short flute length, which allows for greater rigidity and less deflection during the cutting of difficult-to-machine materials.

One of the newer designs for a multiple-flute *roughing end mill* is illustrated in Figure 40–4B. This cobalt high-speed steel end mill is designed for machining low-alloy, high-strength steels and die steels at accelerated feeds and speeds.

Rapid chip removal allows roughing end mills to be set for deeper depths of cut. The rugged design and chip-breaking characteristics also mean that less power is required to remove large amounts of stock than is required with conventional (HSS) end mills. Since the feed/speed/stock removal relationships are less critical than for regular end mills, roughing end mills are widely used for NC machining operations.

Cobalt high-speed steel roughing end mills have the following advantages:

- High material removal rate,
- Minimum horsepower requirement,
- Ability to take heavier cuts at high speeds and with less chatter and vibration,
- Less deflection for the same stock removal rate than conventional end mills,
- Excellent heat dissipation (because of the tooth-form design) under heavy machining operations,
- Fast chip-breaking characteristics,
- Simplified cutter resharpening without affecting the tooth profile.

TYPES OF END TEETH

Two-flute end mills have *center-cutting* end teeth (Figure 40–5A). These end mills are used for plunge and traverse cuts. The end may be ground to a ball (radius) shape or other form.

Multiple-flute end mills (with more than two teeth) are designed with a gashed end or a center-cutting end. *Gashed-end* end teeth (Figure 40–5B) are cut to the center hole or counterbore. The end may be ground square, to a radius, or to another form.

Gashed-end multiple-flute end mills are recommended for the conventional milling of slots and pockets where plunge cutting is not required. Center-cutting, multiple-flute end mills are designed for plunge cutting where a fine surface finish is required. The two center-cutting teeth (Figure 40–5C) cut to the center. Plunge cutting is used as the feeding method in slabbing, pocketing, and die sinking operations.

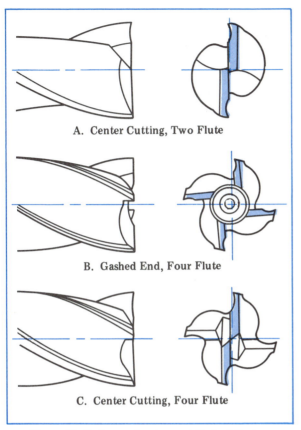

A. Center Cutting, Two Flute

B. Gashed End, Four Flute

C. Center Cutting, Four Flute

Figure 40–5 Center-Cutting and Gashed-End End Teeth on End Mills

CUTTING FLUIDS FOR END MILLING

Mineral, sulphur-base, and soluble oils, as well as water-base synthetics, are common cutting fluids. Water-base cutting fluids have excellent cooling qualities. Oil-base cutting fluids produce a high-quality surface finish.

Cutting fluids are especially necessary to clear the large volume of chips produced by carbide cutters.

Note: The teeth on a carbide end mill will chip when milling cast iron or steel unless there is a *continuous supply* of cutting fluid *under pressure.*

In the case of cast iron, compressed air is used. Brass and plastics are also machined dry. The air supply disperses the chips from the cutting area and cools the end mill.

Note: Safety guards or protective chip shields must be placed around the table and cutting area to prevent chips and other foreign particles from flying.

INTERPRETING CUTTING SPEED AND FEED DATA FROM HANDBOOK TABLES

Efficient machining with end mills and other hole-forming cutting tools requires considerable judgment on the operator's part. The cutting speed for end milling processes is established from tables of technical data. The data have been gathered over many years by cutting and machine tool manufacturers, designers, and engineers. The cutting speeds found in tables are *suggested starting points.* The speeds may be increased or decreased depending on the following variable conditions:

- Material being machined,
- Material of which the cutter is made,
- Type and condition of the machine itself,
- Nature of the machining process,
- Rigidity of the setup and design of the workpiece,
- Composition and application of the cutting fluid,
- Required quality of surface texture.

After the cutting speed has been selected for the diameter of cutter to be used, the feet per minute (or equivalent mm/min or M/min) is converted to RPM. The cutting process is then carefully observed. The operator determines whether a cutting speed adjustment is to be made.

The nature of the technical information contained in reference tables of starting cutting speeds, RPM, and feeds for end mills for selected materials is illustrated by Table 40–1.

Table A–19 in the Appendix includes a full range of diameters from 1/16″ through 3″ (1.6mm through 76mm). The separate columns identify different materials, recommended types and styles of end mills, cutting speeds in sfpm, and feed.

In Table 40–1, the low value of the cutting speed range (in sfpm) is primarily for roughing cuts; the high value, for finish cuts. Corresponding ranges are given for chip load per tooth. For example, assume a part made of machine steel in

Table 40-1 Examples of Suggested Starting Speeds and Feeds
for High-Speed Steel End Mill Applications

Diameter of End Mills (in Inches)	Materials				
	Machine Steel, Hard Brass and Bronze, Electrolytic Copper, Mild Steel Forgings (20-30C)		Brass, Bronze, Alloyed Aluminum, Abrasive Plastics		
	Types and Styles of End Mills				
	High-Speed Steel End Mills, 2 or More Flutes		High-Speed Steel End Mills of High-Helix Type, 1 or 6 Flutes		
	Speed 60–80 sfpm	Feed	Speed 100–200 sfpm	Feed	
	RPM	Chip Load per Tooth	RPM	Chip Load per Tooth	
	∠1 ∠2	∠2 ∠1	∠1 ∠2	∠2 ∠1	
1/16	3667–4888	.0002–.0005	6111–12222	.0002–.0005	
3/32	2750–3259	.0002–.0005	4073–8146	.0002–.0005	
. . .	. . .	. . .	. . .	. . .	
1/2	458–611	.001–.003	764–1528	.0005–.003	
9/16	412–543	.001–.004	678–1356	.0005–.004	
. . .	. . .	. . .	. . .	. . .	
2	115–153	.001–.004	191–382	.0005–.004	
2 1/8	108–144	.001–.004	179–358	.0005–.004	
. . .	. . .	. . .	. . .	. . .	

∠1 Roughing cuts, ∠2 Finishing cuts

the hardness range of 20–30 on the Rockwell C scale is to be rough and finish machined with a 1/2″ (12.5mm) end mill. The information the operator obtains from the table is as follows:

- A high-speed steel end mill with two or more flutes is recommended;

- The cutting speed for end milling ranges from 60 sfpm for roughing cuts to 80 sfpm for finish cuts;

- The feed rate (chip load per tooth) ranges from 0.001″ (0.02mm) for finish cuts to 0.003″ (0.08mm) for roughing cuts;

- The spindle RPM for the roughing cuts, which equals the surface feet per minute (60 × 12) divided by the circumference of the cutter in inches (1/2 × π), is 458 RPM; for finishing cuts, 611 RPM.

It must be emphasized that each speed, feed, and RPM in the table is a *starting point*. The operator determines when each value should be increased or decreased.

ESTABLISHING THE FEED RATE

Efficient machining also requires appropriate feed rates. Again, the operator uses tables that provide suggested starting feed rates. The feed is expressed as a *chip load per tooth*. The feed is expressed in thousandths of an inch or hundredths of a millimeter. The feed rate for end mills varies for different materials, cutting speeds, and depths of cut.

The feed rate (F) for milling operations is the product of the chip load per tooth (f) multiplied by the number of teeth (n) in the cutter and by the RPM. The feed rate is in inches per minute (ipm) or mm/min.

Example: If the chip load per tooth on a 1/4″ (6mm) diameter end mill is 0.0005″ (0.0125mm) and the spindle speed is 4,000 RPM, the feed rate is 8 ipm (200 mm/min):

$$F = f \times n \times RPM$$
$$= 0.0005 \times 4 \times 4,000 = 8 \text{ ipm}$$

in metric units,

$$F = 0.0125 \times 4 \times 4,000 = 200 \text{ mm/min}$$

The depth of cut for end milling operations, as a general rule, should be limited to approximately one-half the diameter of the end mill. When this depth is exceeded, the feed rate must be reduced. Otherwise, the cutter teeth and cutter may fracture due to the excess force beyond the breaking point of the end mill. The depth of cut is limited by: (1) the amount of material to be removed; (2) the power available at the machine spindle; and (3) the rigidity of the setup, cutting tool, and workpiece design.

Accuracy in Groove Milling. Inaccurately formed grooves are produced when excessive feed rates are used with end mills. The added cutting force causes the end mill to be deflected from its axis. The sides of the slot that is milled are not perpendicular to the spindle and work axis. The problem is corrected by reducing the feed rate. The distance the end mill projects from the spindle should also be checked and reduced, if possible. An inaccurate groove or slot is also produced when a dull cutter or a worn spindle is used.

BASIC VERTICAL MILLING MACHINE PROCESSES

The greatest application of the vertical milling machine is in the machining of flat external and internal surfaces. These surfaces may be milled parallel, perpendicular, or at a single or compound angle to the face of the table. Grooving, slotting, and the cutting of T-slots, dovetails, and keyways are common vertical milling processes. The setups and procedures for each of these basic processes follow.

How to Mill a Flat Surface with an End Mill

End Milling a Horizontal Plane Surface

STEP 1 Select an end mill with a diameter and cutter direction appropriate to the job requirements.

Note: It may be necessary to take more than one cut if the width of the surface is greater than the cutter diameter. Large-width surfaces are usually milled with a shell end mill.

STEP 2 Secure the end mill in an end mill holder.
STEP 3 Set the spindle head at the 0° vertical position.
STEP 4 Set the spindle RPM and the horizontal table feed travel rate.
STEP 5 Position the workpiece and cutter so that a minimum number of cuts are taken.
STEP 6 Lock the spindle for maximum rigidity. Start the spindle and flow of cutting fluid.
STEP 7 Raise the knee until the cutter just grazes the workpiece. Set the micrometer collar on the knee at zero.
STEP 8 Clear the end mill and workpiece. Raise the table and take a trial cut.
STEP 9 Move the cutter away from the workpiece. Measure the overall height to check the dimension.
STEP 10 Make whatever adjustment is needed to cut to the required depth. Then, engage the power feed and take the first cut.
STEP 11 Take successive cuts until the surface is machined to the required dimension.

End Milling a Vertical Plane Surface

STEP 1 Select an appropriate side-cutting regular, long, or extra-long end mill.

Note: The cutter length should permit machining the required width in one pass.

STEP 2 Mount the end mill so that it is held as close to the spindle as possible (Figure 40–6).

Note: The direction of the end mill teeth must be checked. The hand must correspond with the spindle and feed directions.

STEP 3 Set the spindle RPM to produce the recommended cutting speed (sfpm or mm/min). Set the feed rate.

STEP 4 Position the cutter vertically at 0° to mill the required area of the workpiece.

Note: The setup should be checked for minimum tool overhang and adequate clearance between the setup, quill, and workpiece.

STEP 5 Move the revolving end mill into position for the first roughing cut.

STEP 6 Hand-feed the end mill to take a cut for about 1/16" (1mm to 2mm) past the center of the cutter. Move the end mill clear of the workpiece.

STEP 7 Stop the spindle. Take a measurement. Make whatever size adjustment is necessary.

STEP 8 Feed the end mill by hand. When all machining conditions have been checked, engage the power feed.

STEP 9 Stop the spindle. Return the cutter to the starting point. Move the cutter into position for other roughing or a finishing cut.

STEP 10 Follow standard procedures for cleaning chips from the machine and workpiece, removing the end mill, and burring.

How to End Mill An External Stepped Surface

Milling a Right-Angle Step

STEP 1 Lay out the stop or multilevel surface that is to be milled.

STEP 2 Select an appropriate end-cutting end mill.

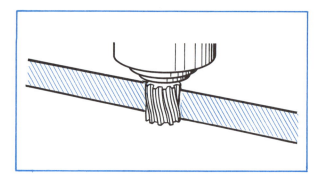

Figure 40–6 End Mill Mounted Close to Spindle Head for Side Milling

Note: A stub or regular-length end mill is used to obtain maximum cutting efficiency and cutter strength.

STEP 3 Position the end mill to cut to the required width and position it vertically for depth.

Note: Wide, deep steps may require one or more roughing cuts and a finish cut.

STEP 4 Take a trial cut. Move the end mill horizontally away from the workpiece. Stop the machine. Measure the width and depth.

STEP 5 Make whatever width or height adjustments are needed. Lock the quill and knee.

STEP 6 Engage the power feed and take the cut.

STEP 7 Repeat steps 3 through 6 for subsequent roughing and finish cuts.

Note: When more than one surface is to be milled, repeat the steps for setting the end mill to depth and width.

Milling a Section (Pad)

STEP 1 Lay out the length, width, and depth of the section to be end milled. (Figure 40–7 shows two typical examples of workpiece sections that are commonly end milled.)

STEP 2 Mount the workpiece securely. Check the layout lines to be sure the setup permits the section to be milled to depth.

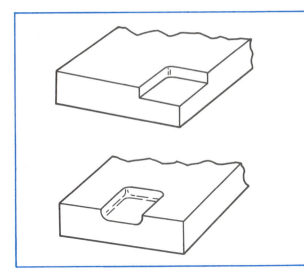

Figure 40–7 Two Examples of End Milled
Stepped Sections

STEP Select the shortest- and largest-diameter
3 end mill for maximum strength. Mount
the end mill. Check to see that the cutter and spindle will clear the workpiece
when the final cut is taken.

STEP Position the end mill to depth to take
4 the first roughing cut.

Note: Locate and/or feed the end mill to
mill to the required depth, width, or
length by using the layout lines or the micrometer collar graduations on the quill.

STEP Move the table transversely so that the
5 cutter is positioned to cut to width at
the layout line for the length.

Note: Allowance is made if considerable
material is to be removed and roughing
and finish cuts are to be taken.

STEP Check the cutter depth and the length of
6 the milled area. Then, continue to feed
the end mill in until the width is reached.

STEP Proceed to hand-feed the table longitudi-
7 nally to the next layout line. This line
marks the length of the surface to be
milled. Lock the table in position at this
point.

STEP Feed the end mill out to produce an L-
8 shaped groove when milling a corner pad

(stepped surface). A U-shaped groove is
milled when the pad location is between
the ends of a workpiece.

STEP Remove the excess material from the area
9 between the grooves.

STEP Repeat steps 4 through 9 when the depth
10 of the stepped area requires roughing
and finish cuts.

How to End Mill an Angular Surface

Positioning the Workpiece at an Angle

STEP Position the workpiece so that the angular
1 surface may be milled with a square-end
end mill held at the 0° vertical axis.

Note: When the angle is laid out, the workpiece is positioned in a vise so that the
layout line is horizontal with the table.
A surface gage is usually used. Sine plates
and tables are used for positioning a
workpiece at a precise angle.

STEP Select an end mill that is appropriate for
2 the operation and material.

STEP Mill the angle surface by using the same
3 procedures as for milling a flat surface.

Note: Care must be taken when the first
cut is made. If it is started at the small
end of the triangle formed by the tapered
surface and the horizontal plane, the cut
gradually increases. The cutter depth
and feed may need to be changed to compensate for this condition.

STEP Take successive cuts to rough and finish
4 mill the angle to size.

Setting the Spindle Head at an Angle

STEP Position and lock the spindle head at the
1 required angle (Figure 40–8).

STEP Move the spindle head so that the center
2 line of the cutting edges of the end mill
falls near the center line of the angular
surface to be milled.

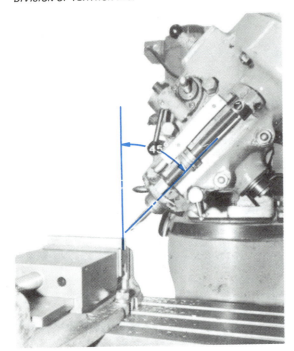

Figure 40–8 Spindle Head Setup for End Milling a Required (45°) Angle

Figure 40–9 Elongated Slot Milled to Size with an End Mill

STEP 3 Turn the quill handwheel to feed the end mill to depth.

STEP 4 Engage the power feed. Take the first cut. Measure the angle. Make angle adjustments when needed.

STEP 5 Take successive cuts until the angular surface is machined to the required dimension.

How to End Mill a Slot or Keyway

STEP 1 Select a keyway end mill.

Note: The standard, slightly undersize keyway cutter compensates for the usual oversize milling of the end mill.

STEP 2 Mount the end mill. Check the setup for minimum overhang. The setup must clear the workpiece when the full depth of cut is reached.

STEP 3 Center the keyway end mill (or regular end mill for slotting) with the centerline location on the workpiece.

STEP 4 Feed the end mill to depth.

Note: The amount and force of the cutting fluid must be sufficient to flow chips out of the slot before the end mill is repositioned for subsequent cuts.

STEP 5 Take a trial cut just long enough to permit the slot to be gaged for its width measurement. Make adjustments as needed.

STEP 6 Proceed to end mill the keyway or slot to the required dimensions (Figure 40–9).

How to Cut a T-Slot on a Vertical Miller

STEP 1 Select an end mill for first milling a groove to the width of the T-slot opening.

STEP 2 Replace the straight end mill with the T-slot cutter. Center the cutter with the slot. Figure 40–10 shows the cutter and workpiece setup for machining a T-slot on a vertical milling machine.

STEP 3 Move the cutter down to the required dimension for the depth of the T-slot.

STEP 4 Feed the cutter into the workpiece by hand to be sure the cutting action is correct. Engage the power feed.

Figure 40–10 Setup for the Vertical Milling of a T-Slot

How to Mill a Dovetail on a Vertical Miller

STEP 1 Lay out the dovetail. Mount the work securely in a work-holding device or clamp it to the table.

STEP 2 Select a dovetail cutter that is ground to the required included angle.

STEP 3 Mill out the area between the two side dimensions of the dovetail with an end mill.

STEP 4 Replace the end mill with a dovetail cutter. Position this cutter for depth to within 0.010″ to 0.015″ (0.2mm to 0.4mm) of the base surface of the dovetail.

STEP 5 Position the cutter to take a series of roughing cuts. Rough machine one side of the dovetail.

STEP 6 Raise the workpiece until the dovetail cutter just grazes the base surface. Then, move the cutter to the layout line for the angle side. Take the final finish cut.

STEP 7 Machine the second angular side of the dovetail. Repeat steps 3 through 6 for roughing out and finish milling.

STEP 8 Measure the depth and width of the dovetail. Burr the edges.

STEP 9 Clean the machine, cutter, and workpiece. Place all tools and accessories in proper storage areas.

Safe Practices for Basic Vertical Milling Processes

• Check the direction of rotation of the spindle and the hand of the end mill to prevent damage to the cutter teeth.

• Select the shortest possible length of end mill to reduce tool overhang and to provide adequate clearance for the operation to be performed.

• Supply a continuous flow of cutting fluid to carbide end mills when cast iron or steel is milled.

• Feed dovetail cutters slowly into a workpiece. Any sharp contact between the fragile ends of the cutter teeth and workpiece may cause the teeth edges to fracture.

• Place safety guards on the table around the workpiece area if air is used as a coolant.

• Use a stiff brush and avoid handling the long, needle-like sharp chips that are produced by an end mill.

• Use safety goggles and/or a transparent shield as protection against flying chips and particles.

• Stop the spindle before cleaning a workpiece. Remove burrs before measuring.

TERMS USED WITH VERTICAL MILLING CUTTING TOOLS AND PROCESSES

Center cutting (end milling)	Cutting across the end face with center-cutting end mill teeth. End mill cutting capability from the outside diameter to the center.
Side cutting (end milling)	Milling with the side-cutting edges (teeth) of an end mill.
Square-end end mill	An end mill with cutting teeth that are ground at a right angle to the cutter axis.
Ball-end end mill	An end mill having end-cutting teeth that are ground to a ball (radius) shape.
Keyway cutter (end milling)	A specially ground end mill having a slightly smaller diameter than the nominal width of a keyway.
Long or extra-long, multiple-flute end mill	An end mill with flutes that are longer than the flutes of a regular-length end mill. An end mill designed to mill to a greater depth that a regular-length end mill.
High-helix end mill	A general-purpose end mill with teeth cut at a steeper helix angle than a standard spiral-tooth end mill.
Stub-length end mill	An end mill designed with short-length teeth that provide greater cutter rigidity and efficient milling.
Gashed-end teeth	End teeth on an end mill that extend from the outside diameter to the center hole.
Collet chuck end mill holder	A vertical milling machine cutter holder and collet combination for holding end mills securely to the spindle.

SUMMARY

- End mills are commercially available with straight or spiral teeth, positive or negative radial rake angle, right- or left-hand helixes, and standard or high helix angles.
  - High-speed steel, cobalt HSS, and carbide cutters are commonly used materials for end mills.
- Single-end and double-end end mills are available in stub, regular, long, and extra-long lengths.
  - Stub-length, small-diameter end mills provide added strength to operate at the higher speeds required for economical milling.
- Square-end, four-flute end mills are adapted to general-purpose end milling, slotting, machining stepped surfaces, slabbing, shallow pocketing, and die sinking processes.
  - High-helix end mills produce a shear cutting action and a fine surface finish.
- Cobalt high-speed steel end mills permit heavy-duty milling at higher speeds, feeds, and depths of cut than conventional HSS mills.
  - The cutting ends on end mills are either center cutting for plunge milling or gashed.

- End mills are used for machining right-angle steps and for milling stepped areas.

  - The end milling of a beveled or other angular surface may be done by either positioning the workpiece or setting the spindle head at the required angle.

- End milling elongated or through slots and keyways requires roughing and finish cuts.

  - T-slots are milled by first cutting a slot and then using a T-slot cutter to form mill the rectangular area.

- The outside diameter of keyway-cutting end mills permits the milling of keyways to precise, nominal sizes.

  - Dovetails are milled on vertical milling machines by finish milling the angular sides with an angle cutter.

UNIT 40 REVIEW AND SELF-TEST

1. Name six design features of standard end mills.

2. a. Identify three design features of stub-length, small-diameter end mills that permit them to cut efficiently at the required higher speeds.
   b. Give the feed rate for general-purpose end milling with stub-length, small-diameter end mills.

3. Give three advantages of high-helix end mills over standard-helix end mills.

4. Explain why a square-end, multiple-flute end mill is a higher production milling cutter than a two-flute end mill.

5. Tell why a keyway-cutting end mill is preferred over a standard end mill for machining a keyway to a precise, nominal dimension.

6. Give two reasons for using a cobalt high-speed steel end mill instead of a conventional high-speed steel end mill for production milling a high-tensile material.

7. Differentiate between a center-cutting and a gashed-end end mill.

8. Describe briefly how a collet chuck holder operates to hold an end mill solidly in the spindle.

9. Use a table to establish (a) the recommended starting spindle speed range for taking roughing and finish cuts and (b) the corresponding feed range for end milling a step in a brass/bronze block. A 1/4″ (6mm) diameter high-speed end mill is used.

10. Give two personal safety precautions to take when performing end milling operations.

Structure and Classification of Industrial Materials

The reduction of iron ores to produce pig iron and processes for producing cast irons and cast steels are described first in this section. Consideration is then given to the characteristics, properties, applications, and manufacture of common ferrous and nonferrous metals and alloys, cemented carbides, and ceramic and diamond cutting tools. Classification systems for iron, steel, aluminum, and alloys are also covered.

UNIT 41

Manufacture, Properties, and Classification of Metals and Alloys

OBJECTIVES

After satisfactorily completing this unit, you will be able to:

- Describe the open hearth, Bessemer converter, electric furnace, and basic oxygen processes of making steels, the composition and grades of carbon steels, and the mechanical properties of metals.
- Classify low-, medium-, and high-carbon steels and tool and die steels according to the Unified Numbering System (UNS), SAE, and AISI systems.
- Establish the characteristics and applications of gray, white, chilled, alloy, and nodular cast iron, and the production of cupola and pearlitic cast irons.
- Describe the effect of basic metallic alloying elements on the properties of alloy steels and copper-, aluminum-, nickel-, zinc-, and magnesium-base alloys.
- Identify grades and uses of cast alloys, cemented carbide, ceramic cutting tools, and polycrystalline diamond cutting tools.
- Read and apply handbook and manufacturer's tables on characteristics, composition, and physical properties of alloying elements for ferrous and nonferrous metals and alloys.
- Follow each recommended *Safe Practice* for handling and machining industrial materials.
- Correctly use each related *Term.*

BASIC STEEL MANUFACTURING PROCESSES

Steel-making processes require the impurities within pig iron to be *burned out*. The removal of impurities is done in one of four types of furnaces:

- Open hearth furnace,
- Bessemer converter,
- Electric furnace,
- Oxygen process furnace.

OPEN HEARTH PROCESS

As much as 80% of all steel is produced by the open hearth process, which was introduced in 1908. The dish-shaped *hearth* is charged with limestone and scrap steel. When the scrap is melted, molten pig iron is poured in. The impurities are burned out by sweeping fuel and hot gas over the molten metal. The burned gases are drawn off. The direction of the flames in the furnace is changed at regular intervals. The limestone in the furnace unites with the impurities and rises to the surface of the molten metal.

The process continues for a number of hours. The molten steel is drawn off into a ladle that has the capacity to take the *heat* (the batch of steel to be poured). The slag floating on the metal is drawn off through a spout in the ladle.

Alloying is done by adding fixed amounts of the required alloying metals—for example, tungsten, molybdenum, and manganese—to the molten steel. After mixing, the alloyed steel is poured into *ingot molds*. When the steel within a mold solidifies, the mold is stripped from the solid steel mass, which is called an *ingot*. The ingot is moved into a *soaking pit* so that the complete ingot is slowly brought to a uniform temperature. Next, the heated ingot is rolled into specific sizes and shapes.

BESSEMER CONVERTER PROCESS

The Bessemer converter today produces a very small percent of the steel manufactured as compared to the earlier production of over 90% of all steel manufactured. The converter consists of a lined circular shell that is swiveled to receive a charge of molten iron. The trunnion mountings are designed to provide a blast of air at high velocity. The blast flows through holes in the bottom when the furnace is moved to vertical position. The oxygen in the air burns out the impurities.

After being subjected to the air blast for 10 to 15 minutes, the converter is tilted to pour the metal into a mixing ladle in which alloying elements are added. The alloyed steel is then poured into ingots, soaked, and rolled in much the same manner as the open hearth steel.

ELECTRIC FURNACE PROCESS

The electric furnace permits precise control over the steel-making process. The furnace operates by lowering three carbon electrodes to strike an arc with the scrap metal loaded in the furnace. The quality of scrap steel, the alloying elements, the amount of oxygen used, and the heat are carefully regulated. The heat generated by the electric arc produces the molten steel.

Chromium, tungsten, nickel, or other alloying materials are added to produce the required alloy steel. After the heat (molten steel) is complete, the furnace is tilted in a pouring position. The molten metal is *teemed* (poured from the ladle) into ingots for subsequent soaking and rolling operations.

BASIC OXYGEN PROCESS

The oxygen process is one of the newer steel-making processes. A high-pressure stream of pure oxygen is directed over the top of the molten metal. The furnace is charged with about 30% scrap metal. Then, molten pig iron is poured into the furnace. Finally, the required fluxes are added.

The making of steel in a basic oxygen process furnace is illustrated in Figure 41–1. The furnace is covered with a water-cooled hood. The oxygen lance is lowered to within 5 to 8 feet above the surface of the molten metal. The force of the oxygen at a pressure of 140 to 160 pounds per square inch at a flow rate of 5,000 to 6,000 cubic feet per minute creates a churning high-temperature movement. This action continues until the impurities are burned out. Then, the flame drops, the oxygen is shut off, and the lance is removed. The entire process takes almost 50 minutes at a steel production rate of about 300 tons per hour.

The furnace is tapped by tilting it. After alloying materials are added to the mixing ladle, the molten metal is poured into ingots. The same processing is followed to form metal stock.

CHEMICAL COMPOSITION OF METALS

Metals are identified according to a natural element. Copper, silver, aluminum, nickel, and chromium are a few examples. Pure metals are alloyed with one or more other elements to provide qualities in a resulting product to meet engineering requirements.

In metallurgical terms, an *alloy* is composed of two or more elements and has metallic properties. One of the elements must be metallic; the other, either metallic or nonmetallic. Commercially, the term *alloy* is used to denote a metallic substance of two of more metallic elements. One of these elements must be intentionally added.

Ferrous alloys contain iron as the base metal and one or more metallic elements. Ferrous alloys such as molybdenum steel, vanadium steel, and nickel steel are all steels that have metallic elements added to change their properties.

Nonferrous alloys such as brass, bronze, and monel metal do not contain iron (except as an impurity). Brass and bronzes are copper-base alloys. Brass is an alloy of copper and zinc; bronze, of copper and tin.

COMPOSITION OF PLAIN CARBON STEEL

The principal elements in plain carbon steel include: *carbon*, *manganese*, *phosphorous*, *silicon*, and *sulphur*. Each element has a decided effect on the properties of steel.

Carbon is the hardening element and exerts the greatest influence on the physical properties. As the percent of carbon increases within a specific range, there is an increase in hardness, hardenability, wear resistance, and tensile strength. The higher the carbon content, the lower the melting point. Any additional carbon above the 0.85% point has limited effect on hardness, although wear resistance increases. The graph shows four general ranges of steel with carbon content within the 0 to 1.3% range. Low-carbon steels contain from 0.02% to 0.30% carbon by weight. Medium-carbon steels contain from 0.30% to approximately 0.60% carbon. High-carbon steels fall within the 0.60% to 0.87% carbon content range. The tool steels begin at the 0.87% carbon content point.

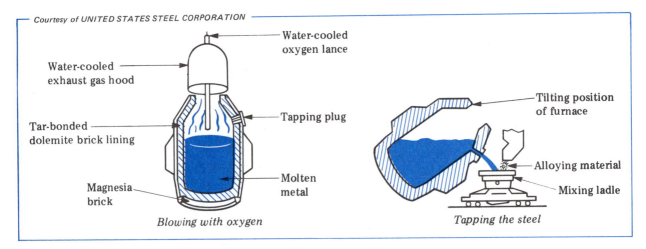

Water-cooled oxygen lance

Water-cooled exhaust gas hood

Tar-bonded dolemite brick lining

Magnesia brick

Tapping plug

Molten metal

Blowing with oxygen

Tilting position of furnace

Alloying material

Mixing ladle

Tapping the steel

Figure 41–1 Making Steel in a Basic Oxygen Process Furnace

A small quantity (0.30% to 0.60%) of manganese is usually added during manufacture to serve as a purifier in removing oxygen from the steel. Oxygen in steel makes it weak and brittle. Manganese also mixes with sulphur, which is an impurity. Manganese is added to steel to increase the tensile strength, toughness, hardenability, and resistance to shock. Manganese is added in amounts from 0.06% to over 15% to produce certain physical properties. A 1.5% to 2.0% addition to high-carbon steel produces a deep-hardening steel. Such a steel must be quenched in oil. The addition of 15% manganese produces hard wear-resistant steels.

The addition of small amounts of phosphorous and sulphur to low-carbon steels increases machinability and decreases shrinkage. Amounts above 0.60% are considered undesirable because the phosphorous causes failures due to shock and vibration.

The presence of from 0.10% to 0.30% silicon makes steel sound as it is cast or hot rolled. Between the 0.60% to 2.0% range, silicon becomes an alloying element. Silicon is usually used in combination with manganese, molybdenum, or chromium. As an alloying element, silicon increases tensile strength, toughness, and hardness penetration.

The addition of from 0.08% to 0.30% sulphur to low-carbon steel increases its machinability. Screw stock for automatic screw machine work is a sulphurized free-cutting steel. Otherwise, sulphur is considered an impurity in steel. As an impurity, sulphur causes cracking during rolling or forging at high temperatures.

MECHANICAL PROPERTIES OF METALS

Mechanical properties are associated with the behavior of a metal as it is acted upon by an external force. The properties of a metal determine the extent, if any, to which it can be hardened, tempered, formed, pulled apart, fractured, or machined.

Hardness is defined as the property of a metal to resist penetration. This property is controlled by heat treating in which a machined or forged tool may be shaped while soft and then hardened.

Hardenability represents the degree to which a metal hardens through completely to its center when heat treated. A low hardenability means the surface layer hardens but the metal is softer toward the center. *Brittleness* is related to hardness. Brittleness refers to the degree to which a metal part breaks or cracks without deformation. *Deformation* is the ability of a material to flex and bend without cracking or breaking.

Ductility refers to the ability of a metal to be bent, twisted, drawn out, or changed in shape without breaking. For example, a deep draw shell requires sheet metal with high ductility to permit drawing without fracturing the metal. The ductility of a metal is usually expressed as a percentage of reduction in area or elongation.

Malleability relates to the ability of a metal to be permanently deformed by rolling, pressing, or hammering. *Toughness* of a metal refers to the property that enables it to withstand heavy impact forces or sudden shock without fracturing.

Fusibility is the ease with which two metals may be joined together when in a liquid state. A high-fusibility metal can be easily welded. In such instances, the term *weldability* is used.

Machinability relates to the ease or difficulty with which a given material may be worked with a cutting tool. After extensive testing of materials under control conditions, machinability ratings have been established. These ratings are published in handbook tables as percentages. AISI 1112 cold-drawn steel is assigned a rating of 100%. Other metals are rated in percentages in comparison to the machinability of AISI 1112 steel. Difficult-to-machine metals are rated below 100%. Easy-to-machine metals are rated above 100%.

Machinability properties need to be considered in terms of the following factors:

- Physical properties of the material (tensile strength, rigidity, hardness, grain structure) and chemical properties;
- Type of chip formation produced during the machining process;
- Cutting characteristics that are influenced by the use or absence of a particular type of cutting fluid;
- Power required in taking a cut.

CLASSIFICATION OF STEELS

There are two general categories of steel: plain carbon steels and alloy steels. The plain carbon

Table 41-1 Effects of Selected Alloying Elements on Certain Properties of Steel

(Examples:) Effect on Properties of Steel	Carbon (C)	Chromium (Cr)	Manganese (Mn)	Molybdenum (Mo)	Nickel (Ni)	Tungsten (T)	Vanadium (V)
Increases							
tensile strength	X	X	X	X	X		
hardness	X	X					
wear resistance	X	X	X		X	X	
hardenability	X	X	X	X	X		X
ductility			X				
elastic limit		X		X			
rust resistance		X			X		
abrasion resistance		X	X				
toughness		X		X	X		X
shock resistance		X			X		X
fatigue resistance							X

steels are grouped in three ranges of carbon content as low-carbon steel, medium-carbon steel, and high-carbon steel.

PLAIN CARBON STEELS

Low-Carbon Steel. The range of carbon by weight is from 0.02% to 0.30%. Low-carbon steels may not be hardened except by adding carbon to permit hardening the outer case while the area below remains soft. The range of carbon is from 0.08% to 0.30% in low-carbon steels that are commonly used for manufacturing parts that do not require hardening. Low-carbon steels such as machine steel and cold-rolled steels are used in manufacturing bolts and nuts, sheet steel products, bars, and rods.

Medium-Carbon Steel. The carbon content is from 0.30% to 0.60%. With this amount of carbon, it is possible to harden tools, such as hammers, screw drivers, and wrenches, that may be drop forged or machined.

High-Carbon Steel. The carbon content range for high-carbon and tool steels is from 0.60% to 1.50%. These steels are adapted for edge cutting tools such as punches, dies, taps, and reamers.

ALLOY STEELS

The addition of alloying elements to steel increases the tensile strength, hardenability, toughness, wear abrasion, red hardness, and corrosion resistance. The most important effects of alloying selected elements on certain properties of steel are indicated in Table 41-1.

SAE AND AISI SYSTEMS OF CLASSIFYING STEELS

Two main systems of classifying steels have been devised. These systems provide for standardization in producing each alloy and in designating each alloy. The systems were developed by the Society of Automotive Engineers (SAE) and the American Iron and Steel Institute (AISI).

FOUR- AND FIVE-DIGIT DESIGNATIONS

A four-digit series of code numbers is used in both the SAE and the AISI systems. Certain alloys are identified by a five-digit series of code numbers. One of two letters is used in cases where an element has been varied from the normal content of the steel. The prefix X denotes a variation of manganese or sulphur. The prefix T denotes a variation of manganese in the 1300-range steels.

The first digit of the SAE and AISI classification systems indicates the *basic type of steel*, as follows:

1. carbon
2. nickel
3. nickel-chrome
4. molybdenum
5. chromium
6. chromium-vanadium
7. tungsten
8. nickel-chromium-molybdenum
9. silicon-manganese

The second digit classifies the steel or alloy within a particular *series*. Table 41-2 shows

Table 41–2 Examples of SAE/AISI Classifications of Standard Carbon
and Alloy Steels by Series

Classification of Carbon Steels		
Series	Type	Composition or Special Treatment*
10XX	Plain carbon	Nonsulphurized
11XX	Free-machining	Resulphurized
12XX		Rephosphorized and resulphurized

Classification of Alloy Steels		
13XX	Free-machining, manganese	Mn 1.75%
23XX	Nickel	Ni 3.50%
25XX		Ni 5.00%
31XX	Nickel-chromium	Ni 1.25% Cr 0.65%
30XXX		Corrosion- and heat-resisting
40XX	Molybdenum	Mo 0.20% or 0.25%
. . .	. . .	. . .
93XX	Triple-alloy	Ni 3.25% Cr 1.20% Mo 0.12%
94XX		Ni 0.45% Cr 0.40% Mo 0.12%
98XX		Ni 1.00% Cr 0.80% Mo 0.25%

*Percents represent the average of the range of a particular metal element. Only the predominant alloying elements (excluding carbon) are identified.

the SAE/AISI classifications of standard carbon and alloy steels by series. A steel designated with a 10XX number is in the nonsulphurized plain carbon steel series. An alloy steel numbered 40XX is in the molybdenum steel series that has 0.20% or 0.25% molybdenum content. A zero in the second digit indicates there is no major alloying element.

The third and fourth digits indicate the middle of the carbon content range—that is, the *average percent of carbon*—in the steel. The percent of carbon is indicated by the third and fourth digits as *points*. A point is the same as 0.01% carbon. Thus, a 0.20% carbon content is expressed by the third and fourth digits as 20 points. Figure 41–2 provides two examples of the use of SAE and AISI code numbers.

In a few instances, a five-digit numerical code is used. For example, in the SAE system, a designation of 71250 means the steel is a tungsten steel (7 as the first digit) with 12% tungsten (12 as the second and third digits) and 0.50% carbon (50 points as the fourth and fifth digits).

The letter H after the SAE or AISI code number specifies that the alloy steel meets specific hardenability standards. Steel manufacturers and handbook tables provide craftspersons with additional information about H designations, composition, properties, and applications of steels designated by SAE and AISI codes.

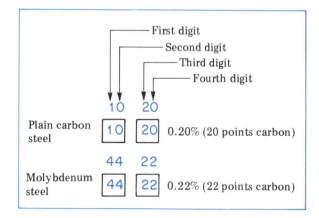

Figure 41–2 Examples of SAE and AISI Code Numbers

Table 41-3 Identification of Major Groups of Tool and Die Steels

Letter Code	Type of Tool and Die Steels		
W	Water-hardening tool steel		
S	Shock-resisting tool steel		
O	Cold-worked tool steel		Oil-hardening
A			Medium alloy, air-hardening
D			High-carbon, high-chromium
H	Hot-worked tool steel	1-19	Chromium types
		20-39	Tungsten types
		40-59	Molybdenum types
M	High-speed tool steel		Molybdenum types
T			Tungsten types
L	Special-purpose tool steel		Low-alloy types
F			Carbon-tungsten types
P	Mold Steel	1-19	Low-carbon types
		20-39	Other types

CODING OF TOOL AND DIE STEELS

There are eleven major groups of tool and die steels. These groups are identified by a letter. In some cases, subgroups are designated by using a numeral following a letter. Table 41-3 provides the codes for identification of the tool and die steel groups.

ADDITIONAL FEATURES OF AISI SYSTEM

A prefix letter is sometimes used with the code number to identify the method used to produce the steel. The four basic prefix letters are: B, for Bessemer carbon steel; C, for general open hearth carbon steel; D, for acid open hearth steel; and E, for electric furnace alloy steel. Where there is no letter prefix to the AISI number given in a table, the steel is predominantly open hearth.

UNIFIED NUMBERING SYSTEM (UNS)

While only the SAE and the AISI systems have been discussed here, various trade associations, professional societies, standards organizations, and private industries have developed different numbering and coding systems. The Unified Numbering System (UNS) represents a joint effort by the American Society for Testing and Materials (ASTM), the SAE, and others to correlate the different numbering systems being used for commercial metals and alloys. The important point to remember about a UNS number is that it provides an *identification, not a specification.* The specifications are to be found in manufacturers' literature, trade journals, and handbooks.

Sixteen code letters are used in the Unified Numbering System. Each letter is followed by five digits. The digits run from 00001 through 99999. Each of the sixteen code letters identifies certain metals and alloys. The letters D, F, G, H, J, K, S, and T designate different cast irons, cast steels, and other ferrous metals and alloys. The letters A, C, E, L, M, P, R, and Z are used for nonferrous metals and alloys.

The numbers in the UNS groups conform, wherever possible, to the code numbers of the other systems. For example, an AISI or SAE 1020 plain carbon steel has a corresponding UNS number of G10200. An SAE or AISI 4130 alloy steel has a UNS identification of G41300. An M33 alloy has a UNS code of T11330. Handbook tables provide the full range of UNS numbers for plain carbon, alloy, and tool steels in the SAE and AISI systems.

ALUMINUM ASSOCIATION (AA) DESIGNATION SYSTEM

A system similar to the SAE, AISI, and UNS classifications for steels and alloys has been developed by the Aluminum Association (AA) for wrought aluminum and aluminum alloys. A four-digit numerical code is used. The first digit identifies the alloy type. For example, a code of 1 indicates an aluminum of 99.00% or greater purity. The numerical code 2 indicates a copper-type alloy; 3, manganese; 4, silicon; 5, magnesium; 6, magnesium and silicon; 7, zinc; and 8, an element other than identified by codes 1 through 7. The number 9 is unassigned at the present time.

The second digit shows the control over one or more impurities. A second digit of zero indicates there is no special impurities control. The third and fourth digits in the 1 series give the amount of aluminum above 99.00%, to the nearest hundredth of a percent. The same digits in the 2 through 8 series are used to identify different alloys in the group. The prefix X is used for experimental alloys. When standardized, the prefix is dropped.

The letters F, O, H, W, and T are used as *temper designations.* These letters follow the four digits (separated by a dash) to designate the temper or degree of hardness, as follows:

F   hardness as fabricated
O   annealed (for wrought alloys only)
H   strain-hardened (for wrought alloys only)
W   solution heat-treated
T   thermally treated

A temper designation may include a numerical code. For example, H designations with one or more digits indicate strain-hardened only (H1), strain-hardened and then annealed (H2), and strain-hardened and then stabilized (H3). A second digit is used to indicate the final degree of strain hardening from 0 to 8—for example, H16. A third digit, when used, gives the variation of a two-digit H temper —for example, H254.

Numerals 2 through 10 have been assigned in the AA system to indicate specific sequences of annealing, heat treating, cold working, or aging, as follows:

T2   annealed (cast products only)
T3   solution heat-treated and then cold-worked
T4   solution heat-treated and naturally aged to a stable condition
T5   artificially aged only
T6   product heat-treated and then artificially aged
T7   product heat-treated and then stabilized
T8   product heat-treated, cold-worked, and then artificially aged
T9   product heat-treated, artificially aged, and then cold-worked
T10   product artificially aged and then cold-worked

VISIBLE IDENTIFICATION OF STEELS

There are two common methods of identifying steels. One method involves the use of the manufacturer's *color code markings.* The other method requires *spark testing.*

COLOR CODE MARKINGS

Some steel producers paint at least one end of steel bars. A color code is supplied by the steel manufacturer. Stock is cut from the unpainted end in order to preserve the painted identification color code.

SPARK TEST

A simple identification test is to observe the color, spacing, and quantity of sparks produced by grinding. The visible characteristics of steel, cast iron, and alloy sparks depend largely on carbon content. A high-carbon tool steel, when spark tested, produces a large quantity of fine, repeating spurts of sparks. These sparks are white in color at the beginning and ending of the stream. By contrast, a spark test of a piece of wrought iron stock produces a few spurts of forked sparks. The color varies from straw color close to the wheel to white near the end of the stream. The spark patterns provide general information about the type of steel, cast iron, or alloy steel.

CHARACTERISTICS OF IRON CASTINGS

Cast iron is an alloy of iron and carbon. The carbon content varies from 1.7% up to 4.5%, with varying amounts of silicon, manganese, and sulphur. The physical properties of cast iron depend on the amount of one of two forms of carbon that is present. *Graphite or free carbon* is one form. *Combined carbon or cementite* (iron carbide) is the second form. The five broad classes of cast iron in general use are as follows:

- Gray cast iron,
- White cast iron,
- Chilled cast iron,
- Alloy cast iron,
- Malleable iron castings.

Ductile iron (nodular cast iron) is a newer type of cast iron used in automotive and industrial applications where factors of high tensile strength and ductility are important.

GRAY CAST IRON

Gray iron castings are widely used in machine tool, farm implement, automotive, and other industries. The American National Standard (ANS) specifications G25.1 and the American Society for Testing and Materials (ASTM) standards A48-64 consider cast iron castings in two groups. The first group of gray iron castings (classes 20, 25, 30, and 35 A, B, and C) have excellent machinability, are comparatively easy to manufacture, and have a low elasticity and a high damping capacity. The second group of gray iron castings (classes 40, 45, 50, and 60 B and C) are more difficult to machine and manufacture, have a lower damping capacity than the first group and have a higher elasticity.

High-strength cast iron castings are produced by the *Meehanite-controlled process*. Some of the more important properties of Meehanite castings were cited earlier in connection with wear-resisting properties on applications such as the cast bed, frame, and column of jig borers, jig grinders, and other machine tools. Meehanite castings may also be produced with heat-resisting, corrosion-resisting, and other combinations of physical properties.

WHITE CAST IRON

White cast iron has a silvery-white fracture. Castings of white cast iron are very brittle, with almost a zero ductility. White cast iron castings have less resistance to impact loading and a comparatively higher compressive strength per square inch than gray cast iron. Nearly all of the carbon in the casting is in the chemically combined carbon (cementite) form. White cast iron is used principally for the production of malleable iron castings and for applications that require a metal with high wear- and abrasive-resistance properties.

A 500X magnification photomicrograph of white cast iron is shown in Figure 41-3 and displays cementite (iron carbide) in the light areas and fine pearlite in the dark areas. Pearlite is comprised of alternating layers of pure iron (ferrite) and cementite.

CHILLED CAST IRON

Chilled cast iron is used for products that require wear-resisting surfaces. The surfaces are designated as chilled cast iron. The hard surfaces are produced in molds that have metal chills for the rapid cooling of the outer surface. Rapid cooling causes the formation of cementite and white cast iron.

ALLOY CAST IRON

Alloy cast iron results when sufficient amounts of chromium, nickel, molybdenum, manganese, and copper are added to cast iron castings to change the physical properties. The alloying elements are added to increase strength; to produce higher wear resistance, corrosion resistance, or heat resistance; or to change other physical properties.

Extensive use is made of alloy cast irons for automotive engine, brake, and other systems; for machine tool castings; and for additional applications where high tensile strength and resistance to scaling at high temperatures are required.

MALLEABLE IRON CASTINGS

Grades and properties of malleable iron castings are specified according to ANS G48.1- and ASTM A47- specifications. These specifications relate to tensile strength, yield strength, and elongation.

Malleable iron is produced by heat treating. In the process, the cast iron is annealed or *graphitized*. Graphitization produces graphite (temper

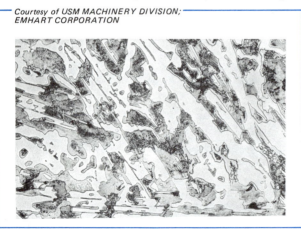

Courtesy of USM MACHINERY DIVISION; EMHART CORPORATION

Figure 41-3 Cementite (Light Areas) and Fine Pearlite (Dark Areas) in White Cast Iron (500X Magnification Photomicrograph)

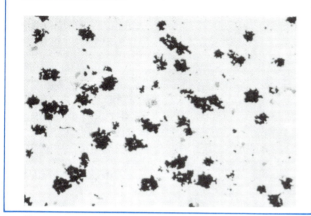

Figure 41–4 Temper Carbon (Graphite) Aggregates in Malleable Cast Iron (100X Magnification)

carbon aggregates). A 100X magnification of temper carbon aggregates in malleable cast iron is shown in Figure 41–4.

Hard, brittle, white cast iron castings are first produced from pig iron and scrap. These castings are then given an annealing heat treatment during which time the temperature is slowly increased over as much as a two-day period up to 1650°F (898.9°C). Then, the temperature is dropped slowly over an equal cooling time.

Malleable iron castings are widely used in industrial applications that require a highly machinable metal, great strength (as compared with other cast irons and nonferrous metals) and ductility, and good resistance to shock.

Cupola Malleable Iron. Malleable iron is also produced by the cupola method. The metal has good fluidity, produces sound castings, and possesses the property of being suited to galvanizing. Therefore, cupola malleable iron is used for making valves, pipe fittings, and similar parts.

Pearlitic Malleable Iron. Pearlitic malleable irons are used where a greater strength or wear resistance is needed than is provided by general malleable iron or in place of steel castings or forgings. There are a number of forms to pearlitic malleable iron. Some forms are engineered to resist deformation. Other forms permit deformation before breaking and are used for crankshafts, camshafts, and differential housings; ordinance equipment; and for other machine parts.

DUCTILE IRON (NODULAR CAST IRON)

Ductile iron is a relatively new type of cast iron. It is also known as *spheroidal graphite* because the graphite is present in ball-like form in contrast to the flake-like form in regular gray cast iron. The spheroidal graphite structure produces a casting of high tensile strength and ductility. The spheroidal structure results from the addition of small amounts of magnesium or cerium-bearing alloys and special processing. A few other advantages of nodular cast iron are as follows:

- Toughness is between the toughness of cast iron and steel;
- Shock resistance is comparable to ordinary grades of carbon steel;
- Melting point and fluidity are similar to the properties of high-carbon cast irons;
- Pressure tightness under high stress is excellent;
- Castings can be softened by annealing;
- Castings can be hardened by normalizing and air cooling or by oil quenching and drawing;
- Wear-resisting surfaces can be cast in molds containing metal chills;
- Surface hardening by flame or induction methods is feasible;
- Parts may be machined with the same ease as gray iron castings.

Nodular cast iron is used in the automotive industry for cylinder heads, crankshafts, and pistons; in the heavy machinery field; and in the machine tool industry for chuck bodies, forming dies, and other industrial applications.

CHARACTERISTICS OF STEEL CASTINGS

Steel castings are used for parts that must withstand shock and heavy loads and must be tougher and stronger than cast iron, malleable iron, or wrought iron.

There are two general classes of steel castings: (1) carbon steel and (2) alloy steel. Carbon steel castings are broken up into three groups: (1) low carbon, (2) medium carbon, and (3) high carbon.

CARBON STEEL CASTINGS

Low-carbon steel castings have a carbon content below 0.20% with most of the castings produced in the 0.16% to 0.19% range. Medium-carbon steel castings have between 0.20% and 0.50% carbon. High-carbon steel castings have a carbon content above 0.50%.

ALLOY STEEL CASTINGS

Alloy steel castings are distinguished from carbon steel castings by the addition of sufficient amounts of special alloying elements to obtain or increase specific desirable qualities. There are two groups of alloy steels: (1) low-alloy steels in which the alloy content is less than 8% and (2) high-alloy steels with more than 8% alloying elements. Alloying elements such as manganese, molybdenum, vanadium, chromium, and nickel are added to steel to affect one or more of the following physical properties:

- Hardenability,
- Distortion reduction and prevention of cracks by decreasing the rate of cooling during the hardening operation,
- Resistance to hardness reduction when being tempered,
- Material strength that may be increased through heat treatment,
- Abrasion resistance at regular and increased temperatures,
- Corrosion resistance at regular and high temperatures,
- Machinability,
- Toughness.

Alloy steels require heat treatment in order to utilize the potential properties of the alloy. There are three categories of alloy steels: (1) construction, (2) special, and (3) tool.

GROUPS OF NONFERROUS ALLOYS

The base metal is used to identify a group of nonferrous metals. A few important groups of nonferrous metals that are commonly used in jobbing shops are described next. The five groups include copper-, aluminum-, zinc-, magnesium-, and nickel-base alloys.

COPPER-BASE ALLOYS

The brasses and bronzes widely used in industry are copper-base alloys. Brasses are pre-dominantly made of copper and zinc. Bronzes are a mixture of copper and tin. Both alloys may be cast, rolled, wrought, or forged. Other alloying elements such as aluminum, manganese, silver, nickel, and antimony are added. High-strength, high-hardness, and age-hardened bronzes are more difficult to machine than free-cutting brasses.

Silicon, manganese, and beryllium are often combined with copper-base alloys. Beryllium in combination with copper forms a hard compound that increases hardness and wear resistance. However, tool wear life is noticeably decreased. Beryllium possesses age-hardening properties.

ALUMINUM-BASE ALLOYS

Alloying elements and heat-treating processes are added to aluminum to increase its tensile strength. Other properties such as machinability, weldability, and ductility are also developed by adding specific alloying elements. Copper, manganese, chromium, iron, nickel, zinc, and titanium are some of the common alloying elements.

NICKEL-BASE ALLOYS

Nickel is noted for its resistance to corrosion from natural elements and many acids. Nickel-base alloys have from 60% to 99% nickel. Nickel is used for parts that are subjected to and must resist corrosive action. In its pure state, nickel is used for plating other metal parts.

One of the common nickel alloys is known as *Monel metal*. This nickel alloy is composed of approximately 65% nickel, 30% copper, and 5% of a few other elements. High-temperature-resistant space age alloys such as Inconel and Waspaloy, depend on nickel as the base metal. Copper, aluminum, iron, manganese, chromium, tungsten, silicon, and titanium are different elements that are alloyed with nickel-base alloys.

ZINC-BASE ALLOYS

Die castings are usually made of zinc-base alloys. The low melting temperature of zinc-base die cast metal alloys of from $725°F$ to $788°F$ ($385°C$ to $420°C$), the ability to flow easily under pressure into die casting dies, and the quality to which details may be produced make zinc-base alloys exceptionally practical for die casting. Zinc alloys are also rolled into sheets for manufacturing sheet metal products.

Zinc is alloyed with aluminum or small amounts of copper, iron, magnesium, tin, lead, and cadmium to produce the sixteen common zinc-base alloys. Die cast metal alloys fall into six groups: zinc-base; tin-base; lead-base; aluminum-base; magnesium-base; and copper, bronze, or brass alloys. The physical properties of each alloy is affected by the combination and quantity of each element.

APPLICATIONS OF INDUSTRIAL MATERIALS

CAST ALLOY CUTTING TOOLS

Cast alloys are considered as one of the special alloy classifications. The other classification includes *wrought alloys*, which are generally not as hard as the cast alloys and contain up to almost 35% iron. Cast alloys require less than 1% iron.

Because of the extreme hardness of cast alloys, they are used as cutting materials. Cast alloys are known by trade names such as *Stellite*, *Tantung*, and *Rexalloy*. Cast alloys have a cobalt base and are nonferrous.

One of the wide fields of application of cast alloys is in cutting tools for machining metals. Cast alloys are produced as cutting tool tips to be brazed on tool shanks, as removable tool bits, and as disposable cutting tool inserts. The extreme hardness of cast alloys requires that they be ground to size with aluminum oxide abrasive wheels. The principal elements in cast alloy cutting tools are cobalt (35% to 55%), chromium (25% to 35%), tungsten (10% to 25%), carbon (1.5% to 3%), and nickel (from 0 to 5%).

One big advantage of cast alloy cutting tools is their ability to cut at high cutting speeds and high cutting temperatures. Due to the high red-hardness property, cast alloy cutting tools perform better at temperatures above the temperatures for high-speed cutters. The temperature range is from 1100°F to 1500°F (593°C to 815°C). In terms of cutting speeds, cast alloy cutting tools are capable of machining at speeds above the highest speeds used with high-speed cutters and the lowest practical speeds for carbide cutting tools.

Tool shank support is important. Most cast alloys are brittle and are not capable of withstanding heavy impact forces as are carbon steel and high-speed cutters. There are, however, special grades of cast alloys that have impact-rupture strength comparable to high-speed steel cutters. Cast alloys are also used in noncutting applications as turbine blades, wear surfaces on machines, and for conveyors in heat-treating furnaces.

CEMENTED CARBIDE CUTTING TOOLS

Cemented carbide cutting tools are made of two main materials: tungsten carbide and cobalt. Special properties are obtained by the addition of titanium and titanium carbides. The term *cemented* means that during production cobalt is used to cement the carbide grains together.

Cemented carbides are cast to required shapes. Cemented carbides do not require further heat treatment. The cast form is very hard. The cast shapes are used as disposable inserts, chip breakers, cutting tool tips that are brazed on shanks, and solid smaller cutting tools such as end mills.

The properties of cemented carbides are affected by cobalt as a principal ingredient. Cobalt affects hardness, wear resistance, and tool life. Increasing the amount of cobalt increases the hardness of the cutting tool. The brittleness is also increased, producing a decreased ability to resist shock and a lower impact toughness. The alloying elements also affect grain structure. A finer grain structure increases the hardness; a coarser structure decreases hardness.

Carbide manufacturers have established a classifying system for cemented carbide machining applications. Of eight general designations, three are used for machining conditions for cast iron and nonferrous metals and five classifications are used for steel and steel alloys. In addition, the carbide industry has six classifications related to wear and impact applications (Table 41–4).

CERAMIC CUTTING TOOLS

Ceramic cutting tools are produced from metal oxide powders. These powders are formed into shape by cold pressing and sintering or by hot pressing. Regardless of the method of forming, the pressed blanks are not as strong as the carbide-shaped cutting tools. Ceramic cutting

Table 41–4 Carbide Industry Classification System

Material to Be Machined	Cemented Carbide Classification	Application
Cast iron and nonferrous metals	C–1	Medium roughing to finishing cuts
	C–2	Roughing cuts
	C–3	High-impact dies
Steels and steel alloys	C–4	Light finishing cuts at high speeds
	C–5	Medium cuts at medium speeds
	C–6	Roughing cuts
	C–7	Light finish cuts
	C–8	Heavy roughing cuts and general-purpose machining cuts

tools are made of aluminum oxide with a metallic binder to improve impact strength. Silicon oxide and magnesium oxide are added.

Ceramic cutting tools are brittle, have low impact resistance, and shatter easily. The hardness of ceramic cutting tools lies below the hardness of a diamond and above the hardness of a sapphire.

Two great advantages of ceramic cutting tools are: (1) the cutter is not affected by the temperature at the cutting edge and (2) hot metal chips do not tend to fuse to the cutter. Thus, cutting fluids are not required for cutting purposes. However, cutting fluids may be needed to prevent distortion of the workpiece. When cutting fluids are used, a liberal and continuous flow must be provided. Otherwise, any intermittent cooling may cause the cutter to fracture or shatter.

The hardness and wear life properties of ceramic cutting tools make them ideal for machining hard and hardened steels. Cutting speeds up to from two to four times faster than the speeds used for cemented carbides may be used for ceramic cutting tools. The surface finish that may be produced by taking light finishing cuts at high speeds is of a quality that eliminates the need for grinding.

Aluminum oxide/titanium carbide ceramic cutting tools use a newer cutting tool material that combines many features of regular oxide cutting tools with its ability to resist thermal and impact shock. This combination of properties permits this oxide family of cutting tools to be used for milling, turning, and other machine

tool operations that are not possible with other cutting tools.

DIAMOND (POLYCRYSTALLINE) CUTTING TOOLS

Industrial-quality natural diamonds and manufactured diamonds provide the hardest known cutting tool material. Hardness and extreme resistance to heat make the diamond ideal for machining soft, low-strength, and highly abrasive materials. The diamond has low strength and shock resistance. Another limitation is to be found in the geometry of the diamond as a single crystal or as a cluster of single-crystal cutting edges. Since the strength, hardness, and wear resistance is related to the orientation of each diamond crystal, the performance may not be as accurately controlled as desired.

This problem of orientation of the crystal is eliminated by bonding fine diamond crystals to form solid tool shapes. The cutting tools in this diamond family are known as *sintered polycrystalline tools.* They are available in forms for brazing on solid shanks, as tips for bonding to carbide tool inserts, and as solid inserts.

Polycrystalline cutting tool inserts have the advantage over the single-crystal cutting edge in that the diamond crystals are randomly oriented. This feature provides for uniform cutting. Interestingly, the wear-resistance property of diamond cutting tools provides a wear life, based on wear resistance alone, in a ratio of from 10:1 to over 400:1 over carbide cutting tools.

Safe Practices in the Handling and Use of Industrial Materials and Cutting Tools

- Grind sharp fins and edges on castings and tumble if the molding sand is not removed from outside and inside surfaces.
- Check the machining impact conditions against the cutting tool manufacturer's recommendations, especially when carbide, ceramic, cast alloy, and other cutting tool materials that may fracture under severe impact are used.
- Avoid quenching a cast alloy cutting tool after dry grinding. Quenching shock may cause the cutter to fracture or shatter.
- Use a liberal, constant flow of coolant when machining with ceramic cutting tools. Inter-mittent cooling may cause the cutter to fracture or shatter.
- Select a rigid tool shank on which carbide, ceramic, and cast alloy inserts may be correctly seated and securely held.
- Cut under the hard outer scale of cast iron castings on the first cut.
- Have a chloride-base fire extinguisher available for any possible fire when machining magnesium and zirconium parts that may ignite.
- Take machining cuts at the fpm, depths, and feeds recommended by the manufacturers of carbide, ceramic, cast alloy, diamond, and other cutting tool materials.
- Make provisions in advance—that is, before start-up—for the desired chip formation (continuous or discontinuous), chip disposal (particularly when operating at high speeds), and positioning of guards and protective devices.

TERMS USED WITH INDUSTRIAL AND CUTTING TOOL MATERIALS

Alloyed cast iron castings	The addition of an alloying element such as chromium or molybdenum to cast iron.
Open hearth, Bessemer, electric, and basic oxygen processes	Four main types of furnaces and processes for reducing pig iron and scrap iron to steel. Processes for burning off impurities and removing slag from a molten iron mixture.
Element (steel and alloy making)	A pure metallic or nonmetallic material found in nature or laboratory-produced. A material combined with one or more metallic or nonmetallic materials to produce desired properties in an alloy.
Hardenability, machinability, ductility	Physical properties of industrial materials and cutting tools.
Alloying	The addition of one or more pure metallic or nonmetallic elements to a base element for purposes of improving properties such as red hardness, corrosion resistance, toughness, and strength.
Metal classification systems	Identification of steels, nonferrous metals, and alloys according to standards established by SAE, AISI, ASTM, AA, and UNS.
Aluminum temper designation	Specific letters used to provide information about the strain hardening, cold working, aging, and heat treating of aluminum.
Graphite structure	Flake and spheroidal forms of carbon. A form of carbon that affects grain structure and hardness. Flake graphite in cast iron improves machinability. Spheroidal graphite adds tensile strength and ductility, as in the case of nodular iron.

Cementite	An iron carbide. A form of combined carbon that produces hardness, toughness, brittleness, and low ductility in white cast iron.
Alloy steel castings	Steel castings with special metallic alloying elements other than carbon.
Martensitic, ferritic, and austenitic stainless steels	Three basic types of stainless steels whose properties are established by the percent of carbon and other alloying elements.
Magnesium, chromium, molybdenum, tungsten, and cobalt	Metal elements added to steel to form alloys that possess specific properties. Metal elements that increase physical properties of steels and alloys such as wear resistance, toughness, red hardness, and hardenability.
Copper-, aluminum-, magnesium-, zinc- and nickel-base alloys	Five important groups of nonferrous metal alloys that are widely used in the machine and metal trades and in industrial manufacturing.
Ceramic cutting tools	A cutting material harder than cemented carbides but softer than the diamond. Metallic oxide powders that are pressed to shape and sintered or hot pressed.
Cast alloys	Cobalt-base alloys with limited traces of iron as an impurity. Extremely hard disposable cutting tool inserts, brazed tips, and removable tool bits.
Polycrystalline diamond cutting tools	Bonding of fine diamond crystals to provide a many-sided, randomly oriented crystal structure to control and improve cutting efficiency and wear life.

SUMMARY

- Pig iron and scrap iron are reduced in a cupola to produce gray, white, chilled, malleable, and alloyed iron castings.

  - Open hearth, Bessemer converter, electric, and oxygen process furnaces are the principal producers of steels and alloys. Steel ingots are soaked, rolled, and formed into specific sizes and shapes.

- Ferrous alloys have iron as the base metal and one or more metallic elements.

  - Nonferrous alloys do not contain iron (except as an impurity).

- Carbon steels are made of iron, a fixed percent of carbon, and other principal elements, including manganese, phosphorous, silicon, and sulphur.

  - The carbon content in tool steels start at 0.87%.

- Manganese, phosphorous, silicon, and sulphur are added to carbon steel to improve mechanical properties such as machinability, hardenability, resistance to shock, grain structure, ductility, and fusibility.

  - Low-, medium-, and high-carbon steels are the three basic groupings of plain carbon steels.

- SAE and AISI classifications are used to identify the composition and properties of steels.

    - The Unified Numbering System (UNS) correlates the major steel and alloy classification systems. Code letters identify metals and alloys. The code numbers indicate the quantity of the base metal and other added metal elements in the steel or alloy.

- The Aluminum Association (AA) has developed a similar classification to SAE/AISI for wrought aluminum and aluminum alloys. The letters F, O, H, W, and T (separated by a dash) follow the four-digit system. The letters relate to temper and heat-treating processes.

    - Two simple, common shop methods are used to identify the kind of steel. One method is to use the manufacturer's color marking on the end of drawn or rolled bars. The other method requires spark testing to identify a steel by the color, spacing, and quantity of sparks in a pattern produced by grinding.

- Five broad, commercially manufactured, cast iron castings include gray, white, chilled, alloy, and malleable cast irons.

    - Malleable iron is produced by heat treating white cast iron. The grain structure changes in the process to produce temper carbon aggregates.

- Pearlitic forms of malleable iron have great strength and wear resistance and may be engineered to permit deformation without fracture.

    - Nodular cast irons contain graphite in a spheroidal structure. Nodular cast irons possess physical properties between the properties of cast iron and steel.

- Steel castings are identified as low-, medium-, or high-carbon and alloy steel.

    - Important physical properties of steel include: hardenability, distortion reduction, machinability, strength, abrasion resistance, corrosion resistance, ductility, and toughness.

- Widely used nonferrous alloys have as the base metal: copper, aluminum, nickel, zinc, and magnesium.

    - The properties of alloy steels are affected by eight major elements: carbon, manganese, nickel, chromium, molybdenum, vanadium, cobalt, and tungsten.

- Stainless steels contain up to 22% nickel alloyed with chromium.

    - Zinc-base alloys are widely used in die casting. The low melting temperature, flowability under force, and ability to reproduce fine cast details makes this group ideal for production die casting of intricate parts.

- Cast alloys have a cobalt base, are nonferrous, and contain additions of chromium, tungsten, and carbon and possess the ability to cut at high speeds and high temperatures of 1100°F to 1500°F (593°C to 815°C).

    - Special grades of alloys have impact-rupture strength. Cast alloys are used for cutting tools and other applications requiring high resistance to abrasion and heat.

- Cobalt increases hardness, wear resistance, tool life, and brittleness.

    - Carbides are classified by manufacturers for machining conditions governing three groups for cast iron and ferrous metals and five groups for steel and steel alloys.

- Ceramic cutting tools are harder than cemented carbides. They are produced from metal oxide powders by cold pressing and sintering or hot forming into cutter inserts and other blank forms.

    - Polycrystalline diamond cutting tools utilize the physical properties of fine diamond crystals. These crystals are bonded in a randomly oriented pattern where the cutting action is not affected by the features of a single crystal.

UNIT 41 REVIEW AND SELF-TEST

1. Describe briefly how pig iron is produced.

2. State three factors, exclusive of the composition and methods of production, that affect the machinability of metals.

3. Refer to a steel manufacturer's catalog or handbook tables of standard carbon steels.
   a. Give the percent range of the elements for carbon steels (1) 1020 (C1020), (2) 1040 (C1040), and (3) 1095 (C1095).
   b. Rate each of the three carbon steels according to the factors of (1) hardness, (2) hardenability, (3) ductility, and (4) wear resistance. Use A for lowest, B for medium, and C for highest.

4. Identify three different properties of steel that are affected by the addition of (1) chromium, (2) manganese, and (3) tungsten.

5. a. List three different types of tool and die steels.
   b. Indicate the letter (and where applicable the number range) of the selected steels.
   c. Give the (1) hardening temperature range and (2) tempering range for a T4 tungsten high-speed tool steel. Refer to a steel manufacturer's data sheets or a handbook table.

6. a. State two differences between cupola and pearlitic malleable iron.
   b. Give an application of each of the two different malleable irons.
   c. List three physical properties of cast iron castings that are changed by alloys to produce alloy cast irons.

7. State two advantages of using alloy tool steels over plain carbon tool steels.

8. State three safety precautions to take when machining with carbide, ceramic, and cast alloy cutting tools.

Metallurgy and Heat Treating: Technology and Processes

This section deals with concepts of basic metallurgy and heat treatment of metals. The structure of steel, its behavior, and the transformation products that result from heating and cooling are described. Phase diagrams and transformation curves are used to explain changes in the microstructure and properties of steels and alloys. Heat treating equipment and heat-treating processes are then covered in detail.

UNIT 42

Heat Treating and Metals Technology

OBJECTIVES

After satisfactorily completing this unit, you will be able to:

- Describe hardening temperature ranges and other iron-carbon phase diagram information.
- Identify common heat treating problems, probable causes, and corrective steps.
- Interpret the functions and use of liquid, oil, water, and molten salt quenching baths.
- Identify basic toolroom and production heat-treating furnaces, controls, and temperature measurement.
- Explain cooling rates and hardening and tempering of carbon tool steels and high-speed steels.
- Describe each of the following processes by giving step-by-step procedures.
  - Heat treatment of tungsten high-speed steels.
  - Annealing, normalizing, and spheroidizing.
  - Liquid, gas, and pack carburizing casehardening.
  - Cyaniding, carbonitriding, and nitriding as surface hardening processes.
  - Flame and induction hardening.
  - Subzero temperatures for stabilizing precision parts.
- Follow *Safe Practices* relating to personal safety and the use of furnaces and quenching baths.
- Correctly use each new related *Term*.

Metallurgy is concerned with the technology of producing metals or alloys from raw materials and alloying elements and preparing metals for use. *Heat treatment* is the combination of heating and cooling operations applied to a metal or alloy in the solid state. The purpose of heat treatment is to produce desired conditions or physical properties in the metal or alloy.

469

HEATING EFFECTS ON CARBON STEELS

Fully annealed carbon steel consists of the element iron and the chemical compound iron carbide. In metallurgical terms, the element iron is known as *ferrite;* the chemical compound *iron carbide* is *cementite.* Cementite is made up of 6.67% carbon and 93.33% iron. There are also traces in carbon steel of impurities such as phosphorous and sulphur.

Depending on the amount of carbon in the steel, the ferrite and cementite are present in certain proportions as a *mechanical mixture.* The mechanical mixture consists of alternate layers or bands of ferrite and cementite. When viewed under a microscope, the mechanical mixture looks like mother-of-pearl. The mixture is known as *pearlite.* With only traces of impurities, pearlite contains about 0.85% carbon and 99.15% iron. Thus, a fully annealed 0.85% carbon steel consists entirely of pearlite. This steel is known as a *eutectoid steel.* A steel having less than 0.85% carbon is a *hypo-eutectoid steel.* This steel has an excess of cementite over the amount that is required to mix with the ferrite to form pearlite. A steel with 0.85% or more carbon content is known as a *hyper-eutectoid steel.* In the fully annealed state, a hyper-eutectoid steel has an excess of cementite so that both cementite and pearlite are present.

CRITICAL POINTS AND HEATING/COOLING RATES

Variations in hardness and other mechanical properties occur when the atomic cell structure of steel is changed through heating and cooling within certain *critical points and rates.* There are two types of atomic structures in steel: *alpha* and *gamma.* Steels that are at a temperature below the critical point have an atomic cell structure of the alpha type. Steels heated above the critical point have a gamma atom structure. Alpha iron *mixes* with carbide. Gamma iron *absorbs* the carbon-iron compound into solid solution. The critical points at which changes take place vary according to the composition of the steel. Hardness and strength variations are produced by changing the atoms and cell structure of the iron atoms from alpha to gamma and then back again to alpha.

The lower critical point for steels is considered to be 1333°F (722.8°C). The upper critical point increases to 1670°F (910°C) for the very low-carbon steels. The lower critical point is used for steels with carbon contents within the 0.8% to 1.7% range. Heat treating takes place when the temperature of a steel is raised above the critical point and the temperature is then reversed and controlled for the required degree rate of cooling.

EFFECTS OF TEMPERATURE AND TIME

Stresses occur in metals at the critical points when a part, or section of a part that is not of equal mass or shape, is heated unevenly or too fast. Usually a part is *preheated* to around 800°F (427°C). Complicated and intricately shaped parts are preheated again to 1250°F (677°C). Preheating permits a better transition from alpha to gamma iron at the critical point.

Steels are heated from 50°F to 200°F (10°C to 93°C) above the critical point to cause all the constituents to go into solid solution. The added temperature also allows the solution to remain in this state a few additional seconds during the cooling cycle between the time the workpiece is removed from the furnace and quenched. This added temperature and short interval of time permit a complete phase change in the metal.

THE IRON-CARBON PHASE DIAGRAM

An *iron-carbon phase diagram* contains general information about the full range of hardening temperatures for carbon steels and is used to show the changes in the microstructure of steel that occur during heat treating. Figure 42–1 is an iron-carbon phase diagram with critical temperatures, heat-treating processes, and grain structures. The diagram shows that when carbon steel in a fully annealed state is heated above the lower critical point—between 1335°F to 1355°F (724°C to 735°C)—the alternate bands of ferrite and pearlite that make up the pearlite begin to merge into one earlier.

The process continues above the critical point until the pearlite completely dissolves to form an iron in an *austenite state.* If an excess of ferrite or cementite is present in the steel, they begin to dissolve into the austenite if the temperature of the steel continues to rise. Ultimately, only austenite is present. The diagram shows that above certain temperatures the excess ferrite or cementite is completely dissolved in the austenite. The temperature at which this transformation occurs is known as the *upper critical point.* This temperature is affected by the carbon content of the steel.

EFFECTS OF SLOW COOLING ON CARBON STEELS

The transformation from alpha to gamma iron to the point where the iron is completely austen- itic is reversed upon cooling. However, the upper and lower critical points occur at slightly lower temperatures to change from gamma to alpha iron. At room temperature, the structure of the original fully annealed carbon steel part will return to the same proportions of ferrite or cementite and pearlite that were present before heating and cooling. No austenite will be present.

EFFECTS OF RAPID COOLING ON CARBON STEELS

As the rate of cooling from an austenitic state is increased, the temperature at which the austenite changes to pearlite decreases below

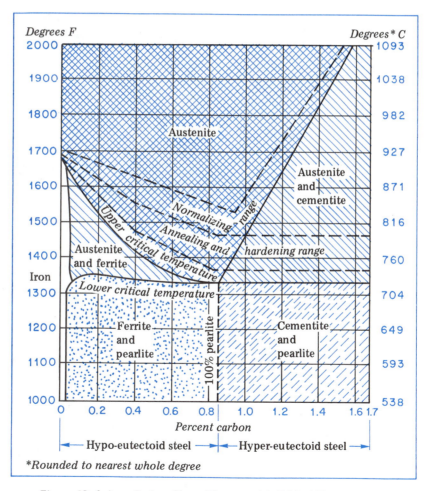

Figure 42–1 Iron-Carbon Phase Diagram with Critical Temperatures, Heat-Treating Processes, and Grain Structures

the slow transformation temperature of $1300°F$ ($704°C$). As the cooling rate is increased, the laminations of the pearlite (formed by the austenite as it is being transformed), become finer.

When a carbon steel is quenched at the *critical cooling rate* or faster, a new *martensite structure* is formed. The critical cooling rate is the cooling rate at which there is a sudden drop in the transformation temperature. The martensite structure formed has angular, needle-like crystals and is very hard.

APPLICATION OF IRON-CARBON PHASE DIAGRAM INFORMATION

CHARACTERISTICS OF MARTENSITE

Maximum hardness is produced in steel when the grain structure after quenching is transformed from austenite to martensite in a matter of seconds. It is important to quench carbon steels from hardening temperature to below the $1000°F$ ($538°C$) temperature in one second or less to prevent austenite from beginning to transform to pearlite: The hardening process avoids the formation of pearlite grains. If the cooling is slower, any pearlite that is formed mixes with the martensite and reduces hardenability throughout the steel. Martensite is the hardest and most brittle form of steel. Figure 42–2 shows the grain structure of martensite (820X magnification). Martensite in pure form is a supersaturated solid solution of carbon in iron and contains no cementite.

DEPTH OF HARDNESS AND HEAT-TREATING PROBLEMS

Hardenability has been considered in relation to the uniformity of hardness from the outer surface to the center of a part. Hardenability may be improved by adding alloying elements.

When the rate of cooling during quenching is too fast, a number of problems emerge, such as warpage, internal stresses, and different types of fractures. Common heat-treating problems and their probable causes are given in Table 42–1. Usually no problems are encountered with the hardening of workpieces of uniform cross section up to 1/2″ (12.7mm) thickness. These workpieces

may be hardened throughout with complete transformation from austenite to martensite. As the thickness increases, additional attention must be paid to the selection of the steel, quenching medium, and cooling time to ensure that there will be uniform hardness or stabilizing of the grain transformation to martensite.

WORKING TEMPERATURES FOR STEEL HEAT-TREATING PROCESSES

Information normally contained on an iron-carbon phase diagram is interpreted into a practical guide in Figure 42–3, which shows the working temperatures for carbon steel heat-treating processes.

Note that the black heat range is from $0°F$ to $1000°F$ ($538°C$); red heat, from $1000°F$ to $2050°F$ ($1121°C$); and white heat, from $2050°F$ to $2900°F$ ($1593°C$). Heat-treating processes for carbon steels are carried on from the subzero range for stabilizing grain structure through normalizing processes starting around $1700°F$ ($927°C$) for steels with a minimum of 0.02% carbon. The higher carburizing temperatures are shown graphically above the normalizing range.

AGING AND GROWTH OF STEEL

A small amount of the austenite is sometimes retained when carbon steels are subjected to severe, extremely rapid cooling. All austenite is not transformed into martensite through quench-

Figure 42–2 Grain Structure of Martensite (820X Magnification)

Table 42-1 Common Heat-Treating Problems and Probable Causes

Heat-Treating Problem	Probable Cause
Circular cracks	—Uneven heating in hardening
Vertical cracks; dark-colored fissures	—Steel burned beyond use
Hard and soft spots	—Uneven or prolonged heating; uneven cooling
Hard and soft spots with tendency to crack	—Tool not moved about continuously in quenching fluid
Soft places	—Tool dropped to bottom of fluid tank
Surface scale; rough decarburized surface	—Heated steel surfaces exposed to effects of an oxidizing agent
Excessive strains (workpiece with cavities and holes)	—Formation of steam or gas bubbles as an insulating film in cavities and pockets
Coarse grain; quench cracking; tool fracture in use	—Overheating of tool
Area around tongs softer than remainder of part	—Tongs not preheated
Quench cracks (straight-line fractures from surface to center)	—Overheating during austenite stage —Incorrect quenching medium and/or nonuniform cooling —Incorrect selection of steel —Time delays in a high-stress state between hardening and tempering
Surface cracks formed by high internal stresses	—Unrelieved stresses produced by high surface temperature generated by prior machining operations such as grinding
Stress cracking	—Failure to temper a part before grinding —Reducing surface hardness through subsequent machining —Forming a hardened crust by machining at a high temperature and immediately quenching area with cutting fluid (coolant)
Undersized machining to clean up decarburized surfaces	—Failure to allow sufficient stock to permit grinding all required surfaces to clean up

ing. The tendency over a period of time is for the austenite to be transformed into martensite without further heating or cooling. This process, called *aging*, results in the *growth* of the steel with the increase in volume. Additional stresses are also introduced. Therefore, control rates are established for preheating, heating to the lower and upper critical points, cooling, and quenching to provide optimum hardening conditions.

CRITICAL POINTS IN HARDENING

The two important stages in hardening steel are heating and quenching. The steel is heated above its transformation point to produce an entirely austenitic structure. The steel is then quenched at a rate faster than the critical rate required to produce a martensitic structure. It should be noted that the critical rate depends on the carbon content, the amount of other alloying elements, and the grain size of the austenite. The hardness of steel depends on the carbon content of the martensitic steel.

Pearlite is transformed into austenite as steel is being heated at a transformation point called the *decalescence point*. At this point, while the steel continues to be heated and the surrounding temperature becomes hotter, the steel continues

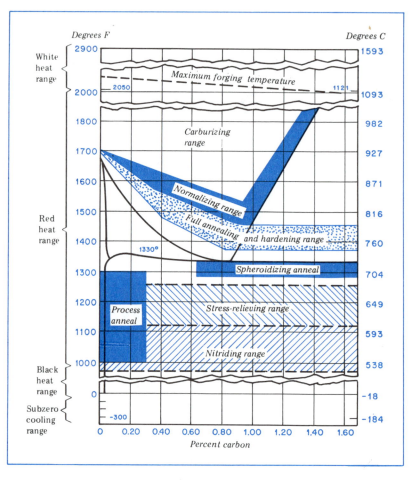

Figure 42–3 Working Temperatures (Interpreted from Iron-Carbon Diagram Data) for Carbon Steel Heat-Treating Processes

to absorb heat without any significant change in temperature.

The process is then reversed in cooling when the austenite is transformed into pearlite at the *recalescence point*. The steel gives out heat at this point so that the temperature rises momentarily instead of continuing to drop.

The control points of decalescence and recalescence have a direct relationship to the hardening of steel. These critical points vary for different kinds of steel, requiring different hardening temperatures. For hardening, the temperature must be sufficient to reach the decalescence point to change the pearlite to austenite. Also, the steel must be cooled rapidly enough before it reaches the recalescence point to prevent the transformation from austenite to pearlite.

HARDENING TEMPERATURES FOR CARBON TOOL STEELS

Tool steel manufacturers' charts contain hardening temperature specifications. As a rule-of-thumb, based on the carbon content in the steel, the following general guidelines are provided:

Carbon Content	Hardening Temperature Range
0.65 to 0.80%	1450 to 1550°F (788 to 843°C)
0.80 to 0.95%	1410 to 1460°F (766 to 793°C)
0.95 to 1.10%	1390 to 1430°F (754 to 777°C)
over 1.10%	1380 to 1420°F (749 to 771°C)

In general, the highest temperature in the range produces deeper hardness penetration and increases strength. Conversely, the lowest temperature in the range decreases the hardness depth but increases the ability of the steel to resist splitting forces.

LIQUID BATHS FOR HEATING

Heating baths are generally used for the following purposes:

- To control the temperature to which a workpiece may be heated,
- To provide uniform heating throughout all sections of a workpiece,
- To protect the finished surfaces against oxidation and scale.

The molten liquid baths that are widely used for steel hardening and tempering operations include sodium chloride, barium chloride, and other metallic salt baths, as well as lead baths. Lead baths are extensively used for quantity heat treating of small tools and for heating below 1500°F (815°C).

QUENCHING BATHS FOR COOLING

Quenching baths serve to remove heat from the steel part being hardened. The rate of cooling must be faster than the critical cooling rate. Hardness partly depends on the rate at which the heat is extracted. The composition of the quenching bath determines the cooling rate. Therefore, different kinds of baths are used, depending on the steel and the required heat treatment.

The two most common quenching baths are fresh, soft water baths; oils of different classes, or oil-water solutions. Brine and caustic soda solutions are also used. High-speed steels are generally cooled in a lead or salt bath. Air cooling serves as a quenching medium for high-speed steel tools that require a slow rate of cooling. Oil quenching is usually used for applications requiring rapid cooling at the highest temperature and slower cooling at temperatures below 750°F (399°C).

OIL QUENCHING BATHS

Oil quenching permits hardening to depth while minimizing distortion and the possibility of cracking of standard steels. Alloy steels are normally oil quenched. Prepared mineral oils have excellent quenching qualities and are chemically stable and cost effective. Vegetable, animal, and fish oils are also used alone or in combination.

One advantage of quenching oils is that they provide fast cooling in the initial stages, followed by slower cooling during the final stages and lower temperatures. This action prevents the steel from cracking. Quenching oils are maintained within a given temperature range of from 90°F to 140°F (32°C to 60°C).

WATER QUENCHING BATHS

Carbon steels are hardened by quenching in a bath of fresh, soft water. The bath temperature must be maintained within a 70°F to 100°F (21°C to 38°C) range so that subsequent workpieces are cooled at the same rate. The temperature of the water must be kept constant within the range because the water temperature seriously affects the cooling rate and hardness penetration.

Design features in workpieces, such as holes, cavities, pockets, and other internal corners, result in uneven cooling when quenched in fresh water. Gas bubbles and an insulating vapor film in cavities produce uneven cooling, excessive internal stresses, and greater danger of cracking.

The addition of rock salt (8% to 9%) or caustic soda (3% to 5%) to a fresh water quenching bath prevents gas pockets and vapor films from forming. Care must be taken to use a clean, uncontaminated, soft water bath. The quantity must be sufficient to dissipate the heat rapidly and permit the workpiece to be agitated (moved) within the bath. As a general rule, thicker sections of a heated part are immersed first.

MOLTEN SALT QUENCHING BATHS

High-speed steels are generally quenched in a molten salt bath in preference to oil quenching. A molten salt bath produces maximum hardness and minimum cooling stresses. Such stresses result in distortion and possible cracking.

Molten salt quenching baths for high-speed steel are maintained at temperatures of 1100°F to 1200°F (593°C to 649°C). After quenching, the hardened part is tempered or drawn in another molten salt bath within a temperature range of 950°F to 1100°F (510°C to 593°C). A general-purpose tempering temperature for high-speed cutting tools is 1050°F (566°C).

QUENCHING BATH CONDITIONS AFFECTING HARDENING

Mention has been made several times about the need to maintain the quenching bath within a specific temperature range. Another condition to consider is the need for movement between the bath and the workpiece so that the temperature throughout the bath remains constant. Under these conditions, cooling proceeds uniformly on all exposed surfaces and completely through the part.

A more desirable practice than agitating the workpiece in the quenching medium is to reverse the process. If the tank permits, the bath is thoroughly agitated while the workpiece is held still. This kind of fluid motion lessens the danger of warping the workpiece during heat treating.

INTERRUPTED QUENCHING PROCESSES

Interrupted quenching processes are used to obtain greater toughness and ductility for a given hardness and to overcome internal stresses that result in quench cracks and distortion. These problems are encountered in general hardening practices.

Austempering, martempering, and *isothermal quenching* are three interrupted methods. The quenching begins at a temperature above the transformation point and proceeds at a rate that is faster than the critical rate. The important fact is that the cooling is interrupted at a temperature above the one at which martensite starts to form.

The steel is maintained at a constant temperature for a fixed time to permit all sections within a part and the external surfaces to reach the same temperature. Transformation of the structure of the steel takes place uniformly for temperature and time throughout the workpiece. Interrupted quenching requires a larger quantity of heat to be absorbed and dissipated without increasing the temperature of the bath.

AUSTEMPERING

Austempering is a patented heat-treating process in which steels (chiefly with 0.60% or higher carbon content) are quenched in a bath at a constant temperature between 350°F to 800°F (176°C to 427°C) at a higher rate than the critical quenching rate. The quenching action is interrupted when the steel temperature reaches the bath temperature.

The steel is held for a specified time at this temperature. The austenitic structure changes to a *bainite structure*, which resembles a tempered martensite structure normally produced by quenching a steel and drawing its temper at 400°F (204°C) or more. The austempered part will have much greater ductility and toughness. In austempering, the steel is quenched rapidly so that there is no formation of pearlite. The steel is also held at the same transformation temperature to ensure that all austenite is transformed to bainite.

MARTEMPERING

Martempering is a heat-treating process that is especially adapted to higher alloyed steels. Martempering produces a high-hardness structure without the problems of internal stresses, which are accompanied in some cases by quench cracks and dimensional changes in the part.

The first rapid quench takes place in martempering at a specific temperature above the transformation point to a temperature above the one where martensite forms. The temperature is held at this point in order to equalize it throughout the part. The workpiece is then removed from the bath and is cooled in air. Martensite begins to form at a uniform rate in a matrix of austenite. The soft austenite tends to absorb some of the internal stresses produced by the formation of martensite.

ISOTHERMAL QUENCHING

Like austempering, isothermal quenching is a process in which steel is rapidly quenched from above the transformation point down to a temperature that is above the one at which martensite forms. The temperature is held constant, usually at 450°F (232°C) or above, until all the austenite is transformed to bainite.

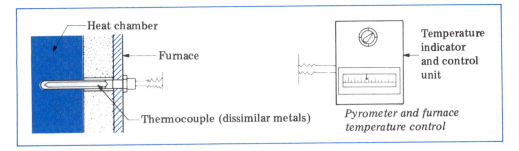

Figure 42-4 Thermocouple and Pyrometer for Measuring, Indicating, and Controlling Temperatures

At this stage, the steel is immersed in another bath and the temperature is raised to a higher specified temperature. After being held at this higher temperature for a definite period, the workpiece is cooled in air. In isothermal quenching, tempering takes place immediately after the steel structure is changed to bainite and before the workpiece is air cooled to normal temperature.

HEAT-TREATING EQUIPMENT

Heat-treating processes require gas, oil-fired, or electrical furnaces that are especially regulated for temperature and, in certain cases, atmosphere. Safety devices are built into the equipment. Exhaust ducts, hoods, close-down valves, and other protective units are requirements of each installation.

Caution: Safety precautions are to be observed for furnace lighting, safety shut off valves and switches, exhausting fumes, performance of all heat treatment processes, and the use of personal protective devices.

HEAT-TREATING FURNACE CONTROLS

Furnace temperatures are usually controlled by a *thermocouple* and a *pyrometer*. A simple installation of an activated thermocouple and an indicator dial pyrometer for measuring, indicating, and controlling temperatures is illustrated in Figure 42-4. The effect of heat in the furnace on the dissimilar metal parts in the thermocouple is to produce small amounts of electrical energy. This varying amount of energy, depending on changes in temperature, is translated on a calibrated temperature scale of the pyrometer.

The pyrometer controls the furnace temperature by setting the instrument at a required temperature.

Pyrometers may have direct temperature sensing and reading controls or a combination of sensing and reading controls and a recording instrument. Control and temperature measurement information provides an important record of heating and cooling processes, particularly when continuous and prolonged temperature controls are required.

BASIC HEAT-TREATING FURNACES

General toolroom hardening, tempering, other heat treatment, forging, and heating processes are performed with small gas-fired or electric heat treatment furnaces. Figure 42-5 shows these two basic types of furnaces with temperature-regulating/indicating controls. The *gas-fired furnace* (Figure 42-5A) is available either for manual regulation or it may be secured with temperature-indicating controls. The *electric furnace* (Figure 42-5B) has its temperatures regulated by electronic controls and readouts on the digital unit. A general temperature control range for these furnaces is from 300°F to 2300°F (148°C to 1260°C). In some furnaces, temperatures are established by using marking crayons, pellets, and liquid coatings. These materials melt at known temperatures and are used to identify the furnace and/or workpiece temperature.

LIQUID HARDENING FURNACES

The *pot-type furnace* is used to heat salt, lead, cyanide, and other baths to a molten state.

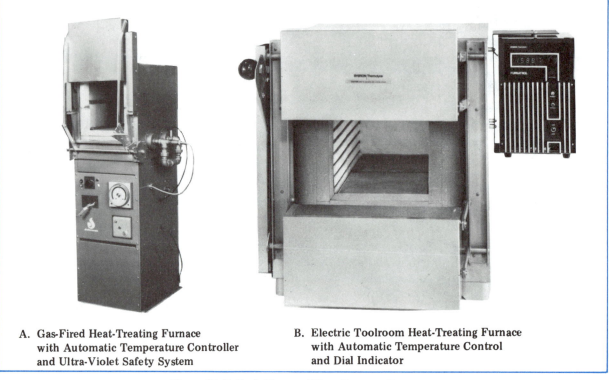

A. Gas-Fired Heat-Treating Furnace
    with Automatic Temperature Controller
    and Ultra-Violet Safety System

B. Electric Toolroom Heat-Treating Furnace
    with Automatic Temperature Control
    and Dial Indicator

Figure 42-5 Basic Types of Heat-Treating Furnaces

This furnace is used for heat-treating processes that require a part to be immersed and heated to a specified temperature. In the case of surface hardening, the outer core that is to be carburized is held in the molten bath for a specified period of time. The time is established by the required depth of the casehardening.

PRODUCTION HEAT-TREATING FURNACES

Two broad groupings of production heat-treating furnaces provide for batch and continuous processing. *Batch furnaces* provide for the heat treatment of quantity parts and process them as a batch. Batch processing is programmed for mass production.

Conveyor furnaces are used for continuous cycling operations. There are different zones of temperature in the furnace to regulate the rate for both heating and cooling. Production furnaces are usually atmospherically controlled—that is, the furnaces are flooded with varying combinations of carbon dioxide, carbon monoxide, nitrogen, and hydrogen gases. The gases prevent decarburization, scale, and surface rust.

HARDENING AND TEMPERING CARBON TOOL STEELS

HARDENING TEMPERATURE RANGE

The hardening temperature is the maximum temperature to which a steel is heated before being quenched for purposes of hardening. The hardening temperature is above the lower critical point of a given steel.

TEMPERING TEMPERATURES FOR CARBON AND HIGH-SPEED STEELS

Tempering, or *drawing*, reduces the brittleness in hardened steel and removes internal strains produced by the sudden cooling in a quenching bath. A cutting tool or part is tempered by reheating it when the steel is in a fully hardened

Table 42–2 Selected Tempering Temperatures, Colors, and Typical Applications of Carbon Steels

Tempering Degrees		Temper Color	Typical Tool Applications
°F	°C		
380	193	Very light yellow	Single-point cutting tools and machine centers requiring maximum hardness
430	221	Light straw	Multiple-point milling cutters, drills, reamers, hollow mills; forming tools
450	232	Pale straw-yellow	Twist drills and screw machine centering tools
470	243	Dark straw	Thread rolling dies, punches, stamping and forming dies, hacksaw blades
490	254	Yellow-brown	Taps over 1/2″ (12.7mm); threading dies for tool steels; shearing blades
510	266	Spotted red-brown	Machine taps under 1/4″ (6.4mm); general threading dies
530	277	Light purple	Hand and pneumatic punches; scribing tools
550	288	Dark purple	Cold chisels and blunt wedge-shape cutting tools
570	299	Dark blue	Cutting/forming tools requiring minimum hardness
590	310	Pale blue	Torque tools: wrenches, screwdrivers, hammer faces
640	338	Light blue	Generally, noncutting tools requiring a minimum hardness

condition and then cooling. Reheating changes the grain structure to one with a reduced hardness.

The general range of temperatures for tempering carbon tool steel is from 300°F to 1050°F (149°C to 565°C). Tempering tables are available for recommended hardness requirements for steel parts, hardening temperatures and time, quenching baths, and tempering temperatures. Table 42–2 gives the tempering temperatures for carbon steels that have been hardened at temperatures from 1350°F to 1550°F (732°C to 843°C).

Alloy tool steels are hardened by heating to temperatures between 1500°F to 1900°F (815°C to 1065°C). High-speed tool hardening temperatures are between 2150°F to 2450°F (1193°C to 1343°C). High-speed parts and tools are tempered at 1000°F to 1100°F (538°C to 593°C).

TEMPERING COLORS

A film of oxide forms on the surface of steel that is heated in an oxydizing atmosphere. As the temperature increases above about 400°F (204°C), the surface of the steel starts to change to a light straw color at 425°F (218°C). As different temperatures are reached, the temper colors change, as listed in Table 61–2. At 640°F (338°C), the temper color of steel is a light blue. Above the tempering color range, the heat colors of steel change from a black red to a white heat around 2400°F (1315°C).

TEMPERING BATHS

Oil Baths. Many tools are tempered in an oil bath that is heated uniformly to a required temperature. Heavy tempering oils may be heated to temperatures between 650°F to 700°F (343°C to 371°C). A partially heated steel part is immersed in the oil. Then, both the steel part and oil are heated to the tempering temperature. The part is next dipped in a tank of caustic soda, followed by quenching in a hot water bath. Oil bath tempering is generally limited to temperatures of 500°F to 600°F (260°C to 316°C).

Salt Baths. Salt baths are recommended for high-speed steel tempering that occurs between 1000°F to 1050°F (538°C to 593°C). Specifications often recommend the use of salt baths above 350°F (173°C) for efficiency and economy. The furnace temperature and part are increased gradually to the tempering range.

Lead Baths. Parts may be tempered by using a lead or lead alloy bath to heat the steel to the required tempering temperature. The workpiece is preheated and immersed in the bath, which has already been brought to the tempering temperature. The part is left in the bath until it reaches the tempering temperature. It is then removed and cooled.

DOUBLE TEMPERING

The tempering operation is often repeated on high-speed steel tools. *Double tempering* ensures that an untempered martensite, which remains in the steel after the first tempering operation, becomes tempered after double tempering. The martensite structure is also relieved of internal strains.

In double tempering, the high-speed steel part is brought to its tempering temperature and held at this temperature for a period of time. The part is then cooled to room temperature. Next, the part is reheated to and held at the original tempering temperature for another period and again cooled to room temperature.

How to Harden and Temper Carbon Tool Steels

Preliminary Steps to Heat Treatment
STEP 1 Check the blueprint for the classification of the steel part and any other hardening and tempering specifications.
STEP 2 Refer to a table of hardening and tempering data to establish the hardening temperature range, the tempering temperature range, the quenching medium, and any other special heat-treating conditions.

Note: See Handbook or Manufacturer's Tables for data on heat-treating temperatures for selected grades and kinds of steel.

STEP 3 Check the temperature of the quenching medium.

Note: Water-hardening steels are cooled best when the water temperature is around 60°F (16°C). The cooling temperature for oil baths is held between 100°F to 140°F (38°C to 60°C).

STEP 4 Determine the desired hardness and toughness from the specifications or technical table.
STEP 5 Determine the tempering temperature.
STEP 6 Establish the soaking periods at the hardening and the tempering temperatures.

Note: Parts that are 1/4" (6mm) thick normally do not require soaking. Parts from 1/4" to 1" (25.4mm) thick require a soaking period of from three minutes to one hour. Thicker parts require longer periods of time for the internal grain structure transformation to occur.

Hardening Carbon Tool Steels
STEP 1 Set the furnace controls to the required hardening temperature.
STEP 2 Light the furnace if it is gas- or oil-fired. Otherwise, turn the electrical furnace on.
STEP 3 Use a pair of tongs and place the part as near the center of the furnace as practical.
STEP 4 Heat the part uniformly and slowly bring it up to the hardening temperature range.

Note: The fuel and air mixture are adjusted so that the furnace atmosphere is neutral to partly oxidizing.

STEP 5 Allow the part to soak at the hardening temperature as required.

Note: If the part is heated above the hardening temperature range, it is necessary to cool the part in air to a temperature below 1000°F (538°C). The part is then brought back up to the correct hardening temperature. Unless this precaution is taken, a coarse-grain hardened steel is produced with decreased toughness.

STEP 6 Remove the part. Quickly quench it in the appropriate quenching bath.

Note: The part must be moved continuously in an up-and-down motion so that the bath is agitated. Prior to removing the part from the furnace, the tong jaws are heated to prevent soft spots in the area where the part is handled. A slicing motion is used with thin, flat parts to prevent uneven cooling and warping.

STEP 7 Check the hardness with hardness testing equipment. A simple file test may also be used.

Note: The hardened part is very hard and brittle at this stage and is under tremendous internal strains. Therefore, the part is tempered as soon as possible to avoid fracturing.

Tempering Carbon Tool Steels

Using Furnace Equipped with Temperature-Indicating Controls

STEP 1 Bring the furnace to the preset tempering temperature.
STEP 2 Place the workpiece in the center of the furnace.
STEP 3 Heat the workpiece uniformly and slowly bring it to tempering temperature.

Note: Small cutting tools and parts are usually tempered by using an auxiliary heated plate. The tool is held against the plate so that the heat travels from an end or side opposite the cutting edge. Once the correct color is reached, the small cutting tool is quenched to prevent the cutting edge from being tempered to a softer degree than required.

STEP 4 Keep the workpiece at this tempering temperature for the required soaking time.
STEP 5 Remove the part from the furnace. Allow it to cool in air or quench it in an appropriate bath.
STEP 6 Determine the hardness either by a simple file test or more exactly by using a hardness tester.

Using Manually Controlled Furnace

STEP 1 Clean the surface coating or discoloration produced while hardening the part.

Note: Generally, an abrasive cloth is used to remove the scale and produce a bright steel surface.

STEP 2 Use a slow heating flame. Place the part in the furnace so that it is positioned for uniform heating. At the same time, the colors as they start to appear must be seen clearly.

Caution: Since the tong jaws are heated before handling the workpiece, wear safety gloves.

STEP 3 Pay special attention to the formation of the temper color just preceding the required temper color.

Note: The temper colors are found in steel suppliers' catalogs and trade handbooks.

STEP 4 Remove the part from the furnace when it reaches the required color.

Note: The part may be cooled either in still air or by quenching.

STEP 5 Test for hardness.

HEAT TREATMENT OF TUNGSTEN HIGH-SPEED STEELS

The hardening temperature for tungsten high-speed steels is from 2200°F to 2500°F (1204°C to 1371°C). Hardening at such a high temperature usually requires one preheating stage. Preheating is particularly important for cutters with thick bodies as compared to the amount of material around cutting teeth. Preheating helps to avoid internal strains. Preheating is done at temperatures below the critical point of the steel, within the 1500°F to 1600°F (816°C to 871°C) range.

At such a temperature, the tool may safely be left in the furnace to heat through uniformly and to bring it up to hardening temperature.

Cutters and other cutting tools that are thicker than 1″ (25.4mm) often require a second pre-heating.

QUENCHING TUNGSTEN HIGH-SPEED STEELS

High-speed steel tools are generally quenched in oil. A tool is moved in the bath to prevent a poor, heat-conducting gas film from forming on the tool. An oil quench permits uniform cooling at the required rate.

Salt baths are also used for quenching. Salt baths are particularly adapted for tools and parts that have complex sections or areas where hardening cracks may develop.

High-speed tools are sometimes quenched in a lead bath or by air cooling. Small sections may be cooled in still air; large, heavier sections require a stream of dry compressed air.

TEMPERING HIGH-SPEED STEEL TOOLS

The drawing temperature for high-speed tools is from 900°F to 1200°F (482°C to 649°C). The temperature is higher for single-point cutting tools used on lathe work than for multiple-tooth milling cutters and form tools. Cobalt high-speed steel tools are tempered between 1200°F and 1300°F (649°C to 704°C).

Once the tool or part is heated to the required drawing temperature, it is held at this temperature until heated uniformly throughout its mass. The tool is allowed to cool in dry, still air away from drafts. The tool is not quenched for tempering. Quenching produces internal strains that may later cause the part to fracture.

ANNEALING, NORMALIZING, AND SPHEROIDIZING METALS

ANNEALING

Annealing is a heat-treating process that requires heating and cooling to:

- Induce softening,
- Remove internal strains and gases,
- Reduce hardening resulting from cold working,
- Produce changes in mechanical properties such as ductility and toughness and magnetic characteristics,
- Form definite grain structures.

The process is also applied to softer metals to permit additional cold working.

The steel is heated to a temperature near the critical range. It is held at this elevated temperature for a period of time and then cooled at a slow rate. Carbon steels are fully annealed by heating *above the upper critical point* for steels with less than 0.85% carbon content (hypo-eutectoid steels). Steels with more than 0.85% carbon content (hyper-eutectoid steels) are heated slightly *above the lower critical point*.

The heated steel is held at this temperature until the part is uniformly heated throughout. It is then slowly cooled to 1000°F (538°C) or below. The result is the formation of pearlite and a layer-like grain structure.

NORMALIZING

The purpose of *normalizing* is to put the grain structure of a steel part into a uniform, unstressed condition of proper grain size and refinement to be able to receive further heat treatment. The iron-base alloy is heated above the transformation range and then cooled in still air.

Depending on the composition of the steel, normalizing may or may not produce a soft, machinable part. By normalizing low-carbon steel parts, the steel is usually placed in best condition for machining. Also, distortion due to carburizing or hardening is decreased.

SPHEROIDIZING

The *spheroidizing of steels* is defined as a heating and cooling process that produces a rounded or globular form of carbide. Steels are spheroidized to increase their resistance to abrasion and to improve machinability, particulary of high-carbon steels that require continuous cutting operations. Low-carbon steels are spheroidized to increase certain properties—for example, strength—before other heat treatment.

Spheroidizing requires heating steel *below the lower critical point*. The part is held at this temperature for a time and then cooled slowly to around 1000°F (538°C) or below. High-carbon steels are spheroidized by heating to a temperature that alternately rises between a temperature inside the critical range and one that is outside the critical range. Tool steels are spheroidized by heating the part slightly above the critical range. The part is held at this temperature for a period of time and then cooled in the furnace.

How to Anneal, Normalize, and Spheroidize

Annealing Steel

STEP 1 Determine the upper critical temperature of the steel.

STEP 2 Adjust the pyrometer from 25°F to 30°F (–4°C to –1°C) above the upper critical temperature.

STEP 3 Prepare the furnace.

STEP 4 Place the part in the furnace so that it may be heated uniformly.

Caution: Use personal safety protective devices for all furnace and heat treatment processes.

STEP 5 Bring the workpiece to the annealing temperature.

STEP 6 Allow the workpiece to soak in the furnace.

Note: A general practice is to soak the part at the upper critical temperature an hour for each 1″ (25.4mm) of thickness.

STEP 7 Shut down the furnace.

Note: The part may be slowly cooled in the furnace. Another method is to remove the part and immediately pack it in ashes or lime for slow cooling.

Normalizing Steel

STEP 1 Follow the same procedures for setting the furnace temperature and placing the workpiece in the furnace for uniform heating as for annealing (steps 1 through 6).

STEP 2 Bring the workpiece to the upper critical temperature.

STEP 3 Soak the workpiece at the normalizing temperature one hour for each 1″ (25.4 mm) of thickness.

STEP 4 Remove the part from the furnace. Allow the part to slowly cool in still air. Avoid drafts.

Note: Thin workpieces generally are packed in ashes or lime to prevent possible hardening in air and to retard the cooling rate.

Spheroidizing Steel

STEP 1 Determining the lower critical temperature of the part.

STEP 2 Set the pyrometer to about 30°F (–1°C) lower than the lower critical temperature.

STEP 3 Start the furnace. Position the part for heating.

STEP 4 Bring the workpiece up to temperature. Allow it to soak for one or more hours depending on the thickness.

STEP 5 Shut down the furnace. Allow the part to slowly cool at a rate of 50°F to 100°F (10°C to 38°C) per hour. Bring the part temperature down to around 1000°F (538°C).

STEP 6 Remove the part at this temperature from the furnace. Bring the temperature down by cooling the part in still air.

CASEHARDENING PROCESSES

Casehardening relates to the process of increasing the carbon content to produce a thin *outer case* that can be heat treated to harden. Impregnating the outer surface with sufficient amounts of carbon to permit hardening is called *carburizing*. When the carburized part is heat treated, the outer case is hardened, leaving a tough, soft inner core. Casehardening refers to the carburizing and hardening processes. The three general groups of casehardening processes are (1) carburizing, (2) carbonitriding, and (3) nitriding.

CARBURIZATION

Iron or steel is carburized when heated to a temperature below the melting point in the presence of solid, liquid, or gaseous carbonaceous materials. These materials liberate carbon when heated. The steel gradually takes on the carbon by penetration, diffusion, or absorption around the outer surface. This action produces a *case or zone* that has a higher carbon content at the outer surface. When a carburized part is heated to hardening temperature and quenched, the outer core acts like a high-carbon steel and becomes hard and tough.

Casehardening produces a steel having surface properties of a hardened high-carbon steel while the steel below the case has the properties of a low-carbon steel. Thus, there are two heat treating processes. One is suitable for the case; the other, for the core. Following an initial heating and slow cooling, a casehardened part is reheated to $1400°F$ to $1500°F$ ($760°C$ to $816°C$). It is then quenched in oil or water and given a final tempering.

PACK HARDENING (CARBURIZING)

The purpose of *pack hardening* is to protect the delicate edges or finished surfaces of workpieces and to encourage uniform heating and contact with a carbonaceous material. Pack hardening also prevents scale formation and minimizes the danger from warping or cracking.

Pack carburizing requires the steel part to be enclosed in a box. Carbonaceous material such as carbonates, coke, and hardwood charcoal and oil, tar, and other binders are packed around the part in the box. Since the carburizing materials are inflammable, the box is sealed with a refractory cement to permit the gases generated within the box to escape while air is prevented from entering.

Penetration Time. The box is heated to carburizing temperature between $1500°F$ to $1800°F$ ($816°C$ to $962°C$), usually within the average temperature range of $1650°F$ to $1700°F$ ($899°C$ to $927°C$). Naturally, the rate of carbon penetration increases at the higher temperature in the range. The approximate *penetration time* for depths of $0.030''$ to $0.045''$ ($0.75mm$ to $1.14mm$) is four hours. It generally takes eight hours to penetrate to $1/16''$ ($1.6mm$) and 24 hours to penetrate $1/8''$ ($3.2mm$).

When the parts are taken out of the box, they are cleaned by wire brush, tumbled, or sand blasted.

Heat Treating Carburized Parts. Once cooled from the carburizing temperature, the parts are reheated to the hardening temperature of the outer case (approximately $1430°F$ or $777°C$) and quenched. This treatment produces a steel with a hardened case and a tough, soft low-carbon steel core.

A finer surface grain structure is produced by *double quenching*, which requires heating to the hardening temperature of the low-carbon steel core ($1650°F$ or $899°C$) and quenching. The coarse grain structure produced is then refined by reheating. The part is brought up to the hardening temperature of the case (about $1430°F$ or $777°C$) and quenched. Double quenching combines the good wearing qualities of a hard case and toughness.

LIQUID CARBURIZING

Liquid carburizing has the advantage of not requiring carbonaceous materials to be packed around the workpiece. Liquid baths have a faster and more uniform penetration with minimum distortion. Where sections of a part are to be selectively carburized, the remaining portions may be copper plated. When the whole piece is immersed in the liquid bath, the copper plate inhibits carburization.

Salt bath furnaces are usually designed to be heated by electrodes that are immersed in the bath. The bath is stirred to ensure uniform temperature. The liquid carburizing baths are molten mixtures of cyanides, chlorides, and carbonates. The composition of the mixture depends on the quantity of carbon and/or nitrogen to be absorbed by the steel.

Liquid carburizing temperatures range from $1550°F$ to $1700°F$ ($843°C$ to $927°C$). In general, it takes about two hours at an average temperature of $1650°F$ ($899°C$) to penetrate to a depth of $0.020''$ ($0.5mm$). Deeper case depths require proportionally greater periods of time to penetrate.

After carburizing in the molten salt bath, parts may be quenched directly in water, brine, or oil, depending on the job requirements. The parts are then tempered as required.

How to Carburize

One commercially prepared carbonaceous substance used in the *pack method* of carburizing is Kasenit®. While nonpoisonous and noninflammable, the fumes must still be exhausted. The part to be carburized is packed in a container with the carbonaceous substance. The box is then covered with a well-vented cover.

A *dip method* is also used where only a shallow case of a few thousandths of an inch (several hundredths of a millimeter) is required. Step-by-step procedures follow for both the pack and dip methods of carburizing.

Pack Method of Carburizing for Case Depths to 0.015″ (0.4mm)

STEP 1 Cover the part, which is placed in a container, with at least a 1″ (25.4mm) or thicker layer of carbonaceous material. Place a well-vented cover on the container.

Note: About 1 1/2″ (38mm) is left between parts that are packed for carburizing.

STEP 2 Place the container in the furnace so that it may heat uniformly. Heat the container and part to 1650°F (899°C).

STEP 3 Soak the part at this temperature for the length of time recommended by the manufacturer of the carbonaceous material, depending on the required depth of case.

STEP 4 Remove the cooled parts and clean them.

STEP 5 Heat each part to its correct critical temperature.

STEP 6 Quench in oil or water, depending on the steel and part specifications.

Note: The tongs should be dry and preheated.

STEP 7 Temper the part, if required.

Dip Method for Shallow Depths

STEP 1 Heat the workpiece to approximately 1650°F (899°C).

STEP 2 Roll the workpiece in a casehardening compound such as Kasenit® until the material fuses around the outer surfaces.

STEP 3 Reheat the part to 1650°F (899°C). Hold at this temperature for a short time. Quench the part immediately in clean cool water.

STEP 4 Reheat to temper, if required.

Note: Parts that are casehardened to just a few thousandths of an inch are not tempered.

Greater depths of casehardening may be obtained by bringing the workpiece up to 1650°F (899°C), redipping the part in the compound, and quenching.

How to Carburize in a Liquid Bath

STEP 1 Preheat the parts to be carburized to about 800°F (410°C). Also heat the jaws of the tongs.

STEP 2 Bring the carbonaceous material to carburizing temperature.

Caution: Strictly observe all safety precautions. Use personal safety equipment. Follow regulations concerning working around liquid salt baths and quenching tanks.

STEP 3 Suspend the dry parts in the liquid bath for the period of time required to produce the depth of case.

STEP 4 Remove the carburized part with a dry pair of tongs. Immediately quench in water (or oil, if oil hardening steel).

STEP 5 Temper the part as required.

GAS CARBURIZING

Methane (natural gas), propane, and butane are three gaseous hydrocarbons (carbon-bearing gases) used for *gas carburizing.* In continuous carburizing furnaces, the parts are heated to carburizing temperature in a horizontal rotary type or vertical pit type of gas carburizer. The carbon-bearing gases are mixed with air and other specially prepared dilutent gases. The carburizing gases are fed continuously to the carburizing retort of the furnace. The spent gases are also exhausted.

The parts are soaked in the carburizing chamber. The soaking temperature and time depend on the required depth of case. Gas carburizing temperatures of around 1700°F (927°C) produce an absorption rate for the first 0.020″ to 0.030″ (0.5mm to 0.75mm) depth of case during a four-hour period.

CYANIDING (LIQUID CARBONITRIDING)

Cyaniding is a casehardening process that is used generally for limited depth case hardening to about 0.020″ (0.5mm). The amount of carbon required in the surface case establishes the properties of cyanide salts in the salt bath (cyanide, chlorate, and chloride salts).

Cyaniding requires temperatures above the critical range of 1400°F to 1600°F (760°C to 871°C) for steels. Lower temperatures in the range are used for a limited case depth. As a general rule, case depths of from 0.003″ to 0.005″ (0.08mm to 0.13mm) require a half-hour soaking. Depths of 0.005″ to 0.010″ (0.13mm to 0.25mm) take from 60 to 70 minutes. The soaking time at cyaniding temperatures for depths of 0.015″ (0.38mm) is two hours.

Parts are quenched in an appropriate water, brine, or oil bath. Tempering where required, is done at temperatures between 250°F to 300°F (120°C to 150°).

Caution: A number of safety precautions must be strictly observed. Most liquid carburizing, cyaniding, and nitriding salts are hazardous. Cyanides are extremely poisonous internally and in open wounds and scratches. When heated, cyanide fumes are toxic. Salt baths require careful, direct venting to outdoors.

CARBONITRIDING (GAS PROCESS)

Carbon and nitrogen are introduced into the surface of steel by a dry (gas) cyaniding process known as *carbonitriding.* Carbonitrided surfaces possess greater hardenability and are harder and more wear resistant than carburized surfaces.

Parts are soaked at between 1350°F to 1650°F (732°C to 899°C) in a gaseous atmosphere composed of carburizing gas and ammonia. The ammonia produces the nitrogen. Carbon and nitrogen are introduced into the heated parts. Case depths of 0.030″ (0.74mm) require soaking at temperatures of 1600°F (871°C) for four to five hours. A lower temperature is used when a higher proportion of nitrogen is required or for a shallower case. In such instances, a 0.005″ to 0.010″ (0.13mm to 0.25mm) case depth may be produced at a temperature around 1450°F (788°C) within 90 minutes. Generally, carbonitrided parts are quenched in oil to prevent distortion and to gain maximum hardness.

NITRIDING

Nitriding is a casehardening process of producing surface hardening by absorption of nitrogen, without quenching. Special alloy steels are heated in an atmosphere of ammonia or in contact with a nitrogenous material. An exceptionally hard surface is produced on machined and heat-treated parts. Nitriding permits hardening carbon alloy steels beyond conventional hardness readings.

The alloy steels contain chromium, vanadium, molybdenum, and aluminum as nitride-forming elements. Nitriding requires heating the part below the lower critical temperature in a protected nitrogenous atmosphere or a salt bath.

Gas Nitriding. Ammonia gas is circulated through a gas furnace chamber. The parts are heated in an air-tight drum to a temperature between 900°F to 1150°F (482°C to 621°C). The nitrogen in the ammonia gas, which decomposes at this temperature, combines with the alloying elements in the steel. The hard nitrides that form produce a harder surface than may be obtained from other heat treatment processes.

Gas nitriding is particularly adapted to increase the hardness of parts that are hardened and ground. Many high-speed steel cutting tools are nitrided with shallow case depths of 0.001" to 0.003" (0.025mm to 0.075mm). The core of these tools is not affected by nitriding.

There is limited, if any, distortion due to the fact that gas-nitrided parts require no quenching. The process is slower than other casehardening processes. Depths of 0.020" (0.5mm) require two to three days.

Salt Bath Nitriding. Cutting tools such as taps, drills, reamers, milling cutters, and dies are often nitrided. Nitriding increases the surface hardness, fatigue and wear resistance, tool wear life, durability, and corrosive resistance (except on stainless steels).

The liquid salt bath is brought to temperature between 900°F to 1100°F (482°C to 593°C), depending on the tool requirements. The tool (or part) is suspended in the molten nitriding salt for the required period of soaking time.

SPECIAL SURFACE HARDENING PROCESSES

Flame hardening and induction hardening are considered as two special surface hardening processes. These processes require heating without using a furnace. The purpose in each instance is to harden particular surfaces from a skin surface to depths up to 1/4" (6.4mm).

FLAME HARDENING

Flame hardening requires heating the surface layer of an iron-base alloy above the transformation temperature range.

The high temperature flame is concentrated in the specific area to be hardened. The surface is immediately quenched by quenching jets in back of the torch or burner. A water or compressed-air quenching medium is used, depending on the type of steel. Tempering is recommended close to the hardening process, where practical.

Similar methods of flame heating are used for tempering. A special low-temperature flame head follows immediately behind the quench to localize the heating area, which is brought to the required tempering temperature.

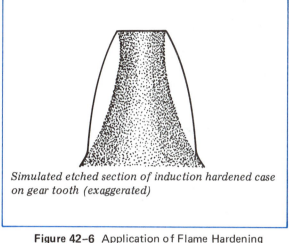

Simulated etched section of induction hardened case on gear tooth (exaggerated)

Figure 42–6 Application of Flame Hardening to Gear Teeth

Medium-carbon and many alloy steels may be flame hardened. Plain carbon steels of from 35 to 60 points of carbon are especially adapted to flame hardening without checking or cracking. The higher-carbon steels are capable of providing greater hardness. However, careful quenching controls are required.

Parts may be flame hardened by the *spinning method*, the *stationary spot method*, or by combining the rotation of the workpiece with the longitudinal movement of the flame head.

Flame hardening is particularly suited to the following applications:

- Steel forgings, castings, and unusually large parts where it is either not practical to heat treat in regular furnaces or the parts must be finish machined prior to hardening. Figure 42–6 shows an application of flame hardening to gear teeth.
- Large parts and irregular surfaces where it is impractical to surface harden selected areas in a furnace efficiently;
- Parts requiring harder wear-resistant surfaces where through-hardening processes produce extensive stresses and distortion;
- Cost-effective applications where carbon steels may be used instead of the more costly high-carbon steels;
- Machine surfaces (such as bed ways, gears, spline shafts, and cams), forming rolls and dies, and other production parts.

INDUCTION HARDENING

Induction hardening is another localized heat-treating process. Electrical heating is required to bring a piece of steel to the required hardening temperature and subsequently quenching it either in a liquid or air bath. Induction heating is especially adapted to parts that require localized and controlled depth of hardening and/or have an irregularly contoured surface. The major advantages to using induction hardening are as follows:

- A short heating cycle produces the same temperature in seconds that normally requires from 30 to 80 times longer to heat in conventional furnaces;
- There is no tendency to produce oxidation or scaling or decarburization;
- The depth and localized zones of hardening may be exactly controlled;
- The automatic heating and quenching cycles permit close controls of the degree of hardness;
- Warpage or distortion are reduced to a minimum;
- Carbon steels may be substituted for higher-cost alloy steels;
- Stress control is possible by localized heating to relieve internal stresses;
- Welded and brazed design features may be added prior to heat treating particular sections of a workpiece;
- Long parts may be heat treated more efficiently than by conventional furnace methods;
- In gear applications of induction hardening, the teeth may be machined and shaved in a soft-annealed or normalized condition (bushings and inserts can be assembled before hardening the gear teeth).

Principle of Induction Heating and Hardening. The process begins when a metal part is placed inside and close to an applicator coil. The part may be held in a fixed position, turned, or fed through the coil. As a high-frequency electrical current passes through the coil, the surface of the steel is raised to a temperature above the critical temperature in a matter of seconds.

The heated part is then quenched in oil, water, or air. The hardness is localized to the surface. The hardness depth is controlled by the length and intensity of the heating cycle. High-frequency currents are used for localized and surface hardening. Low frequencies are employed for through heating, deep hardening, and large workpieces.

One controlling factor on standard types of steels that may be induction hardened is that the carbon content must permit hardening to the required degree by heating and quenching. Low-carbon steels with a carburized case and plain medium-carbon and high-carbon steels may be induction hardened. Cast irons with a percent of carbon in combined form may also be induction hardened. Induction heat treating of alloy steels is generally limited to shallow hardening types that are not affected by high-stressing and possible cracking induced by the required severe quench.

Induction Surface Hardening. To summarize, smaller and thinner section workpieces and shallow hardening depths require a high heating frequency while internal surfaces may readily be heated by shaping the applicator coil to match the cross section. Bored holes of small diameter are often heated by moving an applicator coil or the work so that the heating zone passes through the opening.

The frequency range for induction heating is from 1kHz to 2MHz per second (1,000 cycles to 2,000,000 cycles per second). The higher frequencies permit shallow hardening to depths from a few thousandths of an inch to 1/16" (1.5mm). The lower frequencies are used for hardening depths from 1/16" to 1/4" (1.5mm to 6.4mm). The lower frequencies (180 to 3,000 cycles per second) are used for through applications.

Quenching After Induction Heating. Heated parts may be quenched by immersing in a liquid bath, by liquid spraying, or by self-quenching. *Self-quenching* is associated with rapid absorption of heat by a large mass of surrounding metal, instead of by using a quenching medium. Self-quenching is generally confined to small, simply designed parts. The exact degree of hardness may be automatically timed for heating and quenching. Induction coils or standard furnaces may also be used to temper parts as required.

SUBZERO TREATMENT OF STEEL

Steels are subjected to a *subzero treatment* in order to stabilize a part and to prevent changes in form or size over a period of time.

How to Stabilize Gages and Other Precision Parts by Subzero Cooling

STEP 1 Cool the hardened and ground gage, sine bar, gage blocks, or other precision part uniformly to −120°F (−84°C).

STEP 2 Place the subzero-cooled part in boiling water, oil, or a salt bath.

Note: Precision tools such as thread and plug gages, gage blocks, and other highly accurate parts are immersed for about two hours.

STEP 3 Repeat the cooling and quenching cycle to eventually transform practically all of the austenite to martensite.

STEP 4 Temper according to standard tempering practices for the specific metal in the part.

STEP 5 Proceed to finish grind or lap the subzero-treated part to the required dimensional size and exact form.

Note: Subzero treatment produces a slight increase in size.

Safe Practices in Basic Metallurgy and Heat Treating

- Avoid bringing the part to a higher temperature than the critical point or quenching at a more severe rate than is recommended by the steel manufacturer. Overheating may cause brittleness, fracturing, or shattering a hardened part.

- Use quenching oils with a flash point sufficiently high to permit safe cooling without the danger of fire.

- Agitate the workpiece in the quenching medium to prevent steam or an insulating vapor film from forming.

- Use safety gloves and face and body protection when working around a furnace with hot metals and when quenching. Stand to one side when plunging and agitating a workpiece in the quenching bath, particularly when using an oil.

- Check the furnace system to see that if the exhaust system fails, there is a shutdown of the main gas/oil supply valve.

- Close the loading door on the jacket enclosure on molten bath furnaces to control and direct the fumes and for protection against spattering.

- Make sure that the workpiece and/or tongs that are to come into contact with molten cyanide are clean and dry. Cyanide, mixtures of nitrate, nitrite, and some other salts spatter and explode violently if any one of them in a molten state comes in contact with water.

- Heat a frozen cyanide bath by electrodes from the top down to ensure that any fumes escape as the bath melts from the top on down through the mass. If electrodes are not used, it is important that a removable wedge be inserted in the bath before it is frozen. The *wide side* is on the bottom. The narrow, tapered end extends beyond the top surface.

TERMS USED IN BASIC METALLURGY AND HEAT TREATING

Critical points of steel	Lower and upper temperatures at which transformations of grain structure occur. Upper and lower temperatures affecting hardness and other properties of steels.
Austempering, martempering, and isothermal quenching	Interrupted methods of quenching. Interrupting the cooling that begins above the transformation point.
Subzero, black, red, and white heats	A range of working temperatures from stabilizing subzero –120°F (–84°C) to forging/welding temperatures around 2900°F (1593°C).
Decalescence point	A point in the heat-treating cycle where pearlite is transformed into austenite. The temperature at which steel continues to absorb heat without any significant change in temperature.
Hardening temperature	The temperature at which a steel part is heated or soaked before quenching.
Liquid hardening furnaces	Furnaces equipped to heat salt, lead, carbonaceous, and other materials to molten state. Furnaces used for the immersion of parts to be heat treated in a molten liquid bath.
Transformation range (ferrous metals)	A temperature zone from which a workpiece is raised or lowered to change the grain structure.
Tempering	A transformation temperature range below which a hardened or normalized steel part is reheated, followed by cooling at a specified rate.
Annealing	Heating and cooling metals for purposes of softening, producing a definite microstructure, removing internal stresses, or altering mechanical or physical properties.
Normalizing	A heat-treating process of producing a uniform grain structure for a steel part to be able to receive further heat treatment.
Spheroidizing	Heating and cooling a metal to produce a globular form of carbide. A heat-treating process to increase strength, resistance to abrasion, or other properties.
Casehardening	A series of heat-treating processes to produce a hardenable outer case. The introduction into a steel surface of carbon, carbon-nitrogen, or nitrogen by penetration, diffusion, or absorption. Carburizing, carbonitriding, and nitriding processes affecting the surface hardness of metals.
Liquid carburizing	Introducing carbon into a solid iron-base alloy by heating the workpiece above the transformation temperature range while in contact with a liquid carbonaceous material.
Nitriding	Producing an exceptionally hard surface by heating an alloy steel in a nitrogenous atmosphere or salt bath below the lower critical temperature. Absorption of nitrogen into the outer case and producing a very hard surface without quenching.

Flame hardening	A surface hardening process of localized heating of an iron-base alloy above the transformation temperature range, with an immediate quench.
Induction hardening	Use of an electrical coil and current to quickly raise the surface temperature above the critical temperature of the steel and then quenching in an appropriate bath.
Subzero temperature stabilizing	Reducing the temperature of steel to the $-100°F$ to $-120°F$ ($-73°C$ to $-84°C$) range to transform all austenite to martensite. A series of subzero temperatures to normal cooling temperatures and final tempering.

SUMMARY

- Variations in hardness and other mechanical properties of steel are made possible by changing the cell structure through heating and cooling within certain critical points and rates.

    - Heat-treating processes for steel are carried on from subzero temperatures through the black, red, and white heat ranges. Normalizing processes start at $1700°F$ ($927°C$). Higher temperature ranges are required for carburizing.

- As steels age, unless the grain structure is stabilized, there is a tendency for the steel to grow as the austenite is transformed into martensite.

    - Pearlite is transformed into austenite at the decalescence point. The reverse process takes place at the recalescence point during cooling.

- Steel hardening temperatures before quenching are determined by the amount of ferrite or cementite to be dissolved into austenite, grain size, required physical properties, composition, and tool geometry.

    - The higher temperatures in the hardening range of steel produce deeper hardness penetration.

- Heating baths serve to heat parts uniformly, control the heating temperature, and protect the surface.

    - Quenching baths of air; fresh, soft water; oils; oil-water solutions; lead; and salt are commonly used with ferrous metals.

- Oil baths permit maximum depth hardening with minimum distortion and cracking.

    - Interrupted quenching permits holding the temperature of steel being hardened for a fixed time to provide for the uniform transformation of the grain structure. This process overcomes internal stresses and produces greater toughness and ductility.

- Austempering, martempering, and isothermal quenching are three basic processes of interrupted quenching.

    - The thermocouple and pyrometer provide controls for furnace and workpiece temperatures.

- Gas, oil-fired, and electrical furnaces are generally used for heat-treating and simple forging operations.

    - Pot-type furnaces are used for liquid baths such as salt, lead, cyanide, chlorides, and nitrides.

- The iron-carbon diagram provides basic information about transformation temperatures for changes in grain composition and structure and critical temperatures for hardening, annealing, and normalizing for steels with varying percents of carbon content.

    - Salt baths produce maximum hardness and minimum cooling stresses in high-speed steels.

- Tempering baths may be oil, salt, lead, or some other medium, depending on the hardened material.

    - Tungsten high-speed steels are generally quenched in oil or a salt bath.

- High-speed steel tools are drawn at temperatures from $900°F$ to $1200°F$ ($482°C$ to $649°C$). Cobalt high-speed steel tools are tempered between $1200°F$ to $1300°F$ ($649°C$ to $704°C$).

    - Annealing requires controlled heating and cooling to reduce hardness, improve mechanical properties, and reduce internal stresses.

- Normalizing results in a uniform grain size and refinement and an unstressed structure to receive further heat treatment.

    - Fresh water baths must maintain a constant $70°F$ to $100°F$ ($21°C$ to $38°C$) temperature to uniformly cool a number of workpieces.

- Spheroidizing requires heating the part to its lower critical temperature. The part is allowed to soak for a time and to be brought down to $1000°F$ ($538°C$) before being removed from the furnace and further cooled in still air.

    - Casehardening produces a thin outer case by increasing the carbon content of iron or steel.

- Pack carburizing requires heating a part that is surrounded by a carbonaceous material and binders in a container.

    - Parts to be carburized are preheated to $800°F$ ($410°C$), suspended in a molten liquid bath, brought to hardening temperature, quenched, and tempered.

- Gaseous hydrocarbons are used for gas carburizing. Parts are soaked at temperatures around $1700°F$ ($927°C$) at the rate of absorption required to penetrate to a specific depth.

    - Cyaniding produces a hardenable outer case by heating a steel part in a molten bath at critical temperatures of $1400°F$ to $1600°F$ ($760°C$ to $871°C$) for a stated time and then quenching, followed by tempering.

■ Carbonitriding is a gas process of introducing carbon and nitrogen into the surface of steel for greater hardenability and wear resistance.

■ Nitriding permits surface hardening carbon alloy steels beyond the general hardness range. No quenching is required. The steel core is not affected by nitriding.

■ Flame hardening requires localized heating of the surface layer above the transformation range. The heated area is immediately quenched. Tempering must follow as soon as possible.

■ Induction hardening is especially adapted to localized and controlled depth hardening.

■ Subzero steel treatments accelerate aging and stabilize the grain structure, hardness, and other physical properties of precision tools, gages, and parts.

UNIT 42 REVIEW AND SELF-TEST

1. Tell how a study of the iron-carbon diagram is related to everyday applications in the heat treating of metals.

2. a. Give two probable causes of quench cracks in hardened metals.
   b. Identify two causes of stress cracks (surface failure) produced by machining a hardened workpiece.

3. State three advantages of using molten baths for heat treatment processes.

4. a. Describe martempering.
   b. Cite two desirable features associated with this process.

5. a. State the function of a thermocouple and pyrometer unit.
   b. List three different types of toolroom furnaces that are used for hardening and tempering processes.

6. a. Name two types of tempering baths and give the heat-treating temperature range of each bath.
   b. Identify the bath materials that are used in each case.

7. a. State the purpose of nitriding.
   b. List three properties of cutting tools that are improved by nitriding.

8. Give three advantages of flame hardening over conventional furnace hardening.

9. a. State the function of the subzero temperature treatment of steel.
   b. Give two applications of this process.

SECTION ONE

Nontraditional Tooling and Machines: Technology and Processes

Nontraditional machining and machine tools, which are currently being used in production, are generally calssified within four energy processes groups. These groups relate to (1) mechanical energy, (2) electrical energy, (3) thermal energy, and (4) chemical energy machining. This Section includes examples of commercially available nontraditional machine tools in each group that are becoming widely accepted as economical, desirable manufacturing equipment and processes.

UNIT 43

Nontraditional Machine Tools and Machining Processes

OBJECTIVES

After satisfactorily completing this unit, you will be able to:

- Make comparisons of surface finish ranges of roughness values between conventional and nontraditional machining methods and of metal removal rates.
- Identify features and functions of electrical discharge machines.
- Explain principles and processes of electrochemical machining (ECM), electrochemical deburring (ECD), and electrochemical grinding (ECG).
- Differentiate between chemical blanking and chemical milling processes.
- Describe the unique machining capabilities of ultrasonic machining (USM), plasma beam machining (PBM), and abrasive jet machining (AJM).
- Explain the principles of electron beam machines (EBM) and components and operations.
- Compare laser beam machining (LBM) in terms of solid lasers and gas lasers.
- Use tables of nontraditional machine tools and energy machining processes.
- Follow *Safe Practices* for nontraditional machine tools and processes.
- Correctly use related *Terms*.

Table 43-1 Nontraditional Machine Tools and (Energy) Machining Processes (with Acronyms)

Mechanical Energy Machining	Electrical Energy Machining	Thermal Energy Machining	Chemical Energy Machining
AJM* Abrasive Jet USM* Ultrasonic AFM Abrasive Flow HDM Hydrodynamic LSG Low Stress Grinding	ECD* Electrochemical Deburring ECG* Electrochemical Grinding ECM* Electrochemical Machining ECDG Electrochemical Discharge Grinding ECH Electrochemical Honing ECP Electrochemical Polishing ECT Electrochemical Turning ES Electro-stream STEM Shaped Tube Electrolytic	EBM* Electron Beam EDM* Electrical Discharge Machining LBM* Laser Beam PBM* Plasma Beam EDG Electrical Discharge Grinding EDS Electrical Discharge Sawing EDWC Electrical Discharge Wire Cutting LBT Laser Beam Torch	CHM* Chemical TCM Thermochemical PCM Photochemical ELP Electropolishing

*Identifies machines and processes described in this unit

Principles, processes, and applications of representative machines are provided in this unit. The machines and processes are indicated by an asterisk (*) in Table 43-1.

METAL REMOVAL RATES

Table 43-2 provides technical data for turning and grinding by conventional and for other nontraditional machining processes. Note the tremendous variation from accuracies of 0.01″ (0.25mm) for plastic beam machining (PBM) as contrasted with accuracies within the 0.0002″ to 0.0005″ (0.05mm to 0.0125mm) range for other manufacturing processes.

NONTRADITIONAL MACHINING PROCESSES AND MACHINES

ELECTRICAL DISCHARGE MACHINING (EDM)

The *electrical discharge machine* and process were industrially accepted and marketed beginning in 1946. With an electrical discharge machine, holes and other straight, contour, and embossed surfaces are produced without using a movable

Table 43-2 Examples of Metal Removal Rates for Conventional and Nontraditional Machining Processes

Machining Processes	Maximum Metal Removal Rate (in./min.)	Cutting Speed (ft/min, fpm)	Penetration Rate (in/min, ipm)	Finest Accuracy Attainable	At Maximum Metal Removal Rate	Typical Horsepower (machine input)
				(inch values)		
Conventional						
Turning	200	250	. . .	0.0001	0.005	30
Grinding	50	10	. . .	0.0001	0.003	25
Nontraditional						
EDM	0.3	. . .	0.5	0.0005	0.005	15
ECM	1.0	. . .	0.5	0.0005	0.005	200
LBM	0.0003	. . .	4.0	0.0005	0.005	20
CHM	30	. . .	0.001	0.0005	0.002	. . .
USM	0.05	. . .	0.02	0.0002	0.001	15
EBM	0.0005	200	6	0.0002	0.001	10
PBM	10	50	10	0.01	0.1	200

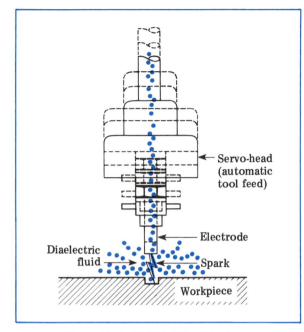

Figure 43-1 EDM Drilling Principle

METAL REMOVING (CUTTING) RATES AND SURFACE FINISH

Since EDM metal removing rates are less than conventional machining processes, as much material as possible is removed by conventional tooling. Three conditions which affect EDM cutting rates and surface finish are: *arcing frequency*, the *amperage setting*, and the properties and condition of the *diaelectric coolant*. The lower the amperage and the lower the arcing frequency, the slower the cutting rate.

Cutting rate, surface finish, and electrode wear life are also affected by the electrode material, the diaelectric fluid, the fast removal of metal chips from the working gap, the filtration of the contaminates, and the continuous removal of heat from the diaelectric fluid. Common diaelectric mediums include silicone and other low-viscosity oils, deionized water, and certain gases.

TRAVELING WIRE (WIRE-CUT) EDM SYSTEM

A second type of EDM general-purpose machine uses a *traveling wire* as an electrode. Extremely accurate internal or external profiles may be produced vertically or at an angle in metals or other materials of any degree of hardness.

Wire EDM machines may be numerically controlled by manual programming or from computer assisted programming (CAP). CAP is used for NC processes involving long or complicated mathematical computations.

CRT, MDI, AND CNC SYSTEM

A number of machine tool and control panel features are illustrated in Figure 43-2. This model has a *CRT display*, *Manual Display Input (MDI) panel*, and *CNC system*.

Part programs, wire position, power settings, simultaneous four-axis positions, and other operator-needed information may be displayed on the CRT. The MDI panel makes it possible to input additional program information during operation or to edit data. The CNC system accepts input from NC tape, memory cassette, or directly from a programming computer.

tool and without tool contact with the workpiece. *Electrical discharge machining* (EDM) uses electrical energy to remove metal of any hardness. Holes and surfaces that have an irregular and intricate contour may be produced. There is no friction-produced heat, so there is no warpage. Tolerances of ±0.002″ (±0.05mm) are practical.

The cutting tool, called an *electrode*, is made of either brass, copper, graphite, or alloys of copper or silver. The tool is hollow to permit pumping a *diaelectric fluid* through it. The diaelectric fluid covers the workpiece during machining. It helps flush the discharge (chips) and acts as a coolant and as the electrical conductor. The principle of EDM drilling is illustrated in Figure 43-1.

Electrical energy is required in EDM, The tool (electrode) is positioned so that there is a gap between it and the work surface. The gap is filled with diaelectric fluid. A high-frequency, pulsating electric current creates sparks that jump the gap. The bombardment vaporizes the material under the tool and produces the desired hole or formed surface. The tool is partly consumed during the process.

Courtesy of ELOX DIVISION, COLT INDUSTRIES

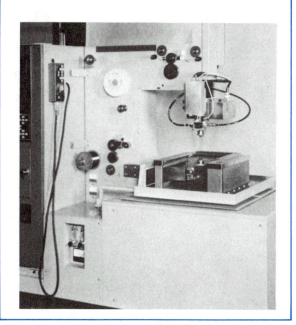

Figure 43-2 High-Speed Wire-Cut Electrical Discharge Machine (EDM) with 4 Simultaneous Controlled Axes and Taper Cutting System

Courtesy of HAMMOND MACHINERY BUILDERS

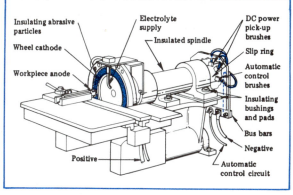

Figure 43-3 Major Features of an ECG Machine

ELECTROCHEMICAL MACHINING (ECM)

There are many similarities between *electrochemical machining* (ECM) and EDM. The workpiece may be of any metal capable of conducting electricity. A tool made of material similar to the material of a tool used for EDM is required. The tool acts as a *cathode* (negative terminal). The workpiece serves as an *anode* (positive terminal). An electrolyte is passed through the gap (from 0.001″ to 0.030″ or 0.025mm to 0.76mm) between the tool and the work at a high rate of flow (velocity). The metal is dissolved and removed by using direct current (DC). A low voltage (30 volts) and high current (2,000 amperes per square inch) are required.

ECM, however, is a faster metal-removing process than EDM. Additional operations, such as etching, deburring, and face milling, may also be performed by ECM.

In electrochemical machining (ECM) metal particles from the workpiece are dissolved into the electrolyte by chemical reaction. The particles are flowed away and are filtered out. Electrolyte temperature affects the electrical conductivity, machining rate, and machining accuracy. The temperature is controlled as it passes through a heat exchanger.

Electrodes (tools) are machined undersize to provide a gap or clearance between a final workpiece dimension and the electrode size which is affected by the *particle size*. The gap or clearance is also known as *overcut*.

ELECTROCHEMICAL DEBURRING (ECD)

Burrs and sharp edges produced by other machining processes may be removed by electrochemical deburring. The ECD machine is an adaptation of ECM. The workpiece is the anode. The tooling (within an insulated coating) forms the cathode. The electrolyte flows through the gap between the tool and the workpiece. The burrs are removed by electrochemical reaction.

ELECTROCHEMICAL GRINDING (ECG)

Electrochemical grinding (ECG) requires an ECM machine, a grinding head, and conventional grinding wheels. These and other major features of an ECG machine are shown in the line drawing (Figure 43-3).

ECG grinding wheels are made with an electrically-conductive bond. The wheel is the cathode; the work is the anode. Electrical current passes between the work and the revolving grinding wheel. An electrochemical reaction is produced for removing metal. The stock removal rate on a standard ECM tool grinder ranges from 0.030" to 0.060" per minute (0.75mm to 1.50mm/min).

The following are major advantages of ECG over conventional grinding.

- Over 75% more metal may be removed in the same period of time.
- Wheel dressing and wheel wear are drastically reduced.
- There is limited loss in dimensional accuracy or the quality of surface finish.
- Cutting edges are smooth and burr free.
- Carbide and other extremely hard alloy cutting tools are ground with greater efficiency and at lower cost, resulting from single-pass grinding.
- Flat surfaces of alloy steel parts may be ground at exceedingly high production rates.

CHEMICAL MACHINING (CHM)

Parts may be *blanked* from thin sheet metals and heavier materials may be *milled* to shape by chemically dissolving unwanted metal. The solutions used for dissolving are generally strongly acid or alkaline

CHEMICAL BLANKING

Chemical blanking is widely used for producing intricately formed, thin sheet metal parts and for the manufacture of printed circuits. Enlarged drawings are made of parts that are to be chemically blanked. The drawing is then reduced to size photographically. The negative produced is used to make a contact printing on the surface of the workpiece which has been given an acid-resisting, photo-sensitive coating.

The light-exposed unwanted portions are dissolved away when the workpiece is placed in an etching solution. The remaining portions form the required blanked workpiece. Less accurate parts are *masked* by using either a silk screen or offset printing process to apply the acid-resistant coatings.

CHEMICAL MILLING

Chemical milling begins when a metal part is thoroughly cleaned in preparation for spraying

or flowing on a masking of vinyl plastic or neoprene rubber. The coating is baked to cure it.

Templates are then used to scribe areas from which the masking is to be removed. The remaining coated metal part is then submerged in a chemical solution to dissolve unwanted metal. At that point, the part is removed, rinsed in water, and *stripped (demasked)* to remove the coating.

ULTRASONIC MACHINING (USM)

Ultrasonic machining (USM) is an *impact grinding* process which requires a high-frequency, low-amplitude vibrating tool head; a ductile, tough material formed tool secured to a tool holder; and an *abrasive slurry*. USM is adaptable to machining tough or hard metal, carbide, or nonconducting ceramic, glass, diamond, and less hard gem stone parts. USM is also used for machining plastics and other soft materials.

The USM process is adapted to blanking uniform or irregular sections in thin, hard conducting and nonconducting metals. The process is also used for forming shallow cavities of simple or intricate shapes.

As an impact grinding process, aluminum oxide, silicon carbide, and boron carbide abrasive grains are suspended in a liquid (usually water) slurry. The slurry is pumped into the gap between the formed tool and the workpiece. The tool vibrates a few thousandths of an inch within a 19,000 to 25,000 Hz range of ultrasonic frequencies. Grinding takes place by the action of the abrasive grains impacting at high velocity on the workpiece as the area immediately under the tool is ground to the shape of the tool.

PLASMA BEAM (ARC) MACHINING (PBM)

Plasma beam (arc) machining (PBM) is primarily a metal cutting process. A high-velocity jet of a super-heated stream of electronically ionized gas is used to form holes and to make other profile cuts in metals. The PBM process is especially adapted to cutting operations on metals which may not be performed efficiently with oxyacetylene equipment (Figure 43–4).

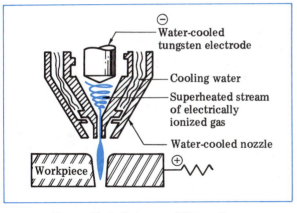

Figure 43-4 Features of Plasma Beam Machining Equipment

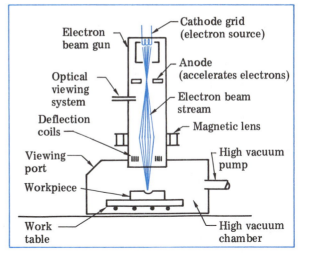

Figure 43-5 Schematic of Major Components in Electron Beam Machining (EBM)

ABRASIVE JET MACHINING (AJM)

Abrasive jet machining (AJM) depends on the cutting force of fine aluminum oxide or silicon carbide abrasive particles. These particles are propelled in a high velocity stream of air to serve as a cutting tool. Accurate AJM cuts as narrow as 0.005" wide may be made.

AJM is an effective cutting method for internal and external deburring and for cutting ceramic, glass, and other hard semiconductor materials.

AJM and PBM machines and processes make it possible to:

• Produce intricate shapes that cannot be economically machined on other regular or nontraditional machines,
• Penetrate hard materials that are difficult to machine,
• Machine holes and other surfaces to high degrees of dimensional accuracy and quality of surface finish.

ELECTRON BEAM MACHINING (EBM)

Electron beam machining (EBM) is used to cut diamonds, carbides, ceramic oxide, and other extremely hard materials. The principles of operation and the major components are pictured in Figure 43-5.

Electron rays from a *cathode (electron) source* are directed by an *electron beam gun*. The electrons are accelerated through an anode to form an electron beam. As the high-speed electron beam passes through a *focusing coil*,

electron energy is transformed into an intense heat energy. This energy (focused on the workpiece) is capable of vaporizing and removing the material before it.

Cutting energy is increased by enclosing the workpiece in a high vacuum chamber. The chamber makes it possible to direct full-energy power electron beams to precise locations. *Magnetic deflection coils* provide for movement of the electron beam (heat energy beam) to generate different cutting patterns. Thus, straight line ╱, circle ◯, square ◈, circle-on-circle ◎, and hole offset ⁚⁚ wave forms may be generated by moving the electron beam off-center. Holes and slots that are larger than a few thousandths of an inch in diameter or width; multiple, closely-spaced holes; and various width slots may be cut from one position. Other longer cuts may be taken by rotating the workpiece or moving the X–Y axes positions of the table.

The disadvantages of the system include: high cost of equipment, slow cutting and machining rates, limitations in terms of workpiece capacity, and the need for extensive machine guarding against harmful x-rays. The same principles and type of equipment (except for increased power requirements) are employed in *electron beam welding (EBW)*.

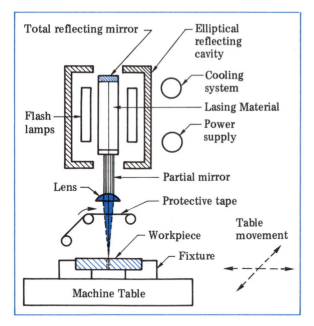

Figure 43–6 Pulsed (Intermittent Bursts) Operation of Solid State Lasers

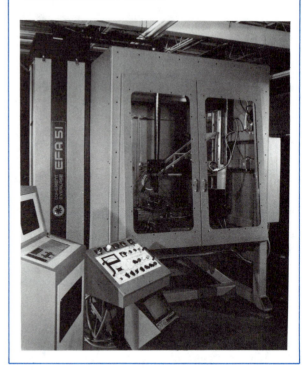

Figure 43–7 Major Components of a Multi-Purpose Machining, Cutting, Welding, and Heat-Treating CO_2 Laser System

LASER BEAM MACHINING (LBM)

The initial source of energy for *laser beam machining* is a low-intensity light beam. The power density of the light beam is amplified by an *optical pump* to become a *high-intensity laser beam*. This beam is focused through a lens system to an exceedingly small area to produce a high level of heat energy capable of vaporizing any material.

Solid Lasers. Solid lasers produce intermittent bursts of high-intensity power (Figure 43–6). The general-purpose system is used in production work to produce small holes, cut out sections of thin materials, and for spot welding and heat treating processes. Tolerances for hole sizes within ±0.0001″ (0.0025mm) are possible.

Laser beams may be directed for straight or angle hole drilling up to 0.050″ (1.25mm) diameter on flat, round, or irregular surfaces, and difficult-to-reach internal areas.

Gas Lasers. One advantage of the gas laser system is its capability to produce a continuous laser beam. Gas lasers require a gas such as carbon dioxide (CO_2) to convert electrical energy into laser energy and to maintain the continuous beam. The energy level of gas lasers is further increased by using oxygen to boost the system.

Figure 43–7 shows the major components of a CO_2 laser focusing optic unit with safety enclosure, a CNC panel, and a CAD/CAM interface system. Table movements are along X, Y, and Z axes. A rotary table accessory permits taper and circular machining.

Mild steel plates 0.400″ (10mm) thick are cut at the rate of 30 ipm (inches per minute) or 12 cm/min. Inconel parts 0.060″ (1.5mm) thick may be butt welded with full penetration at 300 ipm (7.6 meters per minute).

Gas lasers are used for marking, embossing, welding, cutting, and heat treating processes. These are in addition to hole forming and other cutting operations.

Safe Practices for Nontraditional Machines and Processes

- Check to see that an operable class B and C fire extinguisher is readily available near EDM equipment in the unlikely event that a gas flame is ignited at the arc gap.
- Shut down the EDM power supply and wait until the voltmeter reads zero before handling the workpiece, tool holder, or electrode. Electrical shock hazards are associated with all EDM.
- Place the plexiglass guard in position to cover the entire EDM work area.
- Wear a face shield or mask to prevent inhaling dusts, mists, and vapors of chemical compounds which may be released by chemical reaction in ECM.
- Make sure the work enclosure is vented away from the work area to remove any gas produced in ECM.
- Use protective clothing when handling any combustible material which may be injurious to body tissues.
- Wear a full-face shield, acid-resisting apron and clothing, and rubber gloves and boots as protection against personal injury when working with acids and other CHM etching solutions.
- Check for the removal of toxic fumes in work areas around CHM equipment and the availability of an emergency shower.
- Check to see that all x-ray shielding panels are securely in place before using EBM equipment.
- Observe all personal, machine, and area safe operating directions for nontraditional equipment which may generate harmful rays and noise levels.
- Wear laser-proof protective shield or safety glasses for eye protection around LBM systems.
- Check that all laser beam paths are enclosed with safety guards which are secured before the system is turned on.

TERMS RELATED TO NONTRADITIONAL MACHINE TOOLS AND MACHINING PROCESSES

Nontraditional machining processes	Term applied to commercially available emerging material removal processes that have not been extensively used in regular production.
Electrical discharge machining (EDM)	Removal of electrically conductive material. Depends on rapid, repetitive discharges from a d-c pulsating power supply. A continuous flow of dialectic fluid flushes out vaporized material from a gap between the tool and workpiece.
Electrochemical machining (ECM)	Removal of electrically conductive material by anodic dissolution using a rapidly flowing electrolyte and controlling the electrolyte temperature, hydraulic pressure, and flow and current density between the electrode (formed insulated tool) and the workpiece.
Abrasive flow machining (AFM)	A mechanical nontraditional machining method of removing material by flowing a viscous abrasive media under extreme pressure against or through a workpiece.
Laser beam machining (LBM)	A melting and vaporizing process for the removal of material at the point where a highly amplified, focused beam of monochromatic light impinges on the workpiece.

Plasma beam machining (PBM) Materials removal by directing a controlled continuous super-heated stream of electronically ionized gas from a nozzle at near sonic velocity within a cutting area of the workpiece.

Photochemical machining (PCM) Photographic technique of applying and developing a chemically resistant mask to the workpiece before proceding with chemical machining.

Thermochemical machining (TCM) Short-time exposure of a workpiece to extremely hot corrosive gases causing a thermal wave to vaporize workpiece material.

Electrochemical grinding (ECG) Material removal by flowing an electrolyte in the electrical gap between a rotating conductive abrasive wheel and a conductive workpiece.

SUMMARY

- Comparative performance tables for conventional and nontraditional machining processes generally include data on metal removal rates, cutting speed or penetration, and degree of accuracy attainable by each process.

  - Electrical discharge machining (EDM) may be used for producing holes, irregular shaped surfaces, and intricate contours.

- Cutting rates and surface finish for EDM operations are affected by the properties and condition of the dielectric coolant, amperage, and arcing frequency.

  - CNC electric discharge machines have the capacity to accept additional computer-generated input during operation.

- The formed tool used in electrochemical machining (ECM) serves as the negative terminal (cathode); the workpiece, the positive terminal (anode). The high velocity flow of the temperature-controlled electrolyte in the gap between the workpiece and the tool controls machining accuracy and rate of machining.

  - Electrochemical grinding (ECG) depends on an electrochemical reaction. A major advantage of ECG relates to the ability to grind carbides and other extremely hard alloys efficiently and cost effectively.

- Chemical blanking involves the photographic printing of a required form on a workpiece whose surface has been treated with an acid-resisting photo-sensitive coating.

  - Ultrasonic machining (USM) is an impact grinding process. While USM is used for machining hard, tough metals, carbides, ceramic, and diamond products, the technique is also applied to plastic and other soft materials.

- Plasma beam (arc) machining (PBM) is adapted to hole forming, profile cutting, and other metal cutting operations that are difficult to produce by oxyacetylene.

■ Abrasive jet machining (AJM) is a functional technique for producing intricately machined shapes and other surfaces to a high degree of dimensional accuracy and surface finish.

■ Straight line, circle, square, circle-on-circle, and hole offset cutting patterns for electron beam machining (EBM) are controlled by magnetic deflection coils.

■ Laser beam machining (LBM) depends on the focusing of a high intensity laser beam through a lens system to an exceedingly small area.

UNIT 43 REVIEW AND SELF-TEST

1. State two design features or properties of regular cutting tools which are not present in nontraditional machining processes.

2. List three materials which may be more readily machined by nontraditional systems than by conventional practices.

3. Identify two functions of digital readouts on an electrical discharge machine (EDM).

4. State three conditions which govern the cutting rate and surface finish in EDM.

5. List three advantages of electrochemical grinding (ECG) of cutting tools in comparison with conventional tool grinding.

6. a. Give a brief explanation of chemical machining (CHM).
   b. Differentiate between the blanking and the milling CHM processes.

7. a. Describe ultrasonic machining (USM).
   b. Determine the effect the use of coarser abrasive grains has on cutting rate and surface quality.

8. List four metals that may be more efficiently cut by plasma beam machining (PBM) than with other gas-cutting equipment.

9. a. Explain briefly the process of abrasive jet machining (AJM).
   b. Give two examples of AJM.

10. Describe how an electron beam in EBM is moved to (a) generate different cutting patterns and (b) how longer cuts may be taken.

11. a. Name two basic systems of lasers used in laser beam machining (LBM).
    b. Identify general applications of each laser system.
    c. State the principles of operation of each system.

12. List one safety precaution which must be followed in operating nontraditional equipment which require: (a) the direction and control of high energy beams (like EBM and LBM); (b) the application of electrical and chemical energy (as for ECM and CHM); and (c), the use of abrasive particles (as for AJM and USM).

APPENDIX

- **HANDBOOK TABLES**
- **INDEX**

HANDBOOK TABLES

Table A–1 Decimal Equivalents of Fractional, Wire Gage (Number), Letter, and Metric Sizes of Drills

Decimal	Inch	Wire	mm
.0059		97	.15
.0063		96	.16
.0067		95	.17
.0071		94	.18
.0075		93	.19
.0079		92	.20
.0083		91	.21
.0087		90	.22
.0091		89	.23
.0095		88	.24
.0098			.25
.0100		87	
.0102			.26
.0105		86	
.0106			.27
.0110		85	.28
.0114			.29
.0115		84	
.0118			.30
.0120		83	
.0122			.31
.0125		82	
.0126			.32
.0130		81	.33
.0134			.34
.0135		80	
.0138			.35
.0145		79	
.0156	1/64		
.0158			.40
.0160		78	
.0177			.45
.0180		77	
.0197			.50
.0200		76	
.0210		75	
.0217			.55
.0225		74	
.0236			.60
.0240		73	
.0250		72	
.0256			.65
.0260		71	
.0276			.70
.0280		70	
.0292		69	
.0295			.75
.0310		68	
.0312	1/32		
.0315			.80
.0320		67	
.0330		66	
.0335			.85
.0350		65	
.0354			.90
.0360		64	
.0370		63	
.0374			.95
.0380		62	
.0390		61	
.0394			1.00
.0400		60	
.0410		59	
.0413			1.05
.0420		58	
.0430		57	
.0433			1.10
.0453			1.15
.0465		56	
.0469	3/64		
.0472			1.20
.0492			1.25
.0512			1.30
.0520		55	
.0532			1.35
.0550		54	
.0551			1.40
.0571			1.45
.0591			1.50
.0595		53	
.0610			1.55
.0625	1/16		
.0630			1.60
.0635		52	
.0650			1.65
.0669			1.70
.0670		51	
.0689			1.75
.0700		50	
.0709			1.80
.0728			1.85
.0730		49	
.0748			1.90
.0760		48	
.0768			1.95
.0781	5/64		
.0785		47	
.0787			2.00
.0807			2.05
.0810		46	
.0820		45	
.0827			2.10
.0847			2.15
.0860		44	
.0866			2.20
.0886			2.25
.0890		43	
.0906			2.30
.0925			2.35
.0935		42	
.0938	3/32		
.0945			2.40
.0960		41	
.0965			2.45
.0980		40	
.0984			2.50
.0995		39	
.1015		38	
.1024			2.60
.1040		37	
.1063			2.70
.1065		36	
.1083			2.75
.1094	7/64		
.1100		35	
.1102			2.80
.1110		34	
.1130		33	
.1142			2.90
.1160		32	
.1181			3.00
.1200		31	
.1221			3.10
.1250	1/8		
.1260			3.20
.1280			3.25
.1285		30	
.1299			3.30
.1339			3.40
.1360		29	
.1378			3.50
.1405		28	
.1406	9/64		
.1417			3.60
.1440		27	
.1457			3.70
.1470		26	
.1476			3.75
.1495		25	
.1496			3.80
.1520		24	
.1535			3.90
.1540		23	
.1562	5/32		
.1570		22	
.1575			4.00
.1590		21	
.1610		20	
.1614			4.10
.1654			4.20
.1660		19	
.1673			4.25
.1693			4.30
.1695		18	
.1719	11/64		
.1730		17	
.1732			4.40
.1770		16	
.1772			4.50
.1800		15	
.1811			4.60
.1820		14	
.1850		13	4.70
.1870			4.75
.1875	3/16		
.1890		12	4.80
.1910		11	
.1929			4.90
.1935		10	
.1960		9	
.1969			5.00
.1990		8	
.2008			5.10
.2010		7	
.2031	13/64		

Decimal	Inch	Wire	mm
.2040		6	
.2047			5.20
.2055		5	
.2067			5.25
.2087			5.30
.2090		4	
.2126			5.40
.2130		3	
.2165			5.50
.2188	7/32		
.2205			5.60
.2210		2	
.2244			5.70
.2264			5.75
.2280		1	
.2284			5.80
.2323			5.90
.2340		A	
.2344	15/64		
.2362			6.00
.2380		B	
.2402			6.10
.2420		C	
.2441			6.20
.2460		D	
.2461			6.25
.2480			6.30
.2500	1/4	E	
.2520			6.40
.2559			6.50
.2570		F	
.2598			6.60
.2610		G	
.2638			6.70
.2656	17/64		
.2658			6.75
.2660		H	
.2677			6.80

Decimal	Inch	Letter	mm
.2717			6.90
.2720		I	
.2756			7.00
.2770		J	
.2795			7.10
.2810		K	
.2812	9/32		
.2835			7.20
.2854			7.25
.2874			7.30
.2900		L	
.2913			7.40
.2950		M	
.2953			7.50
.2969	19/64		
.2992			7.60
.3020		N	
.3032			7.70
.3051			7.75
.3071			7.80
.3110			7.90
.3125	5/16		
.3150			8.00
.3160		O	
.3189			8.10
.3228			8.20
.3230		P	
.3248			8.25
.3268			8.30
.3281	21/64		
.3307			8.40
.3320		Q	
.3347			8.50
.3386			8.60
.3390		R	
.3425			8.70
.3438	11/32		

Decimal	Inch	Letter	mm
.3445			8.75
.3465			8.80
.3480		S	
.3504			8.90
.3543			9.00
.3580		T	
.3583			9.10
.3594	23/64		
.3622			9.20
.3642			9.25
.3661			9.30
.3680		U	
.3701			9.40
.3740			9.50
.3750	3/8		
.3770		V	
.3780			9.60
.3819			9.70
.3839			9.75
.3858			9.80
.3860		W	
.3898			9.90
.3906	25/64		
.3937		X	10.00
.3970		Y	
.4040			
.4062	13/32		
.4130		Z	

Decimal	Inch	mm
.4134		10.50
.4219	27/64	
.4331		11.00
.4375	7/16	
.4528		11.50
.4531	29/64	
.4688	15/32	
.4724		12.00
.4844	31/64	
.4921		12.50
.5000	1/2	
.5118		13.00
.5156	33/64	
.5312	17/32	
.5315		13.50
.5469	35/64	
.5512		14.00
.5625	9/16	
.5709		14.50
.5781	37/64	
.5906		15.00
.5938	19/32	
.6094	39/64	
.6102		15.50
.6250	5/8	
.6299		16.00
.6406	41/64	
.6496		16.50
.6562	21/32	
.6693		17.00
.6719	43/64	
.6875	11/16	
.6890		17.50
.7031	45/64	
.7087		18.00
.7188	23/32	
.7283		18.50
.7344	47/64	
.7480		19.00

Decimal	Inch	mm
.7500	3/4	
.7656	49/64	
.7677		19.50
.7812	25/32	
.7874		20.00
.7969	51/64	
.8071		20.50
.8125	13/16	
.8268		21.00
.8281	53/64	
.8438	27/32	
.8465		21.50
.8594	55/64	
.8661		22.00
.8750	7/8	
.8858		22.50
.8906	57/64	
.9055		23.00
.9062	29/32	
.9219	59/64	
.9252		23.50
.9375	15/16	
.9449		24.00
.9531	61/64	
.9646		24.50
.9688	31/32	
.9843		25.00
.9844	63/64	
1.0000	1	

Table A–2 Conversion of Metric to Inch-Standard Units of Measure

mm Value	Inch (decimal) Equivalent	mm Value	Inch (decimal) Equivalent	mm Value	Inch (decimal) Equivalent	mm Value	Inch (decimal) Equivalent
.01	.00039	.34	.01339	.67	.02638	1	.03937
.02	.00079	.35	.01378	.68	.02677	2	.07874
.03	.00118	.36	.01417	.69	.02717	3	.11811
.04	.00157	.37	.01457	.70	.02756	4	.15748
.05	.00197	.38	.01496	.71	.02795	5	.19685
.06	.00236	.39	.01535	.72	.02835	6	.23622
.07	.00276	.40	.01575	.73	.02874	7	.27559
.08	.00315	.41	.01614	.74	.02913	8	.31496
.09	.00354	.42	.01654	.75	.02953	9	.35433
.10	.00394	.43	.01693	.76	.02992	10	.39370
.11	.00433	.44	.01732	.77	.03032	11	.43307
.12	.00472	.45	.01772	.78	.03071	12	.47244
.13	.00512	.46	.01811	.79	.03110	13	.51181
.14	.00551	.47	.01850	.80	.03150	14	.55118
.15	.00591	.48	.01890	.81	.03189	15	.59055
.16	.00630	.49	.01929	.82	.03228	16	.62992
.17	.00669	.50	.01969	.83	.03268	17	.66929
.18	.00709	.51	.02008	.84	.03307	18	.70866
.19	.00748	.52	.02047	.85	.03346	19	.74803
.20	.00787	.53	.02087	.86	.03386	20	.78740
.21	.00827	.54	.02126	.87	.03425	21	.82677
.22	.00866	.55	.02165	.88	.03465	22	.86614
.23	.00906	.56	.02205	.89	.03504	23	.90551
.24	.00945	.57	.02244	.90	.03543	24	.94488
.25	.00984	.58	.02283	.91	.03583	25	.98425
.26	.01024	.59	.02323	.92	.03622	26	1.02362
.27	.01063	.60	.02362	.93	.03661	27	1.06299
.28	.01102	.61	.02402	.94	.03701	28	1.10236
.29	.01142	.62	.02441	.95	.03740	29	1.14173
.30	.01181	.63	.02480	.96	.03780	30	1.18110
.31	.01220	.64	.02520	.97	.03819		
.32	.01260	.65	.02559	.98	.03858		
.33	.01299	.66	.02598	.99	.03898		

Table A–3 Conversion of Fractional Inch Values to Metric Units of Measure

Fractional Inch	mm Equivalent	Fractional Inch	mm Equivalent	Fractional Inch	mm Equivalent	Fractional Inch	mm Equivalent
1/64	0.397	17/64	6.747	33/64	13.097	49/64	19.447
1/32	0.794	9/32	7.144	17/32	13.494	25/32	19.844
3/64	1.191	19/64	7.541	35/64	13.890	51/64	20.240
1/16	1.587	5/16	7.937	9/16	14.287	13/16	20.637
5/64	1.984	21/64	8.334	37/64	14.684	53/64	21.034
3/32	2.381	11/32	8.731	19/32	15.081	27/32	21.431
7/64	2.778	23/64	9.128	39/64	15.478	55/64	21.828
1/8	3.175	3/8	9.525	5/8	15.875	7/8	22.225
9/64	3.572	25/64	9.922	41/64	16.272	57/64	22.622
5/32	3.969	13/32	10.319	21/32	16.669	29/32	23.019
11/64	4.366	27/64	10.716	43/64	17.065	59/64	23.415
3/16	4.762	7/16	11.113	11/16	17.462	15/16	23.812
13/64	5.159	29/64	11.509	45/64	17.859	61/64	24.209
7/32	5.556	15/32	11.906	23/32	18.256	31/32	24.606
15/64	5.953	31/64	12.303	47/64	18.653	63/64	25.003
1/4	6.350	1/2	12.700	3/4	19.050	1	25.400

Table A–4 Recommended Saw Blade Pitches, Cutting Speeds, and Feeds for Power Hacksawing Ferrous and Nonferrous Metals*

Material	Pitch (teeth per inch)	Cutting Speed (feet per minute)	Feed (force in pounds)
Ferrous Metals			
Iron			
Cast	6 to 10	120	125
Malleable	6 to 10	90	125
Steel			
Drill rod	10	90	125
Forging stock, mild	3, 4, or 6	120	125
Alloy	4 to 6	90	125
Carbon tool	6 to 10	75	125
High speed	6 to 10	60	125
Machine	6 to 10	120	150
Stainless	6 to 10	75	125
Structural	6 to 10	120	125
Pipe	10 to 14	120	125
Tubing			
Thick-wall	6 to 10	120	90
Thin-wall	14	120	60
Nonferrous Metals			
Aluminum			
Alloy	4 to 6	150	60
Heat-treated	4 to 10	120	60
Soft	4 to 6	180	60
Brass			
Hard	10 to 14	90	60
Free-machining	6 to 10	150	60
Bronze			
Castings	4-6 to 10	110	125
Manganese	6 to 10	80	60
Copper			
Bars	3, 4, or 6	130	150
Tubing	10	120	60

*Coarser pitches are used for sizes over two inches and for thick-walled sections.

Table A–5 Recommended Drill Point and Lip Clearance Angles for Selected Materials

Drill with flat face on cutting lips*

Workpiece Material	Included Drill Point Angle	Lip Clearance Angle
—General-purpose, plain carbon steels		
—Gray cast iron		
—Annealed alloy steels	118°	8–12°
—Medium-hardness, pearlitic, malleable iron		
—Stainless steel		
—Heat-treated steel		
—Hard-to-cut materials	125°	12°
—Alloy steel		
—Drop Forgings		
—Soft cast iron		
—Aluminum		
—Wood	90° to 100°	15°
—Ferritic, malleable iron		
—Plastics		
—Hard rubber		
—Soft aluminum		
—Die casting	60°	20°
—Plastics		
—Wood		
—Soft and medium copper		
—Bronze*	100° to 118°	15°
—Brass*		
—Magnesium alloys*	60° to 118°	10°
—7 to 13 percent manganese steel*		
—Armor plate		
—Tough alloy steels	136° to 150°	10°
—Extra-hard materials		

*Grind flat face on cutting lips.

Table A–6 Conversion of Cutting Speeds and RPM for All Machining Operations
(metric and inch-standard diameters)

		Cutting Speeds																	
Meters per Minute*		6.1	9.1	12.2	15.2	18.2	21.3	24.4	27.3	30.4	38	45.6	60.8	91.2	121.6	152	182.4	243.2	304
Surface Feet per Minute		20	30	40	50	60	70	80	90	100	125	150	200	300	400	500	600	800	1000

Diameter		Revolutions per Minute																	
mm	***Inches**																		
1.6	1/16	1222	1833	2445	3056	3667	4278	4889	5500	6112	7639	9167							
3.1	1/8	611	917	1222	1528	1833	2139	2445	2750	3056	3820	4584	6112	9167					
4.7	3/16	407	611	815	1019	1222	1426	1630	1833	2037	2546	3056	4074	6112	8149				
6.3	1/4	306	458	611	764	917	1070	1222	1375	1528	1910	2292	3056	4584	6112	7639	9167		
7.8	5/16	244	367	489	611	733	856	978	1100	1222	1528	1833	2445	3667	4889	6112	7334	9778	
9.4	3/8	204	306	407	509	611	713	815	917	1019	1273	1528	2037	3056	4074	5093	6112	8149	
10.9	7/16	175	262	349	437	524	611	698	786	873	1091	1310	1746	2619	3492	4365	5238	6985	8731
12.7	1/2	153	229	306	382	458	535	611	688	764	955	1146	1528	2292	3056	3820	4584	6112	7639
15.6	5/8	122	183	244	306	367	428	489	550	611	764	917	1222	1833	2445	3056	3667	4889	6112
18.8	3/4	102	153	204	255	306	357	407	458	509	637	764	1019	1528	2037	2546	3056	4074	5093
21.9	7/8	87	131	175	218	262	306	349	393	437	546	655	873	1310	1746	2183	2619	3492	4365
25.4	1	76	115	153	191	229	267	306	344	382	477	573	764	1146	1528	1910	2292	3056	3820
28.1	1-1/8	68	102	136	170	204	238	272	306	340	424	509	679	1019	1358	1698	2037	2716	3395
31.7	1-1/4	61	92	122	153	183	214	244	275	306	382	458	611	917	1222	1528	1833	2445	3056
34.8	1-3/8	56	83	111	139	167	194	222	250	278	347	417	556	833	1111	1389	1667	2222	2778
38.1	1-1/2	51	76	102	127	153	178	204	229	255	318	382	509	764	1019	1273	1528	2037	2546
41.0	1-5/8	47	71	94	118	141	165	188	212	235	294	353	470	705	940	1175	1410	1880	2351
44.2	1-3/4	44	66	87	109	131	153	175	196	218	273	327	437	655	873	1091	1310	1746	2183
47.3	1-7/8	41	61	82	102	122	143	163	183	204	255	306	407	611	815	1019	1222	1630	2037
50.8	2	38	57	76	96	115	134	153	172	191	239	286	382	573	764	955	1146	1528	1910
63.5	2-1/2	31	46	61	76	92	107	122	138	153	191	229	306	458	611	764	917	1222	1528

*The meters per minute and mm diameters are approximately equivalent to the corresponding sfpm and the diameters given in inches.

Table A-7 Recommended Cutting Speeds for Drilling with High-Speed Drills*

Material	Hardness (Brinell)	Cutting Speed (sfpm)	Material	Hardness (Brinell)	Cutting Speed (sfpm)
Plain Carbon Steels			*Tool Steels*		
AISI-1019, 1020	120–150	80–120	Water-Hardening	150–250	70–80
1030, 1040, 1050	150–170	70–90	Cold-Work	200–250	20–40
1060, 1070, 1080, 1090	170–190	60–80	Shock-Resisting	175–225	40–50
	190–220	50–70	Mold	100–150	60–70
	220–280	40–50		150–200	50–60
	280–350	30–40	High-Speed Steel	200–250	30–40
	350–425	15–30		250–275	15–30
			Cast Iron		
			Gray	110–140	90–140
Alloy Steels				150–190	80–100
AISI-1320, 2317, 2515	125–175	60–80		190–220	60–80
3120, 3316, 4012, 4020	175–225	50–70		220–260	50–70
4120, 4128, 4320, 4620	225–275	45–60		260–320	30–40
4720, 4820, 5020, 5120	275–325	35–55			
6120, 6325, 6415, 8620	325–375	30–40	*Malleable Iron*		
8720, 9315	375–425	15–30	Ferritic	110–160	120–140
			Pearlitic	160–200	90–110
Alloy Steels				200–240	60–90
AISI-1330, 1340, 2330	175–225	50–70		240–280	50–60
2340, 3130, 3140, 3150	225–275	40–60			
4030, 4063, 4130, 4140	275–325	30–50	*Aluminum Alloys*		
4150, 4340, 4640, 5130	325–375	25–40	Cast		
5140, 5160, 5210, 6150,	375–425	15–30	Nonheat-treated		200–300
6180, 6240, 6290, 6340,			Heat-treated		150–250
6380, 8640, 8660, 8740,			Wrought		
9260, 9445, 9840, 9850			Cold-Drawn		150–300
			Heat-treated		140–300
Stainless Steel					
Standard Grades			*Brass, Bronze (ordinary)*		150–300
Austenitic, annealed	135–185	40–50			
Cold-drawn, austenitic	225–275	30–40	*Bronze (high-strength)*		30–100
Ferritic	135–185	50–60			
Martensitic, annealed	135–175	55–70			
	175–225	50–60			
Tempered, quenched	275–325	30–40			
	375–425	15–30			
Free Machining Grades					
Annealed, austenitic	135–185	80–100			
Cold-drawn, austenitic	225–275	60–90			
Ferritic	135–185	100–200			
Martensitic, annealed	135–185	100–130			
Martensitic, cold-drawn	185–240	90–120			
Martensitic, tempered	275–325	50–60			
Quenched	375–425	30–40			

*Use cutting speeds of 50% to 65% of speeds given in the table for high-speed steel reamers.

Table A–8 General NC Tap Sizes and Recommended Tap Drills *(NC standard threads)*

Size (outside diameter inch)	Threads per Inch	Tap Drill Size* (75% thread depth)
1/4	20	#7
5/16	18	"F"
3/8	16	5/16
7/16	14	"U"
1/2	13	27/64
9/16	12	31/64
5/8	11	17/32
11/16	11NS	19/32
3/4	10	21/32
13/16	10NS	23/32
7/8	9	49/64
15/16	9NS	53/64
1	8	7/8
1 1/8	7	63/64
1 1/4	7	1 7/64
1 3/8	6	1 13/64
1 1/2	6	1 11/32
1 5/8	5 1/2NS	1 29/64
1 3/4	5	1 35/64
1 7/8	5NS	1 11/16
2	4 1/2	1 25/32

*Nearest commercial drill size to produce a 75% thread depth.

Table A–9 General NF Tap Sizes and Recommended Tap Drills *(NF standard threads)*

Size (outside diameter inch)	Threads per Inch	Tap Drill Size* (75% thread depth)
1/4	28	#3
5/16	24	"I"
3/8	24	"Q"
7/16	20	"W"
1/2	20	29/64
9/16	18	33/64
5/8	18	37/64
11/16	11NS	19/32
3/4	16	11/16
13/16	10NS	23/32
7/8	14	13/16
15/16	9NS	53/64
1	12	59/64
1	14NS	15/16
1 1/8	12	1 3/64
1 1/4	12	1 11/64
1 3/8	12	1 19/64
1 1/2	12	1 27/64

*Nearest commercial drill size to produce a 75% thread depth.

Table A–10 General Metric Standard Tap Sizes and Recommended Tap Drills *(metric-standard threads)*

Metric Thread Size (nominal outside diameter and pitch) (mm)	Recommended Tap Drill Size			
	Metric Series Drills		Inch Series Drills	
	Size (mm)	Equivalent (inch)	Nominal Size	Diameter (inch)
M1.6×0.35	1.25	.0492		
M2×0.4	1.60	.0630	#52	.0635
M2.5×0.45	2.05	.0807	#45	.0820
M3×0.5	2.50	.0984	#39	.0995
M3.5×0.6	2.90	.1142	#32	.1160
M4×0.7	3.30	.1299	#30	.1285
M5×0.8	4.20	.1654	#19	.1660
M6×1	5.00	.1968	#8	.1990
M8×1.25	6.80	.2677	"H"	.2660
M10×1.5	8.50	.3346	"Q"	.3320
M12×1.75	10.25	.4035	13/32	.4062
M14×2	12.00	.4724	15/32	.4688
M16×2	14.00	.5512	35/64	.5469
M20×2.5	17.50	.6890	11/16	.6875
M24×3	21.00	.8268	53/64	.8281
M30×3.5	26.50	1.0433	1 3/64	1.0469
M36×4	32.00	1.2598	1 1/4	1.2500

Table A–11 Suggested Tapping Speeds (fpm) and Cutting Fluids for Commonly Used Taps and Materials

Material	Carbon Steel Taps				High-Speed Steel Taps				Cutting Fluid
	Hand Tap		Gun Tap		Hand Tap		Gun Tap		
	Coarse	Fine	Coarse	Fine	Coarse	Fine	Coarse	Fine	
	Feet per Minute								
Stainless steel					15	20	20	25	Sulphur-base oil
Drop forgings	20	25	20	25	50	55	60	65	Soluble oil or sulphur-base oil
Aluminum and die castings					70	90	80	100	Kerosene and lard oil
Mild steel	35	40	35	40	70	80	90	100	Sulphur-base oil
Cast iron	35	40	35	40	80	90	100	110	Dry or soluble oil
Brass	70	80	70	80	160	180	200	220	Soluble or light oil

Table A–12 Recommended Cutting Speeds for Reaming Common Materials (High-Speed Steel Machine Reamers)

| Material | Cutting Speed Range | |
	sfpm	m/min*
Steel		
Machinery .2C to .3C	50–70	15–22
Annealed .4C to .5C	40–50	12–15
Tool, 1.2C	35–40	10–12
Alloy	35–40	10–12
Automotive forgings	35–40	10–12
Alloy 300–400 brinell	20–30	6–9
Free-machining stainless	40–50	12–15
Hard Stainless	20–30	6–9
Monel Metal	25–35	8–12
Cast Iron		
Soft	70–100	22–30
Hard	50–70	15–22
Chilled	20–30	6–9
Malleable	50–60	15–18
Brass/Bronze		
Ordinary	130–200	40–60
High tensile	50–70	15–22
Bakelite	70–100	20–30
Aluminum and its alloys	130–200	40–60
Magnesium and its alloys	170–270	50–80

*m/min: meters per minute.

Table A–13 Recommended Cutting Speeds (in sfpm) for Carbide and Ceramic Cutting Tools

| Material | Carbide Cutting Tools (sfpm) | | Ceramic Cutting Tools (sfpm) | |
	Rough Cuts	Finish Cuts	Rough Cuts	Finish Cuts
Nonferrous				
Cast Iron	200–250	350–450	200–800	200–2000
Machine Steel	400–500	700–1000	250–1200	400–1800
Tool Steel	300–400	500–750	300–1500	600–2000
Stainless Steel	250–300	375–500	300–1000	400–1200
Ferrous				
Aluminum	300–450	700–1000	40–2000	600–3000
Brass	500–600	700–800	400–800	600–1200
Bronze	500–600	700–800	150–800	200–1000

Table A–14 Recommended Cutting Speeds and Feeds (in sfpm) for Selected Materials and Lathe Processes (High-Speed Steel Cutting Tools and Knurls)

	Turning, Boring, Facing						
	Rough Cut		*Finish Cut*				
Material	*RPM*	*Feed (")*	*RPM*	*Feed (")*	*Threading*	*Parting*	*Knurling*
Ferrous							
Cast Iron	70	.015 to .025	80	.005 to .012	20	80	30
Machine Steel	100	.010 to .020	110	.003 to .010	40	80	30
Tool Steel	70	.010 to .020	80	.003 to .010	25	70	30
Stainless Steel	75–90	.015 to .030	100–130	.005 to .010	35	70	20
Nonferrous							
Aluminum	200	.015 to .030	300	.005 to .010	50	200	40
Brass	180	.015 to .025	220	.003 to .010	45	150	40
Bronze	180	.015 to .025	220	.003 to .010	45	150	40

Table A–15 Tapers and Included Angles

		Included Angle		
Taper per Foot	*Taper per Inch*	*Deg.*	*Min.*	*Sec.*
1/8	0.010416	0	35	48
3/16	0.015625	0	53	44
1/4	0.020833	1	11	36
5/16	0.026042	1	29	30
3/8	0.031250	1	47	24
7/16	0.036458	2	5	18
1/2	0.416667	2	23	10
9/16	0.046875	2	41	4
5/8	0.052084	2	59	42
11/16	0.057292	3	16	54
3/4	0.06250	3	34	44
13/16	0.067708	3	52	38
7/8	0.072917	4	10	32
15/16	0.078125	4	28	24
1	0.083330	4	46	18
1 1/4	0.104666	5	57	48
1 1/2	0.125000	7	9	10
1 3/4	0.145833	8	20	26
2	0.166666	9	31	36
2 1/2	0.208333	11	53	36
3	0.250000	14	15	0
3 1/2	0.291666	16	35	40
4	0.333333	18	55	28
4 1/2	0.375000	21	14	2
5	0.416666	23	32	12
6	0.500000	28	4	2

Table A–16 Average Cutting Tool Angles and Cutting Speeds for Single-Point, High-Speed Steel Cutting Tools

Material to Be Machined	Side[1] Relief	End[1] Relief	Side Rake[2]	True Back[3] Rake	Suggested Cutting Speeds[4]	
					sfpm	m/min
Steel						
Free-machining	10°	10°	10°–22°	16°	160–350	50–110
Low-carbon (.05%–.30%C)	10°	10°	10°–14°	16°	90–100	27–30
Medium-carbon (.30%–.60%C)	10°	10°	10°–14°	12°	70–90	21–27
High-carbon tool (.60–1.70%C)	8°	8°	8°–12°	8°	50–70	15–21
Tough alloy	8°	8°	8°–12°	8°	50–70	15–21
Stainless	8°	8°	5°–10°	8°	40–70	12–21
Stainless, free-machining	10°	10°	5°–10°	16°	80–140	24–42
Cast Iron						
Soft	8°	8°	10°	8°	50–80	15–24
Hard	8°	8°	8°	5°	30–50	9–15
Malleable	8°	8°	10°	8°	80–100	24–30
Nonferrous						
Aluminum	10°	10°	10°–20°	35°	200–1500	60–460
Copper	10°	10°	10°–20°	16°	100–120	30–36
Brass	10°	8°	0°	0°	150–300	45–90
Bronze	10°	8°	0°	0°	90–100	27–30
Plastics						
Molded	10°	12°	0°	0°	150–300	45–90
Acrylics	15°	15°	0°	0°	60–70	18–21
Fiber	15°	15°	0°	0°	80–100	24–30

[1]The side- and end-relief angles are usually ground to the same angle for general machining operations. Average 8° end-relief and 10° side-relief angles are standard for turning most metals. The end- and side-relief angles for shaper and planer cutting tools generally range from 3° to 5°.

[2]The smaller angle is used without a chip breaker; the larger angle, with a chip breaker.

[3]The rake angles are measured from horizontal and vertical planes.

[4]The slower speeds are used for roughing cuts, machining dry. The faster speeds are used for finishing cuts, using cutting fluids. Cutting speeds for cast alloy bits are 50% to 70% faster; for cemented carbide tips and inserts, two to four times faster.

Table A–17 Recommended Cutting Speeds (sfpm) of Common Cutter Materials for Milling Selected Metals

Material to Be Milled	Milling Cutter Material				
	High-Speed Steel	Super High-Speed Steel	Stellite	Tantalum Carbide	Tungsten Carbide
Nonferrous					
Aluminum	500–1000	. . . .	800–1500	. . . .	1000–2000
Brass	70–175	. . . .	150–200	. . . .	350–600
Hard Bronze	65–130	. . . .	100–160	. . . .	200–425
Very Hard Bronze	30–50	. . . .	50–70	. . . .	125–200
Cast Iron					
Soft	50–80	60–115	90–130	. . . .	250–325
Hard	30–50	40–70	60–90	. . . .	150–200
Chilled	. . . .	30–50	40–60	. . . .	100–200
Malleable	70–100	80–125	115–150	. . . .	350–370
Steel					
Soft (low-carbon)	60–90	70–100	. . . .	150–250	300–550
Medium-carbon	50–80	60–90	. . . .	125–200	225–400
High-carbon	40–70	50–80	. . . .	100–175	150–250
Stainless	30–80	40–90	. . . .	75–200	100–300

Table A–18 Recommended Cutting Fluids for Ferrous and Nonferrous Metals

		Ferrous Metals			Nonferrous Metals	
Group ⟶	I	II	III	IV	V	VI
Machinability:* ⟶	Above 70%	50–70%	40–50%	Below 40%	Above 100%	Below 100%
Materials ⟶	Low-Carbon Steels High-Carbon Steels		Stainless Steels		Aluminum and Alloys Brasses and Bronzes	
Type of Machining Operation)**	Malleable Iron Cast Steel Stainless Steel	Cast Iron	Ingot Iron Wrought Iron	Tool Steels High-Speed Steels	Magnesium and Alloys Zinc	Copper Nickel Inconel Monel
1 Broaching, internal	Em Sul	Sul Em	Sul Em	Sul Em	MO Em	Sul ML
2 Broaching, surface	Em Sul	Em Sul	Sul Em	Sul Em	MO Em	Sul ML
Threading, pipe	Sul	Sul ML	Sul	Sul		Sul †
3 Tapping, plain	Sul	Sul	Sul	Sul	Em Dry	Sul ML
Threading, plain	Sul	Sul	Sul	Sul	Em Sul	Sul †
Gear shaving	Sul L	Sul L	Sul L	Sul L		
4 Reaming, plain	ML Sul	ML Sul	ML Sul	ML Sul	ML MO Em	ML MO Sul
Gear cutting	Sul ML Em	Sul	Sul	Sul ML		Sul ML
5 Drilling, deep	Em ML	Em Sul	Sul	Sul	MO ML Em	Sul ML
6 Milling Plain Multiple cutter	Em ML Sul ML	'Em Sul	Em Sul	Sul Sul ML	Em MO Dry Em MO Dry	Sul Em Sul Em
Boring, multiple-head	Sul Em	Sul HDS	Sul HDS	Sul Em	K Dry Em	Sul Em
7 Multiple-spindle automatic screw machines and turret lathes: drilling, forming, turning, reaming, cutting off, tapping, threading	Sul Em ML	Sul Em ML	Sul Em ML HDS	Sul ML Em HDS	Em Dry ML	Sul
8 High-speed, light-feed screw machines: drilling, forming, tapping, threading, turning, reaming, box milling, cutting off	Sul Em ML	Sul Em ML	Sul Em ML	Sul ML Em	Em Dry ML	Sul
Drilling	Em	Em	Em	Em Sul	Em Dry	Em
9 Planing, shaping	Em Sul ML	Em Sul ML	Sul Em	Em Sul	Em Dry	Em
Turning, single-point tool, form tools	Em Sul ML	Em Sul ML	Em Sul ML	Em Sul ML	Em Dry ML	Em Sul
10 Sawing, circular, hack	Sul ML Em	Sul Em ML	Sul Em ML	Sul Em ML	Dry MO Em	Sul Em ML
Grinding Plain Form (thread, etc.)	Em Sul	Em Sul	Em Sul	Em Sul	Em MO Sul	Em Sul

CUTTING FLUID SYMBOL: K = Kerosene; L = Lard Oil; MO = Mineral oils; ML = Mineral-lard oils; Sul = Sulphurized oils, with or without chlorine; Em = Soluble or emulsifiable oils and compounds; Dry = No cutting fluid needed; HDS = Heavy-duty soluble oil.

*Machinability rating based on 100% for cold-drawn Bessemer screw stock (B1112).

**Operations are ranked according to severity; 1 is the greatest and 10 is the least.

†Palm oil is frequently used to thread copper.

Compiled from *Metals Handbook*, *Machinery's Handbook*, and *AISI Steel Products Manual*.

Table A–19 Suggested Starting Speeds and Feeds for High-Speed Steel End Mill Applications

	Heat-Resistant Cobalt-Base Alloys, High Tensile Steels (50–55C)		Cast Iron, Mild Steel, Half-Hard Brass and Bronze		Aluminum, Plastics, Wood	
Materials						
Types and Styles of End Mills	Premium Cobalt High-Speed Steel 2 or More Flutes		High-Speed Steel End Mills, 2 or More Flutes Surface Treatment Helpful in C. I. Applications		High-Speed Steel End Mills of High-Helix Type, 1 to 6 Flutes	
Diameter of End Mills (in Inches)	Speed 5–10 sfpm RPM	Feed Chip Load per Tooth	Speed 10–15 sfpm RPM	Feed Chip Load per Tooth	Speed 15–20 sfpm RPM	Feed Chip Load per Tooth
1/16	*	*	*	*		
3/32	*	*	*	*	611–815	.0002–.0005
1/8	*	*	*	*	456–611	.0002–.0005
3/16	*	*	204–306	.0002–.0005	306–407	.0002–.0005
1/4	76–153	.0002–.001	153–230	.0002–.001	229–306	.0002–.001
5/16	61–122	.0002–.001	122–183	.0002–.001	183–244	.0002–.001
3/8	51–102	.0002–.001	102–153	.0002–.001	153–203	.0002–.001
7/16	44–88	.0005–.001	88–132	.0005–.001	131–175	.0005–.002
1/2	38–76	.0005–.001	76–115	.0005–.001	115–153	.0005–.002
9/16	34–68	.0005–.002	68–104	.0005–.002	104–136	.0005–.002
5/8	31–61	.0005–.002	61–92	.0005–.002	92–122	.0005–.002
11/16	28–56	.0005–.002	56–84	.0005–.002	84–111	.0005–.002
3/4	26–51	.0005–.002	51–76	.0005–.002	76–102	.001–.004
13/16	24–47	.001–.003	47–71	.001–.003	71–94	.001–.004
7/8	22–44	.001–.003	44–65	.001–.003	65–87	.001–.004
15/16	20–40	.001–.003	40–62	.001–.003	62–81	.001–.004
1	19–38	.001–.003	38–58	.001–.003	58–76	.001–.004
1 1/8	34	.0015–.004	34–51	.0015–.004	51–68	.0015–.005
1 1/4	31	.0015–.004	31–46	.0015–.004	46–61	.0015–.005
1 3/8	28	.0015–.004	28–42	.0015–.004	42–55	.0015–.005
1 1/2	26	.0015–.004	26–38	.0015–.004	38–51	.002 Up
1 5/8	24	.002 Up	35	.002 Up	35–47	.002 Up
1 3/4	22	.002 Up	32	.002 Up	32–43	.002 Up
1 7/8	20	.002 Up	30	.002 Up	30–40	.003 Up
2	19	.002 Up	29	.003 Up	29–38	.003 Up
2 1/8	18	.003 Up	28	.003 Up	36	.003 Up
2 1/4	17	.003 Up	26	.003 Up	34	.003 Up
2 3/8	16	.003 Up	25	.003 Up	32	.003 Up
2 1/2	15	.003 Up	23	.003 Up	30	.003 Up
2 5/8	15	.003 Up	22	.003 Up	29	.003 Up
2 3/4	14	.003 Up	21	.003 Up	28	.003 Up
2 7/8	14	.003 Up	20	.003 Up	27	.003 Up
3	13	.003 Up	19	.003 Up	26	.003 Up

*Use solid carbide end mills in small-diameter applications for materials that are harder than 46 C Rockwell.
Note: The speeds and feeds apply also to cavity and slotting cuts up to the depth equal to the diameter of the end mill. Deeper cuts require these values to be decreased. Table adapted with permission from The Cleveland Twist Drill Co.

INDEX